VOLUME FOUR HUNDRED AND SEVENTY-ONE

METHODS IN ENZYMOLOGY

Two-Component Signaling Systems, Part C

METHODS IN ENZYMOLOGY

VOLUME FOUR HUNDRED AND SEVENTY-ONE

METHODS IN ENZYMOLOGY

Two-Component Signaling Systems, Part C

EDITED BY

MELVIN I. SIMON
Division of Biology, California Institute of Technology
California Boulevard, Pasadena, CA

BRIAN R. CRANE
Cornell University
Department of Chemistry and Chemical Biology
Olin, Ithaca, NY

ALEXANDRINE CRANE
Cornell University
Department of Chemistry and Chemical Biology
Olin, Ithaca, NY

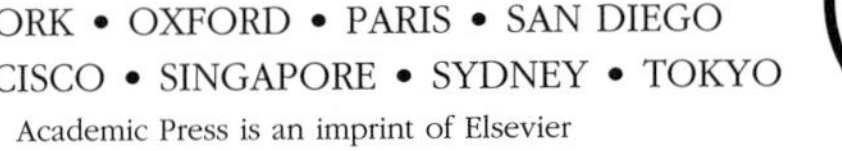

AMSTERDAM • BOSTON • HEIDELBERG • LONDON
NEW YORK • OXFORD • PARIS • SAN DIEGO
SAN FRANCISCO • SINGAPORE • SYDNEY • TOKYO
Academic Press is an imprint of Elsevier

Academic Press is an imprint of Elsevier
525 B Street, Suite 1900, San Diego, CA 92101-4495, USA
30 Corporate Drive, Suite 400, Burlington, MA 01803, USA
32 Jamestown Road, London NW1 7BY, UK

First edition 2010

For information on all Academic Press publications
visit our website at elsevierdirect.com

ISBN: 978-0-12-381347-3
ISSN: 0076-6879

Printed and bound in United States of America
10 11 12 10 9 8 7 6 5 4 3 2 1

Contents

Contributors

Adrián F. Alvarez
Departamento de Genética Molecular, Instituto de Fisiología Celular, Universidad Nacional Autónoma de México, México D.F., México

Babak Andi
Department of Chemistry and Biochemistry, University of Oklahoma, Norman, Oklahoma, USA

Paul V. Attwood
School of Biomedical, Biomolecular and Chemical Sciences (M310), The University of Western Australia, Crawley, Western Australia, Australia

Paul G. Besant
School of Biomedical, Biomolecular and Chemical Sciences (M310), The University of Western Australia, Crawley, Western Australia, Australia

Roberto A. Bogomolni
Department of Chemistry and Biochemistry, University of California, Santa Cruz, California, USA

Katherine A. Borkovich
Department of Plant Pathology and Microbiology, University of California, Riverside, California, USA

Robert B. Bourret
Departments of Microbiology and Immunology, University of North Carolina, Chapel Hill, North Carolina, USA

Winslow R. Briggs
Department of Plant Biology, Carnegie Institution for Science, Stanford, California, USA

Edmundo Calva
Departamento de Microbiología Molecular, Instituto de Biotecnología, Universidad Nacional Autónoma de México AP 510-3, Cuernavaca, Morelos, Mexico

Paul F. Cook
Department of Chemistry and Biochemistry, University of Oklahoma, Norman, Oklahoma, USA

Rachel L. Creager-Allen
Departments of Biochemistry and Biophysics, University of North Carolina, Chapel Hill, North Carolina, USA

Nabanita De
Department of Molecular Medicine, College of Veterinary Medicine, Cornell University, Ithaca, New York, USA

Jan S. Fassler
Department of Biology, University of Iowa, Iowa City, Iowa, USA

Marcus A. Frederickson
Department of Chemistry and Biochemistry, University of California, Santa Cruz, California, USA

Sumire Fujiwara
Department of Plant Cellular and Molecular Biology, Ohio State University, Columbus, Ohio, USA

Dimitris Georgellis
Departamento de Genética Molecular, Instituto de Fisiología Celular, Universidad Nacional Autónoma de México, México D.F., México

Eric Giraud
Laboratoire des Symbioses Tropicales et Méditerranéennes, IRD, CIRAD, AGRO-M, INRA, UM2, TA A-82/J, Campus de Baillarguet, Montpellier Cedex 5, France

Zemer Gitai
Department of Molecular Biology, Princeton University, Princeton, New Jersey, USA

Erin A. Gontang
Department of Microbiology and Molecular Genetics, Harvard Medical School, Boston, Massachusetts, USA

Mark Goulian
Department of Biology, University of Pennsylvania, Philadelphia, Pennsylvania, USA

Penelope I. Higgs
Department of Ecophysiology, Max Planck Institute for Terrestrial Microbiology, Marburg, Germany

Hans-Jörg Hippe
Medizinische Klinik III, Universitätsklinikum, Universität Heidelberg, Heidelberg, Germany

James A. Hoch
Department of Molecular and Experimental Medicine, The Scripps Research Institute, La Jolla, California, USA

Terence Hwa
Center for Theoretical Biological Physics, University of California San Diego, La Jolla, California, USA

Sakthimala Jagadeesan
Department of Molecular Biology, Max Planck Institute for Infection Biology, Berlin, Germany

Carol A. Jones
Department of Plant Pathology and Microbiology, University of California, Riverside, California, USA

Alla O. Kaserer
Department of Chemistry and Biochemistry, University of Oklahoma, Norman, Oklahoma, USA

Woe-Yeon Kim
Department of Plant Cellular and Molecular Biology, Ohio State University, Columbus, Ohio, USA, and Division of Applied Life Science (BK21 Program) and Environmental Biotechnology National Core Research Center, Gyeongsang National University, Jinju, Korea

Susanne Klumpp✠
Institut für Pharmazeutische und Medizinische Chemie, Westfälische Wilhelms-Universität, Münster, Germany

Roberto Kolter
Department of Microbiology and Molecular Genetics, Harvard Medical School, Boston, Massachusetts, USA

Michael Korth
Institut für Pharmakologie für Pharmazeuten, Universitätsklinikum Hamburg-Eppendorf, Hamburg, Germany

Petya V. Krasteva
Department of Molecular Medicine, College of Veterinary Medicine, Cornell University, Ithaca, New York, USA

Jérôme Lavergne
CEA, DSV, IBEB, Laboratoire de Bioénergétique Cellulaire, and CNRS, UMR 6191, Biologie Végétale et Microbiologie Environnementales, and Aix-Marseille Université, Saint-Paul-lez-Durance, France

Bongsoo Lee
Department of Ecophysiology, Max Planck Institute for Terrestrial Microbiology, Marburg, Germany

✠ Deceased June 17, 2009

Jürgen U. Linder
Pharmaceutical Institute, University of Tübingen, Tübingen, Germany

Daniel López
Department of Microbiology and Molecular Genetics, Harvard Medical School, Boston, Massachusetts, USA

Katrin Ludwig
Institut für Pharmazeutische und Medizinische Chemie, Westfälische Wilhelms-Universität, Münster, Germany

Bryan Lunt
Institute for Scientific Interchange, Viale S. Severo 65, Torino, Italy

C. Robertson McClung
Department of Biological Sciences, Dartmouth College, Hanover, New Hampshire, USA

Takeshi Mizuno
Laboratory of Molecular Microbiology, School of Agriculture, Nagoya University, Chikusa-ku, Nagoya, Japan

Aaron M. Moore
Departments of Microbiology and Immunology, University of North Carolina, Chapel Hill, North Carolina, USA

Susumu Morigasaki
Department of Microbiology, University of California, Davis, California, USA, and Graduate School of Biological Sciences, Nara Institute of Science and Technology, Ikoma, Nara, Japan

Marcos V. A. S. Navarro
Department of Molecular Medicine, College of Veterinary Medicine, Cornell University, Ithaca, New York, USA

Jose N. Onuchic
Center for Theoretical Biological Physics, University of California San Diego, La Jolla, California, USA

Ricardo Oropeza
Departamento de Microbiología Molecular, Instituto de Biotecnología, Universidad Nacional Autónoma de México AP 510-3, Cuernavaca, Morelos, Mexico

Stephani C. Page
Departments of Biochemistry and Biophysics, University of North Carolina, Chapel Hill, North Carolina, USA

Andrea Procaccini
Institute for Scientific Interchange, Viale S. Severo 65, Torino, Italy

Patrice A. Salomé
Department of Biological Sciences, Dartmouth College, Hanover, New Hampshire, USA

Andreas Schramm
Department of Ecophysiology, Max Planck Institute for Terrestrial Microbiology, Marburg, Germany

Alexander Schug
Center for Theoretical Biological Physics, University of California San Diego, La Jolla, California, USA

Joachim E. Schultz
Pharmaceutical Institute, University of Tübingen, Tübingen, Germany

Kazuhiro Shiozaki
Department of Microbiology, University of California, Davis, California, USA

Ruth E. Silversmith
Departments of Microbiology and Immunology, University of North Carolina, Chapel Hill, North Carolina, USA

Albert Siryaporn
Department of Biology, University of Pennsylvania, Philadelphia, Pennsylvania, USA

David E. Somers
Department of Plant Cellular and Molecular Biology, Ohio State University, Columbus, Ohio, USA

Holger Sondermann
Department of Molecular Medicine, College of Veterinary Medicine, Cornell University, Ithaca, New York, USA

Hendrik Szurmant
Department of Molecular and Experimental Medicine, The Scripps Research Institute, La Jolla, California, USA

Stephanie A. Thomas
Departments of Microbiology and Immunology, University of North Carolina, Chapel Hill, North Carolina, USA

Tong-Seung Tseng
Department of Plant Biology, Carnegie Institution for Science, Stanford, California, USA

André Verméglio
CEA, DSV, IBEB, Laboratoire de Bioénergétique Cellulaire, and CNRS, UMR 6191, Biologie Végétale et Microbiologie Environnementales, and Aix-Marseille Université, Saint-Paul-lez-Durance, France

Qi Wang
Department of Molecular Medicine, College of Veterinary Medicine, Cornell University, Ithaca, New York, USA

Martin Weigt
Institute for Scientific Interchange, Viale S. Severo 65, Torino, Italy

John N. Werner
Department of Molecular Biology, Princeton University, Princeton, New Jersey, USA

Ann H. West
Department of Chemistry and Biochemistry, University of Oklahoma, Norman, Oklahoma, USA

Thomas Wieland
Institut für Experimentelle und Klinische Pharmakologie und Toxikologie, Medizinische Fakultät Mannheim, Universität Heidelberg, Mannheim, Germany

Takafumi Yamashino
Laboratory of Molecular Microbiology, School of Agriculture, Nagoya University, Chikusa-ku, Nagoya, Japan

Xiao-Bo Zhou
Institut für Pharmakologie für Pharmazeuten, Universitätsklinikum Hamburg-Eppendorf, Hamburg, Germany

Methods in Enzymology

Volume XVI. Fast Reactions
Edited by Kenneth Kustin

Volume XVII. Metabolism of Amino Acids and Amines (Parts A and B)
Edited by Herbert Tabor and Celia White Tabor

Volume XVIII. Vitamins and Coenzymes (Parts A, B, and C)
Edited by Donald B. McCormick and Lemuel D. Wright

Volume XIX. Proteolytic Enzymes
Edited by Gertrude E. Perlmann and Laszlo Lorand

Volume XX. Nucleic Acids and Protein Synthesis (Part C)
Edited by Kivie Moldave and Lawrence Grossman

Volume XXI. Nucleic Acids (Part D)
Edited by Lawrence Grossman and Kivie Moldave

Volume XXII. Enzyme Purification and Related Techniques
Edited by William B. Jakoby

Volume XXIII. Photosynthesis (Part A)
Edited by Anthony San Pietro

Volume XXIV. Photosynthesis and Nitrogen Fixation (Part B)
Edited by Anthony San Pietro

Volume XXV. Enzyme Structure (Part B)
Edited by C. H. W. Hirs and Serge N. Timasheff

Volume XXVI. Enzyme Structure (Part C)
Edited by C. H. W. Hirs and Serge N. Timasheff

Volume XXVII. Enzyme Structure (Part D)
Edited by C. H. W. Hirs and Serge N. Timasheff

Volume XXVIII. Complex Carbohydrates (Part B)
Edited by Victor Ginsburg

Volume XXIX. Nucleic Acids and Protein Synthesis (Part E)
Edited by Lawrence Grossman and Kivie Moldave

Volume XXX. Nucleic Acids and Protein Synthesis (Part F)
Edited by Kivie Moldave and Lawrence Grossman

Volume XXXI. Biomembranes (Part A)
Edited by Sidney Fleischer and Lester Packer

Volume XXXII. Biomembranes (Part B)
Edited by Sidney Fleischer and Lester Packer

Volume XXXIII. Cumulative Subject Index Volumes I-XXX
Edited by Martha G. Dennis and Edward A. Dennis

Volume XXXIV. Affinity Techniques (Enzyme Purification: Part B)
Edited by William B. Jakoby and Meir Wilchek

VOLUME LIII. Biomembranes (Part D: Biological Oxidations)
Edited by SIDNEY FLEISCHER AND LESTER PACKER

VOLUME LIV. Biomembranes (Part E: Biological Oxidations)
Edited by SIDNEY FLEISCHER AND LESTER PACKER

VOLUME LV. Biomembranes (Part F: Bioenergetics)
Edited by SIDNEY FLEISCHER AND LESTER PACKER

VOLUME LVI. Biomembranes (Part G: Bioenergetics)
Edited by SIDNEY FLEISCHER AND LESTER PACKER

VOLUME LVII. Bioluminescence and Chemiluminescence
Edited by MARLENE A. DELUCA

VOLUME LVIII. Cell Culture
Edited by WILLIAM B. JAKOBY AND IRA PASTAN

VOLUME LIX. Nucleic Acids and Protein Synthesis (Part G)
Edited by KIVIE MOLDAVE AND LAWRENCE GROSSMAN

VOLUME LX. Nucleic Acids and Protein Synthesis (Part H)
Edited by KIVIE MOLDAVE AND LAWRENCE GROSSMAN

VOLUME 61. Enzyme Structure (Part H)
Edited by C. H. W. HIRS AND SERGE N. TIMASHEFF

VOLUME 62. Vitamins and Coenzymes (Part D)
Edited by DONALD B. MCCORMICK AND LEMUEL D. WRIGHT

VOLUME 63. Enzyme Kinetics and Mechanism (Part A: Initial Rate and Inhibitor Methods)
Edited by DANIEL L. PURICH

VOLUME 64. Enzyme Kinetics and Mechanism (Part B: Isotopic Probes and Complex Enzyme Systems)
Edited by DANIEL L. PURICH

VOLUME 65. Nucleic Acids (Part I)
Edited by LAWRENCE GROSSMAN AND KIVIE MOLDAVE

VOLUME 66. Vitamins and Coenzymes (Part E)
Edited by DONALD B. MCCORMICK AND LEMUEL D. WRIGHT

VOLUME 67. Vitamins and Coenzymes (Part F)
Edited by DONALD B. MCCORMICK AND LEMUEL D. WRIGHT

VOLUME 68. Recombinant DNA
Edited by RAY WU

VOLUME 69. Photosynthesis and Nitrogen Fixation (Part C)
Edited by ANTHONY SAN PIETRO

VOLUME 70. Immunochemical Techniques (Part A)
Edited by HELEN VAN VUNAKIS AND JOHN J. LANGONE

VOLUME 89. Carbohydrate Metabolism (Part D)
Edited by WILLIS A. WOOD

VOLUME 90. Carbohydrate Metabolism (Part E)
Edited by WILLIS A. WOOD

VOLUME 91. Enzyme Structure (Part I)
Edited by C. H. W. HIRS AND SERGE N. TIMASHEFF

VOLUME 92. Immunochemical Techniques (Part E: Monoclonal Antibodies and General Immunoassay Methods)
Edited by JOHN J. LANGONE AND HELEN VAN VUNAKIS

VOLUME 93. Immunochemical Techniques (Part F: Conventional Antibodies, Fc Receptors, and Cytotoxicity)
Edited by JOHN J. LANGONE AND HELEN VAN VUNAKIS

VOLUME 94. Polyamines
Edited by HERBERT TABOR AND CELIA WHITE TABOR

VOLUME 95. Cumulative Subject Index Volumes 61–74, 76–80
Edited by EDWARD A. DENNIS AND MARTHA G. DENNIS

VOLUME 96. Biomembranes [Part J: Membrane Biogenesis: Assembly and Targeting (General Methods; Eukaryotes)]
Edited by SIDNEY FLEISCHER AND BECCA FLEISCHER

VOLUME 97. Biomembranes [Part K: Membrane Biogenesis: Assembly and Targeting (Prokaryotes, Mitochondria, and Chloroplasts)]
Edited by SIDNEY FLEISCHER AND BECCA FLEISCHER

VOLUME 98. Biomembranes (Part L: Membrane Biogenesis: Processing and Recycling)
Edited by SIDNEY FLEISCHER AND BECCA FLEISCHER

VOLUME 99. Hormone Action (Part F: Protein Kinases)
Edited by JACKIE D. CORBIN AND JOEL G. HARDMAN

VOLUME 100. Recombinant DNA (Part B)
Edited by RAY WU, LAWRENCE GROSSMAN, AND KIVIE MOLDAVE

VOLUME 101. Recombinant DNA (Part C)
Edited by RAY WU, LAWRENCE GROSSMAN, AND KIVIE MOLDAVE

VOLUME 102. Hormone Action (Part G: Calmodulin and Calcium-Binding Proteins)
Edited by ANTHONY R. MEANS AND BERT W. O'MALLEY

VOLUME 103. Hormone Action (Part H: Neuroendocrine Peptides)
Edited by P. MICHAEL CONN

VOLUME 104. Enzyme Purification and Related Techniques (Part C)
Edited by WILLIAM B. JAKOBY

Volume 139. Cellular Regulators (Part A: Calcium- and Calmodulin-Binding Proteins)
Edited by Anthony R. Means and P. Michael Conn

Volume 140. Cumulative Subject Index Volumes 102–119, 121–134

Volume 141. Cellular Regulators (Part B: Calcium and Lipids)
Edited by P. Michael Conn and Anthony R. Means

Volume 142. Metabolism of Aromatic Amino Acids and Amines
Edited by Seymour Kaufman

Volume 143. Sulfur and Sulfur Amino Acids
Edited by William B. Jakoby and Owen Griffith

Volume 144. Structural and Contractile Proteins (Part D: Extracellular Matrix)
Edited by Leon W. Cunningham

Volume 145. Structural and Contractile Proteins (Part E: Extracellular Matrix)
Edited by Leon W. Cunningham

Volume 146. Peptide Growth Factors (Part A)
Edited by David Barnes and David A. Sirbasku

Volume 147. Peptide Growth Factors (Part B)
Edited by David Barnes and David A. Sirbasku

Volume 148. Plant Cell Membranes
Edited by Lester Packer and Roland Douce

Volume 149. Drug and Enzyme Targeting (Part B)
Edited by Ralph Green and Kenneth J. Widder

Volume 150. Immunochemical Techniques (Part K: *In Vitro* Models of B and T Cell Functions and Lymphoid Cell Receptors)
Edited by Giovanni Di Sabato

Volume 151. Molecular Genetics of Mammalian Cells
Edited by Michael M. Gottesman

Volume 152. Guide to Molecular Cloning Techniques
Edited by Shelby L. Berger and Alan R. Kimmel

Volume 153. Recombinant DNA (Part D)
Edited by Ray Wu and Lawrence Grossman

Volume 154. Recombinant DNA (Part E)
Edited by Ray Wu and Lawrence Grossman

Volume 155. Recombinant DNA (Part F)
Edited by Ray Wu

Volume 156. Biomembranes (Part P: ATP-Driven Pumps and Related Transport: The Na, K-Pump)
Edited by Sidney Fleischer and Becca Fleischer

Volume 157. Biomembranes (Part Q: ATP-Driven Pumps and Related Transport: Calcium, Proton, and Potassium Pumps)
Edited by Sidney Fleischer and Becca Fleischer

Volume 158. Metalloproteins (Part A)
Edited by James F. Riordan and Bert L. Vallee

Volume 159. Initiation and Termination of Cyclic Nucleotide Action
Edited by Jackie D. Corbin and Roger A. Johnson

Volume 160. Biomass (Part A: Cellulose and Hemicellulose)
Edited by Willis A. Wood and Scott T. Kellogg

Volume 161. Biomass (Part B: Lignin, Pectin, and Chitin)
Edited by Willis A. Wood and Scott T. Kellogg

Volume 162. Immunochemical Techniques (Part L: Chemotaxis and Inflammation)
Edited by Giovanni Di Sabato

Volume 163. Immunochemical Techniques (Part M: Chemotaxis and Inflammation)
Edited by Giovanni Di Sabato

Volume 164. Ribosomes
Edited by Harry F. Noller, Jr., and Kivie Moldave

Volume 165. Microbial Toxins: Tools for Enzymology
Edited by Sidney Harshman

Volume 166. Branched-Chain Amino Acids
Edited by Robert Harris and John R. Sokatch

Volume 167. Cyanobacteria
Edited by Lester Packer and Alexander N. Glazer

Volume 168. Hormone Action (Part K: Neuroendocrine Peptides)
Edited by P. Michael Conn

Volume 169. Platelets: Receptors, Adhesion, Secretion (Part A)
Edited by Jacek Hawiger

Volume 170. Nucleosomes
Edited by Paul M. Wassarman and Roger D. Kornberg

Volume 171. Biomembranes (Part R: Transport Theory: Cells and Model Membranes)
Edited by Sidney Fleischer and Becca Fleischer

Volume 172. Biomembranes (Part S: Transport: Membrane Isolation and Characterization)
Edited by Sidney Fleischer and Becca Fleischer

VOLUME 173. Biomembranes [Part T: Cellular and Subcellular Transport: Eukaryotic (Nonepithelial) Cells]
Edited by SIDNEY FLEISCHER AND BECCA FLEISCHER

VOLUME 174. Biomembranes [Part U: Cellular and Subcellular Transport: Eukaryotic (Nonepithelial) Cells]
Edited by SIDNEY FLEISCHER AND BECCA FLEISCHER

VOLUME 175. Cumulative Subject Index Volumes 135–139, 141–167

VOLUME 176. Nuclear Magnetic Resonance (Part A: Spectral Techniques and Dynamics)
Edited by NORMAN J. OPPENHEIMER AND THOMAS L. JAMES

VOLUME 177. Nuclear Magnetic Resonance (Part B: Structure and Mechanism)
Edited by NORMAN J. OPPENHEIMER AND THOMAS L. JAMES

VOLUME 178. Antibodies, Antigens, and Molecular Mimicry
Edited by JOHN J. LANGONE

VOLUME 179. Complex Carbohydrates (Part F)
Edited by VICTOR GINSBURG

VOLUME 180. RNA Processing (Part A: General Methods)
Edited by JAMES E. DAHLBERG AND JOHN N. ABELSON

VOLUME 181. RNA Processing (Part B: Specific Methods)
Edited by JAMES E. DAHLBERG AND JOHN N. ABELSON

VOLUME 182. Guide to Protein Purification
Edited by MURRAY P. DEUTSCHER

VOLUME 183. Molecular Evolution: Computer Analysis of Protein and Nucleic Acid Sequences
Edited by RUSSELL F. DOOLITTLE

VOLUME 184. Avidin-Biotin Technology
Edited by MEIR WILCHEK AND EDWARD A. BAYER

VOLUME 185. Gene Expression Technology
Edited by DAVID V. GOEDDEL

VOLUME 186. Oxygen Radicals in Biological Systems (Part B: Oxygen Radicals and Antioxidants)
Edited by LESTER PACKER AND ALEXANDER N. GLAZER

VOLUME 187. Arachidonate Related Lipid Mediators
Edited by ROBERT C. MURPHY AND FRANK A. FITZPATRICK

VOLUME 188. Hydrocarbons and Methylotrophy
Edited by MARY E. LIDSTROM

VOLUME 189. Retinoids (Part A: Molecular and Metabolic Aspects)
Edited by LESTER PACKER

Volume 190. Retinoids (Part B: Cell Differentiation and Clinical Applications)
Edited by Lester Packer

Volume 191. Biomembranes (Part V: Cellular and Subcellular Transport: Epithelial Cells)
Edited by Sidney Fleischer and Becca Fleischer

Volume 192. Biomembranes (Part W: Cellular and Subcellular Transport: Epithelial Cells)
Edited by Sidney Fleischer and Becca Fleischer

Volume 193. Mass Spectrometry
Edited by James A. McCloskey

Volume 194. Guide to Yeast Genetics and Molecular Biology
Edited by Christine Guthrie and Gerald R. Fink

Volume 195. Adenylyl Cyclase, G Proteins, and Guanylyl Cyclase
Edited by Roger A. Johnson and Jackie D. Corbin

Volume 196. Molecular Motors and the Cytoskeleton
Edited by Richard B. Vallee

Volume 197. Phospholipases
Edited by Edward A. Dennis

Volume 198. Peptide Growth Factors (Part C)
Edited by David Barnes, J. P. Mather, and Gordon H. Sato

Volume 199. Cumulative Subject Index Volumes 168–174, 176–194

Volume 200. Protein Phosphorylation (Part A: Protein Kinases: Assays, Purification, Antibodies, Functional Analysis, Cloning, and Expression)
Edited by Tony Hunter and Bartholomew M. Sefton

Volume 201. Protein Phosphorylation (Part B: Analysis of Protein Phosphorylation, Protein Kinase Inhibitors, and Protein Phosphatases)
Edited by Tony Hunter and Bartholomew M. Sefton

Volume 202. Molecular Design and Modeling: Concepts and Applications (Part A: Proteins, Peptides, and Enzymes)
Edited by John J. Langone

Volume 203. Molecular Design and Modeling: Concepts and Applications (Part B: Antibodies and Antigens, Nucleic Acids, Polysaccharides, and Drugs)
Edited by John J. Langone

Volume 204. Bacterial Genetic Systems
Edited by Jeffrey H. Miller

Volume 205. Metallobiochemistry (Part B: Metallothionein and Related Molecules)
Edited by James F. Riordan and Bert L. Vallee

Volume 256. Small GTPases and Their Regulators (Part B: Rho Family)
Edited by W. E. Balch, Channing J. Der, and Alan Hall

Volume 257. Small GTPases and Their Regulators (Part C: Proteins Involved in Transport)
Edited by W. E. Balch, Channing J. Der, and Alan Hall

Volume 258. Redox-Active Amino Acids in Biology
Edited by Judith P. Klinman

Volume 259. Energetics of Biological Macromolecules
Edited by Michael L. Johnson and Gary K. Ackers

Volume 260. Mitochondrial Biogenesis and Genetics (Part A)
Edited by Giuseppe M. Attardi and Anne Chomyn

Volume 261. Nuclear Magnetic Resonance and Nucleic Acids
Edited by Thomas L. James

Volume 262. DNA Replication
Edited by Judith L. Campbell

Volume 263. Plasma Lipoproteins (Part C: Quantitation)
Edited by William A. Bradley, Sandra H. Gianturco, and Jere P. Segrest

Volume 264. Mitochondrial Biogenesis and Genetics (Part B)
Edited by Giuseppe M. Attardi and Anne Chomyn

Volume 265. Cumulative Subject Index Volumes 228, 230–262

Volume 266. Computer Methods for Macromolecular Sequence Analysis
Edited by Russell F. Doolittle

Volume 267. Combinatorial Chemistry
Edited by John N. Abelson

Volume 268. Nitric Oxide (Part A: Sources and Detection of NO; NO Synthase)
Edited by Lester Packer

Volume 269. Nitric Oxide (Part B: Physiological and Pathological Processes)
Edited by Lester Packer

Volume 270. High Resolution Separation and Analysis of Biological Macromolecules (Part A: Fundamentals)
Edited by Barry L. Karger and William S. Hancock

Volume 271. High Resolution Separation and Analysis of Biological Macromolecules (Part B: Applications)
Edited by Barry L. Karger and William S. Hancock

Volume 272. Cytochrome P450 (Part B)
Edited by Eric F. Johnson and Michael R. Waterman

Volume 273. RNA Polymerase and Associated Factors (Part A)
Edited by Sankar Adhya

VOLUME 274. RNA Polymerase and Associated Factors (Part B)
Edited by SANKAR ADHYA

VOLUME 275. Viral Polymerases and Related Proteins
Edited by LAWRENCE C. KUO, DAVID B. OLSEN, AND STEVEN S. CARROLL

VOLUME 276. Macromolecular Crystallography (Part A)
Edited by CHARLES W. CARTER, JR., AND ROBERT M. SWEET

VOLUME 277. Macromolecular Crystallography (Part B)
Edited by CHARLES W. CARTER, JR., AND ROBERT M. SWEET

VOLUME 278. Fluorescence Spectroscopy
Edited by LUDWIG BRAND AND MICHAEL L. JOHNSON

VOLUME 279. Vitamins and Coenzymes (Part I)
Edited by DONALD B. MCCORMICK, JOHN W. SUTTIE, AND CONRAD WAGNER

VOLUME 280. Vitamins and Coenzymes (Part J)
Edited by DONALD B. MCCORMICK, JOHN W. SUTTIE, AND CONRAD WAGNER

VOLUME 281. Vitamins and Coenzymes (Part K)
Edited by DONALD B. MCCORMICK, JOHN W. SUTTIE, AND CONRAD WAGNER

VOLUME 282. Vitamins and Coenzymes (Part L)
Edited by DONALD B. MCCORMICK, JOHN W. SUTTIE, AND CONRAD WAGNER

VOLUME 283. Cell Cycle Control
Edited by WILLIAM G. DUNPHY

VOLUME 284. Lipases (Part A: Biotechnology)
Edited by BYRON RUBIN AND EDWARD A. DENNIS

VOLUME 285. Cumulative Subject Index Volumes 263, 264, 266–284, 286–289

VOLUME 286. Lipases (Part B: Enzyme Characterization and Utilization)
Edited by BYRON RUBIN AND EDWARD A. DENNIS

VOLUME 287. Chemokines
Edited by RICHARD HORUK

VOLUME 288. Chemokine Receptors
Edited by RICHARD HORUK

VOLUME 289. Solid Phase Peptide Synthesis
Edited by GREGG B. FIELDS

VOLUME 290. Molecular Chaperones
Edited by GEORGE H. LORIMER AND THOMAS BALDWIN

VOLUME 291. Caged Compounds
Edited by GERARD MARRIOTT

VOLUME 292. ABC Transporters: Biochemical, Cellular, and Molecular Aspects
Edited by SURESH V. AMBUDKAR AND MICHAEL M. GOTTESMAN

VOLUME 293. Ion Channels (Part B)
Edited by P. MICHAEL CONN

VOLUME 294. Ion Channels (Part C)
Edited by P. MICHAEL CONN

VOLUME 295. Energetics of Biological Macromolecules (Part B)
Edited by GARY K. ACKERS AND MICHAEL L. JOHNSON

VOLUME 296. Neurotransmitter Transporters
Edited by SUSAN G. AMARA

VOLUME 297. Photosynthesis: Molecular Biology of Energy Capture
Edited by LEE MCINTOSH

VOLUME 298. Molecular Motors and the Cytoskeleton (Part B)
Edited by RICHARD B. VALLEE

VOLUME 299. Oxidants and Antioxidants (Part A)
Edited by LESTER PACKER

VOLUME 300. Oxidants and Antioxidants (Part B)
Edited by LESTER PACKER

VOLUME 301. Nitric Oxide: Biological and Antioxidant Activities (Part C)
Edited by LESTER PACKER

VOLUME 302. Green Fluorescent Protein
Edited by P. MICHAEL CONN

VOLUME 303. cDNA Preparation and Display
Edited by SHERMAN M. WEISSMAN

VOLUME 304. Chromatin
Edited by PAUL M. WASSARMAN AND ALAN P. WOLFFE

VOLUME 305. Bioluminescence and Chemiluminescence (Part C)
Edited by THOMAS O. BALDWIN AND MIRIAM M. ZIEGLER

VOLUME 306. Expression of Recombinant Genes in Eukaryotic Systems
Edited by JOSEPH C. GLORIOSO AND MARTIN C. SCHMIDT

VOLUME 307. Confocal Microscopy
Edited by P. MICHAEL CONN

VOLUME 308. Enzyme Kinetics and Mechanism (Part E: Energetics of Enzyme Catalysis)
Edited by DANIEL L. PURICH AND VERN L. SCHRAMM

VOLUME 309. Amyloid, Prions, and Other Protein Aggregates
Edited by RONALD WETZEL

VOLUME 310. Biofilms
Edited by RON J. DOYLE

Volume 311. Sphingolipid Metabolism and Cell Signaling (Part A)
Edited by Alfred H. Merrill, Jr., and Yusuf A. Hannun

Volume 312. Sphingolipid Metabolism and Cell Signaling (Part B)
Edited by Alfred H. Merrill, Jr., and Yusuf A. Hannun

Volume 313. Antisense Technology
(Part A: General Methods, Methods of Delivery, and RNA Studies)
Edited by M. Ian Phillips

Volume 314. Antisense Technology (Part B: Applications)
Edited by M. Ian Phillips

Volume 315. Vertebrate Phototransduction and the Visual Cycle (Part A)
Edited by Krzysztof Palczewski

Volume 316. Vertebrate Phototransduction and the Visual Cycle (Part B)
Edited by Krzysztof Palczewski

Volume 317. RNA–Ligand Interactions (Part A: Structural Biology Methods)
Edited by Daniel W. Celander and John N. Abelson

Volume 318. RNA–Ligand Interactions (Part B: Molecular Biology Methods)
Edited by Daniel W. Celander and John N. Abelson

Volume 319. Singlet Oxygen, UV-A, and Ozone
Edited by Lester Packer and Helmut Sies

Volume 320. Cumulative Subject Index Volumes 290–319

Volume 321. Numerical Computer Methods (Part C)
Edited by Michael L. Johnson and Ludwig Brand

Volume 322. Apoptosis
Edited by John C. Reed

Volume 323. Energetics of Biological Macromolecules (Part C)
Edited by Michael L. Johnson and Gary K. Ackers

Volume 324. Branched-Chain Amino Acids (Part B)
Edited by Robert A. Harris and John R. Sokatch

Volume 325. Regulators and Effectors of Small GTPases
(Part D: Rho Family)
Edited by W. E. Balch, Channing J. Der, and Alan Hall

Volume 326. Applications of Chimeric Genes and Hybrid Proteins
(Part A: Gene Expression and Protein Purification)
Edited by Jeremy Thorner, Scott D. Emr, and John N. Abelson

Volume 327. Applications of Chimeric Genes and Hybrid Proteins
(Part B: Cell Biology and Physiology)
Edited by Jeremy Thorner, Scott D. Emr, and John N. Abelson

Volume 328. Applications of Chimeric Genes and Hybrid Proteins (Part C: Protein–Protein Interactions and Genomics)
Edited by Jeremy Thorner, Scott D. Emr, and John N. Abelson

Volume 329. Regulators and Effectors of Small GTPases (Part E: GTPases Involved in Vesicular Traffic)
Edited by W. E. Balch, Channing J. Der, and Alan Hall

Volume 330. Hyperthermophilic Enzymes (Part A)
Edited by Michael W. W. Adams and Robert M. Kelly

Volume 331. Hyperthermophilic Enzymes (Part B)
Edited by Michael W. W. Adams and Robert M. Kelly

Volume 332. Regulators and Effectors of Small GTPases (Part F: Ras Family I)
Edited by W. E. Balch, Channing J. Der, and Alan Hall

Volume 333. Regulators and Effectors of Small GTPases (Part G: Ras Family II)
Edited by W. E. Balch, Channing J. Der, and Alan Hall

Volume 334. Hyperthermophilic Enzymes (Part C)
Edited by Michael W. W. Adams and Robert M. Kelly

Volume 335. Flavonoids and Other Polyphenols
Edited by Lester Packer

Volume 336. Microbial Growth in Biofilms (Part A: Developmental and Molecular Biological Aspects)
Edited by Ron J. Doyle

Volume 337. Microbial Growth in Biofilms (Part B: Special Environments and Physicochemical Aspects)
Edited by Ron J. Doyle

Volume 338. Nuclear Magnetic Resonance of Biological Macromolecules (Part A)
Edited by Thomas L. James, Volker Dötsch, and Uli Schmitz

Volume 339. Nuclear Magnetic Resonance of Biological Macromolecules (Part B)
Edited by Thomas L. James, Volker Dötsch, and Uli Schmitz

Volume 340. Drug–Nucleic Acid Interactions
Edited by Jonathan B. Chaires and Michael J. Waring

Volume 341. Ribonucleases (Part A)
Edited by Allen W. Nicholson

Volume 342. Ribonucleases (Part B)
Edited by Allen W. Nicholson

Volume 343. G Protein Pathways (Part A: Receptors)
Edited by Ravi Iyengar and John D. Hildebrandt

Volume 344. G Protein Pathways (Part B: G Proteins and Their Regulators)
Edited by Ravi Iyengar and John D. Hildebrandt

Volume 362. Recognition of Carbohydrates in Biological Systems (Part A)
Edited by Yuan C. Lee and Reiko T. Lee

Volume 363. Recognition of Carbohydrates in Biological Systems (Part B)
Edited by Yuan C. Lee and Reiko T. Lee

Volume 364. Nuclear Receptors
Edited by David W. Russell and David J. Mangelsdorf

Volume 365. Differentiation of Embryonic Stem Cells
Edited by Paul M. Wassauman and Gordon M. Keller

Volume 366. Protein Phosphatases
Edited by Susanne Klumpp and Josef Krieglstein

Volume 367. Liposomes (Part A)
Edited by Nejat Düzgüneş

Volume 368. Macromolecular Crystallography (Part C)
Edited by Charles W. Carter, Jr., and Robert M. Sweet

Volume 369. Combinational Chemistry (Part B)
Edited by Guillermo A. Morales and Barry A. Bunin

Volume 370. RNA Polymerases and Associated Factors (Part C)
Edited by Sankar L. Adhya and Susan Garges

Volume 371. RNA Polymerases and Associated Factors (Part D)
Edited by Sankar L. Adhya and Susan Garges

Volume 372. Liposomes (Part B)
Edited by Nejat Düzgüneş

Volume 373. Liposomes (Part C)
Edited by Nejat Düzgüneş

Volume 374. Macromolecular Crystallography (Part D)
Edited by Charles W. Carter, Jr., and Robert W. Sweet

Volume 375. Chromatin and Chromatin Remodeling Enzymes (Part A)
Edited by C. David Allis and Carl Wu

Volume 376. Chromatin and Chromatin Remodeling Enzymes (Part B)
Edited by C. David Allis and Carl Wu

Volume 377. Chromatin and Chromatin Remodeling Enzymes (Part C)
Edited by C. David Allis and Carl Wu

Volume 378. Quinones and Quinone Enzymes (Part A)
Edited by Helmut Sies and Lester Packer

Volume 379. Energetics of Biological Macromolecules (Part D)
Edited by Jo M. Holt, Michael L. Johnson, and Gary K. Ackers

Volume 380. Energetics of Biological Macromolecules (Part E)
Edited by Jo M. Holt, Michael L. Johnson, and Gary K. Ackers

VOLUME 434. Lipidomics and Bioactive Lipids: Lipids and Cell Signaling
Edited by H. ALEX BROWN

VOLUME 435. Oxygen Biology and Hypoxia
Edited by HELMUT SIES AND BERNHARD BRÜNE

VOLUME 436. Globins and Other Nitric Oxide-Reactive Protiens (Part A)
Edited by ROBERT K. POOLE

VOLUME 437. Globins and Other Nitric Oxide-Reactive Protiens (Part B)
Edited by ROBERT K. POOLE

VOLUME 438. Small GTPases in Disease (Part A)
Edited by WILLIAM E. BALCH, CHANNING J. DER, AND ALAN HALL

VOLUME 439. Small GTPases in Disease (Part B)
Edited by WILLIAM E. BALCH, CHANNING J. DER, AND ALAN HALL

VOLUME 440. Nitric Oxide, Part F Oxidative and Nitrosative Stress in Redox Regulation of Cell Signaling
Edited by ENRIQUE CADENAS AND LESTER PACKER

VOLUME 441. Nitric Oxide, Part G Oxidative and Nitrosative Stress in Redox Regulation of Cell Signaling
Edited by ENRIQUE CADENAS AND LESTER PACKER

VOLUME 442. Programmed Cell Death, General Principles for Studying Cell Death (Part A)
Edited by ROYA KHOSRAVI-FAR, ZAHRA ZAKERI, RICHARD A. LOCKSHIN, AND MAURO PIACENTINI

VOLUME 443. Angiogenesis: *In Vitro* Systems
Edited by DAVID A. CHERESH

VOLUME 444. Angiogenesis: *In Vivo* Systems (Part A)
Edited by DAVID A. CHERESH

VOLUME 445. Angiogenesis: *In Vivo* Systems (Part B)
Edited by DAVID A. CHERESH

VOLUME 446. Programmed Cell Death, The Biology and Therapeutic Implications of Cell Death (Part B)
Edited by ROYA KHOSRAVI-FAR, ZAHRA ZAKERI, RICHARD A. LOCKSHIN, AND MAURO PIACENTINI

VOLUME 447. RNA Turnover in Bacteria, Archaea and Organelles
Edited by LYNNE E. MAQUAT AND CECILIA M. ARRAIANO

VOLUME 448. RNA Turnover in Eukaryotes: Nucleases, Pathways and Analysis of mRNA Decay
Edited by LYNNE E. MAQUAT AND MEGERDITCH KILEDJIAN

VOLUME 449. RNA Turnover in Eukaryotes: Analysis of Specialized and Quality Control RNA Decay Pathways
Edited by LYNNE E. MAQUAT AND MEGERDITCH KILEDJIAN

VOLUME 450. Fluorescence Spectroscopy
Edited by LUDWIG BRAND AND MICHAEL L. JOHNSON

VOLUME 451. Autophagy: Lower Eukaryotes and Non-Mammalian Systems (Part A)
Edited by DANIEL J. KLIONSKY

VOLUME 452. Autophagy in Mammalian Systems (Part B)
Edited by DANIEL J. KLIONSKY

VOLUME 453. Autophagy in Disease and Clinical Applications (Part C)
Edited by DANIEL J. KLIONSKY

VOLUME 454. Computer Methods (Part A)
Edited by MICHAEL L. JOHNSON AND LUDWIG BRAND

VOLUME 455. Biothermodynamics (Part A)
Edited by MICHAEL L. JOHNSON, JO M. HOLT, AND GARY K. ACKERS (RETIRED)

VOLUME 456. Mitochondrial Function, Part A: Mitochondrial Electron Transport Complexes and Reactive Oxygen Species
Edited by WILLIAM S. ALLISON AND IMMO E. SCHEFFLER

VOLUME 457. Mitochondrial Function, Part B: Mitochondrial Protein Kinases, Protein Phosphatases and Mitochondrial Diseases
Edited by WILLIAM S. ALLISON AND ANNE N. MURPHY

VOLUME 458. Complex Enzymes in Microbial Natural Product Biosynthesis, Part A: Overview Articles and Peptides
Edited by DAVID A. HOPWOOD

VOLUME 459. Complex Enzymes in Microbial Natural Product Biosynthesis, Part B: Polyketides, Aminocoumarins and Carbohydrates
Edited by DAVID A. HOPWOOD

VOLUME 460. Chemokines, Part A
Edited by TRACY M. HANDEL AND DAMON J. HAMEL

VOLUME 461. Chemokines, Part B
Edited by TRACY M. HANDEL AND DAMON J. HAMEL

VOLUME 462. Non-Natural Amino Acids
Edited by TOM W. MUIR AND JOHN N. ABELSON

VOLUME 463. Guide to Protein Purification, 2nd Edition
Edited by RICHARD R. BURGESS AND MURRAY P. DEUTSCHER

VOLUME 464. Liposomes, Part F
Edited by NEJAT DÜZGÜNEŞ

CHAPTER ONE

Characterizing Cross-Talk *In Vivo*: Avoiding Pitfalls and Overinterpretation

Albert Siryaporn *and* Mark Goulian

Contents

Abstract

Cross-talk between noncognate histidine kinases and response regulators has been widely reported *in vitro* and, in specific mutant backgrounds and conditions, *in vivo*. However, in most cases there is little evidence supporting a physiological role of cross-talk. Indeed, histidine kinases and response regulators show remarkable specificity for their cognate partners. *In vivo* studies of cross-talk have the potential to establish mechanisms that control specificity and, if the cross-talk is observable in wild-type strains, may reveal new levels of cross-regulation. However such studies can be complicated by effects of other regulatory circuits and by the inactivation of mechanisms that would otherwise suppress cross-talk. It is thus easy to mis- or overinterpret the significance of such studies. We address potential complications associated with measuring cross-talk and discuss some methods for identifying and unmasking sources of cross-talk in cells using transcriptional reporters and *in vivo* DNA-binding assays.

Department of Biology, University of Pennsylvania, Philadelphia, Pennsylvania, USA

Methods in Enzymology, Volume 471
ISSN 0076-6879, DOI: 10.1016/S0076-6879(10)71001-6

1. Overview

When sequence comparisons first revealed the generality of two-component signaling, it was immediately appreciated that histidine kinases could potentially phosphorylate noncognate response regulators (Nixon *et al.*, 1986) and that such cross-talk could play a role in signal integration and processing (Stock *et al.*, 1989) (Fig. 1.1). Early studies demonstrated the existence of cross-talk *in vitro*, although the cross-phosphorylation occurred with considerably slower kinetics compared with phosphorylation between cognate partners (Igo *et al.*, 1989; Ninfa *et al.*, 1988). Cross-talk was also observed *in vivo*, but only in modified strains in which various regulators were deleted or overexpressed. In the two decades since these early considerations, cross-talk has not emerged as a common theme in two-component signaling. Despite numerous reports in the literature, in most cases there is little evidence to support the claim that there are detectable effects from cross-talk in wild-type strains (Laub and Goulian, 2007).

There are several factors that make attempts at identifying cross-talk subject to mis- or overinterpretation. *In vitro* studies have demonstrated that numerous histidine kinases can phosphorylate noncognate response regulators, with some histidine kinases showing a considerable level of promiscuity (Skerker *et al.*, 2005; Yamamoto *et al.*, 2005). However, as a rule, histidine kinases show a very strong kinetic preference for their cognate response regulator (Skerker *et al.*, 2005). Thus, a demonstration of cross-talk *in vitro*, without a comparison of the kinetics of phosphotransfer between cognate and noncognate pairs, is not particularly compelling evidence for cross-talk function *in vivo* (Laub and Goulian, 2007; Skerker *et al.*, 2005).

There are additional complications when studying cross-talk *in vivo*. For most systems, it is not feasible to directly measure response regulator

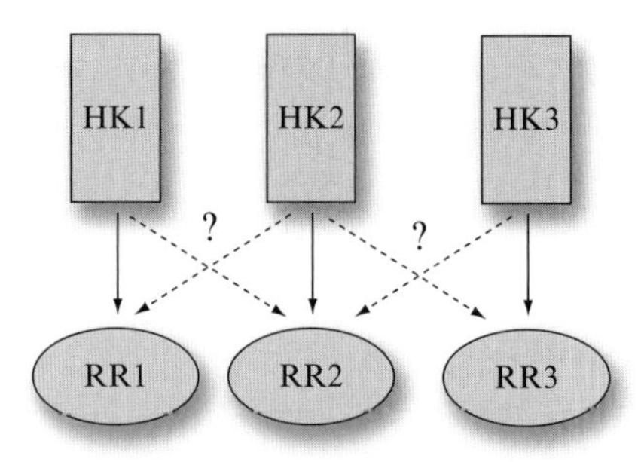

Figure 1.1 Cross-talk between noncognate histidine kinases (HKs) and response regulators (RRs) could potentially enable complex processing of multiple input signals. However, to date there has been relatively little evidence for cross-talk in wild-type strains.

phosphorylation in cells. Instead, phosphorylation is inferred from transcription measurements, e.g., with transcriptional reporters. However, it is often difficult to disentangle the effects of other regulatory factors that may act at the promoter of interest. In addition, studies in mutant strains may identify cross-talk that is irrelevant in the context of wild-type cells. Mechanisms have been identified that suppress cross-talk between noncognate pairs *in vivo* (Groban *et al.*, 2009; Siryaporn and Goulian, 2008). When these mechanisms are eliminated by mutation, significant cross-talk may emerge (Kim *et al.*, 1996; Silva *et al.*, 1998; Siryaporn and Goulian, 2008), which can give the false impression that cross-talk functions in wild-type strains.

Here we describe some of the issues one should take into account when looking for sources or effects of cross-talk. These methods can also be used to explore various mechanisms that limit cross-talk and to explore the molecular determinants of specificity among histidine kinases and response regulators. This chapter covers potential sources of cross-talk, cross-talk suppression mechanisms, and methods for measuring cross-talk in cells. In particular, it provides specific details on how to measure the association of response regulator-fluorescent protein fusions to DNA *in vivo*, a method which can distinguish cross-talk from the effects of other regulatory pathways. We also briefly discuss methods for characterizing phosphatase cross-talk and the effects of signal response on cross-talk networks.

2. Sources of Cross-Talk

A response regulator can potentially be phosphorylated by a cognate histidine kinase, noncognate histidine kinases, and small molecule phosphodonors. It is generally assumed that when genes for a response regulator and a histidine kinase are in the same operon, the encoded proteins are cognate pairs. Situations in which more than one histidine kinase or response regulator is encoded in an operon, however, may be more complex. It is also not unusual to encounter "orphan" histidine kinases and response regulators for which no clear partner has been identified. Many microbial genomes appear to have unequal numbers of histidine kinases and response regulators (Alm *et al.*, 2006), which suggests that some response regulators may have more than one source of phosphorylation and similarly some histidine kinases may phosphorylate multiple response regulators.

In some systems, the histidine kinase is bifunctional and mediates response regulator phosphorylation and dephosphorylation. In this case, there could in principle be cross-talk from either the phosphotransfer or phosphatase activities. However, we are unaware of any reports of phosphatase cross-talk, even in mutant strains where phosphotransfer cross-talk has been observed. For example, the histidine kinase CpxA phosphorylates and dephosphorylates its cognate response regulator CpxR. Cross-talk from CpxA to OmpR can be detected in strains deleted for *envZ* and *cpxR*, however, dephosphorylation of OmpR-P by CpxA is not detected (Siryaporn and Goulian, 2008). Kinetic modeling of the histidine kinase–response regulator interaction suggests that phosphatase activity may emerge for sufficiently strong interaction between a bifunctional histidine kinase and response regulator (Siryaporn *et al.*, in preparation). Thus, the difficulty in observing phosphatase cross-talk may reflect the relatively weak interactions between noncognate partners.

A number of response regulators are phosphorylated by the small molecule phosphodonor acetyl phosphate (McCleary and Stock, 1994; Wanner and Wilmes-Riesenberg, 1992; Wolfe, 2005). The cellular concentration of this high-energy source of phosphoryl groups is strongly dependent on growth conditions, such as carbon source (Wolfe, 2005). Concentrations can be quite high in some cases, which raises the possibility that acetyl phosphate may integrate metabolic status into some two-component systems (Fredericks *et al.*, 2006; Klein *et al.*, 2007; Wanner, 1992). For most two-component systems, however, a role for acetyl phosphate has not been established. There is also the possibility that other small molecule phosphodonors (Lukat *et al.*, 1992) could function to phosphorylate response regulators in some contexts *in vivo*.

The ability of a phosphodonor to affect the steady-state phosphorylation level of a particular response regulator depends on the rate of the phosphorylation reaction relative to the rates of other sources of phosphorylation and dephosphorylation. While phosphorylation rates of cytoplasmic fragments of histidine kinases are readily measured *in vitro* using purified components (Laub *et al.*, 2007), the effects of individual phosphorylation sources on levels of phosphorylated response regulator *in vivo* can be quite sensitive to the strain background and can be difficult to measure. For example, a bifunctional histidine kinase that has high rates of phosphorylation and dephosphorylation can give levels of phosphorylated response regulator that are comparable to those that would arise from a histidine kinase that has weak phosphorylation activity but lacks a mechanism for response regulator dephosphorylation (Fig. 1.2A). However, through a careful and systematic analysis using transcriptional and translational reporter fusions, the contributions from different phosphorylation sources can in some cases be disentangled.

3. Cross-Talk Suppression

Studies have shown that cross-talk can be suppressed by cognate histidine kinases and response regulators (Fig. 1.2B and C) (Groban *et al.*, 2009; Kim *et al.*, 1996; Silva *et al.*, 1998; Siryaporn and Goulian, 2008).

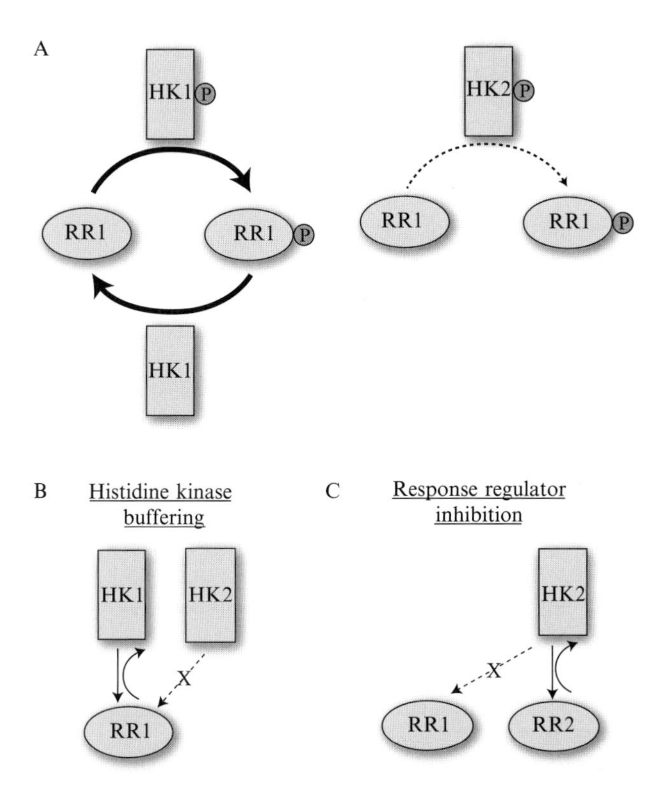

Figure 1.2 (A) Two different scenarios in which very different rates of phosphorylation can give the same steady state level of phosphorylated response regulator (RR-P). (Left) A bifunctional histidine kinase phosphorylates and dephosphorylates a response regulator with very high rates such that, under moderate stimulation, a moderate level of RR-P is obtained at steady state. (Right) A histidine kinase that phosphorylates a response regulator very weakly may, in the absence of any phosphatase activity, also produce moderate levels of RR-P. (B, C) Suppression of cross-talk from a histidine kinase HK2 to response regulator RR1 by cognate partners (Groban *et al.*, 2009; Siryaporn and Goulian, 2008). (B) The high flux of phosphorylation and dephosphorylation of RR1 by the bifunctional histidine kinase HK1 can suppress or buffer against cross-talk from HK2. (C) Suppression of cross-talk by a cognate response regulator. The cognate response regulator RR2 outcompetes the noncognate regulator RR1 for interaction with the histidine kinase HK2. By removing RR2, significant cross-talk from HK2 to RR1 may be detected, which would be otherwise absent in the wild-type strain.

Mutations that relieve these suppression mechanisms could produce signals of cross-talk that would otherwise be squelched in the wild-type strain. To avoid over- or misinterpreting the results from analysis of mutant strains, it is important to take into account the potential effects of these mechanisms.

Bifunctional histidine kinases set the level of phosphorylated response regulator through a balance of phosphorylation and dephosphorylation. This cycle can effectively act as a buffer to suppress the effects of weaker phosphorylation sources (Fig. 1.2B) (Groban *et al.*, 2009; Kim *et al.*, 1996; Silva *et al.*, 1998; Siryaporn and Goulian, 2008). Therefore, deletion of a bifunctional histidine kinase could reveal phosphorylation of the response regulator from a noncognate histidine kinase or other phosphodonor (e.g., Batchelor *et al.*, 2005; Danese and Silhavy, 1998; Hutchings *et al.*, 2006; Wolfe *et al.*, 2008). However, this alternate source may give a much lower flux of phosphoryl groups and may thus be physiologically irrelevant in wild-type strains.

A response regulator can also prevent its cognate histidine kinase from participating in cross-talk (Fig. 1.2C). For example, cross-talk from CpxA to OmpR is suppressed by CpxR (Siryaporn and Goulian, 2008). This is likely to occur through competitive inhibition, in which the cognate response regulator outcompetes noncognate response regulators for interaction with the histidine kinase. Thus, deletion or inactivation of a response regulator could potentially produce cross-talk from the cognate histidine kinase to some other response regulator. Significant overexpression of the histidine kinase could also have a similar result.

4. Transcriptional Reporters

In most two-component systems that have been studied, the phosphorylated response regulator controls transcription of effecter genes. Therefore, transcriptional reporter fusions have been a standard tool for measuring the output of signal transduction systems. A reporter gene such as *gfp* or *lacZ* is placed under control of a promoter of interest and expression of the reporter is measured through fluorescence or enzyme assays.

Cross-talk can be measured using transcriptional reporter fusions. Chromosomal fusions are preferable to constructs on multicopy plasmids since they avoid potential effects on reporter gene expression from titrating out response regulator and changes in plasmid copy number. It is always a good idea to assess the dynamic range of reporter expression since this gives an indication of the sensitivity to changes in output under various conditions. The basal expression level can be determined from measurements in a strain deleted for the response regulator of interest. The range of activation can be determined by making measurements for different levels of input stimulus, if the stimulus is known, or by varying expression of a constitutively active version of the histidine kinase, if one is available. For experiments in which a fluorescent reporter such as *gfp* is used, it is preferable to grow cultures using a growth medium with a low level of autofluorescence such as a minimal salts medium (Miller, 1992), as this generally increases the sensitivity of measurements. Rich media often give moderate levels of autofluorescence, which could reduce the sensitivity of fluorescence measurements.

Evidence for the presence and absence of cross-talk can be followed by monitoring changes in transcriptional reporter expression for different genetic backgrounds and conditions. Depending on the application, it may be necessary to remove cross-talk suppression mechanisms described above by deleting appropriate histidine kinases and response regulators before cross-talk can be detected. The observation of cross-talk in this case should not be used to conclude that there is cross-talk in the wild-type strain. Specific histidine kinases that are potential sources of cross-talk can be tested by changing expression levels or stimulation with signal. However, their effects on cross-talk will depend on details of the histidine kinase and its interaction with the noncognate response regulator, as discussed below.

A histidine kinase sets the level of phosphorylated response regulator through phosphorylation alone (monofunctional behavior) or through a balance of phosphorylation and dephosphorylation (bifunctional behavior). For histidine kinases displaying monofunctional behavior, the level of phosphorylated response regulator is expected to be quite sensitive to histidine kinase expression level (E. Batchelor and M. Goulian, unpublished observations; Miyashiro and Goulian, 2008). On the other hand, when the histidine kinase has bifunctional behavior, modeling and experiments suggest the level of phosphorylated response regulator can be relatively insensitive to histidine kinase expression levels (Batchelor and Goulian, 2003; Miyashiro and Goulian, 2008; Shinar *et al.*, 2007). Thus, cross-talk may be observed in a context of histidine kinase overexpression even though it appears that cognate response regulator phosphorylation is unchanged. In addition, in at least some cases, a histidine kinase can exhibit bifunctional behavior against its cognate partner while displaying monofunctional behavior against a noncognate partner (Siryaporn and Goulian, 2008).

The choice of carbon source may have a significant effect on the level of acetyl phosphate in the cell (McCleary and Stock, 1994) and thus the level of response regulator phosphorylation. The desired level of acetyl phosphate may depend on the experimental condition being tested. In some cases, significant levels of acetyl phosphate may provide a means of increasing the basal level of response regulator phosphorylation and hence increase the sensitivity to cross-talk by other phosphodonors. In other cases, one may want to eliminate acetyl phosphate or keep its level to a minimum in order to distinguish its effects from other sources of response regulator phosphorylation. For *Escherichia coli*, growth using carbon sources such as glucose, pyruvate, and acetate produce high levels of acetyl phosphate. Low levels can be achieved by growing on glycerol. One can also modulate acetyl phosphate levels by deleting one or both of *pta* and *ackA*, depending on the flux of acetate through the Pta (phosphotransacetylase)–AckA (acetate kinase) pathway (Wolfe *et al.*, 2008). Deletion of *pta ackA* abolishes production of acetyl phosphate. However, care should be taken in altering the Pta–AckA pathway as this may have additional indirect effects on some two-component systems (Wolfe *et al.*, 2008).

While transcriptional reporter fusions can provide effective measurements of two-component system output, they are also subject to the effects of other regulatory factors. There are many examples of genes whose expression is controlled by multiple regulatory proteins and small RNAs. Even many relatively well-characterized systems may have additional unidentified regulators whose effects could give the false impression of cross-talk. In some cases, promoters may also be controlled by multiple response regulators (e.g., Batchelor *et al.*, 2005; Mouslim and Groisman, 2003). To distinguish between cross-talk and alternate regulatory pathways, one may need to use additional methods to follow response regulator phosphorylation or binding to promoters.

5. Response Regulator Localization

If a functional fluorescent protein (FP) fusion to the response regulator of interest is available then one can use a fluorescence localization assay to provide further support for cross-talk *in vivo* (Batchelor and Goulian, 2006; Siryaporn and Goulian, 2008). The method takes advantage of the tendency of plasmids with partitioning systems to form clusters in the cell (Pogliano, 2002). In this assay, cells express the response regulator-FP fusion and also contain a plasmid that has one or more binding sites for the response regulator. Plasmid clustering provides a high local density of response regulator binding sites in the cell (Fig. 1.3A). Significant response regulator-FP binding to DNA results in intense localized regions of fluorescence, which appear as bright foci or spots under a fluorescence microscope and can be easily quantified by image analysis (Fig. 1.3B–D). Assuming DNA binding is modulated by response regulator phosphorylation, changes in response regulator-FP colocalization with the plasmid can be used to infer changes in response regulator phosphorylation.

We have been able to construct functional fluorescent protein fusions to the C-termini of several different response regulators in *E. coli*, although in all of the cases that we tested the fusions showed decreased activity. We are unaware of any general rules regarding the choice of C- or N-termini for the fusion, or whether or not to include a flexible linker. It is probably best to try several different constructs. For the plasmid, we have used constructs based on pRS415 (Batchelor and Goulian, 2006; Simons *et al.*, 1987), which is derived from pBR322. Many other plasmids should work as well although they may show differing levels of clustering. Plasmid clustering can

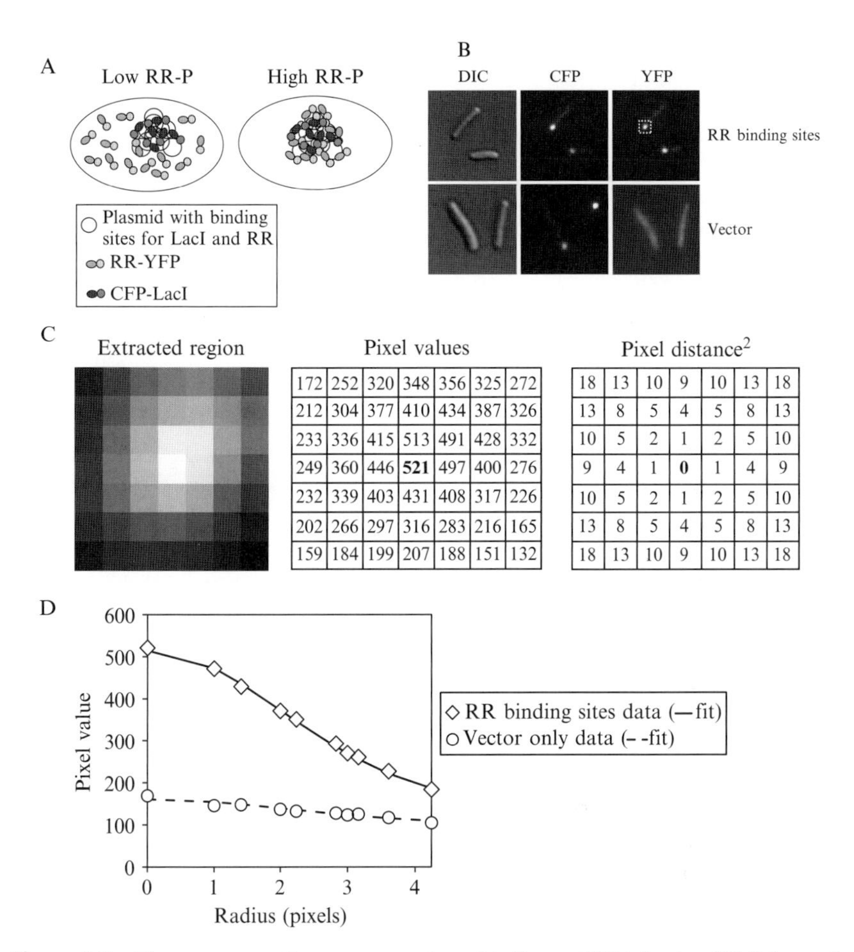

Pixel values

172	252	320	348	356	325	272
212	304	377	410	434	387	326
233	336	415	513	491	428	332
249	360	446	**521**	497	400	276
232	339	403	431	408	317	226
202	266	297	316	283	216	165
159	184	199	207	188	151	132

Pixel distance^2

18	13	10	9	10	13	18
13	8	5	4	5	8	13
10	5	2	1	2	5	10
9	4	1	**0**	1	4	9
10	5	2	1	2	5	10
13	8	5	4	5	8	13
18	13	10	9	10	13	18

Figure 1.3 Measurement of response regulator binding to DNA *in vivo*. (A) Schematic of the method, which uses a functional translational fusion of a fluorescent protein (e.g., YFP) to the response regulator and a plasmid containing response regulator binding sites. The plasmid also contains *lac* operators, which are bound by CFP-LacI. Plasmid clustering in the cell results in a high local concentration of binding sites, which is easily visualized as a bright fluorescent spot in the CFP channel. Response regulator binding to plasmid also results in a bright fluorescent spot in the YFP channel. (B) Images of cells expressing OmpR-YFP and containing a plasmid with OmpR binding sites (top row) or an empty control vector (bottom row). DIC—differential interference contrast image. (C) Example of a YFP spot, corresponding to the dashed square in the upper right image in (B), and the associated pixel values and distances from the center. (D) Gaussian fit to the profile of pixel values from (C). A corresponding fit to the neighborhood of a maximal YFP pixel in a cell containing the vector control is also shown. (See Color Insert.)

be characterized through the use of *lac* operators and a fluorescent protein fusion to *lac* repressor, as discussed below. The ability to detect fluorescent protein localization to plasmids will depend on the binding affinity of the phosphorylated response regulator-FP fusion as well as the extent of phosphorylation. One can increase the sensitivity by including several binding sites in the plasmid as needed.

It is generally advisable to include a means to monitor plasmid clustering independently of response regulator binding to DNA. This controls for potential changes in plasmid clustering as a result of changes in cell physiology. This is easily accomplished by cloning *lac* operators into the plasmid and expressing a fluorescent protein fusion to *lac* repressor. We routinely use a CFP-LacI fusion (Batchelor and Goulian, 2006), which can be simultaneously imaged with a response regulator-YFP fusion. We have used an array of tandem *lac* operator repeats (Robinett *et al.*, 1996) to mark the plasmid. This produces extremely bright foci, which is convenient as it provides increased sensitivity and enables shorter exposures when acquiring images. After plasmid clusters have been identified in the CFP channel, one can then quantify the YFP fluorescence in the neighborhood of this location. Under conditions of very high response regulator phosphorylation one may observe additional foci that are not associated with plasmids. In at least one case (OmpR-YFP), we have determined that these foci are due to response regulator binding to chromosomal loci (E. Libby and M. Goulian, manuscript in preparation).

Fluorescence localization can be visualized using techniques adapted from general single-cell fluorescence microscopy methods (e.g., Batchelor and Goulian, 2006; Miyashiro and Goulian, 2007).

1. Grow cultures in medium with aeration at the appropriate growth temperature to saturation. To maximize the sensitivity of this assay, a culture medium that gives relatively low levels of autofluorescence should be used. A minimal salts medium is ideal in this respect and also allows for choice of carbon source.
2. Dilute cultures at least 1:1000 into the same medium and grow to early exponential phase (optical density of ~0.1–0.3 at 600 nm for *E. coli*).
3. Shortly before cells reach the target density, preheat the microscope stage to the same temperature as that used for growing the cell cultures. Prepare 1% agarose pads made from the same growth medium and placed between a microscope slide and cover glass. For many small-scale experiments, it is sufficient to use 3 in. × 1 in. × 1 mm slides and 22 mm square #1.5 cover glass. Place 50–100 μl of molten agarose on the center of slide and cover immediately with coverslip. Agarose pads should be maintained at the growth temperature of the culture and may need time to dry slightly before use. Cells will not immobilize well on agarose pads that are too moist.
4. Carefully lift the cover glass off of the pad and place ~5 μl of culture onto cover glass. Replace cover glass back onto agarose pad with culture between the cover glass and pad and put the slide on the

microscope stage immediately. The number of cells on the agarose pad should be such that cells are easily found in most fields but are in sufficiently low density that there is little or no contact between neighboring cells. Obtaining proper cell density in images is especially important for the fluorescence analysis that follows.

5. Acquire a phase contrast or differential interference contrast (DIC) image and fluorescence images of the same field on a fluorescence microscope (Fig. 1.3B). Image acquisition times should be long enough so that cell fluorescence is significantly higher than background but brief enough so that none of the pixel values is equal to the maximum allowed value. The maximum pixel values are 255, 1023, 4095, and 65355 for 8-, 10- 12-, and 16-bit images, respectively. Objective magnification and resolution of the digital camera should be sufficient so that the shortest dimension of individual cells is at least approximately 5 pixels. Images should be saved without applying any contrast enhancements or other processing.

Cell fluorescence levels may be relatively weak for response regulator fusions that are expressed at wild-type levels. In addition, fluorescent foci are not always observable through the eyepiece and may photo-bleach within seconds. It is important to acquire images from fields of cells that have not been previously exposed to light from the fluorescence illuminator. In particular, cells should be brought into focus using phase or DIC imaging. The extent of fluorescence localization will vary from cell to cell due to fluctuations in protein expression levels, plasmid copy number, plasmid localization, or other variations in the cellular environment. A representative measure of fluorescence localization for the population can be attained by acquiring images of a large number of cells (e.g., 100–200) and quantifying the extent of localization as described below.

While dramatic changes in fluorescence localization may be sufficient to support a particular conclusion, in some cases it may be necessary to distinguish smaller changes in localization that are not easily assessed by eye. The fluorescence localization of cell populations can be quantified from images using software packages such as ImageJ (National Institutes of Health, Bethesda, MD), LabVIEW (National Instruments, Austin, TX), and MATLAB (The MathWorks, Natick, MA), which provide convenient libraries of imaging tools. Modules and plug-ins provided by these packages can be stitched together through scripts or macros to process large numbers of images quickly and accurately.

The basic assumption behind this analysis is that fluorescent foci can be accurately represented by point-source intensity profiles. The maximum value of the fluorescence in the cell or the maximum value in the neighborhood of the plasmid cluster is taken to be the peak of the fluorescence intensity profile. Pixel values surrounding this maximum are extracted, assigned pixel distances, and fitted to a Gaussian function. Numerical parameters derived from the best-fit intensity profiles are then used to compute an integrated fluorescence intensity, which is a measure of the extent of fluorescence localization. This method assumes that YFP fluorescence and phase contrast or DIC images have been collected. Instructions also describe cases where the investigator has chosen to express CFP-LacI to identify plasmid clusters as described above.

1. A phase contrast or DIC image and a YFP fluorescence image should be available for analysis. If the CFP-LacI system was also used, then a CFP fluorescence image should also be on hand.
2. Determine cell boundaries using the phase or DIC image. Edge detection algorithms, such as the Sobel edge filter (Castleman, 1996), can be used for identifying cell boundaries in phase contrast images. For images which contain relatively low or complex contrast, such as images from DIC microscopy, cell boundaries can be identified by thresholding pixel values (Batchelor and Goulian, 2006; Miyashiro and Goulian, 2007).
3. Once cell boundaries are determined, it is often convenient to construct a binary (black and white) mask, which defines areas where cells are present. Masks can be constructed from cell boundary images by filling in regions marked by the cell boundaries.
4. Extract pixel values from corresponding fluorescence images using the mask as a guide for cell locations. It is convenient to extract pixel values for individual cells so that each cell can be processed individually from this point on.
5. Determine the location of the plasmid cluster within the cell as follows:
   (a) If a CFP-LacI image marking plasmid clusters has been acquired, the brightest pixel within each cell boundary in the CFP image should be identified. If the response regulator-YFP is colocalized with the plasmid cluster, it should be observable in the vicinity of the LacI-CFP peak. Identify the location of the maximum YFP peak within the local pixel neighborhood of the CFP peak. For example, for a cell that is represented by approximately 500 pixels, one might choose to restrict the search area to be within a 3×3 pixel area. This maximum YFP value is taken to be the peak of the YFP (response regulator) intensity profile.
   (b) If a CFP-LacI image is not available, identify the location of the maximum pixel value within the cell boundary of the YFP (response regulator) image. This is taken to be the peak of the YFP (response regulator) intensity profile.
6. Extract pixel values surrounding the peak of the YFP intensity profile (Fig. 1.3B). The extracted area should be large enough to encompass a section representative of the spot and should not include areas outside of the cell. Assign a radial distance to each pixel value with the peak pixel value as the center (Fig. 1.3C).

7. Using a nonlinear fitting algorithm, such as the Levenberg-Marquardt algorithm (Press and Numerical Recipes Software (Firm), 1992), fit the pixel intensities I as a function of radial distance r to the Gaussian form $I = A\ exp(-Br^2) + C$ with fitting parameters A, B, and C (Fig. 1.3D). To compute the integrated peak fluorescence intensity, one can take the integral of the Gaussian using parameters for the best fit. One can also fit to a Gaussian that is not radially symmetric as described in Siryaporn and Goulian (2008). This has the advantage that it allows for asymmetric patterns of fluorescence in the neighborhood of fluorescence maxima near the boundary of the cell.

The parameters describe intensity characteristics of the peak profile: A characterizes the peak intensity value, the inverse of B characterizes the width of the spot, and C characterizes the local background fluorescence. We have found the parameter A as well as the integral of the Gaussian over the fitting area can both provide reasonable measures of response regulator-YFP colocalization with plasmids. In some cases, the fits may give very poor representations of the fluorescence distribution profile. Obviously poor fits (e.g., negative parameters for A or C) can be thrown out using a predetermined cut-off criterion. When analyzing cell images, it is important to spot-check parameters from individual cells and compare these to actual locations in the cell to verify that the program is working properly and that parameters accurately describe characteristics of the fluorescence profile.

6. Phosphatase Cross-Talk

The effects of phosphatase cross-talk in principle should be readily detectable as a decrease in response regulator phosphorylation. However, the mechanisms described above will likely suppress weak phosphatase cross-talk in the same way that phosphorylation cross-talk is suppressed. To look for possible signals of phosphatase activity, it is therefore useful to establish a system for response regulator phosphorylation in the absence of the wild-type cognate histidine kinase. For some response regulators, this can be accomplished by growing cells under conditions of high acetyl phosphate (e.g., growth on glucose or pyruvate). Another approach is to use a mutant of the cognate histidine kinase that lacks phosphatase activity but retains kinase activity. Many of these kinase+ phosphatase- mutants appear to function as relatively weak kinases. However, they nevertheless produce high levels of response regulator phosphorylation because of the absence of a mechanism for rapid response regulator dephosphorylation. If the histidine kinase being tested does show evidence of dephosphorylation activity against the noncognate response regulator, this is at least consistent with some level of phosphatase cross-talk. Of course all of the caveats described above regarding the relevance for wild-type cells still apply. In addition, one must consider the possibility that the decreased phosphorylation is due to an indirect effect, for example, from decreased response regulator expression, decreased expression of the (kinase+ phosphatase-) mutant, or decreased acetyl phosphate levels.

7. Signal Response in Cross-Talk Networks

Most histidine kinases that have been studied are associated with signal detection; they modulate the level of phosphorylated response regulator in response to changes in signal. In general, the signal could affect the rate of autophosphorylation or the rate of response regulator dephosphorylation. In some cases, modulation of the input signal for a specific histidine kinase could provide evidence for cross-talk to a noncognate response regulator. However, the interpretation of these results should be viewed with caution if other physiological effects of the input signal are not well understood. Indeed, the signals for many histidine kinases, such as osmolarity, unfolded proteins, pH, variations in specific ions, etc. are likely to have many effects on the cell through pathways distinct from the specific histidine kinase that one is attempting to stimulate. It is also quite possible that cross-talk is present but is blind to signal. Interestingly, this appears to be the case in some examples where cross-talk has been characterized (Silva *et al.*, 1998; Siryaporn and Goulian, 2008). One model suggests that this is due to the relatively weak interactions between the noncognate histidine kinase and response regulator (Siryaporn *et al.*, in preparation).

8. Concluding Remarks

The level of phosphorylated response regulator is generally expected to be sensitive to perturbations in the rates of the phosphorylation and dephosphorylation reactions. Most wild-type circuits appear to be designed to protect against such sensitivity. However, in mutant backgrounds, in which cognate histidine kinases and response regulators have been deleted, sensitivity can emerge, resulting in cross-talk from noncognate histidine kinases or alternative phosphodonors. This can be useful for studying the specificity of histidine kinase–response regulator interactions, but does not necessarily indicate a physiological role of cross-talk, or even the existence of cross-talk, in wild-type cells. Indirect effects on histidine kinase or response regulator expression level, the level of other phosphodonors, or possibly other regulators can also give an easily misinterpreted signal of cross-talk in mutant backgrounds. Nevertheless, through a careful and systematic dissection of two-component networks, the level of insulation, and flow of phosphorylation between individual components can be disentangled from the complex web of interacting regulatory pathways.

ACKNOWLEDGMENTS

The work was supported by NIH grant R01GM080279 to M. G. A. S. was also supported by the NIH Bacteriology Training Grant T32 AI060516.

REFERENCES

Alm, E., Huang, K., *et al.* (2006). The evolution of two-component systems in bacteria reveals different strategies for niche adaptation. *PLoS Comput. Biol.* **2**(11), e143.

Batchelor, E., and Goulian, M. (2003). Robustness and the cycle of phosphorylation and dephosphorylation in a two-component regulatory system. *Proc. Natl. Acad. Sci. USA* **100**(2), 691–696.

Batchelor, E., and Goulian, M. (2006). Imaging OmpR localization in *Escherichia coli*. *Mol. Microbiol.* **59**(6), 1767–1778.

Batchelor, E., Walthers, D., *et al.* (2005). The *Escherichia coli* CpxA-CpxR envelope stress response system regulates expression of the porins ompF and ompC. *J. Bacteriol.* **187**(16), 5723–5731.

Castleman, K. R. (1996). Digital Image Processing. Upper Saddle River, Prentice Hall, NJ.

Danese, P. N., and Silhavy, T. J. (1998). CpxP, a stress-combative member of the Cpx regulon. *J. Bacteriol.* **180**(4), 831–839.

Fredericks, C. E., Shibata, S., *et al.* (2006). Acetyl phosphate-sensitive regulation of flagellar biogenesis and capsular biosynthesis depends on the Rcs phosphorelay. *Mol. Microbiol.* **61**(3), 734–747.

Groban, E. S., Clarke, E. J., *et al.* (2009). Kinetic buffering of cross talk between bacterial two-component sensors. *J. Mol. Biol.* **390**(3), 380–393.

Hutchings, M. I., Hong, H. J., *et al.* (2006). The vancomycin resistance VanRS two-component signal transduction system of Streptomyces coelicolor. *Mol. Microbiol.* **59**(3), 923–935.

Igo, M. M., Ninfa, A. J., *et al.* (1989). Phosphorylation and dephosphorylation of a bacterial transcriptional activator by a transmembrane receptor. *Genes Dev.* **3**(11), 1725–1734.

Kim, S. K., Wilmes-Riesenberg, M. R., *et al.* (1996). Involvement of the sensor kinase EnvZ in the in vivo activation of the response-regulator PhoB by acetyl phosphate. *Mol. Microbiol.* **22**(1), 135–147.

Klein, A. H., Shulla, A., *et al.* (2007). The intracellular concentration of acetyl phosphate in *Escherichia coli* is sufficient for direct phosphorylation of two-component response regulators. *J. Bacteriol.* **189**(15), 5574–5581.

Laub, M. T., and Goulian, M. (2007). Specificity in two-component signal transduction pathways. *Annu. Rev. Genet.* **41,** 121–145.

Laub, M. T., Biondi, E. G., *et al.* (2007). Phosphotransfer profiling: Systematic mapping of two-component signal transduction pathways and phosphorelays. *Methods Enzymol.* **423,** 531–548.

Lukat, G. S., McCleary, W. R., *et al.* (1992). Phosphorylation of bacterial response regulator proteins by low molecular weight phospho-donors. *Proc. Natl. Acad. Sci. USA* **89**(2), 718–722.

McCleary, W. R., and Stock, J. B. (1994). Acetyl phosphate and the activation of two-component response regulators. *J. Biol. Chem.* **269**(50), 31567–31572.
Miller, J. H. (1992). A Short Course in Bacterial Genetics: A Laboratory Manual and Handbook for *Escherichia coli* and Related Bacteria. Cold Spring Harbor Laboratory Press, Plainview, NY.
Miyashiro, T., and Goulian, M. (2007). Single-cell analysis of gene expression by fluorescence microscopy. *Methods Enzymol.* **423,** 458–475.
Miyashiro, T., and Goulian, M. (2008). High stimulus unmasks positive feedback in an autoregulated bacterial signaling circuit. *Proc. Natl. Acad. Sci. USA* **105**(45), 17457–17462.
Mouslim, C., and Groisman, E. A. (2003). Control of the Salmonella ugd gene by three two-component regulatory systems. *Mol. Microbiol.* **47**(2), 335–344.
Ninfa, A. J., Ninfa, E. G., *et al.* (1988). Crosstalk between bacterial chemotaxis signal transduction proteins and regulators of transcription of the Ntr regulon: Evidence that nitrogen assimilation and chemotaxis are controlled by a common phosphotransfer mechanism. *Proc. Natl. Acad. Sci. USA* **85**(15), 5492–5496.
Nixon, B. T., Ronson, C. W., *et al.* (1986). Two-component regulatory systems responsive to environmental stimuli share strongly conserved domains with the nitrogen assimilation regulatory genes ntrB and ntrC. *Proc. Natl. Acad. Sci. USA* **83**(20), 7850–7854.
Pogliano, J. (2002). Dynamic cellular location of bacterial plasmids. *Curr. Opin. Microbiol.* **5**(6), 586–590.
Press, W. H., and Numerical Recipes Software (Firm) (1992). Numerical recipes in C. Cambridge [England]. Cambridge University Press, New York.
Robinett, C. C., Straight, A., *et al.* (1996). In vivo localization of DNA sequences and visualization of large-scale chromatin organization using lac operator/repressor recognition. *J. Cell. Biol.* **135**(6 Pt 2), 1685–1700.
Shinar, G., Milo, R., *et al.* (2007). Input output robustness in simple bacterial signaling systems. *Proc. Natl. Acad. Sci. USA* **104**(50), 19931–19935.
Silva, J. C., Haldimann, A., *et al.* (1998). In vivo characterization of the type A and B vancomycin-resistant enterococci (VRE) VanRS two-component systems in *Escherichia coli*: A nonpathogenic model for studying the VRE signal transduction pathways. *Proc. Natl. Acad. Sci. USA* **95**(20), 11951–11956.
Simons, R. W., Houman, F., *et al.* (1987). Improved single and multicopy lac-based cloning vectors for protein and operon fusions. *Gene* **53**(1), 85–96.
Siryaporn, A., and Goulian, M. (2008). Cross-talk suppression between the CpxA-CpxR and EnvZ-OmpR two-component systems in *E. coli*. *Mol. Microbiol.* **70**(2), 494–506.
Skerker, J. M., Prasol, M. S., *et al.* (2005). Two-component signal transduction pathways regulating growth and cell cycle progression in a bacterium: A system-level analysis. *PLoS Biol.* **3**(10), e334.
Stock, J. B., Ninfa, A. J., *et al.* (1989). Protein phosphorylation and regulation of adaptive responses in bacteria. *Microbiol. Rev.* **53**(4), 450–490.
Wanner, B. L. (1992). Is cross regulation by phosphorylation of two-component response regulator proteins important in bacteria? *J. Bacteriol.* **174**(7), 2053–2058.
Wanner, B. L., and Wilmes-Riesenberg, M. R. (1992). Involvement of phosphotransacetylase, acetate kinase, and acetyl phosphate synthesis in control of the phosphate regulon in *Escherichia coli*. *J. Bacteriol.* **174**(7), 2124–2130.
Wolfe, A. J. (2005). The acetate switch. *Microbiol. Mol. Biol. Rev.* **69**(1), 12–50.
Wolfe, A. J., Parikh, N., *et al.* (2008). Signal integration by the two-component signal transduction response regulator CpxR. *J. Bacteriol.* **190**(7), 2314–2322.
Yamamoto, K., Hirao, K., *et al.* (2005). Functional characterization in vitro of all two-component signal transduction systems from *Escherichia coli*. *J. Biol. Chem.* **280**(2), 1448–1456.

CHAPTER TWO

Inference of Direct Residue Contacts in Two-Component Signaling

Bryan Lunt,* Hendrik Szurmant,† Andrea Procaccini,* James A. Hoch,† Terence Hwa,‡ *and* Martin Weigt*

Contents

Abstract

Since the onset of the genomic era more than 1000 bacterial genomes have been sequenced and several fold more are expected to be completed in the near future. These genome sequences supply a wealth of information that can be

* Institute for Scientific Interchange, Viale S. Severo 65, Torino, Italy
† Department of Molecular and Experimental Medicine, The Scripps Research Institute, La Jolla, California, USA
‡ Center for Theoretical Biological Physics, University of California San Diego, La Jolla, California, USA

Methods in Enzymology, Volume 471
ISSN 0076-6879, DOI: 10.1016/S0076-6879(10)71002-8

exploited by statistical methods to gain significant insights into cellular processes. In Volume 422 of *Methods in Enzymology* we described a covariance-based method, which was able to identify coevolving residue pairs between the ubiquitous bacterial two-component signal transduction proteins, the sensor kinase and the response regulator. Such residue position pairs supply interaction specificity in the light of highly amplified but structurally conserved two-component systems in a typical bacterium and are enriched with interaction surface residue pairings. In this chapter we describe an extended version of this method, termed "direct coupling analysis" (DCA), which greatly enhances the predictive power of traditional covariance analysis. DCA introduces a statistical inference step to covariance analysis, which allows to distinguish coevolution patterns introduced by direct correlations between two-residue positions, from those patterns that arise via indirect correlations, that is, correlations that are introduced by covariance with other residues in the respective proteins. This method was shown to reliably identify residue positions in spatial proximity within a protein or at the interface between two interaction partners. It is the goal of this chapter to allow an experienced programer to reproduce our techniques and results so that DCA can soon be applied to new targets.

1. Introduction

Proteins serve to execute most biochemical functions within all cellular organisms. Equally important as the individual activity of a particular protein is its ability to specifically interact with partner proteins. Cases in point are multiprotein machineries such as the ribosome, RNA polymerase holoenzyme, or the bacterial motility apparatus, the flagella. Most of the individual components of these macromolecular complexes are without use to a cell when not in contact with the other components of the complex. For this reason, protein–protein interaction interfaces are considered as potential, yet relatively unutilized drug targets (Wells and McClendon, 2007).

High-resolution X-ray structures have provided significant insights into many macromolecular protein complexes and identified their protein–protein interfaces. One of the great success stories in X-ray crystallography was the resolution of the entire ribosome (Ramakrishnan, 2008; Yusupov *et al.*, 2001). Still, not all protein complexes are as inherently stable as the above-mentioned examples (Cusick *et al.*, 2005). Indeed, there is a requirement for many protein interactions to be transient to allow a single protein to travel in cellular space and to interact with different partners, while utilizing overlapping interaction surfaces. An example, well known to the bacterial signal transduction community, is the chemotaxis response regulator protein CheY, which utilizes overlapping surfaces to interact with either the P2 domain of the kinase CheA, the C-terminal signature peptide of the phosphatase CheZ or the N-terminal signature peptide of the flagella switch protein FliM (and FliY in *Bacillus subtilis*) (Dyer *et al.*, 2004; Szurmant *et al.*, 2003; Welch *et al.*, 1998; Zhao *et al.*, 2002; Zhu *et al.*, 1997). In general protein–protein interactions in signal transduction are expected to be transient, for above-mentioned reasons. Capturing such transient interactions in X-ray crystals has proven challenging.

The common signaling cascade utilized by the bacteria is the two-component system (Hoch, 2000). These systems transform a signal into an appropriate response via two proteins, a signal detecting sensor histidine kinase and a response regulator protein, typically a transcription factor. The message between the proteins is passed by transfer of a phosphoryl group from the kinase to the regulator. Not due to lack of effort, as of August 2009, there was no published structure of the complex of a true sensor kinase/response regulator trapped in the act of phosphotransfer. A close structural and functional homologue and the only representative crystal structure of such a complex is that of the sporulation phosphorelay proteins Spo0B with Spo0F (Zapf *et al.*, 2000). Based on structural similarity, Spo0B is an evolutionary divergent kinase, having lost the ability to autophosphorylate but having retained the ability to interact with and transfer phosphate to response regulator proteins Spo0F and Spo0A (Burbulys *et al.*, 1991; Varughese, 2002).

In the light of the obvious difficulty of capturing transient interactions by experimental means, we recently developed a covariance-based method utilizing the mutual information (MI) measure with the aim of identifying residue/residue contacts at protein/protein interfaces from sequence alone. This method, applied to two-component signaling proteins, was described in some detail in Volume 422 of *Methods in Enzymology* (White *et al.*, 2007). This chapter is an extension of the previous work featuring an additional step, which vastly improves the predictive power of covariance analysis for reasons outlined below.

Covariance-based methods have been extensively applied to gain insights into protein tertiary structure (Altschuh *et al.*, 1987; Atchley *et al.*, 2000; Göbel *et al.*, 1994; Suel *et al.*, 2003) and more recently to identifying protein interaction surfaces (Burger and van Nimwegen, 2008; Kass and Horovitz, 2002; Thattai *et al.*, 2007; White *et al.*, 2007). These methods are based on the underlying assumption that structural details of the interaction are conserved across homologous proteins, and that the residue positions at the contact surface between two protein interaction partners (or in contact within a protein fold) are constrained. Not all amino acid combinations are equally acceptable for positions in contact; the statistical properties of pairs of contact positions thus differ from arbitrarily chosen residue position pairs. To measure such constraints one needs to have a large protein sequence database of homologous paired interaction partners. In the light of the number of bacterial genomes that have been sequenced ($\sim$1000) or whose sequence is in the works (~3500) (Liolios *et al.*, 2008), such methods are starting to become amenable for those proteins, which are encoded by the majority of bacterial genomes. Some protein systems are amplified in bacterial genomes, for example, two-component signaling systems. While these provide more statistically relevant sequence data for analysis, an added difficulty is that the actually interacting protein pairs cannot be identified merely by being found in the same genome. Instead additional information is necessary to identify, which two proteins are interaction partners. For two-component systems, this tends to be unproblematic since a large fraction of these systems are organized into operons. Hence, chromosomal adjacency is utilized to infer interaction partners.

In this context, MI measures the amount of information provided by the knowledge of the amino acid present in one position (in the first protein) about the one present in the other position (in the second protein). When applying covariance analysis to two-component signaling proteins to infer how the two proteins interact with each other during phosphotransfer, it becomes apparent that within the highest MI residue position pairs (i.e., the most constrained ones), many were found to be in close proximity in the above mentioned Spo0B/Spo0F cocrystal structure, demonstrating that covariance analysis is able to strongly enrich surface contact pairings (White *et al.*, 2007). Other high-MI residue pairings, however, were distant from the interface and involved cluster of residues connecting buried residues in the four-helix bundle core of the kinase to a highly dynamic region of the response regulator. The importance of these correlations for SK/RR recognition has been described (McLaughlin *et al.*, 2007; Szurmant *et al.*, 2008). In the light of identifying protein interaction surfaces, however, these highly correlated pairings have to be considered false positives.

A shortcoming of covariance analysis is that correlations between a pair of residue positions might arise from direct as well as indirect effects. Indirect effects can occur, for example, when a given residue has a conformational effect on the placement of residues at the protein interface. Similarly, a highly connected net of weak direct interactions will lead to inflation of covariance values due to multiple correlation chains. To reduce the effect of correlation chains we previously applied a so-called "best-friend" transformation (White *et al.*, 2007). Within the set of highly correlated residue positions only those pairings are considered relevant, which display the highest MI value for a particular residue positions. While such a transformation reduces the effect of correlation chains it certainly does not eliminate them, and in addition relevant information is discarded.

Covariance analysis cannot distinguish between direct and indirect correlations since it is a local measure, that is, correlations between residue positions are calculated individually without the context of the other residue positions in the proteins of interest. To distinguish between direct and indirect correlations, each residue position pair has to be investigated in the context of all other positions in the proteins. To achieve this we developed an improved method, called direct coupling analysis (DCA), which adds a global model-inference step using message passing to covariance analysis (Weigt *et al.*, 2009). In this step, the MI values derived by covariance analysis are split into direct and indirect contributions. The approach is based on the premise that only these strong direct interactions are an indicator for correlated substitutions caused by functional residue contacts. The procedure produces a new measure termed direct information (DI), which represents the contribution to the MI that is estimated to derive due to direct correlation of two residue positions.

When, applying DCA to two-component signaling proteins, it becomes apparent that the predictive power of covariance analysis is greatly enhanced by the message-passing step. This has been described in detail (Weigt *et al.*, 2009). Here we highlight some of the results. Positions of the top 15 MI and DI pairings in structural HisKA and RR models are compared in Fig. 2.1A. Within the set of highly correlated residue positions less than 40% are found at the interaction surface between Spo0F and Spo0B, whereas the remainder of connections is distal to the interface (Fig. 2.1B and C). After applying the message-passing filter, the set of directly correlated positions involves 10 pairings that are at the interaction surface in the Spo0B/Spo0F cocrystal structure.[1] Such information is sufficient to generate high-resolution structural models of protein complexes, where the structure of the individual proteins is known, as described in the following chapter in this edition of *Methods in Enzymology*.

In the following we describe the step-by-step procedure of how to extract coupled interaction surface residue positions from databases of interacting proteins. The dataflow of the entire process is given in Fig. 2.2. The procedure is presented in two major sections. The first section, Section 2, describes how a database of interacting protein sequences is build. Genomic data is downloaded from NCBI's RefSeq database (described in Section 2.1), analyzed to create a database of predicted operons (described in Section 2.2), and searched with the HMMER package to extract and align protein domains of interest to the user (described in Section 2.3). Extracted protein domains are joined into predicted pairs (i.e., pairs of sample sequences of interacted proteins) using the Operon database (described in Section 2.4). Utilizing the assembled database, the second section, Section 3, describes how coupled residue positions are identified by MI analysis, and subsequently how direct correlation is distinguished from

[1] The other five parings involve HisKA helix α2 residues 291, 294, and 298, which are ignored here, since they cannot be reliably mapped to Spo0B, but likely represent interface contact pairings in SK/RR complexes.

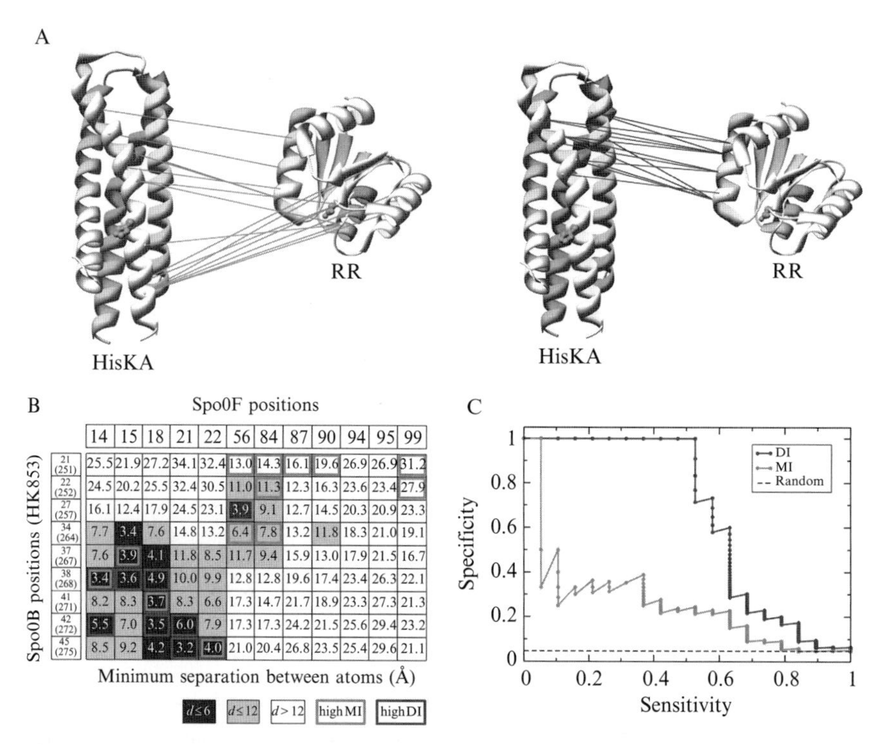

Spo0B positions (HK853)	14	15	18	21	22	56	84	87	90	94	95	99
21 (251)	25.5	21.9	27.2	34.1	32.4	13.0	14.3	16.1	19.6	26.9	26.9	31.2
22 (252)	24.5	20.2	25.5	32.4	30.5	11.0	11.3	12.3	16.3	23.6	23.4	27.9
27 (257)	16.1	12.4	17.9	24.5	23.1	3.9	9.1	12.7	14.5	20.3	20.9	23.3
34 (264)	7.7	3.4	7.6	14.8	13.2	6.4	7.8	13.2	11.8	18.3	21.0	19.1
37 (267)	7.6	3.9	4.1	11.8	8.5	11.7	9.4	15.9	13.0	17.9	21.5	16.7
38 (268)	3.4	3.6	4.9	10.0	9.9	12.8	12.8	19.6	17.4	23.4	26.3	22.1
41 (271)	8.2	8.3	3.7	8.3	6.6	17.3	14.7	21.7	18.9	23.3	27.3	21.3
42 (272)	5.5	7.0	3.5	6.0	7.9	17.3	17.3	24.2	21.5	25.6	29.4	23.2
45 (275)	8.5	9.2	4.2	3.2	4.0	21.0	20.4	26.8	23.5	25.4	29.6	21.1

Figure 2.1 Comparison of results derived by covariance analysis with direct coupling analysis. (A) The top 15 residue pairings identified by covariance analysis (left in green) and the top 15 residue pairings identified by direct coupling analysis (DCA), which includes an additional statistical inference step are mapped on exemplary structures for the HisKA domain (HK853 from *Thermotoga maritima*: PDBID 2C2A) and RR domain (Spo0F from *B. subtilis*: PDBID 1PEY). It becomes apparent that the additional inference step increases strongly the specificity of the contact residue prediction. (B) The table shows minimal atom distances in the Spo0B/Spo0F cocrystal structure between all residues that are identified by covariance analysis or DCA. The top pairings identified by covariance analysis are framed in green and identified by DCA are framed in red. Since the Spo0B helix $\alpha2$ is oriented different and cannot be aligned with regular HisKA helix $\alpha2$, residue positions 291, 294, and 298 (HK853 numbering) are ignored for this analysis. Out of the 15 pairings displayed in each of the figures in Panel (A), 14 high-MI and 10 high-DI pairings do not involve these residues, and are included into the figure. (C) Comparison of specificity/sensitivity curves of covariance analysis versus DCA, where distances below 6 Å in the Spo0B/Spo0F cocrystal structure are considered as real contacts. (See Color Insert.)

indirectly coupled positions by applying a message-passing algorithm. In Section 3.1, we explain a reweighting procedure to compensate for unequal sampling of the space of possible protein sequences. The main part of the section deals with the extraction of correlation measures, in particular

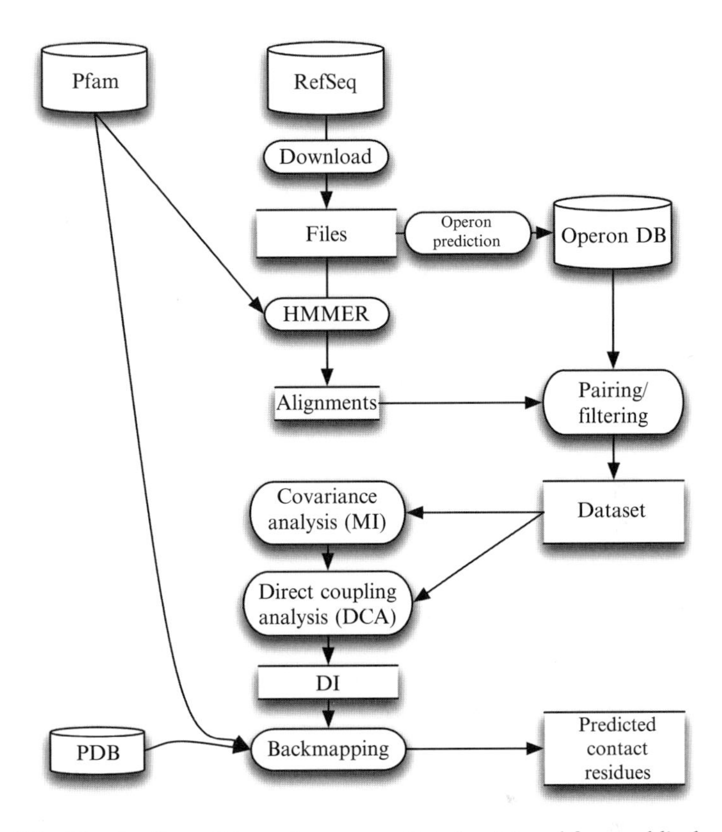

Figure 2.2 The dataflow of the entire process. Data is retrieved from public databases, analyzed to predict operons and populate the Operon database, searched with HAMMER and paired and filtered with the Operon database to create the dataset. In the "direct coupling analysis" step, the statistical correlations between columns in the protein alignments are analyzed to determine which are direct and indirect. Finally, "direct information" is used to predict interfacing residues, and this prediction is expressed on a molecular model.

MI and the novel DI, which will provide candidate pairs for interprotein residue contacts. As the derivation of the inference algorithm in the second section is given in detail in Weigt *et al.* (2009) and based on standard Belief Propagation methods described in Kschischang *et al.* (2001), Mezard and Mora (2009), and Yedidia *et al.* (2001), we will dispense with the derivation and describe only implementation in concrete terms. It is the ultimate goal that using this document, an experienced programer should be able to reproduce our techniques and results. Furthermore, it is greatly hoped that these techniques can then be extended to new problems, or added to by the reader.

2. Extraction Tools

Before it is possible to make any detailed inference about a pair of interacting protein domain families, it is necessary to generate a paired dataset of those domains in question. In the case where the domains appear on the same protein, this is relatively straightforward. However, interprotein interactions are of as much or more interest, and the generation of these datasets presents significantly more difficulty. The dataset generation phase is of crucial importance, because larger datasets will provide better estimates of the statistical properties of the domains in questions. However, if a large dataset is generated at the expense of fidelity, the estimates generated will be of little value. In this section, we present the tools and techniques for extracting the largest possible high-fidelity datasets.

2.1. Data sources

2.1.1. RefSeq

The NCBI RefSeq database (Pruitt *et al.*, 2009) provides nonredundant, curated sequence information for a variety of organisms, its bacterial database is available from ftp://ftp.ncbi.nih.gov/genomes/Bacteria. As of July 2009, it contains 905 unique taxonomy IDs for genome projects of bacteria and their substrains. This data is made available in a variety of common file-formats.

2.1.2. Pfam

The Protein Families Database (Finn *et al.*, 2008) contains HMMER (Eddy, 1998) format profile hidden Markov models (HMM(s)) for 10,340 protein domains and families. These models are instrumental in the extraction and alignment phase of dataset generation.

2.2. Operon database

To go beyond the limitations of earlier protein-pair prediction schemes based on GI number adjacency (and presumed genetic adjacency), in this work we have introduced the use of a database of predicted operons. Genes that function together are often transcribed on a single mRNA in bacteria, and in particular this is the case for many of the TCS, which are at the center of interest in this publication. We take advantage of this fact to increase the number and accuracy of our protein-pair predictions.

2.2.1. Database structure

The database is a simple relational database consisting of two necessary and two optional tables that represent the relationship between genes, predicted operons, and optionally chromosomes and genomes. The Entity Relationship diagram of the database is given in Fig. 2.3.

From the various files provided in the RefSeq directory indicated in Section 2.1, a unified list of all predicted protein and RNA coding genes can be extracted for each chromosome. This list is then broken into regions of contiguous genes with the same coding sense. Finally, each of these contiguous regions are broken into predicted operons at any intergenic region larger than a specified threshold, in our case chosen to be any distance larger than 200 base pairs (bp). Operons are predicted solely with an intergenic distance cutoff of same-sense genes inspired by (Moreno-Hagelsieb and Collado-Vides, 2002). Brouwer *et al.* (2008) concluded that Moreno-Hagelsieb and Collado-Vides' method is more effective than many more sophisticated contributions involving considerable time investment.

2.2.2. Comparison to known operons

Of 876 *Escherichia coli* operons contained in RegulonDB (Gama-Castro *et al.*, 2008) whose descriptions contain either the word "experimental" (experimentally identified) or that are predicted based on having no adjacent genes on the same strand, 576 are predicted identically with the 200 bp cutoff, 219 are joined with others, but have none of their own genes separated, and 19 are split apart, with 62 unaccounted for. Since it is more

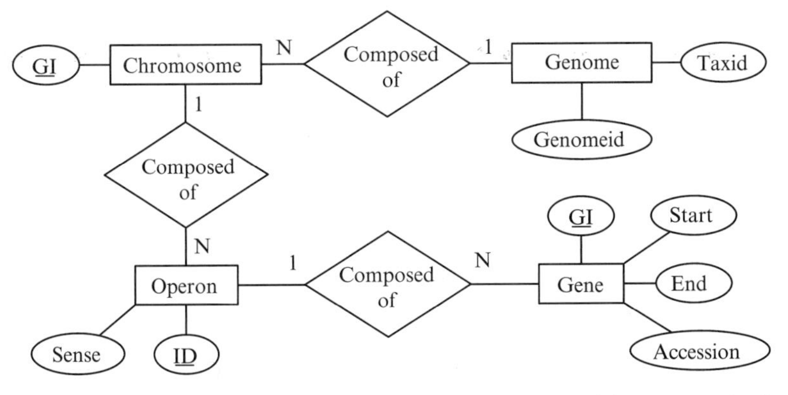

Figure 2.3 Entity relationship diagram of a simple realization of the Operon database. Genes store the starting and ending position, GI number, an Accession, and the Operon ID of the operon that contains them. Operons contain a system generated ID and the coding sense of the genes contained thereon. If the Chromosome and Genome tables are impelmented, Operons will also have a reference to the GI of the chromosome that contains them, Chromosomes make reference to the Genome that contains them. Any of these tables can be extended with other information.

important in our application to maintain the grouping of nearby genes while providing a reasonable cutoff for pairing searches, we consider identically predicted and joined operons as "correct." This gives us a 91% accuracy for maintaining operon grouping on the experimentally identified *E. coli* operons.

2.3. Extraction and alignment

The HMMER suite of tools is used to create alignments to the HMMs of those domains from which one wishes to produce matched pairs. Additionally, any domain that will be used in the logical filtering step must also be extracted and aligned. For example, in addition to the HisKA and Response_reg domains, the HATPase_c domain is also extracted to be used as a filter to improve the extraction specificity of HisKA-containing proteins. All extracted domains are automatically aligned to their HMMs, and we use this alignment to join all domains into large multiple-sequence alignments (MSA).

As insert positions in the alignment will be of varying length and introduce MSA columns that are predominantly gap characters, insert positions are discarded, resulting in alignments only of those residues from match and gap states in the HMM.

It is essential at this stage that the GI number, accession, subsequence location of the match, and other data be stored for use in the next processes. In our implementation, this is stored in the description line of a FASTA-like format file, for example:

```
>gi|16131282|ref|NP_417864.1|/5-117 E=2.4e-41
  [Escherichia_coli_K12_substr_MG1655]
```

Note that it may be practical to include the *E*-value of the alignment to the HMM for further processing into this description line, as will be discussed below.

2.4. Pairing and filtering

The processes in this section describe the actual creation of protein pairs. These steps need not be executed in the order presented here, and can be mixed together into dataflows appropriate for the system under scrutiny.

2.4.1. Single protein architecture filtering

Members of each input alignment can be filtered based on the presence or absence of other domains on the same protein, for example, HisKA containing proteins without an occurrence of HATPase_c may be discarded, as can hybrid proteins containing both HisKA and Response_reg, and proteins containing multiples of either, etc.

The fraction of false positive HisKA containing proteins extracted by the HMM is drastically reduced by requiring the presence of the HATPase_c domain. No similar constraint exists for Response_reg domains (the corresponding proteins may exist as single-domain proteins). We have used the E-value provided by HMMer to reduce the false positive rate, including only domains with $E < 0.01$. Similarly, models from the Pfam database include various cutoff scores that have been well tuned by the Pfam team.

2.4.2. Operon pairing

All protein domains provided to the operon pairing function should have the ID of the operon that they appear on looked up in the Operon database, and be put into groups accordingly. It is important to mention that at this stage, these groupings are more flexible than being a simple pair. For example, two or more of any of the target domains may appear on one operon, or other filtering domains may or may not appear. This is handled in the next stage.

2.4.3. Operon architecture filtering

Once the domains of interest have been grouped by predicted operon, these operons can be filtered according to the presence or absence of other filter domains or even whole protein architectures appearing in the same operon. For example, in the TCS system, an operon containing multiple Response_reg domains will be discarded, whereas those with an unambiguous HisKA/Response_reg pair will be immediately added to the dataset.

2.4.4. Orphan datasets

Finally, though the handling of unpaired "Orphans" is beyond the scope of this document, it should be noted in passing that domains that could not be placed in paired proteins can be saved for use in other analyses.

2.5. Final dataset

After being extracted according to the description in the preceding sections, data takes the form of concatenated strings of length N from an alphabet consisting of the standard IUPAC amino acid codes, and an additional character ("-") to represent alignment gaps. This results in a $Q = 21$ letter alphabet. Letting M be the number of domain pairs in the dataset, this gives an MSA in the form of a $M \times N$ matrix (A_i^a), $i = 1, \ldots, N$, $a = 1, \ldots, M$ over $\{1, \ldots, Q\}$, of data from which we will compile the statistics that will be used for inference.

In the specific case of TCS, we obtained $M = 8998$ pairs. The single Pfam domains have length 87 (HisKA) resp. 117 (Response_reg), resulting in $N = 204$ letters per MSA row.

3. DCA: Direct Coupling Analysis

In this section, we present the algorithm for modeling the statistical properties of protein domain pairs. As our ultimate goal is to predict physical interprotein residue contacts in protein multimers, we will first measure the statistical correlation between residue positions in the domain-pairs using Shannon's MI (Shannon, 1948). However, while MI can help determine which residue pairs show a statistically significant correlation, MI alone cannot reveal which pairs interact directly. Significant values of MI can be the result not only of strong direct couplings, but also of multitudinous couplings through intermediate residues. Thus without some way to discriminate between the contributions to MI from direct coupling and induced coupling, we would be at a loss to identify physically interacting residue positions. In this section, we introduce a global inference method that will lead to the notion of DI, as a measure of those contributions to MI, which result only from direct interaction. The main idea of disentangling direct and indirect statistical coupling is illustrated in Fig. 2.4, and the dataflow for the full statistical analysis of our dataset is given in Fig. 2.5.

In this section, we describe our step-by-step approach to statistical analysis of the MSA. First we describe how to correct for uneven sampling effects by a simple reweighting procedure, and introduce reweighted frequency counts for single residue positions and position pairs. Based on these counts we calculate MI as a total correlation measure. The major part of this section is dedicated to disentangling direct and indirect statistical coupling: First the global statistical model is introduced together with compatibility

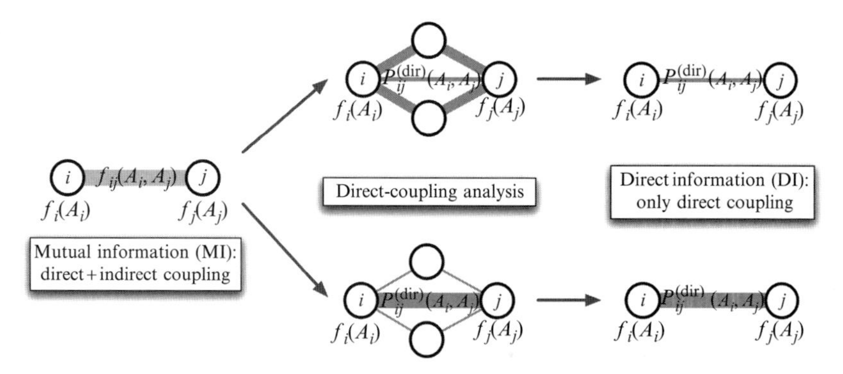

Figure 2.4 The main idea of *direct coupling analysis* (DCA). The MI between MSA columns in our dataset measures the total statistical coupling between two-residue positions in the protein domains under consideration. However, as illustrated in the figure, high MI can result from direct and indirect couplings, which are disentangled by DCA. *Direct information* (DI) measures the correlation between the two positions due to direct coupling alone, by pruning all indirect effects including intermediate positions.

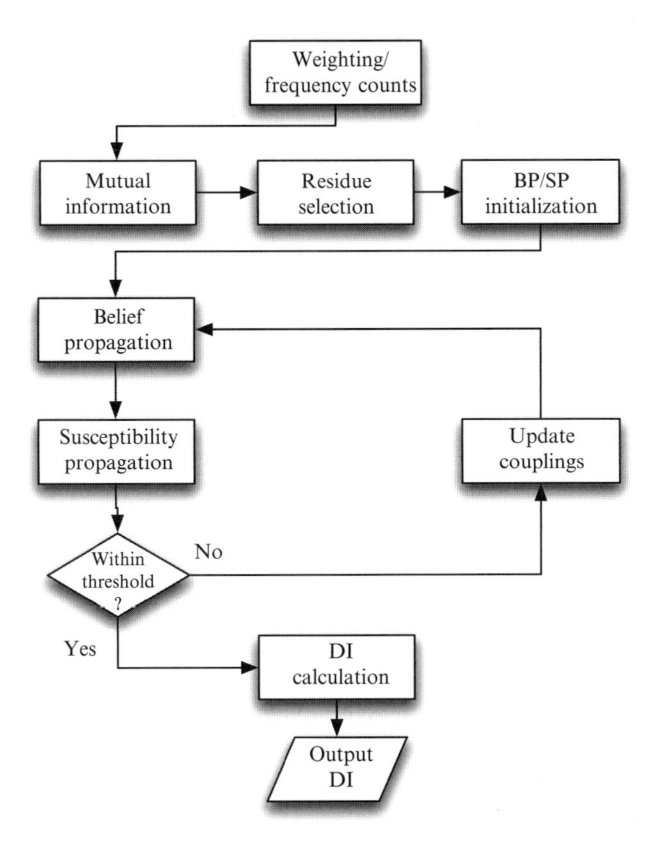

Figure 2.5 The dataflow of the *direct coupling analysis* (DCA) segment. The dataset is read and reweighted, "mutual information" is calculated and used to select residues for DCA, Belief Propagation and Susceptibility Propagation are used to calculate two-site marginal values, calculated values are compared to observed values and couplings are updated until overall convergence. Finally, direct information (DI) is calculated and output.

contraints to the empirical genomic data, then our approach of extracting model parameters using message passing is discussed. It results directly in a description of how to calculate DI, which is used to predict residue contacts. At the very end of this section, a backmapping procedure is discussed, which allows for translation of the results obtained in terms of MSA columns to actual representatives of the protein families.

3.1. Weighting

For a number of reasons, the dataset at this point may not represent an even sampling of the space of possible functional pairs. These reasons include the effects of phylogeny, paralogy, and oversampling of given pairs caused by duplication within a single genome or duplicates derived from an abundance of very similar substrains of a particular bacterial species. Such pairs may represent a disproportionately large part of the dataset, and must be downweighted.

Each of the concatenated paired sequences in the dataset is taken as a single sequence. Each one is compared to all others according to a user defined distance metric. For each entry (i.e., sequence), the number of entries closer than a chosen threshold for the chosen distance metric is recorded. Thus we define the list of weights W^a for each entry a in the dataset as

$$\begin{aligned} W^a &= \frac{1}{d^a} \\ d^a &= |\{x \in \text{entries} \mid \text{dist}(x, a) \leq \text{threshold}\}| \end{aligned} \tag{2.1}$$

Note that the elements x will always include a as dist$(a, a) \equiv 0$, thus a functional pair with no similar pair receives a weight of 1, each copy of a pair that occurs twice receives a weight of 0.5 and so forth.

In our case, we chose to use the Hamming distance between the two strings, and the cutoff value of 80% sequence identity was chosen, however, for systems showing much higher conservation than the TCS system, this would not be appropriate.

3.2. Frequency counts

Both parts of the inference task are dependent solely on single and dual-site frequency counts, which are loaded from the dataset described in the previous section.

Thus we define:

$$\begin{aligned} f_i(A_i) &= \frac{1}{\lambda Q + \sum_a W^a}\left[\lambda + \sum_{a=1}^{M} \delta(A_i, A_i^a) W^a\right] \\ f_{ij}(A_i, A_j) &= \frac{1}{\lambda Q + \sum_a W^a}\left[\frac{\lambda}{Q} + \sum_{a=1}^{M} \delta(A_i, A_i^a)\delta(A_j, A_j^a) W^a\right] \\ \delta(x, y) &= \begin{cases} 1 & x = y \\ 0 & x \neq y \end{cases} \end{aligned} \tag{2.2}$$

for the frequency of occurrence of amino acid A_i in column i, and the frequency of co-occurrence of (A_i, A_j) in residue pair (i, j). These formulae contain a pseudocount $\lambda > 0$, which helps to regularize frequency counts for finite-sample effects. It prevents zero counts, which would lead to divergent couplings in the following inference task. In our implementation we choose a pseudocount of one, which becomes less and less relevant in cases where $M \gg \lambda Q$. It is essential to note that the pseudocount terms in Eq. (2.2) are chosen such that consistency is preserved,

$$\sum_{A=1}^{Q} f_{ij}(A_i, A_j) \equiv f_i(A_i)$$

for all $i, j \in \{1,\ldots,N\}$ and all $A_i \in \{1,\ldots,Q\}$

The frequency counts f_i and f_{ij} are the only inputs to the calculation of MI and to DCA. Indeed, this has important implications for the efficiency of these algorithms, as an increase in the size of the dataset will give better estimates of occurrence and co-occurrence frequencies, without affecting the speed of inference.

3.3. Mutual information

One of the simplest methods to detect correlation between column couplets is Shannon's MI (Shannon, 1948):

$$\mathrm{MI}_{ij} = \sum_{A_i, A_j \in \{1,\ldots,Q\}} f_{ij}(A_i, A_j) \ln \frac{f_{ij}(A_i, A_j)}{f_i(A_i) f_j(A_j)} \qquad (2.3)$$

MI measures the Kullback–Leibler divergence of the joint distribution $f_{ij}(A_i, A_j)$ and the factorized term $f_i(A_i)f_j(A_j)$. MI is the amount of information in nats that are available about the identity of amino acid A_i by knowing A_j, and *vice versa*. One nat is equivalent to $\sim$1.44 bits = 1 bit/ln 2 = 1 nat. When $f_{ij}(A_i, A_j)$ is in fact factorized, and the joint frequency shows nothing more than what would be expected of two independent random variables, MI is zero, otherwise it is positive. Mutual information MI_{ij} is calculated for every residue pair $i, j|i < j$ and these values are stored.

3.4. Global statistical modeling

As illustrated in Figs. 2.1 and 2.4, high MI can result both from direct and indirect couplings including intermediate residues. Hence some of the high-MI pairings are found at the interaction surface of the proteins of interest whereas others are not. To improve the predictive power of covariance analysis, the direct coupling effect alone needs to be estimated, which may be connected to physical interprotein contacts; it is necessary to consider not only a single residue pair at a time (as in MI), but also to model the global statistical properties of entire protein sequences. In principle, we would thus wish to construct a full joint-probability distribution for the entire concatenated protein string $P(A_1,\ldots,A_N)$, however, in order to accurately sample this distribution directly, $\Omega(21^N)$ sequence pairs would be needed. With current dataset sizes, accurate direct estimates are only available for pairs of residues. Even triplets of residues would need substantially more than $Q^3 = 9261$ samples, which goes beyond the currently still limited number of data available from fully sequenced genomes.

Because we are only capable of directly measuring single and pairwise joint distributions, the only consistency-check that we can make on our statistical model is that it renders those same distributions when marginalized.

$$\begin{aligned} f_i(A_i) &\equiv P_i(A_i) = \sum_{A_k|k\neq i} P(A_1, \ldots, A_N) \\ f_{ij}(A_i, A_j) &\equiv P_{ij}(A_i, A_j) = \sum_{A_k|k\neq i,j} P(A_1, \ldots, A_N) \end{aligned} \tag{2.4}$$

The principle of maximum entropy (Jaynes, 1949) provides the motivation for the otherwise minimally constrained model that we shall construct. As discussed in more detail in Weigt *et al.* (2009), this principle leads to a simple form of the statistical model in terms of pairwise residue interactions and single residue biases:

$$\begin{aligned} P(A_1, \ldots, A_N) &= \frac{1}{Z} \prod_{i<j} \exp\{-e_{ij}(A_i, A_j)\} \prod_i \exp\{h_i(A_i)\} \\ Z &= \sum_{\{A_i\}} \prod_{i<j} \exp\{-e_{ij}(A_i, A_j)\} \prod_i \exp\{h_i(A_i)\} \end{aligned} \tag{2.5}$$

This distribution includes (still unknown) local biases (fields) $h_i(A_i)$ and two-residue couplings (interactions) $e_{ij}(A_i, A_j)$, which will ultimately be used to estimate the direct interactions between i and j. Z serves to guarantee the normalization $\sum_{\{A_i\}} P(A_i, \ldots, A_j) = 1$. Readers with a statistical-physics background will recognize Eqs. (2.5) as the Boltzmann–Gibbs distribution of a disordered Q-state Potts model.

The model parameters $\{h_i(A_i)\}$ and $\{e_{ij}(A_i, A_j)\}$ have to be determined such that Eqs. (2.4) are fulfiled. To do this exactly would require summation over 21^N terms and is therefore computationally infeasible. To overcome this barrier, we first have to restrict the number of residues under consideration (i.e., to reduce N), and second to use semiheuristic approaches like message passing providing beliefs for the single-residue and pair marginals of $P(A_1, \ldots, A_N)$ (i.e., to go from an exact procedure of exponential time complexity to a semiheuristic polynomial-time algorithm).

3.5. Residue selection

The main idea of selecting potentially relevant residues is that the set of position pairs i and j with considerable direct coupling is included in the set of position pairs of high total statistical coupling as measured by MI. In our model inference, we therefore include only those columns of our dataset, which show high MI with at least one other residue in the other protein domain.

Interdomain mutual information values MI_{ij} (where $i \leq N_1 < j$, with $N_1 = 87$ denoting the length of the HisKA domain) calculated in the previous step are therefore sorted, and those residues participating in position pairs with highest MI are progressively selected until the requisite number of residues is attained.

As all subsequent calculations will only operate on this subset of residues, we will reuse the symbol N, which from now on will be taken to be the size of the subset in consideration. The value of N can be selected by the implementor based on time and hardware constraints. In our implementation, we chose to use 60 residues. Results are found to depend only weakly on this number (Weigt *et al.*, 2009).

3.6. Initialization

The following data structures will be necessary during the execution of the program, some are optional caches of derived values, which are used often; omitting them results in a loss of speed in exchange for the smaller memory-footprint.

$$e_{ij} \in \mathbb{R}_{Q\times Q} \,|\, \forall A_i : \sum_{A_j} e_{ij}(A_i, A_j) = 0 \wedge \forall A_j : \sum_{A_i} e_{ij}(A_i, A_j) = 0$$

- The statistical "residue couplings" for every residue pair $i, j | i < j$.
- Initialized to zeros.
- For all pairs, $e_{ji} \equiv e_{ij}^{\mathrm{T}}$ will also be needed, but as computer linear algebra systems (Blackford *et al.*, 2002; Jones *et al.*, 2001) provide matrix multiplication functions that accept a transposed matrix, it is unnecessary to store these values in memory. Furthermore, specifically not doing so ensures consistency.

$$G_{ij} \in \mathbb{R}_{Q\times Q} \,|\, G_{ij}(A_i, A_j) \equiv \mathrm{e}^{-e_{ij}(A_i,\, A_j)}$$

- Optional caches of the exponentiated values of e_{ij} (Boltzmann weights), as these values are used quite often, this offers a significant performance improvement.

$$P_{i\to j} \in \mathbb{R}_{Q\times 1} | \sum_{A_i} P_{i\to j}(A_i) = 1$$

- Probability vector messages for the Belief Propagation step. This represents the belief that i has that it should take on values A_i in the absence of the direct influence of j.

- These messages are initialized randomly, but normalized according to their definition.
- In pseudocode, the datastructure containing these vectors is listed as P_ij and subscripted with the source and target, for example, P_ij[x, y] for the message from x to y.

$$M_{i\to j;k} \in \mathbb{R}_{Q\times Q} | \forall A_k : \sum_{A_i} M_{i\to j;k}(A_i, A_k) = 0$$

- Susceptibility messages giving the partial derivatives $\partial P_{i\to j}(A_i)/\partial h_k(A_k)$ for the Susceptibility Propagation step.
- Initialized to zeros.
- In pseudocode, the datastructure containing these matrices is listed as M_ijk and subscripted with the source, target, and influencing field, for example, M_ijk[x, y, z] for the message from x to y with respect to variation of the field in z.

3.7. Belief Propagation

Standard Belief Propagation (BP) is an efficient method for estimating the marginal values of unobserved nodes in Markov Random Fields. In our inverse problem, the single-site marginal values are fixed to known values (to the frequency counts f_i), but the messages from this step are necessary for the later Susceptibility Propagation step.

BP acts by passing "beliefs" around a graph representing a Random Markov Field, from one node to another, providing to the recipient information about what values the sender would be likely to take on in the absence of the direct influence of the recipient. While BP is exact on trees, it is also possible to send messages around a loopy graph several times until these messages converge to a fixed point (i.e., until no message is updated more than a given threshold). At the end of this process, it is possible to calculate beliefs for the marginal values of every node in the graph.

Because of the inverse nature of our inference, it is possible to realize a great improvement in efficiency over standard BP: already knowing the marginal values, we wish to calculate the fields and interaction terms. While standard BP passes messages globally around a graph, in this case, solving standard BP equations with known marginal probabilities renders single edge BP equations only dependent on the two messages sent back and forth between the two nodes:

$$P_{i\to j} \sim \frac{f_i}{G_{ij} \cdot P_{j\to i}} \tag{2.6}$$

where the proportionality indicates that the message is normalized to 1 according to its definition. See Appendix for the definition of nonstandard matrix and vector operations used in this section.

Belief Propagation

```
for every residue i:
  for every residue j > i:
    until convergence:
      update message P_ij[i, j] according to Eq. (2.6)
      update message P_ij[j, i] according to Eq. (2.6)
```

3.8. Susceptibility Propagation

Because of the fully connected nature of our graph, it is a difficult problem to determine two-point distributions from the provided interaction terms. However, Mézard and Mora have made great progress in this area with the technique of Susceptibility Propagation (SP), introduced in (Mezard *et al.*, 2009). A detailed description of SP is beyond the scope of this paper, but it allows for the efficient calculation of two-point distributions.

SP is executed in a fashion more similar to traditional BP, where all messages are interdependent, and updated until all messages have converged. To obtain efficient convergence we have chosen to use a random sequential update, where the order of the nodes originating SP messages is a random permutation of the set of nodes.

As defined previously, SP messages give the partial derivatives of BP messages with regard to local fields in any position. Their update formula, as derived in the original writing (and updated to this document's matrix notation) is given as

$$M_{i\to j;k} = \mathrm{Diag}(P_{i\to j})\left[\delta(i,k)I_Q + \sum_{l\neq i,j}\frac{G_{il}M_{l\to i;k}}{G_{il}P_{l\to i}} - c_{i\to j;k}\right] \tag{2.7}$$

where $c_{i\to j;k}$ can only be determined after the first terms have been calculated, and is chosen to enforce the normalization to zero in the above definition of $M_{i\to j;k}$. To implement message update directly as given here would be extremely inefficient, but an improvement of order N can be made by caching the summation over all $l \neq i$ and individually subtracting terms j when calculating each message, and updating all messages sent by node i at once. Pseudocode for an efficient implementation of this algorithm is provided below. Please note that some operations can be made even more efficient by using such functions as "elementwise multiplication" provided in many computer linear algebra packages, and are trivial to implement if unavailable (Blackford *et al.*, 2002; Jones *et al.*, 2001).

See Appendix for the definition of nonstandard matrix and vector operations used in this section.

$$\begin{aligned}\mathrm{spA}[l, i, k] &= \frac{G_{il} M_{l \to i;k}}{G_{il} P_{l \to i}} \\ \mathrm{spB}[i, k] &= \sum_{l \neq i} \mathrm{spA}[l, i, k] \\ \mathrm{spC}[i, j, k] &= \mathrm{spB}[i, k] - \mathrm{spA}[j, i, k]\end{aligned} \tag{2.8}$$

Susceptibility Propagation

```
until all susceptibilities converge:
  for every residue i in a random order:
    for every residue k:
      for every residue l!=i:
        cache spA[l, i, k] according to Eq. (2.8)
      spB[i,k] = ∑ spA[l,i,k]
                l≠i
      for every residue j!=i:
        M_ijk[i, j, k] = spB[i, k] - spA[j, i, k]
        if i==k:
          M_ijk[i, j, k] = M_ijk[i, j, k] + I_Q
        M_ijk[i, j, k] = diag(P_ij[i, j]) * M_ijk[i, j, k]
        M_ijk[i, j ,k] = M_ijk[i, j, k] -
                        P_ij[i, j]*ColumnSums
                        (M_ijk[i, j, k])
```

As before, convergence means that no message has been updated by more than a user-selected threshold.

3.9. Parameter update

3.9.1. Does the model describe empirical data within threshold?

After all SP messages have converged, it is possible to calculate the values $\partial P_i(A_i)/\partial h_j(A_j)$ according to Eq. (2.9),

$$\frac{\partial P_i}{\partial h_j} = \mathrm{Diag}(f_i)\left[\sum_{l \neq i} \frac{G_{il} M_{l \to i;j}}{G_{il} P_{l \to i}} - c_{ij}\right] \tag{2.9}$$

Pseudocode for this is analogous to that given in the SP section, above. Finally, with these values and Eq. (2.10),

$$\frac{\partial P_i}{\partial h_j} = P_{ij} - f_i f_j^{\mathrm{T}} \tag{2.10}$$

it is possible to estimate the two-residue marginal probabilities $P_{ij}(A_i, A_j)$, which will be used in this step to determine how closely the model predicts the measured two-point marginal values $f_{ij}(A_i, A_j)$.

If no marginal value $P_{ij}(A_i, A_j)$ differs from the measured marginal $f_{ij}(A_i, A_j)$ by more than the threshold value, in our implementation, gradient descent has finished, and it is now possible to calculate DI according to the subsequent section. If any marginal value differs more than this, couplings e_{ij} must be updated according to the subsequent subsection, and the BP and SP steps must be repeated.

Note that due to the heuristic character of BP on loopy graphs, the two-residue marginals are not exact. This may lead to a situation where the following coupling updates, even if derived via gradient descent in a convex optimization problem, actually increase the distance between the model-derived $P_{ij}(A_i, A_j)$ and the empirical counts $f_{ij}(A_i, A_j)$. In this case, the algorithm is required to halt and pass the best-found parameter values to the calculation of DI.

3.9.2. Update couplings

In any case where the empirical two-residue counts and the marginals calculated via SP differ by more than the threshold value, new coupling values must be chosen according to Eq. (2.11),

$$\Delta e_{ij}(A_i, A_j) = -\Delta_{\mathrm{GD}}\left[f_{ij}(A_i, A_j) - P_{ij}(A_i, A_j)\right] \tag{2.11}$$

where Δ_{GD} is the gradient descent step size. A larger Δ_{GD} will lead to a more rapid approach to the vicinity of the fixed point of the couplings, but will also tend to cause the program to overshoot and possibly to enter an infinite loop in the endgame, thus this value needs to be adjusted to the system under scrutiny.

3.10. Direct information

Once all two-residue coupling parameters have converged to a fixed point where calculated two-site marginal values, P_{ij}, match (as well as possible) the empirically observed two-site (f_{ij}) marginal values (recall that as a result of our inverted BP step, calculated single site marginals, P_i, always match observed single-site marginals, f_i), it is possible to calculate the contribution to the MI (Eq. 2.3) given only by the direct statistical interaction of two residues, introducing the joint probability that arises only from direct interaction as

$$\begin{aligned} P_{ij}^{(\mathrm{dir})} &= \frac{1}{Z_{ij}} \mathrm{Diag}(P_{i\to j})\, G_{ij} \mathrm{Diag}(P_{j\to i}) \\ Z_{ij} &= \sum_{A_i,\, A_j} \mathrm{Diag}(P_{i\to j})\, G_{ij} \mathrm{Diag}(P_{j\to i}) \end{aligned} \tag{2.12}$$

This value allows for the calculation of DI, which is analogous to MI covered earlier in this section:

$$DI_{ij} = \sum_{A_i, A_j} P_{ij}^{(\mathrm{dir})}(A_i, A_j) \ln \frac{P_{ij}^{(\mathrm{dir})}(A_i, A_j)}{f_i(A_i) f_j(A_j)} \tag{2.13}$$

Interdomain residue pairs are sorted according to their DI values. Large DI, that is, strong direct statistical coupling of the two residues under consideration, is taken as a predictor for a physical contact in the protein dimer. Therefore, DI is the most important output of our approach.

3.11. Backmapping

Ultimately, the overall goal of the procedure described herein is to facilitate the prediction and analysis of contact residues in interacting protein domains. This information is most easily understood, visualized, and processed when taken in the context of molecular models. While significantly less involved than the implementations of the above processes, the procedure for backmapping identified residue pairs with high DI onto molecular structures should be mentioned.

While Pfam provides this data in the tables msd_data, pdb, and pdbmap, this necessarily excludes novel domains and data newer than the newest Pfam release. In cases where possible, it is recommended to use this provided data.

Simple backmapping can be achieved by reading sequence data from molecular model files, searching, and aligning this data with HMMER. Making note of the location of the matching subsequence, it is simple to translate columns in the aligned dataset to residues in the model file.

These model files can be visualized with highlighting according to the DI values of residues, or multiple models may be visualized and manipulated simultaneously with links generated by high DI column-pairs. Finally, these linked models may become input to molecular dynamics calculations, or other simulations.

Appendix: Nonstandard Linear Algebra Functions

Because of its usefulness in clarifying equations, we will introduce the concept of matrix division by another matrix of the same size, a column vector with the same number of rows, or a row vector with the same number of columns. In the case of a matrix divisor, each position of the dividend is divided by the corresponding position of the divisor. In the case of column and row vector divisors, respectively; each column or row of the dividend is divided by the corresponding position of the divisor. This function is provided in some computer linear algebra systems, but can be implemented readily if not available.

Elementwise operators

Matrix Division by Matrix:

$$\text{Mat}_{M\times N} \times \text{Mat}_{M\times N} \rightarrow \text{Mat}_{M\times N}$$
$$\text{Quotient}_{\alpha,\beta} = \text{Dividend}_{\alpha,\beta}/\text{Divisor}_{\alpha,\beta}$$

Matrix Division by Row Vector:

$$\text{Mat}_{M\times N} \times \text{Mat}_{1\times N} \rightarrow \text{Mat}_{M\times N}$$
$$\text{Quotient}_{\alpha,\beta} = \text{Dividend}_{\alpha,\beta}/\text{Divisor}_{\beta}$$

Matrix Division by Column Vector:

$$\text{Mat}_{M\times N} \times \text{Mat}_{M\times 1} \rightarrow \text{Mat}_{M\times N}$$
$$\text{Quotient}_{\alpha,\beta} = \text{Dividend}_{\alpha,\beta}/\text{Divisor}_{\alpha}$$

ACKNOWLEDGMENTS

This work was supported by the Center for Theoretical Biological Physics (CTBP) sponsored by the NSF (Grant PHY-0822283) with additional support from NIH grant R01GM077298 (T. H.) and by NIH grant R01GM019416 (J. A. H.). B. L. and A. P. acknowledge funding from the EC via the STREP GENNETEC ("Genetic networks: emergence and complexity").

REFERENCES

Altschuh, D., Lesk, A. M., Bloomer, A. C., and Klug, A. (1987). Correlation of coordinated amino acid substitutions with function in viruses related to tobacco mosaic virus. *J. Mol. Biol.* **193,** 693–707.

Atchley, W. R., Wollenberg, K. R., Fitch, W. M., Terhalle, W., and Dress, A. W. (2000). Correlations among amino acid sites in bHLH protein domains: An information theoretic analysis. *Mol. Biol. Evol.* **17,** 164–178.

Blackford, L. S., Demmel, J., Dongarra, J., Duff, I., Hammarling, S., Henry, G., Heroux, M., Kaufman, L., Lumsdaine, A., Petitet, A., Pozo, R., Remington, K., and Whaley, R. C. (2002). An updated set of Basic Linear Algebra Subprograms (BLAS). *Trans. Math. Soft.* **28**(2), 135–151, ISSN 0098-3500.

Brouwer, R. W., Kuipers, O. P., and van Hijum, S. A. (2008). The relative value of operon predictions. *Brief. Bioinform.* **9,** 367–375.

Burbulys, D., Trach, K. A., and Hoch, J. A. (1991). Initiation of sporulation in *B. subtilis* is controlled by a multicomponent phosphorelay. *Cell* **64,** 545–552.

Burger, L., and van Nimwegen, E. (2008). Accurate prediction of protein-protein interactions from sequence alignments using a Bayesian method. *Mol. Syst. Biol.* **4,** 165.

Cusick, M. E., Klitgord, N., Vidal, M., and Hill, D. E. (2005). Interactome: Gateway into systems biology. *Hum. Mol. Genet.* **14** Spec No. 2, R171–R181.

Dyer, C. M., Quillin, M. L., Campos, A., Lu, J., McEvoy, M. M., Hausrath, A. C., Westbrook, E. M., Matsumura, P., Matthews, B. W., and Dahlquist, F. W. (2004). Structure of the constitutively active double mutant CheYD13K Y106W alone and in complex with a FliM peptide. *J. Mol. Biol.* **342,** 1325–1335.

Eddy, S. R. (1998). Profile hidden Markov models. *Bioinformatics* **14,** 755–763.

Finn, R. D., Tate, J., Mistry, J., Coggill, P. C., Sammut, S. J., Hotz, H. R., Ceric, G., Forslund, K., Eddy, S. R., Sonnhammer, E. L., and Bateman, A. (2008). The Pfam protein families database. *Nucleic Acids Res.* **36,** D281–D288.

Gama-Castro, S., Jimenez-Jacinto, V., Peralta-Gil, M., Santos-Zavaleta, A., Penaloza-Spinola, M. I., Contreras-Moreira, B., Segura-Salazar, J., Muniz-Rascado, L., Martinez-Flores, I., Salgado, H., Bonavides-Martinez, C., Abreu-Goodger, C., *et al.* (2008). RegulonDB (version 6.0): Gene regulation model of *Escherichia coli* K-12 beyond transcription, active (experimental) annotated promoters and Textpresso navigation. *Nucleic Acids Res.* **36,** D120–D124.

Göbel, U., Sander, C., Schneider, R., and Valencia, A. (1994). Correlated mutations and residue contacts in proteins. *Proteins Struct. Funct. Genet.* **18,** 309–317.

Hoch, J. A. (2000). Two-component and phosphorelay signal transduction. *Curr. Opin. Microbiol.* **3,** 165–170.

Jaynes, E. T. (1949). Information theory and statistical mechanics. *Physiol. Rev.* **106,** 620–630.

Jones, E., Oliphant, T., Peterson, P., *et al.* (2001–2005). Open source scientific tools for Python. http://www.scipy.org/.

Kass, I., and Horovitz, A. (2002). Mapping pathways of allosteric communication in GroEL by analysis of correlated mutations. *Proteins Struct. Funct. Genet.* **48,** 611–617.

Kschischang, F. R., Frey, B. J., and Loeliger, H.-A. (2001). Factor graphs and the sum-product algorithm. *IEEE Trans. Inf. Theory* **47,** 498–519.

Liolios, K., Mavromatis, K., Tavernarakis, N., and Kyrpides, N. C. (2008). The genomes on line database (GOLD) in 2007: Status of genomic and metagenomic projects and their associated metadata. *Nucleic Acids Res.* **36,** D475–D479.

McLaughlin, P. D., Bobay, B. G., Regel, E. J., Thompson, R. J., Hoch, J. A., and Cavanagh, J. (2007). Predominantly buried residues in the response regulator Spo0F influence specific sensor kinase recognition. *FEBS Lett.* **581,** 1425–1429.

Mezard, M., and Mora, T. (2009). Constraint satisfaction and neural networks: A statistical-physics perspective. *J. Physiol. Paris* **103,** 107–113.

Moreno-Hagelsieb, G., and Collado-Vides, J. (2002). A powerful non-homology method for the prediction of operons in prokaryotes. *Bioinformatics* **18**(Suppl. 1), S329–S336.

Pruitt, K. D., Tatusova, T., Klimke, W., and Maglott, D. R. (2009). NCBI Reference Sequences: Current status, policy and new initiatives. *Nucleic Acids Res.* **37,** D32–D36.

Ramakrishnan, V. (2008). What we have learned from ribosome structures. *Biochem. Soc. Trans.* **36,** 567–574.

Shannon, C. E. (1948). A mathematical theory of communication. *Bell Syst. Tech. J.* **27,** 379–423.

Suel, G. M., Lockless, S. W., Wall, M. A., and Ranganathan, R. (2003). Evolutionarily conserved networks of residues mediate allosteric communication in proteins. *Nat. Struct. Biol.* **10,** 59–69.

Szurmant, H., Bunn, M. W., Cannistraro, V. J., and Ordal, G. W. (2003). *Bacillus subtilis* hydrolyzes CheY-P at the location of its action, the flagellar switch. *J. Biol. Chem.* **278,** 48611–48616.

Szurmant, H., Bobay, B. G., White, R. A., Sullivan, D. M., Thompson, R. J., Hwa, T., Hoch, J. A., and Cavanagh, J. (2008). Co-evolving motions at protein-protein interfaces of two-component signaling systems identified by covariance analysis. *Biochemistry* **47,** 7782–7784.

Thattai, M., Burak, Y., and Shraiman, B. I. (2007). The origins of specificity in polyketide synthase protein interactions. *PLoS Comput. Biol.* **3,** 1827–1835.

Varughese, K. I. (2002). Molecular recognition of bacterial phosphorelay proteins. *Curr. Opin. Microbiol.* **5,** 142–148.

Weigt, M., White, R. A., Szurmant, H., Hoch, J. A., and Hwa, T. (2009). Identification of direct residue contacts in protein-protein interaction by message passing. *Proc. Natl. Acad. Sci. USA* **106,** 67–72.

Welch, M., Chinardet, N., Mourey, L., Birck, C., and Samama, J. P. (1998). Structure of the CheY-binding domain of histidine kinase CheA in complex with CheY. *Nat. Struct. Biol.* **5,** 25–29.

Wells, J. A., and McClendon, C. L. (2007). Reaching for high-hanging fruit in drug discovery at protein-protein interfaces. *Nature* **450,** 1001–1009.

White, R. A., Szurmant, H., Hoch, J. A., and Hwa, T. (2007). Features of protein-protein interactions in two-component signaling deduced from genomic libraries. *Methods Enzymol.* **422,** 75–101.

Yedidia, J. S., Freeman, W. T., and Weiss, Y. (2001). Generalized belief propagation. *NIPS* **13,** 689–695.

Yusupov, M. M., Yusupova, G. Z., Baucom, A., Lieberman, K., Earnest, T. N., Cate, J. H., and Noller, H. F. (2001). Crystal structure of the ribosome at 5.5 A resolution. *Science* **292,** 883–896.

Zapf, J., Sen, U., Madhusudan, X., Hoch, J. A., and Varughese, K. I. (2000). A transient interaction between two phosphorelay proteins trapped in a crystal lattice reveals the mechanism of molecular recognition and phosphotransfer in signal transduction. *Structure* **8,** 851–862.

Zhao, R., Collins, E. J., Bourret, R. B., and Silversmith, R. E. (2002). Structure and catalytic mechanism of the *E. coli* chemotaxis phosphatase CheZ. *Nat. Struct. Biol.* **9,** 570–575.

Zhu, X., Volz, K., and Matsumura, P. (1997). The CheZ-binding surface of CheY overlaps the CheA- and FliM-binding surfaces. *J. Biol. Chem.* **272,** 23758–23764.

CHAPTER THREE

Computational Modeling of Phosphotransfer Complexes in Two-Component Signaling

Alexander Schug,* Martin Weigt,† James A. Hoch,‡ Jose N. Onuchic,* Terence Hwa,* *and* Hendrik Szurmant‡

Contents

Abstract

Two-component signal transduction systems enable cells in bacteria, fungi, and plants to react to extracellular stimuli. A sensor histidine kinase (SK) detects such stimuli with its sensor domains and transduces the input signals to a response regulator (RR) by *trans*-phosphorylation. This *trans*-phosphorylation reaction requires the formation of a complex formed by the two interacting proteins. The complex is stabilized by transient interactions. The nature of the transient interactions makes it challenging for experimental techniques to gain structural information. X-ray crystallography requires stable crystals, which are difficult to grow and stabilize. Similarly, the mere size of these systems proves problematic for NMR. Theoretical methods can, however, complement existing data. The statistical direct coupling analysis presented in the previous chapter reveals the interacting residues at the contact interface of the SK/RR pair. This information can be combined with the structures of the individual proteins in molecular dynamical simulation to generate structural models of the complex. The general approach, referred to as MAGMA, was tested on the

* Center for Theoretical Biological Physics, University of California San Diego, La Jolla, California, USA
† Institute for Scientific Interchange, Viale S. Severo 65, Torino, Italy
‡ Department of Molecular and Experimental Medicine, The Scripps Research Institute, La Jolla, California, USA

Methods in Enzymology, Volume 471
ISSN 0076-6879, DOI: 10.1016/S0076-6879(10)71003-X

sporulation phosphorelay phosphotransfer complex, the Spo0B/Spo0F pair, delivering crystal resolution accuracy. The MAGMA method is described here in a step-by-step explanation. The developed parameters are transferrable to other SK/RR systems.

1. Introduction

Two-component signal transduction systems (TCS) enable bacteria, fungi, and plants to respond to stimuli and changes of environments like nutrients, light, or pressure (for a recent review see Mascher *et al.*, 2006). They consist of two proteins, a sensor histidine kinase (SK) and a response regulator (RR). The multidomain SK consists of intra- and/or extracellular sensor domains and a catalytic histidine kinase core (reviewed in Szurmant *et al.*, 2007). The latter can be subdivided into the phosphorylatable histidine containing HisKA domain and the catalytic ATP-binding ATPase domain. In response to a stimulus, the phosphoryl flux between the SK and the RR is modulated. As a first step, the histidine on the HisKA domain is autophosphorylated. In a second step, this phosphoryl group is transferred to an aspartate residue on the RR protein, which most commonly serves as a transcription factor (Galperin, 2006).

Extended versions of the TCS signal transduction pathway are the phosphorelays (Hoch, 2000). In these systems, the phosphoryl flux between the SK and the RR is mediated by a second single domain RR and a phosphotransferase in a His-Asp-His-Asp phosphotransfer cascade. The phosphotransferase protein can feature either a monomeric four-helix bundle Hpt domain (also utilized by the chemotaxis histidine kinase, CheA) or a second dimeric four-helix bundle HisKA-like domain (Hoch and Varughese, 2001). The latter form of phopsphorelay is exemplified by the well-described sporulation phosphorelay of the *Bacilli*, which connects five SK KinA–KinE with the sporulation RR Spo0A via the single domain RR Spo0F and the HisKA-like phosphotransferase Spo0B (Burbulys *et al.*, 1991; Jiang *et al.*, 2000).

The phosphotransfer complex between HisKA and RR domains is ruled by transient interactions. While many individual TCS proteins have been structurally resolved, these transient interactions result in short-lived and unstable complexes, which have proven resistant to structural resolution by experimental means. For this reason, as of August, 2009, the complex of above described Spo0F and Spo0B proteins of the *Bacillus subtilis* sporulation phosphorelay remains as the only published structural example of a HisKA–RR pair, trapped in the act of phosphotransfer (Zapf *et al.*, 2000).

This lack of structural templates hampers theoretical structure-prediction methods. Established methods like homology modeling (Eswar *et al.*, 2008) rely on the presence of such templates with high sequential similarity. Physics-based approaches (Fujitsuka *et al.*, 2004; Hardin *et al.*, 2002; Schug *et al.*, 2005a) struggle with the accuracy of the underlying force-fields, especially for mixed α-helical/β-sheet structures (Best *et al.*, 2008). Even assuming perfect suitability of a specific force-field, the mere size of TCS protein complexes might make such prediction methods computationally prohibitive by requiring searches of huge conformational spaces with accordingly high numbers of degrees of freedom ("problem of sampling") (Schug and Wenzel, 2006; Schug *et al.*, 2005b). Currently, no reliable "gold standard" for the prediction of protein complexes has been found as demonstrated by the CAPRI competitions, in which different approaches at protein–protein docking are compared (Janin *et al.*, 2003).

Not all is bad news, however. The current ongoing "genomic revolution" provides scientists with a wealth of sequential information and complete genomes for an exponentially growing number of systems. As TCS are ubiquitously used and highly amplified in bacteria, fungi, and plants, one can take advantage of the abundance of sequential information. The *direct coupling analysis* (DCA) (Weigt *et al.*, 2009), presented in the previous chapter, investigates the mutational patterns of coevolving protein like the SK/RR pair in TCS and has three subsequent steps. The first step is a homology search of the two target sequences in as many bacterial genomes as possible. The second step constitutes aligning the sequences and performing a covariance analysis. This reveals pairs of amino acids with high mutual information in the two proteins. In a third step, a statistical inference analysis distinguishes between pairs, which directly interact, from such, which are indirectly correlated, that is, for which the correlation is mediated by additional residue positions. These three steps result in a set of directly interacting pairs of amino acids between the SK and the RR, that is residue positions, which lie on the interface of the two proteins (here thereto referred as DI contacts). This information describes the evolutionary most crucial interactions of the surface and is therefore by itself of high scientific interest. It becomes, however, even more valuable when combined with molecular dynamical (MD) simulations.

MD-simulations approximate the physics of an entire system of interacting biomolecules over a defined period of time. Typically, one has to balance the accuracy of description/physics with the available computational resources (Adcock and McCammon, 2006). Here, we employ native structure-based simulations (SBS), which are highly successful in describing protein folding (Clementi *et al.*, 2001; Onuchic and Wolynes, 2004; Onuchic *et al.*, 1997), conformational transitions related to protein function (Schug *et al.*, 2007; Whitford *et al.*, 2007), RNA folding and function (Thirumalai and Hyeon, 2005; Whitford *et al.*, 2009b), and protein–protein interactions (Levy and Onuchic, 2006; Levy *et al.*, 2007), and make prediction in agreement with experimental measurements (Clementi and Plotkin, 2004; Gambin *et al.*, 2009). They describe protein/RNA dynamics based on a specific structural conformation, which is usually the native state. This results in a concise Hamiltonian and allows adopting the level of coarse-graining to the individual scientific question (Lammert *et al.*, 2009; Oliveira *et al.*, 2008; Schug *et al.*, 2009a; Whitford *et al.*, 2009a).

We previously demonstrated that combining complementary independent information, the coevolutionary information obtained from DCA and structural data of the unbound monomers, by SBS can predict a TCS complex in agreement with experimental data (see Fig. 3.1) (Schug *et al.*, 2009b). The approach will hereafter be referred to as MAGMA (Molecular dynamics And Genomic information for Macromolecular Assembly). The MAGMA method along with some of its results is described here in some detail. Relying on the DCA analysis of subtle *mutational patterns* is an orthogonal approach to typical structure-prediction methods like homology modeling, which rely on *highly conserved residues* in sequential and structural libraries. The ultimate aim is, however, not stopping at protein structure prediction. Right now, we only use DCA to determine interacting residues

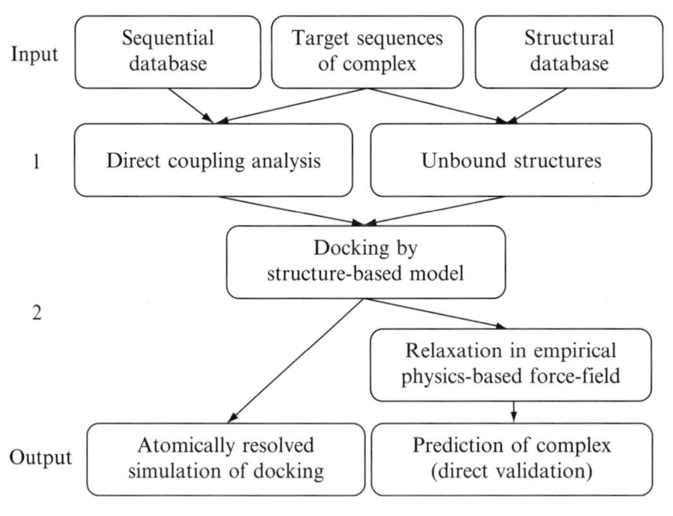

Figure 3.1 Flow-chart of the MAGMA approach. Given the target sequence of an unknown protein complex, direct-coupling analysis (DCA, see previous chapter) investigates statistical fluctuations of mutational pattern in sequential homologues and suggests pairwise contacts defining an interaction surface. Similarly, unbound structures for the given target sequences can be either directly extracted from a structural database or generated by structure-prediction methods like homology modeling. This information of the unbound structures and interaction surface contacts is sufficient information for docking simulations in computationally efficient structure-based models, providing both insight into the mechanism of docking and making a prediction of the protein complex. To improve the quality of the prediction, it can be additionally relaxed in physics-based empirical force fields.

at the intermolecular docking interface of RR and SK. We hope to combine the additional DCA intramolecular information with molecular simulations to generate realistic descriptions of the conformational changes underlying the biological function of biomolecules. In the context of TCS, that would be the simulation of conformational changes during autophosphorylation, docking of the RR to the SK, phosphoryl-transfer, and the final dissociation of RR and SK in the absence of detailed experimental structural data for each step.

2. Methods

2.1. Structure-based simulations

Structure-based simulations are used in protein folding simulations based on the funneled energy landscape and the principle of minimal frustration. Accordingly, evolution shapes and concurrently smoothens the energy landscape of proteins by ensuring a dominance of interactions present in the native state during the entire folding process (Bryngelson *et al.*, 1995; Frauenfelder *et al.*, 1991; Onuchic and Wolynes, 2004). This guiding bias prevents entrapment in minima representing nonnative folds. It also provides a degree of robustness, permitting protein folding and function despite moderate environmental changes or mutations.

In folding simulations, native structure-based models[1] represent the ideal case of a perfectly funneled energy landscape where only interactions present in the native state are taken into account and no energetic frustration occurs. They have shown to be in high agreement with experimental measurements (Chavez *et al.*, 2004; Cheung *et al.*, 2003; Clementi *et al.*, 2000). In a typical mathematical description, each amino acid is represented as a single C_α-bead. Bridging these methods toward empirical all-atom force fields, variants using multiwelled Gaussians for the contacts (Lammert *et al.*, 2009), $C_\alpha C_\beta$ (Finke *et al.*, 2004; Oliveira *et al.*, 2008) or all-atom representations have been developed (Linhananta and Zhou, 2002; Shimada *et al.*, 2001; Whitford *et al.*, 2009a; Zhou *et al.*, 2003). As the latter incorporate the details of packing best while maintaining computational tractability, we choose (Whitford *et al.*, 2009a) as a basis for our docking simulations (the Hamiltonian is given as Eq. (3.1)):

[1] Structure-based models are often referred to as Go-models.

$$E = \sum_{\text{bonds}} K_r(r - r_0)^2 + \sum_{\text{angles}} K_\theta(\theta - \theta_0)^2 + \sum_{\text{impropers/planar}} K_\chi(\chi - \chi_0)^2$$
$$+ \sum_{\text{backbone}} K_{\text{BB}} F_{\text{D}}(\varphi)^2 + \sum_{\text{sidechain}} K_{\text{SC}} F_{\text{D}}(\varphi)^2$$
$$+ \sum_{\text{contacts}} \left\{ \varepsilon_{\text{C}} \left[\left(\frac{\sigma_{ij}}{r_{ij}} \right)^{12} - 2 \left(\frac{\sigma_{ij}}{r_{ij}} \right)^{6} \right] + \varepsilon_{\text{NC}} \left(\frac{\sigma_{\text{NC}}}{r_{ij}} \right)^{12} \right\},$$
$$F_{\text{D}}(\varphi) = [1 - \cos(\varphi - \varphi_0)] + \frac{1}{2}[1 - \cos(3(\varphi - \varphi_0))] \tag{3.1}$$

with $K_r = 100 k_B T$ (Å^2), $K_\theta = 20 k_B T$, $K_\chi = 20 k_B T$, and $\varepsilon_{\text{NC}} = 0.01 k_B T$. The values for r_0, θ_0, χ_0, Φ_0, and σ_{ij} are given by the native conformation. An illustration of the different terms can be found in Figs. 3.2 and 3.3. Both impropers/planar and backbone denote dihedral terms, pending on how rigid they are in the structure (for a thorough discussion see Oliveira *et al.*, 2008). i and j run over all atoms and r_{ij} is the distance between any two atoms. An attractive interaction with $\varepsilon_{\text{C}}(i, j) = 1 k_B T$ and $\varepsilon_{\text{NC}}(i, j) = 0 k_B T$ is assigned to natively interacting residues, while $\varepsilon_{\text{C}}(i, j) = 1 k_B T$ and $\varepsilon_{\text{NC}}(i, j) = 0.01 k_B T$ enforce an excluded volume for noninteracting residues. σ_{ij} is the native distance between interacting residues and set to 2.5 Å for noninteracting residues. The dihedral strengths K_{BB} and K_{SC} are assigned in a way that the interaction energy between the sidechains and the backbones is balanced 2:1 and the total contacts energy (determined by the total number of contacts) is balanced 2:1 against the total sidechain energy (Whitford *et al.*, 2009a).

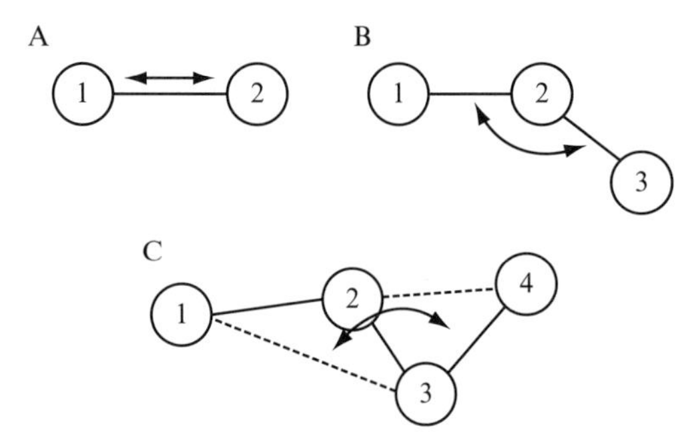

Figure 3.2 Illustration of the different interactions. The backbone interactions of native structure-based simulations have three contributions. A harmonic pair interaction or 1–2 interaction (A) involves two atoms and describes the vibrations around a harmonic bond. The angle term or 1–3 interaction (B) is given by the angle between the bonds 1–2 and 2–3. Finally, the dihedral term (C) or 1–4 interaction is defined as the angle between two planes. In the example, atoms 1–3 span the first plane and atoms 2–4 the second plane.

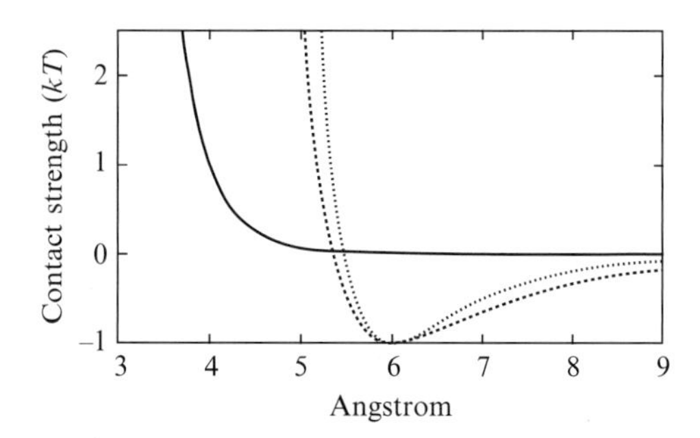

Figure 3.3 Illustration of the contact potential. The Lennard-Jones-type contact potential of structure-based models biases against distances much shorter than the contact distance σ_{ij} while providing an attractive basin for values around the distance. This results from the repulsive $(\sigma_{ij}/r_{ij})^{12}$-term, which dominates for short distances (solid line, $\sigma_{ij} = 4$ Å) and prevents overlap of the electron shells of any two atoms. Typically, structure-based models have an attractive $(\sigma_{ij}/r_{ij})^{6}$ or $(\sigma_{ij}/r_{ij})^{10}$-term for natively interacting residues (dashed lines) with the former being "softer" (longer bars) and the latter being more localized (shorter bars).

2.1.1. Deriving parameters for molecular docking simulations

SBS allow direct modeling of the independent protein monomers, which shall be docked. For docking, however, we need to introduce *a priori* unknown interprotein forces. A weak harmonical center-of-mass force for all atoms ($k = 0.25 \times 10^{-6} k_B T$, Å^2) simulates a sufficient molecular concentration of both molecules to bring the proteins into frequent contact with each other (Schug *et al.*, 2007).

Another crucial term in MAGMA is the specific inclusion of the DCA-predicted residue pairs (see previous chapter and Weigt *et al.*, 2009). These predicted direct interactions at the surface will determine the exact orientation of the two molecules with respect to each other ("docking pose"). The most natural implementation of these interactions is as additional contacts. To prevent any bias or artifacts from overfitting to a dataset, we choose a homogenous distance for these contacts between the corresponding C_α atoms. As typically two amino acids, which are in contact with each other interact by 3–6 contacts between individual atoms on the all-atom level, the contact strength between the C_α atoms should account for the total interaction between the amino acids and was hence increased fivefold. We tested contact distances between 5.5 and 7 Å, which is the range of average contact distances[2] over a range of reduced temperatures (1/3–2/3), all well below folding temperature to ensure fast convergence to a docked complex ("kinetic simulations").

[2] It is important not to overestimate the contact distance. Otherwise the repulsive part of the van der Waals contact potential will dominate the interaction and prevent close approach of the interacting residues (see Fig. 3.3).

2.1.2. Test: Docking of the Spo0B/Spo0F complex

To develop and test our parameters, we choose the described Spo0B/Spo0F system, since both, structures of the individual proteins (pdb-codes: 1pey, Mukhopadhyay *et al.*, 2004; and 1ixm, Varughese *et al.*, 1998) and a structure of the complexed crystal (1f51; Zapf *et al.*, 2000) have been determined (see also Figs. 3.4 and 3.5).

In the case of several copies of a protein in a pdb-file, for example, as a result from crystal packing, one has to choose a representative conformation. We choose as a consistent but somewhat arbitrary choice the first representation in each pdb-file. The two proteins are then combined into one pdb-file with consecutively numbered amino acids and atoms.[3] It is important to check that the two proteins do not overlap, as the resulting atomic clashes will stop the MD-simulations. If the proteins overlap or if one wants to speed up the subsequent docking simulations, the two proteins can be brought into spatial vicinity of each other with the docking interfaces close to each other and without overlap of the atoms by using, for example, the VMD software (Humphrey *et al.*, 1996) (Mouse → Move → Fragment and Mouse → Rotate).

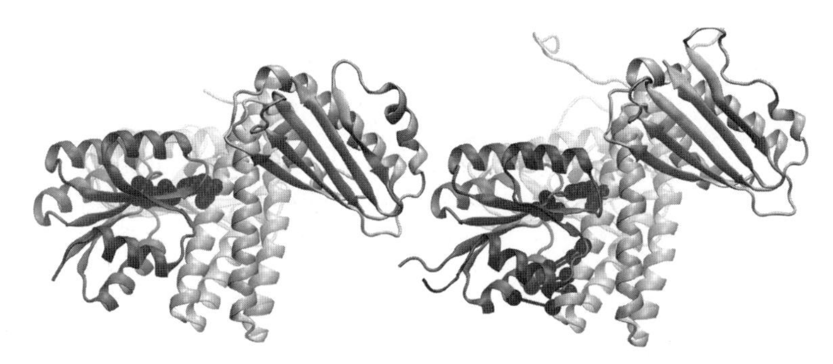

Figure 3.4 The Spo0B/Spo0F complex. The phosphotransferase Spo0B/Spo0F system is part of a phosphorelay in the sporulation pathway of *Bacillus subtilis*. (Left) The crystal structure (PDB-ID 1f51) shows Spo0B (light gray) and Spo0F (dark gray). The residues His30 and Asp54 responsible for the phosphoryl-transfer are highlighted in black. (Right) DCA identifies 6 residue pairs, which are highly directly correlated (black) at the interface of the two proteins. Docking simulations using this information, information about the spatial vicinity of the His-Asp pair (also black), and the unbound protein structures generate a structural model in high agreement with the experimental structure (Cα-RMSD < 3 Å).

[3] The consecutive numbering avoids ambiguities, for example, when software packages ignore the chain identifier.

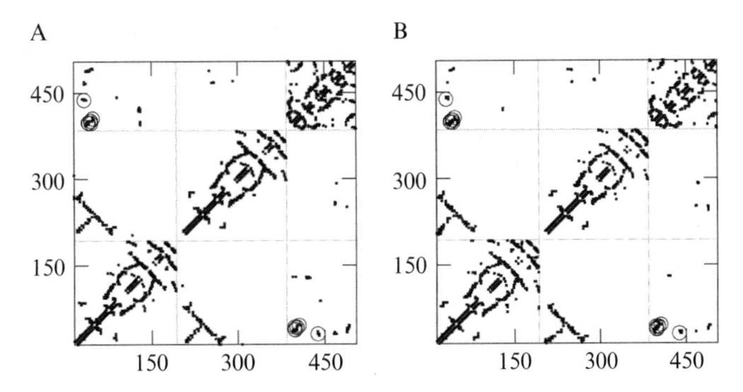

Figure 3.5 Contact maps of the Spo0B/Spo0F complex. The contact maps of (left) the crystal structure and (right) the prediction agree well. The axes denote the consecutively numbered residues of Spo0B-A (1–192), Spo0B-B (193–384), and Spo0F (385+). The DCA-contacts and the Asp-His pair are highlighted as circles. The quadrant, which contains the contacts of the interface region of the complex shows in addition to these explicitly included contacts other contacts formed in the crystal structure which have not been found by DCA. Some of these "missing" contacts are reconstituted in our docking simulations. It seems therefore possible that this subset of contacts consists of the crucial contacts for docking based on two facts: (A) DCA identifies them to have statistically strongly linked coevolution compared to all possible interface contact pairs and (B) they are sufficient information for successful reconstitution of the protein complex in docking by MAGMA.

The next step is preparing the docking simulations. Here, we use the GROMACS software package (Kutzner *et al.*, 2007; Van Der Spoel *et al.*, 2005). The required files for the simulations can be created by, for example, the webpage http://sbm.ucsd.edu (prepare a simulation, default parameters except contact map: Cut-off instead of shadow-map; the same webpage also contains a small tutorial into structure-based simulations). One receives two files, the gro-file with the atomic coordinates and a top-file with simulation parameters. In order to add the DCA-contacts, one has to edit the top-file using a text-editor right after the [pairs]-entry. The format is:

[pairs]
$i \quad j \quad C(\sigma_{ij})^6 \quad C(\sigma_{ij})^{12}$

i and j are the atom numbers between which a Lennard-Jones-type contact potential is introduced, σ_{ij} designates the desired contact distance in nm (we suggest 0.7 nm, see below), and C is the contact strength. For Spo0B/Spo0F, DCA predicts six contacts with significant direct information (see Table 3.1) (Weigt *et al.*, 2009).[4] The crucial His30–Asp54 interaction

[4] As discussed in the previous chapter the top 10 DI pairings are contacts that could have been included for docking analysis. For the present analysis only the six high DI contacts that also showed above threshold mutual information, as published in Weigt et al. (2009) were considered. We do not anticipate that results would change much if the additional four pairings are included since they involve the same response regulator and sensor kinase residues.

Table 3.1 Variation of C_α–C_α contact distances in docking simulations of Spo0F/Spo0B

Spo0B	Spo0F	Native distance (Å)	C_α–C_α distance (Å), 6 Å	C_α–C_α distance (Å), 6.5 Å	C_α–C_α distance (Å), 7 Å	C_α–C_α distance (Å), 7/11 Å
GLN 37	ILE 15	7.7	7.4 ± 0.2	7.4 ± 0.1	7.8 ± 0.2	7.7 ± 0.1
LEU 38	GLY 14	6.7	6.0 ± 0.2	6.4 ± 0.1	6.8 ± 0.1	6.9 ± 0.1
GLY 41	LEU 18	7.0	6.0 ± 0.1	6.5 ± 0.1	6.9 ± 0.2	6.9 ± 0.1
ASN 42	GLY 14	9.7	6.0 ± 0.1	6.5 ± 0.1	7.0 ± 0.1	6.9 ± 0.1
ASN 42	LEU 18	7.0	5.9 ± 0.1	6.4 ± 0.1	7.0 ± 0.1	6.9 ± 0.1
LEU 45	VAL 22	8.2	6.0 ± 0.2	6.5 ± 0.1	6.9 ± 0.1	7.8 ± 0.2
HIS 30	ASP 54	12.2	16.5 ± 0.5	15.9 ± 0.4	16.4 ± 0.3	11.6 ± 0.1

When varying the C_α–C_α distance of the DI contacts (first six rows) and the His-Asp pair (last row) in docking simulations, the deviations to the native distances are small for all choices of parameters. Each number represents 10 docking simulations at $T = 1/3$ with contact strengths for the DI predicted contacts of $5kT$. The parameters in the last column have an additional contact added between His30 and Asp54 of $10kT$.

responsible for the phosphoryl transfer cannot be detected by DCA due to perfectly conserved amino acids (covariance needs variance). We therefore test including an additional contact (see Fig. 3.6 and Table 3.1). To accommodate for the size of the phosphoryl, we add 4 Å to the contact distance, close to the typical size of a phosphate group in empirical force fields. This additional contact improves the quality of the prediction.

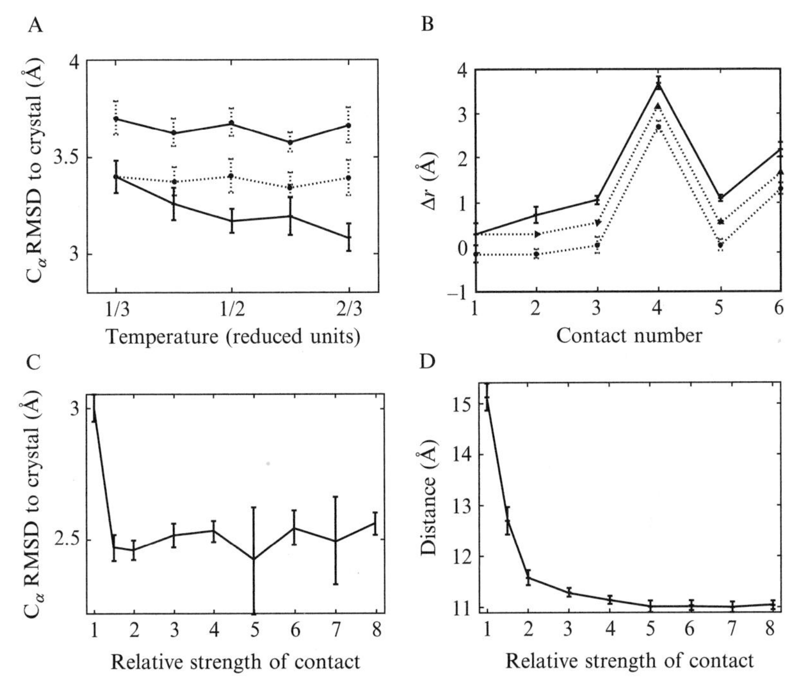

Figure 3.6 Robustness of parameters for the Spo0F/Spo0B system. Ten docking simulations using the DCA predicted interface contacts (full: contact strength $5kT$ and 6 Å, striped: $5kT$ and 6.5 Å and thinly striped: $5kT$ and 7 Å) lead to comparably good results. (A) We find successful docking simulations (all contacts are formed in the docked conformation) for different temperatures with a RMSD difference to the crystal around 3–3.5 Å Cα-RMSD. (B) The differences Δr of the six contact distances with the crystal distance show comparable deviations (see also Table 3.1). While (A) suggests a contact distance of 6 Å optimal, (B) suggests 7 Å to be a slightly better choice. We arbitrarily choose 7 Å as default value for further simulations. (C, D) Due to the perfect conservation of the His-Asp pair in the two proteins, DCA in principle cannot detect them as an interactions (covariance requires variance). The prediction quality improves when including this contact additionally. The strength is relative to the DCA-contacts ($5kT$, 7 Å) and the distance 11 Å allows to accommodation of the phosphoryl group (assumed to be roughly 4 Å large). It shows that this additional contact needs to be strong enough ($>7.5kT$) to compete with the DI contacts to improve the prediction quality.

For the docking simulations the temperature is kept constant by the Berendsen algorithm (Berendsen *et al.*, 1984) with a coupling constant of 1. Each docking simulation runs 2.5 Mio time steps of 0.0025 using the described center-of-mass force and subsequent 0.5 Mio time steps without the center-of-mass force, running for a total of roughly 20 h on a typical CPU (Spo0F/Spo0B system, 4164 atoms). Having tested various sets of parameters (see Fig. 3.6 and Table 3.1), we suggest $T = 2/3$ and a combination of a contact strength of $5kT$ with a contact distance of 7 Å for the DCA-contacts and $15kT/11$ Å for the His-Asp contact for simulations.

2.2. Relaxation in an empirical force field

It is possible to relax the docked complexes additionally in an empirical all-atom force field for refinement. Here, we use AmberF99 (Wang *et al.*, 2000) (http://chemistry.csulb.edu/ffamber) with explicit Tip3p solvent and counterions (Jorgensen *et al.*, 1983), a time step of 0.002 fs, and particle mesh Eswalds electrostatics (Essmann *et al.*, 1995). This refinement aims at removing artifacts from different physical environments for the isolated and docked proteins.

There are several pitfalls/common problems when starting such simulations. First, the input pdb-file must be modified to accommodate the possibility of charged amino acids. Here we treat all LYS as LYP, CYS as CYN, and HIS as HID. Also, the start and end of each chain has to be identified (e.g., ASP to CASP or NASP indicating it being on the C- or N-terminal). For these changes, it might be necessary to add/remove some atoms. This can be done manually or by using homology modeling software (Eswar *et al.*, 2008). After that, it is necessary to minimize the structure for later simulations. We find it useful to first minimize the structure, then add solvent molecules (commands: editconf, genbox), minimize again, add counterions (genion), and minimize again. Afterward the simulation can be started.

While we see some relaxation of the sidechains in this simulation, the backbone shows only minor movements. The resulting structures are in high agreement (RMSD ≤ 3 Å excluding the mobile C-termini) with highly similar contact maps to the complexed crystal structure (1f51) (see Figs. 3.4 and 3.5) (Schug *et al.*, 2009b).

3. Summary

We described the detailed MAGMA method that exemplified the feasibility of integrating sequence-based genomic analysis with molecular simulation to generate structural models of a signal transduction complex at a resolution matching experimental accuracy (Schug *et al.*, 2009b). DCA described in the previous chapter was shown to give sufficient information to successfully dock the Spo0B/Spo0F system. The parameters for the SBS simulations are robust toward slight variations without significant changes of the resulting structure. This allows tuning and refining them for new specific systems or questions. We are confident MAGMA will successfully introduce other TCS or, more general, short-lived complex structures ruled by transient interactions, and allow concurrent simulation of the conformational and functional motions of the complex, such as those during the autophosphorylation reaction or phosphoryl-transfer reaction.

ACKNOWLEDGMENTS

This work was supported by the Center for Theoretical Biological Physics (CTBP) sponsored by the NSF (Grant PHY-0822283) with additional support from NSF grant MCB-0543906 (J. N. O.), NIH grant R01GM077298 (T. H.), and by NIH grant R01GM019416 (J. A. H.).

REFERENCES

Adcock, S. A., and McCammon, J. A. (2006). Molecular dynamics: Survey of methods for simulating the activity of proteins. *Chem. Rev.* **106,** 1589–1615.

Berendsen, H. J. C., Postma, J. P. M., van Gunsteren, W. F., DiNola, A., and Haak, J. R. (1984). Molecular dynamics with coupling to an external bath. *J. Chem. Phys.* **81,** 3684–3690.

Best, R. B., Buchete, N. V., and Hummer, G. (2008). Are current molecular dynamics force fields too helical? *Biophys. J.* **95,** L07–L09.

Bryngelson, J. D., Onuchic, J. N., Socci, N. D., and Wolynes, P. G. (1995). Funnels, pathways, and the energy landscape of protein-folding—A synthesis. *Proteins* **21,** 167–195.

Burbulys, D., Trach, K. A., and Hoch, J. A. (1991). Initiation of sporulation in *B. subtilis* is controlled by a multicomponent phosphorelay. *Cell* **64,** 545–552.

Chavez, L. L., Onuchic, J. N., and Clementi, C. (2004). Quantifying the roughness on the free energy landscape: Entropic bottlenecks and protein folding rates. *J. Am. Chem. Soc.* **126,** 8426–8432.

Cheung, M. S., Finke, J. M., Callahan, B., and Onuchic, J. N. (2003). Exploring the interplay between topology and secondary structural formation in the protein folding problem. *J. Phys. Chem. B* **107,** 11193–11200.

Clementi, C., and Plotkin, S. S. (2004). The effects of nonnative interactions on protein folding rates: Theory and simulation. *Protein Sci.* **13,** 1750–1766.

Clementi, C., Nymeyer, H., and Onuchic, J. N. (2000). Topological and energetic factors: What determines the structural details of the transition state ensemble and "en-route" intermediates for protein folding? An investigation for small globular proteins. *J. Mol. Biol.* **298,** 937–953.

Clementi, C., Jennings, P. A., and Onuchic, J. N. (2001). Prediction of folding mechanism for circular-permuted proteins. *J. Mol. Biol.* **311,** 879–890.

Essmann, U., Perara, L., Berkowity, M. L., Darden, T., Lee, H., and Pedersen, L. G. (1995). A smooth particle Eswald method. *J. Chem. Phys.* **103,** 8577.

Eswar, N., Eramian, D., Webb, B., Shen, M.-Y., and Sali, A. (2008). Protein structure modeling with MODELLER. *Methods Mol. Biol.* **426,** 145–159.

Finke, J. M., Cheung, M. S., and Onuchic, J. N. (2004). A structural model of polyglutamine determined from a host-guest method combining experiments and landscape theory. *Biophys. J.* **87,** 1900–1918.

Frauenfelder, H., Sligar, S. G., and Wolynes, P. G. (1991). The energy landscapes and motions of proteins. *Science* **254,** 1598–1603.

Fujitsuka, Y., Takada, S., Luthey-Schulten, Z. A., and Wolynes, P. G. (2004). Optimizing physical energy functions for protein folding. *Proteins* **54,** 88–103.

Galperin, M. Y. (2006). Structural classification of bacterial response regulators: Diversity of output domains and domain combinations. *J. Bacteriol.* **188,** 4169–4182.

Gambin, Y., Schug, A., Lemke, E. A., Lavinder, J. J., Ferreon, A. C., Magliery, T. J., Onuchic, J. N., and Deniz, A. A. (2009). Direct single-molecule observation of a protein living in two opposed native structures. *Proc. Natl. Acad. Sci. USA* **106,** 10153–10158.

Hardin, C., Pogorelov, T. V., and Luthey-Schulten, Z. (2002). Ab initio protein structure prediction. *Curr. Opin. Struct. Biol.* **12,** 176–181.

Hoch, J. A. (2000). Two-component and phosphorelay signal transduction. *Curr. Opin. Microbiol.* **3,** 165–170.

Hoch, J. A., and Varughese, K. I. (2001). Keeping signals straight in phosphorelay signal transduction. *J. Bacteriol.* **183,** 4941–4949.

Humphrey, W., Dalke, A., and Schulten, K. (1996). VMD: Visual molecular dynamics. *J. Mol. Graph.* **14**(33–8), 27–28.

Janin, J., Henrick, K., Moult, J., Eyck, L. T., Sternberg, M. J., Vajda, S., Vakser, I., and Wodak, S. J. (2003). CAPRI: A critical assessment of predicted interactions. *Proteins* **52,** 2–9.

Jiang, M., Shao, W., Perego, M., and Hoch, J. A. (2000). Multiple histidine kinases regulate entry into stationary phase and sporulation in *Bacillus subtilis*. *Mol. Microbiol.* **38,** 535–542.

Jorgensen, W. L., Chandrasekhar, J., Madura, J. D., Impey, R. W., and Klein, M. L. (1983). Comparison of simple potential functions for simulating liquid water. *J. Chem. Phys.* **79,** 926–935.

Kutzner, C., van der Spoel, D., Fechner, M., Lindahl, E., Schmitt, U. W., de Groot, B. L., and Grubmuller, H. (2007). Speeding up parallel GROMACS on high-latency networks. *J. Comput. Chem.* **28,** 2075–2084.

Lammert, H., Schug, A., and Onuchic, J. N. (2009). Robustness and generalization of structure-based models for protein folding and function. *Proteins* **77,** 881–891.

Levy, Y., and Onuchic, J. N. (2006). Mechanisms of protein assembly: Lessons from minimalist models. *Acc. Chem. Res.* **39,** 135–142.

Levy, Y., Onuchic, J. N., and Wolynes, P. G. (2007). Fly-casting in protein-DNA binding: Frustration between protein folding and electrostatics facilitates target recognition. *J. Am. Chem. Soc.* **129,** 738–739.

Linhananta, A., and Zhou, Y. Q. (2002). The role of sidechain packing and native contact interactions in folding: Discontinuous molecular dynamics folding simulations of an all-atom G(o)over-bar model of fragment B of Staphylococcal protein A. *J. Chem. Phys.* **117,** 8983–8995.

Mascher, T., Helmann, J. D., and Unden, G. (2006). Stimulus perception in bacterial signal-transducing histidine kinases. *Microbiol. Mol. Biol. Rev.* **70,** 910–938.

Mukhopadhyay, D., Sen, U., Zapf, J., and Varughese, K. I. (2004). Metals in the sporulation phosphorelay: Manganese binding by the response regulator Spo0F. *Acta Crystallogr. D Biol. Crystallogr.* **60,** 638–645.

Oliveira, L. C., Schug, A., and Onuchic, J. N. (2008). Geometrical features of the protein folding mechanism are a robust property of the energy landscape: A detailed investigation of several reduced models. *J. Phys. Chem. B* **112,** 6131–6136.

Onuchic, J. N., and Wolynes, P. G. (2004). Theory of protein folding. *Curr. Opin. Struct. Biol.* **14,** 70–75.

Onuchic, J. N., Luthey-Schulten, Z., and Wolynes, P. G. (1997). Theory of protein folding: The energy landscape perspective. *Annu. Rev. Phys. Chem.* **48,** 545–600.

Schug, A., and Wenzel, W. (2006). An evolutionary strategy for all-atom folding of the 60-amino-acid bacterial ribosomal protein L20. *Biophys. J.* **90,** 4273–4280.

Schug, A., Fischer, B., Verma, A., Merlitz, H., Wenzel, W., and Schoen, G. (2005a). Biomolecular structure prediction stochastic optimization methods. *Adv. Eng. Mat.* **7,** 1005–1009.

Schug, A., Herges, T., Verma, A., Lee, K. H., and Wenzel, W. (2005b). Comparison of Stochastic optimization methods for all-atom folding of the Trp-cage protein. *Chemphyschem* **6,** 2640–2646.

Schug, A., Whitford, P. C., Levy, Y., and Onuchic, J. N. (2007). Mutations as trapdoors to two competing native conformations of the Rop-dimer. *Proc. Natl. Acad. Sci. USA* **104,** 17674–17679.

Schug, A., Hyeon, C., and Onuchic, J. (2009a). Coarse-grained structure-based simulations of proteins and RNA. *In* "Coarse-Graining of Condensed Phase and Biomolecular Systems," (G. A. Voth, ed.), pp. 123–140. CRC Press, Boca Raton, FL.

Schug, A., Weigt, M., Onuchic, J. N., Hwa, T., and Szurmant, H. (2009b). High resolution complexes from integrating genomic information with molecular simulation. *Proc. Natl. Acad. Sci. USA* **106**, 22124–22129.

Shimada, J., Kussell, E. L., and Shakhnovich, E. I. (2001). The folding thermodynamics and kinetics of crambin using an all-atom Monte Carlo simulation. *J. Mol. Biol.* **308,** 79–95.

Szurmant, H., White, R. A., and Hoch, J. A. (2007). Sensor complexes regulating two-component signal transduction. *Curr. Opin. Struct. Biol.* **17,** 706–715.

Thirumalai, D., and Hyeon, C. (2005). RNA and protein folding: Common themes and variations. *Biochemistry* **44,** 4957–4970.

Van Der Spoel, D., Lindahl, E., Hess, B., Groenhof, G., Mark, A. E., and Berendsen, H. J. (2005). GROMACS: Fast, flexible, and free. *J. Comput. Chem.* **26,** 1701–1718.

Varughese, K. I., Madhusudan, Zhou, X. Z., Whiteley, J. M., and Hoch, J. A. (1998). Formation of a novel four-helix bundle and molecular recognition sites by dimerization of a response regulator phosphotransferase. *Mol. Cell.* **2,** 485–493.

Wang, J., Cieplak, P., and Kollman, P. A. (2000). How well does a restrained electrostatic potential (RESP) perform in calculating conformational energies of organic and biological molecules? *J. Chem. Phys.* **21,** 1049–1074.

Weigt, M., White, R. A., Szurmant, H., Hoch, J. A., and Hwa, T. (2009). Identification of direct residue contacts in protein-protein interaction by message passing. *Proc. Natl. Acad. Sci. USA* **106,** 67–72.

Whitford, P. C., Miyashita, O., Levy, Y., and Onuchic, J. N. (2007). Conformational transitions of adenylate kinase: Switching by cracking. *J. Mol. Biol.* **366,** 1661–1671.

Whitford, P. C., Noel, J. K., Gosavi, S., Schug, A., Sanbonmatsu, K. Y., and Onuchic, J. N. (2009a). An all-atom structure-based potential for proteins: Bridging minimal models with all-atom empirical forcefields. *Proteins* **75,** 430–441.

Whitford, P. C., Schug, A., Saunders, J., Hennelly, S. P., Onuchic, J. N., and Sanbonmatsu, K. Y. (2009b). Nonlocal helix formation is key to understanding S-adenosylmethionine-1 riboswitch function. *Biophys. J.* **96,** L7–L9.

Zapf, J., Sen, U., Madhusudan, Hoch, J. A., and Varughese, K. I. (2000). A transient interaction between two phosphorelay proteins trapped in a crystal lattice reveals the mechanism of molecular recognition and phosphotransfer in signal transduction. *Structure* **8,** 851–862.

Zhou, Y., Zhang, C., Stell, G., and Wang, J. (2003). Temperature dependence of the distribution of the first passage time: Results from discontinuous molecular dynamics simulations of an all-atom model of the second beta-hairpin fragment of protein G. *J. Am. Chem. Soc.* **125,** 6300–6305.

CHAPTER FOUR

Kinetic Studies of the Yeast His-Asp Phosphorelay Signaling Pathway

Alla O. Kaserer, Babak Andi, Paul F. Cook, *and* Ann H. West

Contents

Abstract

For both prokaryotic and eukaryotic His-Asp phosphorelay signaling pathways, the rates of protein phosphorylation and dephosphorylation determine the stimulus-to-response time frame. Thus, kinetic studies of phosphoryl group transfer between signaling partners are important for gaining a full understanding of how the system is regulated. In many cases, the phosphotransfer reactions are too fast for rates to be determined by manual experimentation. Rapid quench flow techniques thus provide a powerful method for studying rapid reactions that occur in the millisecond time frame. In this chapter, we describe experimental design and procedures for kinetic characterization of the yeast SLN1–YPD1–SSK1 osmoregulatory phosphorelay system using a rapid quench flow kinetic instrument.

Abbreviations

ATP	adenosine-5′-triphosphate
DTT	dithiothreitol
EDTA	ethylenediaminetetraacetic acid

Department of Chemistry and Biochemistry, University of Oklahoma, Norman, Oklahoma, USA

Methods in Enzymology, Volume 471
ISSN 0076-6879, DOI: 10.1016/S0076-6879(10)71004-1

GST	glutathione-*S*-transferase
HK	histidine kinase domain
HPt	histidine-containing phosphotransfer
MAP	mitogen-activated protein
PAGE	polyacrylamide gel electrophoresis
RR	response regulator
SLN1-R1	C-terminal response regulator domain of SLN1
SSK1-R2	C-terminal response regulator domain of SSK1
SKN7-R3	C-terminal response regulator domain of SKN7
RQF	rapid quench flow
SDS	sodium dodecyl sulfate

1. Introduction

Kinetic studies have contributed to our understanding of many two-component signaling systems. For example, the rate of histidine kinase (HK) autophosphorylation and subsequent phosphotransfer to the downstream response regulator (RR) protein determines how quickly the cell can respond to changes in the environment. Likewise, the rate of hydrolysis of the aspartyl phosphate on the RR (due to its intrinsic stability or phosphatase-catalyzed rate) will determine the duration of the cellular response and the return to a prestimulus state.

Two-component regulatory systems and the expanded multistep His-Asp phosphorelay systems are essential for adaptation to a variety of environmental stresses in bacteria and, to a more limited extent, in eukaryotic organisms such as fungi and plants. The number of proteins participating in these His-Asp phosphorelay systems can vary from a minimum of two components, an HK and RR, to three or more signaling molecules that comprise multistep phosphorelay systems (Appleby *et al.*, 1996; Parkinson and Kofoid, 1992; Stock *et al.*, 2000; West and Stock, 2001). These systems are regulated by sequential phosphoryl group transfer and hydrolysis reactions as a means of information transfer.

In cases where the phosphotransfer reaction is too rapid to capture manually, a rapid quench flow (RQF) instrument can be used for monitoring phosphotransfer reactions that occur in the millisecond time frame, measuring the rate constants and other kinetic parameters. This approach has been used to study two-component phosphorelay systems in bacteria (Fisher *et al.*, 1996; Grimshaw *et al.*, 1998; Stewart, 1997) and has been more recently applied to the study of the multistep phosphorelay system from *Saccharomyces cerevisiae* (Janiak-Spens *et al.*, 2005; Kaserer *et al.*, 2009). In *S. cerevisiae*, a branched multistep phosphorelay system is responsible for adaptation to hyperosmotic, oxidative, and other environmental stresses (Hohmann *et al.*, 2007; Posas *et al.*, 1996; Saito and Tatebayashi, 2004). The SLN1–YPD1–SSK1 branch controls the downstream HOG1 MAP kinase cascade that allows cells to adapt to hyperosmotic stress. Under nonosmotic stress conditions, the transmembrane hybrid SLN1 kinase is active and transfers phosphoryl groups from its central kinase domain to its C-terminal receiver domain (referred to as the SLN1-R1 domain). Subsequently, phosphoryl groups are transferred to YPD1, a cytoplasmic histidine-containing phosphotransfer (HPt) protein and then to the RR domain on SSK1 (referred to as the SSK1-R2 domain), thereby maintaining SSK1 in a constitutively phosphorylated state. Hyperosmotic stress leads to dephosphorylation of SSK1 and activation of the downstream HOG1 MAP kinase cascade resulting in an increase in intracellular glycerol, a compatible osmolyte that restores homeostasis (Horie *et al.*, 2008; Posas and Saito, 1998). The SLN1–YPD1–SKN7 branch responds primarily to cell wall perturbations (Li *et al.*, 2002; Lu *et al.*, 2003; Shankarnarayan *et al.*, 2008). The SKN7 RR is a nuclear localized transcription factor, thus its function is to modulate gene expression in response to environmental conditions (Brown *et al.*, 1993, 1994; Krems *et al.*, 1996).

Previous studies of the multistep phosphorelay system from *S. cerevisiae* from our laboratory have centered on structural and functional characterization of the YPD1 HPt protein. Specifically, important information regarding YPD1-RR recognition and binding (Porter and West, 2005; Porter *et al.*, 2003) and X-ray structures of YPD1/SLN1-R1 complexes (Xu *et al.*, 2003; Zhao *et al.*, 2008) have been obtained. *In vitro* data had shown that YPD1 can form a complex with the phosphorylated SSK1-R2 domain and it was suggested that YPD1 shields the phosphoryl group on SSK1 preventing it from hydrolysis (Janiak-Spens *et al.*, 2000). Further studies focused on the interaction of YPD1 with the RR domains associated with SLN1, SSK1, and SKN7 (R1, R2, and R3, respectively) with respect to phosphotransfer, protein binding affinity, specificity of interaction, and characterization of YPD1 mutants (Janiak-Spens and West, 2000; Janiak-Spens *et al.*, 1999, 2000, 2005; Porter and West, 2005).

The phosphotransfer reactions that comprise the SLN1–YPD1–SSK1 phosphorelay reach steady-state levels within 8 s (Janiak-Spens and West, 2000), thus necessitating utilization of rapid quench kinetics as a means of studying phosphotransfer rates. The individual phosphoryl transfer reactions between YPD1 and the RR domains have been examined kinetically and the individual phosphotransfer rates and dissociation constants were determined and analyzed (Janiak-Spens *et al.*, 2005). The data demonstrated that phosphotransfer from YPD1 to SSK1 was strongly favored over phosphotransfer to SKN7, and phosphotransfer from YPD1 to SSK1 was irreversible. These findings were consistent with the concept that SSK1 is constitutively phosphorylated under normal osmotic conditions.

Moreover, these data led to the hypothesis that upon hyperosmotic stress, when water rapidly effluxes the cell, the increasing ion/solute concentrations inside the cell might disrupt the YPD1·SSK1~P complex. Therefore, the effect of osmolyte concentrations on the half-life of phosphorylated SSK1-R2 in the presence and absence of YPD1 and the kinetics of the individual phosphorelay reactions was examined (Kaserer *et al.*, 2009). Our findings suggest that as intracellular osmolyte concentrations increase, the YPD1·SSK1~P complex dissociates thereby facilitating dephosphorylation of SSK1 and activating the HOG1 MAP kinase cascade. Later, when glycerol and other ions reach their highest concentration in the cell, attenuation of the pathway is achieved, in part, because the kinetics of the phosphorelay favor production of SSK1~P and inhibition of the HOG1 pathway.

In this chapter, we provide a description of the application of RQF analysis in order to measure kinetic parameters of the SLN1–YPD1–SSK1 osmoregulatory phosphorelay system from *S. cerevisiae*.

2. Materials and Methods

2.1. Protein purification and phosphorylation

Expression and purification of the *S. cerevisiae* SLN1-HK, SLN1-R1, YPD1, SSK1-R2, and SKN7-R3 proteins has been described in detail in another chapter in this volume (Fassler and West, 2010). Here, we will summarize preparation of the radiolabeled phosphorylated protein donors for the purpose of rapid quench kinetic experiments. The phosphorylated SLN1-R1 domain acts as the phosphoryl donor in the first half-reaction between SLN1-R1~P and YPD1. GST-linked SLN1-HK (7 μM) bound to glutathione-Sepharose 4B resin is incubated with 7 μM [γ-^{32}P]-ATP for 30 min. Unincorporated [γ-^{32}P]-ATP is then washed from SLN1-HK~P with 50 mM Tris–HCl, pH 8.0, 100 mM KCl, 15 mM $MgCl_2$, 2 mM DTT, and 20% glycerol using three consecutive centrifugations (1 min at 1000×g). The SLN1-R1 protein (18.6 μM) is then added in the same buffer and incubated for 10 min at room temperature in a total volume of 300 μl. Phospho-SLN1-R1 is recovered in the supernatant after gently pelleting the GST–SLN1–HK bound to the resin. EDTA is added to the supernatant to a final concentration of 30 mM to prevent dephosphorylation. Phosphorylated SLN1-R1 is diluted to 0.45 μM in S2 buffer containing 50 mM Tris–HCl, pH 8.0, 1 mM EDTA, 1 mM DTT.

For the second half-reaction, phosphotransfer from YPD1~P to SSK1-R2, a similar protocol can be followed with the following modifications.

Incubation of GST-tagged SLN1-HK-R1 (7 *μM*) and [γ-^{32}P] ATP (7 *μM*) is for 60 min. The YPD1 protein (18.6 *μM*) is then added to the reaction mixture.

2.2. The RQF instrument

There are several commercially available rapid quench instruments[1] and they all consist of a control unit, a drive mechanism, and a mixing chamber. In order to cover a broader timescale of enzymatic reactions, different quench flow devices can be used; however, the principles remain the same. For example, an RQF device (Fig. 4.1A) is applicable for times in the millisecond range from 5 to 300 ms and a time-delay quench flow (Fig. 4.1B) for times greater than 300 ms (Barman *et al.*, 2006).

An RQF apparatus makes use of continuous liquid flow, as shown in Fig. 4.1A, a drive system pushes the plungers at constant speed, enzyme and substrate are mixed in a mixing chamber, and the reaction mixture fills and passes through the capillary tube at a constant speed, *S*. The reaction time, or age of the mixture, is calculated as $t = V/S$, where V is the volume of the capillary. At the end of the capillary tube, the reaction is quenched (alternatively, the quenching solution can also be injected into a second mixing chamber with a syringe or by the drive mechanism, Fig. 4.1B). The quenched sample is then collected and analyzed. Changing the capillary tube (V), and or the drive speed (S), provides for flexibility in the reaction time. Accurate calculation of the rates and other kinetic parameters requires knowledge of reaction time (t) and the dilution factor of the reaction mixture by the quenching solution.

In this chapter, attention is focused on the SFM-4/Q rapid quench instrument[2] from Bio-Logic used in our laboratory (Janiak-Spens *et al.*, 2005; Kaserer *et al.*, 2009). As shown in Fig. 4.2A, the instrument can be operated with a total of four syringes,[3] including two reagent syringes (S2 and S3), one syringe for the buffer or optionally the third reagent of the reaction (S1), and one syringe for the quenching solution (S4). The standard syringe volumes for the instrument are 5 ml for S2 and S3, and 20–30 ml for S1 and S4. Syringes are interchangeable, which allows custom adjustment of the system. Four independently programmable stepping motors are used to actuate syringes S1, S2, S3, and S4. Motor drive rates are independent, so variable-mixing ratios can be obtained by simply programming the drive sequence.

[1] TgK Scientific (www.tgkscientifc.com), Bio-Logic (www.bio-logic.info), KinTek (www.kintek-corp.com), and AppliedPhotophysics (www.photophysics.com).

[2] The SFM400 is a newer model from Bio-Logic that has replaced the SFM-4/Q instrument, but the basic design and principle of use is the same.

[3] The instrument can also be used with two to three syringes (although the fourth syringe cannot be empty; one can fill it with buffer) and a delay line with either single or double mixing.

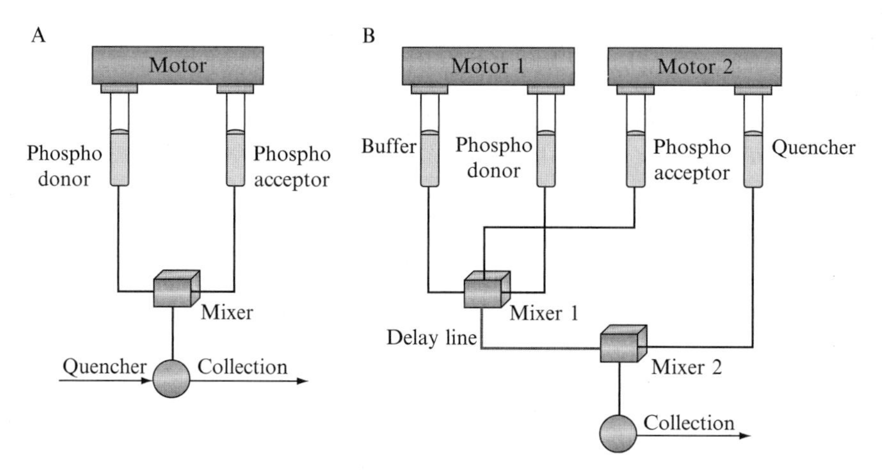

Figure 4.1 (A) Continuous or rapid quench flow apparatus with a single drive mode. By activation of the drive, phosphodonor and phosphoacceptor are mixed in the mixing chamber. The reaction mixture is aged. The age of the reaction mixture corresponds to $t = V/S$ where V is the volume of the capillary and S is the rate of the flow of the reaction mixture down the capillary tube. The value of t can be controlled by varying V and S which is in the range of 30–300 ms. The reaction mixture is quenched upon sample collection. (B) Time-delay mode for the quench flow device. In this case, there are two drives. One is for the buffer and phosphodonor and the second one is for the phosphoacceptor and quencher. By activation of the first drive, phosphodonor and buffer are mixed in the first mixing chamber. Activation of the second drive fills the mixing chamber with phosphoacceptor. The reaction mixture is driven in the delay line and stays there for a certain amount of time. After a specific time delay, the reaction mixture is expelled from the delay line into a second mixing chamber where it is quenched by the injected quencher and collected for analysis. For this method, the reaction time is in the range of 150 ms to minutes and the reaction mixture is less than 50 μl (figure adapted from Barman *et al.*, 2006).

The reaction mixture can be aged in the delay line allowing various delay times. The volume of the delay line is fixed, however, a variety of delay line volumes are available; the volume can be as low as a few microliters. Aging in the delay line can be set by varying the mean flow rate of the syringes (effectively from 1 μl/s to 5 ml/s). The reaction product is mixed with the quenching solution in the mixer and the final mixture (product) is collected via the exit line. The final product is collected through the exit purge port using buffer or the next reaction mixture.

2.3. Experimental design

The following aspects of experimental design should be considered in preparation for rapid quench-flow kinetic analysis:

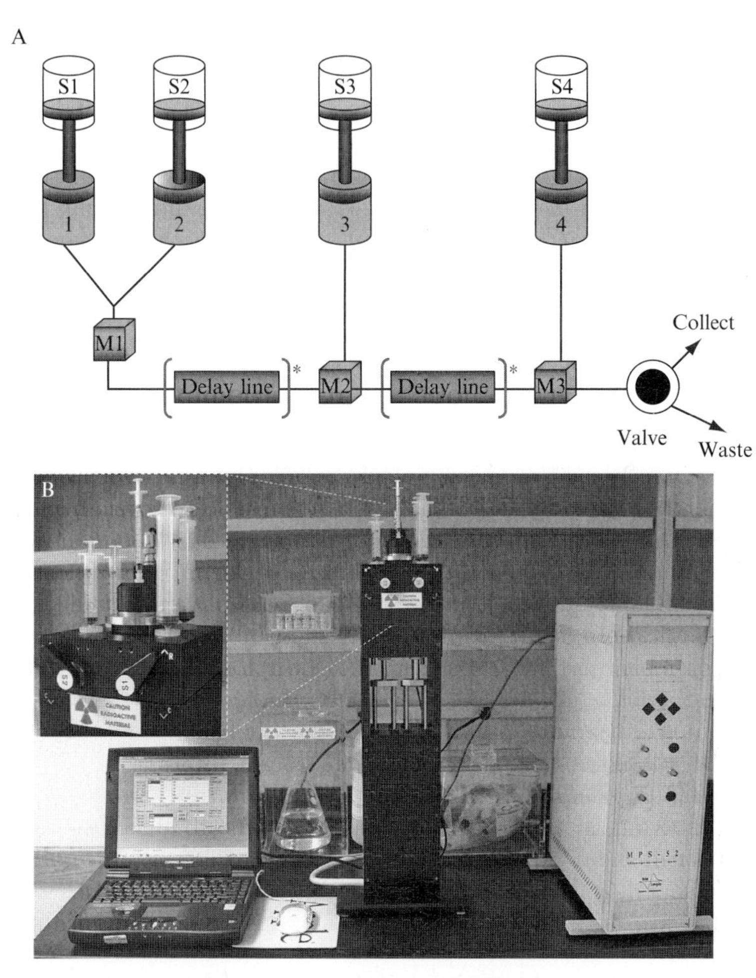

Figure 4.2 (A) A schematic representation of the RQF experiment. S1–S4, syringes; M1–M3, mixers; 1–4, motors; [delay line]*, possible location of the delay line. Figure is reproduced from the SFM-4/Q instrument manual with permission from Bio-Logic (www.bio-logic.info). (B) The SFM-4/Q rapid quench-flow instrument from Bio-Logic (newer model is the SFM400) showing placement of syringes and valves (inset). Note that users should follow proper safety procedures for working with radioactivity, including shielding while working with the instrument and properly disposing of both liquid and solid radioactive waste.

(a) What are the major components of the reaction mixture?

For example, the SLN1–YPD1–SSK1 multistep phosphorelay pathway was subdivided for the RQF analysis into two half-reactions of SLN1-R1~P to YPD1 (Scheme 4.1) and YPD1~P to SSK1-R2 (Scheme 4.2).

$$\text{SLN1-R1}{\sim}\text{P} + \text{YPD1} \underset{k_{-1}}{\overset{k_1}{\rightleftharpoons}} (\text{SLN1-R1}{\sim}\text{P} \cdot \text{YPD1}) \underset{k_{-2}}{\overset{k_2}{\rightleftharpoons}} (\text{SLN1-R1} \cdot \text{YPD1}{\sim}\text{P}) \underset{k_{-3}}{\overset{k_3}{\rightleftharpoons}} \text{SLN1-R1} + \text{YPD1}{\sim}\text{P}$$

Scheme 4.1

$$\text{YPD1}{\sim}\text{P} + \text{SSK1-R2} \underset{k_{-4}}{\overset{k_4}{\rightleftharpoons}} (\text{YPD1}{\sim}\text{P} \cdot \text{SSK1-R2}) \underset{k_{-5}}{\overset{k_5}{\rightleftharpoons}} (\text{YPD1} \cdot \text{SSK1-R2}{\sim}\text{P}) \underset{k_{-6}}{\overset{k_6}{\rightleftharpoons}} \text{YPD1} + \text{SSK1-R2}{\sim}\text{P}$$

Scheme 4.2

(b) What protein concentrations are required?

This will depend on the specific activity of the radiolabeled phosphodonor protein preparation. One could determine empirically what the minimal detectable concentration of the protein phosphodonor is based on the amount of radioactivity detectable by phosphorimager analysis (Molecular Dynamics). In the case of SLN1-R1~P, we estimate ~10% of the protein is radiolabeled (Janiak-Spens *et al.*, 2005) and a protein concentration of ~0.45 μM is suitable for the rapid quench experiments.

(c) What quenching solution will be used?

The quenching solution may vary depending on the reaction components used for the experiment. In the case of phosphotransfer analysis of the *S. cerevisiae* multistep phosphorelay system, the quenching solution contained EDTA to chelate Mg^{2+} and SDS for protein denaturation for subsequent SDS–PAGE analysis.

(d) What enzyme to substrate ratios will be assayed?

For example, in the case of SLN1–YPD1–SSK1 phosphotransfer reactions, the following concentration ratios of SSK1-R1 to YPD1 and YPD1 to SSK1-R2 were used: 1:0.5, 1:1, 1:2, 1:5, 1:10, and 1:20.

(e) What are the timescales?

Different timescales are expected for different phosphotransfer reactions and this should be determined empirically. The earliest manually detectable time point for the phosphotransfer reaction between SLN1-R1~P and YPD1 was 8 s (Janiak-Spens and West, 2000). Thus, the following time points were used for both half-reactions: 30, 40, 60, 80, 100, 150, and 300 ms.

(f) Will a delay line be used?

The RQF instrument does not require installation of the delay lines if the reaction timescale is within 30–300 ms. However, if the reaction timescale is longer, delay lines should be installed and used for proper aging time.

The RQF final result will always depend on an accurate and precise chemical analysis of the product or intermediate.

2.3.1. A typical experiment

For kinetic analysis of the SLN1–YPD1–SSK1 phosphorelay, the RQF technique was employed and the two half-reactions, SLN1-R1 to YPD1 and YPD1 to SSK1-R2, were analyzed in detail. The following operational mode is given here as an example adapted for the *S. cerevisiae* SLN1–YPD1–SSK1 multistep phosphorelay.

- Only syringes S2, S3, and S4 are used for the typical experiment (phosphotransfer from SLN1-R1~P to YPD1 or YPD1~P to SSK1-R2), while S1 is filled with buffer but is not used for the reaction.[4]
- Reactants are in S2 (5 ml) and S3 (5 ml) and the quenching reagent is in S4 (20 ml).
- A delay line is not necessary.
- For collection of the final mixture, the SFM-4/Q instrument has a programmable exit electrovalve, which also provides sample economy. After the final solution is accumulated in the mixer and pushed to the exit valve, the sample is collected in a pipette or syringe. Use of a plastic syringe is recommended for collection of radioactive samples.[5]

2.4. Rapid quench flow experiments

The SFM-4/Q instrument (Bio-Logic) was used for the RQF experiments to determine the phosphotransfer rates of the SLN1–YPD1–SSK1 pathway (Fig. 4.2). The instrument was calibrated via monitoring the base-catalyzed hydrolysis of *p*-nitrophenylacetate (Gutfreund, 1969) as recommended by the instrument manufacturer (Bio-Logic). The reaction mixture contained 500 μl of 0.625 mM 2,4-dinitrophenyl acetate and 0.237 M NaOH, quenched in 500 μl of 4 M HC1. The base-catalyzed hydrolysis of 2,4-dinitrophenylacetate can be examined over a wide range of rates, generated by changing the concentration of the excess reagent.

After calibration, the SFM-4/Q should be flushed thoroughly with reaction buffers (S1–S3, see below); the syringe drive motors should be run in forward and reverse to release any air bubbles trapped in the system. Blank preruns can be conducted to ensure proper volume dispensing during sample collection. The minimal reaction mixture volume is approximately 180 μl for the SLN1–YPD1–SSK1 phosphotransfer reactions. To estimate minimal

[4] The reason for filling syringe S1 is to avoid software interface problems in communicating with the instrument for this specific instrument model.

[5] During a typical experiment, be aware that backpressure from a purge port can push out the collection syringe if it is not held by hand during the collection mode.

protein concentration, nonradioactive preruns were conducted with 60 μl of SLN1-R1 (1, 0.75, and 0.45 μM) rapidly mixed with 60 μl of YPD1 (0.9 μM) and quenched with 60 μl of S4 buffer (see below) for the first half-reaction; optimal concentrations were 0.45 μM SLN1-R1 or YPD1.

Using the same procedure, the second half-reaction YPD1 to SSK1-R2 was analyzed in similar manner. Blank nonradioactive preruns were conducted (60 μl of SSK1-R2 (1, 0.6, and 0.45 μM), 60 μl of YPD1 (0.9 μM) and 60 μl of quenching S4 buffer.

The datasets for the blank experiments were collected in a time-dependent manner (30, 40, 60, 80, 100, 150, and 300 ms) and each experiment included three phosphodonor to phosphoacceptor concentration ratios (SLN1-R1 to YPD1 or YPD1 to SSK1-R2): 1:2, 1:5, 1:10. The concentration of the phosphodonor (SLN1-R1 or YPD1) was kept constant throughout the experimental procedure. Once the blank experiments were completed, the collected samples were inspected by SDS–PAGE to verify even dispensing by the instrument.

When the preliminary analysis was completed, the instrument was filled with the following buffers as designed for the His-Asp phosphorelay system from *S. cerevisiae* (Fig. 4.2):

S1:20 ml of 50 m*M* Tris–HCl, pH 8.0, 1 m*M* EDTA, 1 m*M* DTT;
S3:3 ml of the phosphoaccepting protein at concentrations of 0.45–20 μ*M* were diluted into 50 m*M* Tris–HCl, pH 8.0, 20 m*M* $MgCl_2$, 1 m*M* DTT;
S4:20 ml of stop buffer (8% SDS, 80 m*M* EDTA).

The radiolabeled phosphodonor protein (SLN1-R1 or YPD1; diluted to 0.45 μM) is immediately transferred into the S2 syringe (Fig. 4.2). The diluted phosphorylated SLN1-R1 (60 μl) is then mixed with 60 μl of phosphoaccepting protein (YPD1 or SSK1-R2 at 0.45–20 μM). Reactions are quenched with 60 μl of the stop buffer after a specified time. To further prevent phosphate hydrolysis, the reaction samples can be frozen or placed on ice prior to gel electrophoresis.

Data sets are collected in the time-dependent manner (30, 40, 60, 80, 100, 150, and 300 ms) and each new experiment contained a different phosphodonor to phosphoaccepting concentration ratio as specified above. However, the concentration of the phosphodonor (SLN1-R1 or YPD1) was kept constant for all experiments.

To analyze the results, 30 μl of the quenched reaction was mixed with 10 μl of 4× SDS–PAGE loading buffer (200 m*M* Tris, pH 6.8, 400 m*M* DTT or β-mercaptoethanol, 8% SDS, 0.4% bromophenol blue, and 40% glycerol), and then 30 μl samples were loaded onto 15% SDS–PAGE gels.[6]

[6] Do not boil samples prior to gel loading, as this increases the rate of phosphate hydrolysis.

After gel electrophoresis, wet gels are wrapped in plastic wrap and analyzed using a Phosphorimager (Molecular Dynamics, Storm 840). The phosphotransfer reaction kinetic parameters are quantified on the basis of disappearance of the ^{32}P-label from the phosphodonor protein and the appearance of ^{32}P-label in the phosphoaccepting protein as described below.

2.5. Data analysis and interpretation

A phosphotransfer reaction profile or time course is used to extract the kinetic parameters. A representative radiograph image of the raw data is shown in Fig. 4.3. The parameters are quantified on the basis of the disappearance of the band corresponding to the ^{32}P-label from the phosphodonor protein or the appearance of the band corresponding to the ^{32}P-label in the phosphoaccepting protein. The phosphotransfer profile is usually time dependent and the time of incubation depends on how fast the reaction occurs, which is usually captured in the range of milliseconds to seconds. After scanning the storage phosphor screen of the gel or membrane containing the proteins of interests, using a pixel processing software (ImageQuant version 5.2 in our case), the total intensities of the pixels corresponding to a specific band is calculated. For volume calculation, ImageQuant subtracts the background value from the intensity of each pixel in the object, and then adds all the values as follows:

$$\text{volume} = \sum_{y=1}^{M} \sum_{x=1}^{N} [f(X, Y) - \text{background}] \tag{4.1}$$

where the value of the background is calculated as a local average of all the pixel values in the object outline (ImageQuant reference, version 5.0, Molecular Dynamics, Inc., USA).

To determine the percent of the remaining phosphodonor protein, the volume of the corresponding band is divided by the sum of the volumes of the bands for phosphodonor and phosphoacceptor proteins and the fraction value is multiplied by 100. The percent of the remaining phosphodonor protein is then plotted as the natural logarithm versus reaction time (Fig. 4.4A) (Janiak-Spens *et al.*, 2005). The observed first-order rate constants (k_{obs}) is obtained by fitting each time course (obtained at a particular fixed [S]) to the linear relationship as given in Eq. (4.2):

$$\ln A_t = \ln A_0 - k_{obs} t \tag{4.2}$$

where A is the amount of the phosphorylated phosphodonor protein at times t and 0, and k_{obs} is the observed first-order rate constant obtained using least-squares fitting in Excel (Microsoft Office). The purpose of the data plotting as the natural logarithm is to assess data quality and obtain a

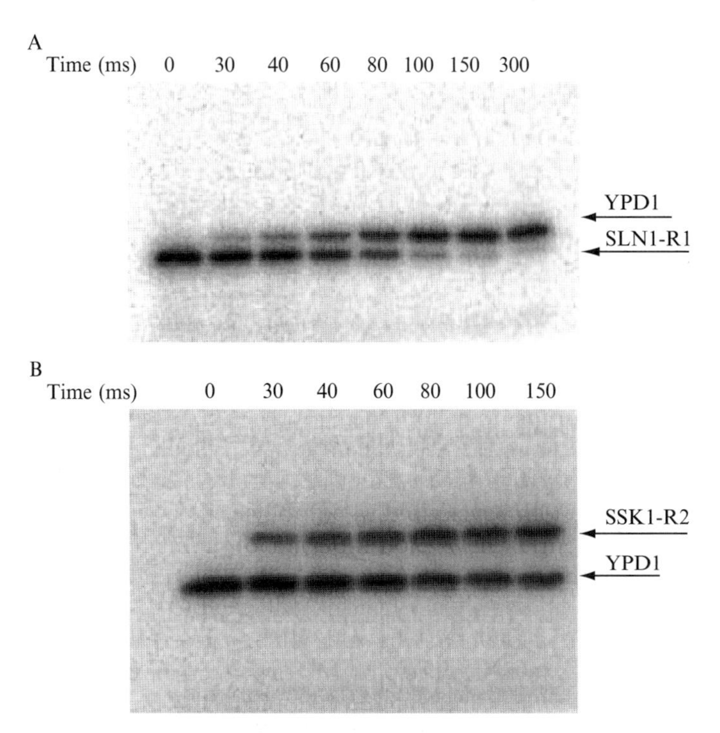

Figure 4.3 Phosphotransfer analysis of the multistep phosphorelay pathway from *S. cerevisiae* using RQF kinetics. (A) The resulting gel radiograph image of the phosphotransfer reaction between phosphorylated receiver domain of SLN1-R1~P (phosphodonor) and histidine phosphotransfer protein YPD1 (phosphoacceptor) as a function of time. Phosphorylated SLN1-R1~P (0.45 μM) was incubated with YPD1 (0.9 μM) for 30, 40, 60, 80, 100, 150, and 300 ms. Each reaction mixture was quenched with quench buffer (8% SDS, 80 mM EDTA) at the specified time point. Each quenched reaction mixture (30 μl) was mixed with 4× SDS–PAGE loading buffer (10 μl) and separated on a 15% SDS–PAGE gel. The radiolabeled bands were quantified using phosphorimager analysis. (B) The radiograph image of the phosphotransfer reaction between phosphorylated YPD1 (phosphodonor) and the response regulator domain of SSK1-R2 (phosphoacceptor) as a function of time. Phosphorylated YPD1 (0.45 μM) was incubated with SSK1-R2 (0.9 μM) for 30, 40, 60, 80, 100, and 150 ms. Each reaction mixture was quenched with quench buffer and analyzed as described in (A).

linear relationship between the dependent (A_t) and independent variables (t). However, time course data can also be fitted to the exponential form of Eq. (4.2) given as follows:

$$A_t = A_0 \, e^{-k_{obs} t} \tag{4.3}$$

However, the fit does not change the value of k_{obs} in comparison to Eq. (4.2). The advantage of fitting to Eq. (4.2) is the feasibility of the linear regression as well as better data assessment.

The general procedures for the subsequent data fitting are to take the slope of the line fitted to the raw data using Eq. (4.2) (the slope is equal to k_{obs} and is obtained at a particular fixed phosphoacceptor concentration). Then, graph the slopes or k_{obs} (obtained at each fixed substrate concentration) versus substrate concentration to assess data quality and evaluate the mathematical trend of the data and or the shape of the graph in order to find the proper model and equation for data fitting.

As shown in the Fig. 4.4B, k_{obs} is a rectangular hyperbolic function of phosphoacceptor concentration with a finite value on the y-axis. Data were fitted to Eq. (4.4) (three-parameter rectangular hyperbola), which adheres to the following phosphotransfer reaction model (Scheme 4.3):

$$k_{obs} = k_{rev} + k_{fwd}\left(\frac{[S]}{K_d + [S]}\right) \tag{4.4}$$

$$\mathrm{A{\sim}P + B} \underset{k_{-1}}{\overset{k_1}{\rightleftharpoons}} \mathrm{(A{\sim}P{\cdot}B)} \underset{k_{-2}}{\overset{k_2}{\rightleftharpoons}} \mathrm{(A{\cdot}B{\sim}P)} \underset{k_{-3}}{\overset{k_3}{\rightleftharpoons}} \mathrm{A + B{\sim}P}$$

Scheme 4.3

where [S] is the concentration of the phosphoaccepting protein, k_{obs} is the observed first-order rate constant for the phosphotransfer reaction at a particular [S], k_{fwd} is the maximal forward net rate constant for phosphoryl transfer from the phosphorylated protein to the phosphoacceptor protein, k_{rev} is the corresponding maximum reverse net rate constant for the reaction between phosphodonor and phosphoacceptor proteins, and K_d is the dissociation constant of the phosphodonor–acceptor complex (Fig. 4.4B). The fit can be obtained using any program with data fitting algorithms. In this case, the Enzfitter program was used (version 2.04, Biosoft, Cambridge, UK).

The limit of Eq. (4.4) when the concentration of the phosphoaccepting protein is very high relative to the K_d is as follows:

$$k_{obs} = k_{max} = k_{rev} + k_{fwd} \tag{4.5}$$

where k_{max} is the maximum observed first-order rate constant. When the concentration of the phosphoaccepting protein is near zero, the limit of Eq. (4.4) is equal to k_{rev}. The limits of Eq. (4.4) are applicable to Fig. 4.4B. The values of the kinetic parameters (k_{fwd}, k_{rev}, and K_d) can also be obtained under different conditions such as different concentrations of osmolytes (Kaserer *et al.*, 2009) or some other variable. In this case, the values of the kinetic parameters (k_{fwd}, K_d, or k_{fwd}/K_d) are plotted against the concentrations of osmolytes and the data interpretation depends on the overall mathematical shape of the plots and their interrelationships.

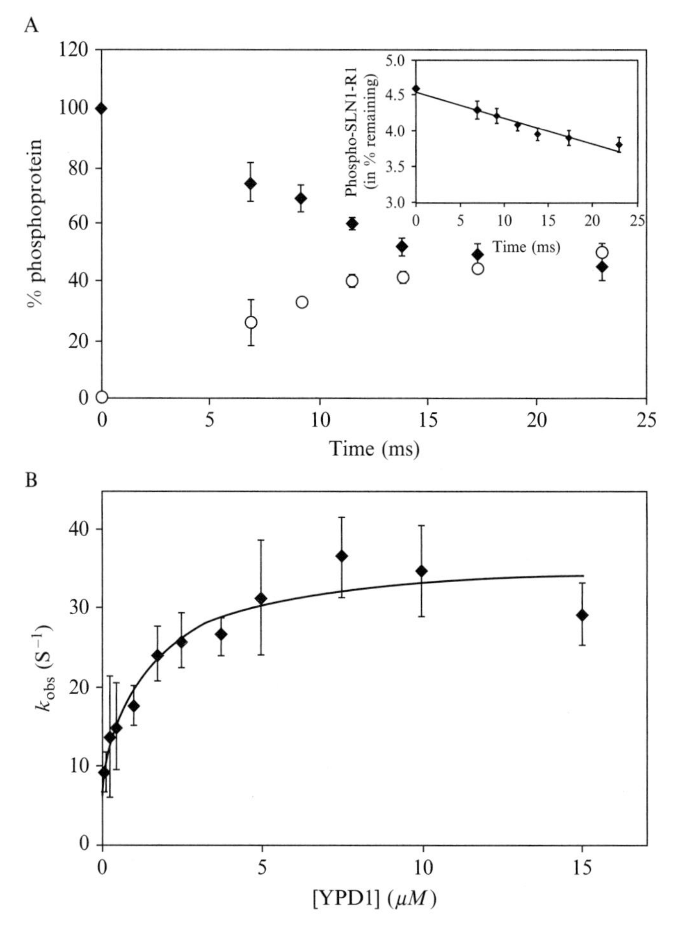

Figure 4.4 (A) A time course of phosphoryl transfer from SLN1-R1~P to YPD1. The percent of the ^{32}P-radiolabeled proteins are shown as follows: (○) YPD1~P, (◆) SLN1-R1~P. Data obtained was from the average of three experiments. The inset shows the percent of the remaining SLN1-R1~P plotted as natural logarithm versus reaction time. The slope of the linear fit (Eq. (4.2)) is k_{obs}. (B) YPD1-dependent rate of the phosphoryl transfer reaction. The plot shows saturation at high YPD1 concentrations and a finite nonzero ordinate intercept (k_{rev}). K_d can be defined as the concentration of YPD1 where $k_{obs} = (1/2)k_{max}$, which is 1.4 μM. Data were fitted to Eq. (4.2) (Janiak-Spens *et al.*, 2005). Figure was reproduced with permission from the American Chemical Society.

In general, the kinetics of the phosphotransfer reactions (see examples below) between biological macromolecules follow the enzyme kinetic model (Michaelis–Menten model) and the data analysis is similar. However, different approaches may be taken for the analysis of the raw data. Here, we give three examples of data obtained using the RQF method:

1. Specificity of the vancomycin resistance kinase VanS for two RRs, VanR and PhoB (Fisher *et al.*, 1996). In this study, the catalytic efficiency (k_{cat}/K_M) was determined for phosphotransfer from VanS to VanR and PhoB with the former reaction showing a 10^4-fold preference. Both reactions follow Michaelis–Menten kinetics with slight modifications. The raw data for the time course experiments were treated with a nonlinear least-squares regression curve fitting to a rectangular hyberbola.
2. Phosphotransfer between CheA and CheY in the bacterial chemotaxis signal transduction pathway (Stewart, 1997). Experiments showed a reversible three step mechanism for CheA to CheY phosphotransfer, which includes binding of CheY to CheA~P, rapid phosphotransfer to CheY, and dissociation of CheA•CheY~P complex. The rate of phosphotransfer shows saturation kinetics similar to enzymatic reactions. Raw data were fitted to the equation for a rectangular hyberbola using a nonlinear least-squares method.
3. Interactions between components of the phosphorelay controlling sporulation in *Bacillus subtilis* (Grimshaw *et al.*, 1998). Divalent-metal dependent autophosphorylation of KinA is the first step in the phosphotransfer reaction. The value of k_{cat}/K_M for Spo0F in the formation of Spo0F~P is 57,000 fold higher than Spo0A in the formation of Spo0A~P. The authors showed the kinetic mechanism was hybrid ping-pong/sequential with a pronounced (>40-fold) substrate synergism by Spo0F in the autophosphorylation of KinA. In this study, reciprocal initial velocities were plotted vs. reciprocal substrate concentrations and fitted by a least-square method for some of the data collected.

The three examples above, as well as recent papers from this lab (Janiak-Spens *et al.*, 2005; Kaserer *et al.*, 2009) are typical of the kinetics of the phosphotransfer reactions obtained using the RQF method. They show the versatility of the method in obtaining the data under different conditions.

3. Conclusion

The highly versatile method of RQF kinetics was described here with regard to its application to the study of His-Asp phosphotransfer reactions in the yeast osmoregulatory signal transduction pathway. Using this method, data on intermediate species can be obtained in the time range of milliseconds to minutes. Finally, treatment and analysis of the data obtained using the RQF method is facilitated by computer operated instrument modes, software, and fitting algorithms.

ACKNOWLEDGMENTS

We gratefully acknowledge funding from the NIH (GM59311 to A. H. W.), the Oklahoma Center for the Advancement of Science and Technology (OCAST) (HR 06-123 to A. H. W.) and the Grayce B. Kerr endowment to the University of Oklahoma (to support the research of P. F. C.) for the research described here. We would also like to thank Dr. Fabiola Janiak-Spens for superb technical advice regarding quench flow experiments and experimental procedures and Vidya Kumar for useful feedback on the manuscript.

REFERENCES

Appleby, J. L., Parkinson, J. S., and Bourret, R. B. (1996). Signal transduction via the multi-step phosphorelay: Not necessarily a road less traveled. *Cell* **86,** 845–848.

Barman, T. E., Bellamy, S. R., Gutfreund, H., Halford, S. E., and Lionne, C. (2006). The identification of chemical intermediates in enzyme catalysis by the rapid quench-flow technique. *Cell. Mol. Life Sci.* **63,** 2571–2583.

Brown, J. L., North, S., and Bussey, H. (1993). *SKN7*, a yeast multicopy suppressor of a mutation affecting cell wall *β*-glucan assembly, encodes a product with domains homologous to prokaryotic two-component regulators and to heat shock transcription factors. *J. Bacteriol.* **175,** 6908–6915.

Brown, J. L., Bussey, H., and Stewart, R. C. (1994). Yeast Skn7p functions in a eukaryotic two-component regulatory pathway. *EMBO J.* **13,** 5186–5194.

Fassler, J. F., and West, A. H. (2010). Genetic and biochemical analysis of the SLN1 pathway in *Saccharomyces cerevisiae*. *Methods Enzymol.* **471,** 293–319.

Fisher, S. L., Kim, S.-K., Wanner, B. L., and Walsh, C. T. (1996). Kinetic comparisons of the specificity of the vancomycin resistance kinase VanS for two response regulators, VanR and PhoB. *Biochemistry* **35,** 4732–4740.

Grimshaw, C. E., Huang, S., Hanstein, C. G., Strauch, M. A., Burbulys, D., Wang, L., Hoch, J. A., and Whiteley, J. M. (1998). Synergistic kinetic interactions between components of the phosphorelay controlling sporulation in *Bacillus subtilis*. *Biochemistry* **37,** 1365–1375.

Gutfreund, H. (1969). Rapid mixing: Continuous flow. *Methods Enzymol.* **16,** 229–249.

Hohmann, S., Krantz, M., and Nordlander, B. (2007). Yeast osmoregulation. *Methods Enzymol.* **428,** 29–45.

Horie, T., Tatebayashi, K., Yamada, R., and Saito, H. (2008). Phosphorylated Ssk1 prevents unphosphorylated Ssk1 from activating the Ssk2 MAP kinase kinase kinase in the yeast HOG osmoregulatory pathway. *Mol. Cell. Biol.* **28,** 5172–5183.

Janiak-Spens, F., and West, A. H. (2000). Functional roles of conserved amino acid residues surrounding the phosphorylatable histidine of the yeast phosphorelay protein YPD1. *Mol. Microbiol.* **37,** 136–144.

Janiak-Spens, F., Sparling, J. M., Gurfinkel, M., and West, A. H. (1999). Differential stabilities of phosphorylated response regulator domains reflect functional roles of the yeast osmoregulatory SLN1 and SSK1 proteins. *J. Bacteriol.* **181,** 411–417.

Janiak-Spens, F., Sparling, D. P., and West, A. H. (2000). Novel role for an HPt domain in stabilizing the phosphorylated state of a response regulator domain. *J. Bacteriol.* **182,** 6673–6678.

Janiak-Spens, F., Cook, P. F., and West, A. H. (2005). Kinetic analysis of YPD1-dependent phosphotransfer reactions in the yeast osmoregulatory phosphorelay system. *Biochemistry* **44,** 377–386.

Kaserer, A. O., Andi, B., Cook, P. F., and West, A. H. (2009). Effects of osmolytes on the SLN1-YPD1-SSK1 phosphorelay system from *Saccharomyces cerevisiae*. *Biochemistry* **48,** 8044–8050.

Krems, B., Charizanis, C., and Entian, K.-D. (1996). The response regulator-like protein Pos9/Skn7 of *Saccharomyces cerevisiae* is involved in oxidative stress resistance. *Curr. Genet.* **29,** 327–334.

Li, S., Dean, S., Li, Z., Horecka, J., Deschenes, R. J., and Fassler, J. S. (2002). The eukaryotic two-component histidine kinase Sln1p regulates OCH1 via the transcription factor, Skn7p. *Mol. Biol. Cell* **13,** 412–424.

Lu, J. M.-Y., Deschenes, R. J., and Fassler, J. S. (2003). *Saccharomyces cerevisiae* histidine phosphotransferase Ypd1p shuttles between the nucleus and cytoplasm for *SLN1*-dependent phosphorylation of Ssk1p and Skn7p. *Eukaryot. Cell* **2,** 1304–1314.

Parkinson, J. S., and Kofoid, E. C. (1992). Communication modules in bacterial signaling proteins. *Annu. Rev. Genet.* **26,** 71–112.

Porter, S. W., and West, A. H. (2005). A common docking site for response regulators on the yeast phosphorelay protein YPD1. *Biochim. Biophys. Acta* **1748,** 138–145.

Porter, S. W., Xu, Q., and West, A. H. (2003). Ssk1p response regulator binding surface on histidine-containing phosphotransfer protein Ypd1p. *Eukaryot. Cell* **2,** 27–33.

Posas, F., and Saito, H. (1998). Activation of the yeast SSK2 MAP kinase kinase kinase by the SSK1 two-component response regulator. *EMBO J.* **17,** 1385–1394.

Posas, F., Wurgler-Murphy, S. M., Maeda, T., Witten, E. A., Thai, T. C., and Saito, H. (1996). Yeast HOG1 MAP kinase cascade is regulated by a multistep phosphorelay mechanism in the SLN1-YPD1-SSK1 "two-component" osmosensor. *Cell* **86,** 865–875.

Saito, H., and Tatebayashi, K. (2004). Regulation of the osmoregulatory HOG MAPK cascade in yeast. *J. Biochem.* **136,** 267–272.

Shankarnarayan, S., Malone, C. L., Deschenes, R. J., and Fassler, J. S. (2008). Modulation of yeast Sln1 kinase activity by the CCW12 cell wall protein. *J. Biol. Chem.* **283,** 1962–1973.

Stewart, R. C. (1997). Kinetic characterization of phosphotransfer between CheA and CheY in the bacterial chemotaxis signal transduction pathway. *Biochemistry* **36,** 2030–2040.

Stock, A. M., Robinson, V. L., and Goudreau, P. N. (2000). Two-component signal transduction. *Annu. Rev. Biochem.* **69,** 183–215.

West, A. H., and Stock, A. M. (2001). Histidine kinases and response regulator proteins in two-component signaling systems. *Trends Biochem. Sci.* **26,** 369–376.

Xu, Q., Porter, S. W., and West, A. H. (2003). The yeast YPD1/SLN1 complex: Insights into molecular recognition in two-component systems. *Structure* **11,** 1569–1581.

Zhao, X., Copeland, D. M., Soares, A. S., and West, A. H. (2008). Crystal structure of a complex between the phosphorelay protein YPD1 and the response regulator domain of SLN1 bound to a phosphoryl analog. *J. Mol. Biol.* **375,** 1141–1151.

CHAPTER FIVE

Purification of MBP-EnvZ Fusion Proteins Using an Automated System

Ricardo Oropeza *and* Edmundo Calva

Contents

Abstract

Bacteria use two-component signal transduction systems to detect and respond to environmental changes. These systems have been studied systematically in *Escherichia coli* as a model organism. Most of the signal transduction systems present in *E. coli* are conserved in related pathogenic bacteria; however, differences in regulation by these systems have been reported from one bacterial species to another [Oropeza, R., and Calva, E. (2009). The cysteine 354 and 277 residues of *Salmonella enterica* serovar Typhi EnvZ are determinants of autophosphorylation and OmpR phosphorylation. *FEMS Microbiol. Lett.* 292, 282–290]. Our laboratory has been interested in studying the OmpR/EnvZ two-component system in *S. enterica*. In *S. enterica* serovar Typhi (Typhi), it regulates the expression of the porin genes, namely *ompC*, *ompF*, *ompS1*, and *ompS2*. OmpR proteins are identical between *E. coli* and Typhi, but several differences exist between the EnvZ proteins. To define whether some differences in porin regulation are due to changes on EnvZ, we decided to overexpress and purify *E. coli*, Typhi, and *S. enterica* serovar Typhimurium (Typhimurium) EnvZ proteins fused to the maltose-binding protein (MBP) as a purification tag. Differences in the autophosphorylation level of these proteins

Departamento de Microbiología Molecular, Instituto de Biotecnología, Universidad Nacional Autónoma de México AP 510-3, Cuernavaca, Morelos, Mexico

Methods in Enzymology, Volume 471
ISSN 0076-6879, DOI: 10.1016/S0076-6879(10)71005-3

were evidenced. Hence, considering the differences at the amino acid level between *E. coli* and Typhi EnvZ proteins, several mutations were introduced in the Typhi EnvZ protein in order to try to find the amino acids affecting the enzymatic activity of the protein. We found that Cys354 plays an important role in defining the enzymatic activity of this histidine kinase. Here, we report the automated purification of a collection of MBP-EnvZ fusions using a mini-chromatography commercial system, but adapting an amylose affinity column packed by ourselves.

1. Introduction

The two-component system EnvZ–OmpR was first identified in *Escherichia coli* as playing a role in porin regulation in response to changes in medium osmolarity (Lan and Igo, 1998; Mizuno and Mizushima, 1990; Russo and Silhavy, 1991; Slauch *et al.*, 1988; Sutcliffe and Foulds, 1983; Taylor *et al.*, 1983). Since then it has been found in a variety of enterobacteria controlling different genes, among them those involved in virulence in species such as *Shigella*, *Yersinia*, *Salmonella*, and pathogenic *E. coli* (Bernardini *et al.*, 1990; Dorrell *et al.*, 1998; Garmendia *et al.*, 2003; Schwan, 2009). Moreover, the importance of understanding the environmental signals that lead to two-component signal transduction in bacterial pathogens has been recently emphasized (Calva and Oropeza, 2006). Most of the work involving two-component system characterization has been done using as a model the nonpathogenic strain *E. coli* K-12. In this bacterium, the *ompR–envZ* genes are encoded by the *ompB* locus, forming an operon. The initial codons of the *envZ* gene overlap with the final ones in *ompR* and presumably, thus, the ratio between these two proteins is controlled (Comeau *et al.*, 1985).

The OmpR and EnvZ proteins from *E. coli* have been studied in great detail. As any typical response regulator OmpR has two domains: the N-terminal encompasses the receiver domain where the Asp 55 residue is located, which plays an important role in signal transduction, as it is phosphorylated by EnvZ resulting in an increase in the DNA-binding activity by the C-terminal domain of the protein (Delgado *et al.*, 1993; Forst *et al.*, 1990; Huang *et al.*, 1997; Kenney *et al.*, 1995; Lan and Igo, 1998; Mizuno and Mizushima, 1990). The OmpR DNA-binding domain was crystallized and the whole protein has been purified and used to perform DNA–protein interaction experiments (Bergstrom *et al.*, 1998; Huang *et al.*, 1997; Martinez-Hackert *et al.*, 1996; Norioka *et al.*, 1986; Oropeza *et al.*, 1999). EnvZ is a typical histidine kinase, which is attached at the bacterial inner membrane by two transmembranal domains and has a periplasmic loop; it is believed that around this region occurs the sensing of some environmental stimulus that is communicated to the cytoplasmic region of the protein to regulate its histidine kinase or phosphatase activity (Khorchid *et al.*, 2005). By the cytoplasmic side, starting at residue 180, EnvZ has several subdomains: the first one is the linker that has been regarded as important for transmitting a signal coming from the N-terminal region of EnvZ; then comes subdomain A, which is involved in protein dimerization and where the His 243 residue for autophosphorylation is located. At the EnvZ carboxy-end is subdomain B, which has several conserved boxes with ATPase proteins involved in ATP-binding and hydrolysis. The crystal structure of the EnvZ A and B subdomains has been reported which, together with genetic and biochemical data, has allowed the conclusion that the autophosphorylation reaction occurs in trans, with the B subdomain of one monomer transferring the phosphate group to the histidine residue at the other monomer (Cai and Inouye, 2003; Hsing *et al.*, 1998; Qin *et al.*, 2003; Tanaka *et al.*, 1998; Tomomori *et al.*, 1999).

Our laboratory has been interested on studying porin regulation in Typhi. As in *E. coli*, OmpR and EnvZ play major roles in the control of the genes coding for outer membrane proteins. However, in *Salmonella*, aside from the *ompC* and *ompF* genes which code for abundant major porins, there are the *ompS1* and *ompS2* genes that code for low-expressed porins under standard laboratory growth conditions (De la Cruz *et al.*, 2007, 2009; Fernandez-Mora *et al.*, 1995, 2004; Flores-Valdez *et al.*, 2003; Martinez-Flores *et al.*, 1999; Oropeza *et al.*, 1999; Puente *et al.*, 1987, 1991). Interestingly, all of them are regulated by this two-component regulator system. Moreover, and in contrast to *E. coli*, the *ompC* gene is not osmoregulated in Typhi, a phenomenon that has been attributed to differences in EnvZ (Martinez-Flores *et al.*, 1999).

To gain further insight on the role of Typhi OmpR and EnvZ on porin regulation, an initial biochemical characterization was performed (Oropeza and Calva, 2009). Interestingly, OmpR is identical in *E. coli*, Typhimurium and Typhi; however, several differences exist among the EnvZ proteins. After performing *in vitro* autophosphorylation experiments on wild type His-tagged derivatives carrying the cytoplasmic domain of the protein, differences on the EnvZ-P yield were evident between *E. coli*, Typhimurium and Typhi EnvZ. Our next goal was to find the amino acid residues defining this biochemical activity on Typhi EnvZ. A technical problem we had to solve was how to purify in a reproducible way the different versions of the Typhi EnvZ protein, because the yield of the His-tagged versions varied depending on whether the tag was at the N- or C-terminal of the protein. The one-step affinity purification of overexpressed fusion proteins with specific tags has proven to be a powerful tool for the biochemical characterization of protein function. There are several purification strategies that make use of different tags, such as a histidine hexamer and glutathione-*S*-transferase (GST). The maltose-binding protein (MalE or MBP) is a cyto- and periplasmic protein of *E. coli* with high binding affinity to maltose, which acts as a receptor in chemiotaxis and signal transduction (Nikaido, 1994). MBP has been successfully used as an N-terminal tag to purify overexpressed proteins, by affinity chromatography based on amylose resins, especially when the protein of interest is difficult to purify because of its biochemical properties (di Guan *et al.*, 1988). Proteins are mildly eluted with 10 m*M* maltose, which does not interfere with the native conformation and properties of most tested proteins. Moreover, MBP increases solubility and stability during purification and storage of the fused protein, a feature that often allows high yields (Ibarra *et al.*, 2003; Perez-Martin *et al.*, 1997).

For the past few years, numerous methods have been described for the purification of MBP-fusion proteins, but the best adopted is affinity chromatography with amylose resin. Thus, proteins have been manually purified with laboratory-made columns because of their simplicity in use and low cost, but this is time-consuming (Riggs, 2000). In this report, we describe the adaptation of an FPLC system, in this case the ÄKTAprime system (Amersham-Pharmacia-Biotech, UK), for use with an amylose affinity column, hence increasing the effectiveness of the purification of MBP-fused proteins. To our knowledge, at the moment there are no protocols or custom prefilled columns designed for the routine purification of MBP fusion proteins in FPLC systems. In our laboratory, we have successfully purified different cytoplasmic MBP fusions following the protocol described below, which is exemplified with the purification of EnvZ cytoplasmic domain. Moreover, the same procedure has been used successfully in some other proteins, especially with transcriptional regulators (Ibarra *et al.*, 2003).

2. Comparison Between the Cytoplasmic Domains of *E. coli* and Typhi EnvZ Proteins

An initial characterization of the *E. coli*, Typhimurium and Typhi EnvZ cytoplasmic domains revealed differences in the autophosphorylation levels, with a higher level for *E. coli* EnvZ, intermediate level for Typhimurium and lower level for Typhi EnvZ (Oropeza and Calva, 2009). To find out which amino acids could be causing these differences, the sequences for the cytoplasmic domain of *E. coli* and Typhi EnvZ proteins were aligned (Fig. 5.1). Eighteen amino acid changes were evidenced. Most of them are located at subdomain B, which is involved in ATP-binding and hydrolysis. However, neither is inside the consensus sequences of this

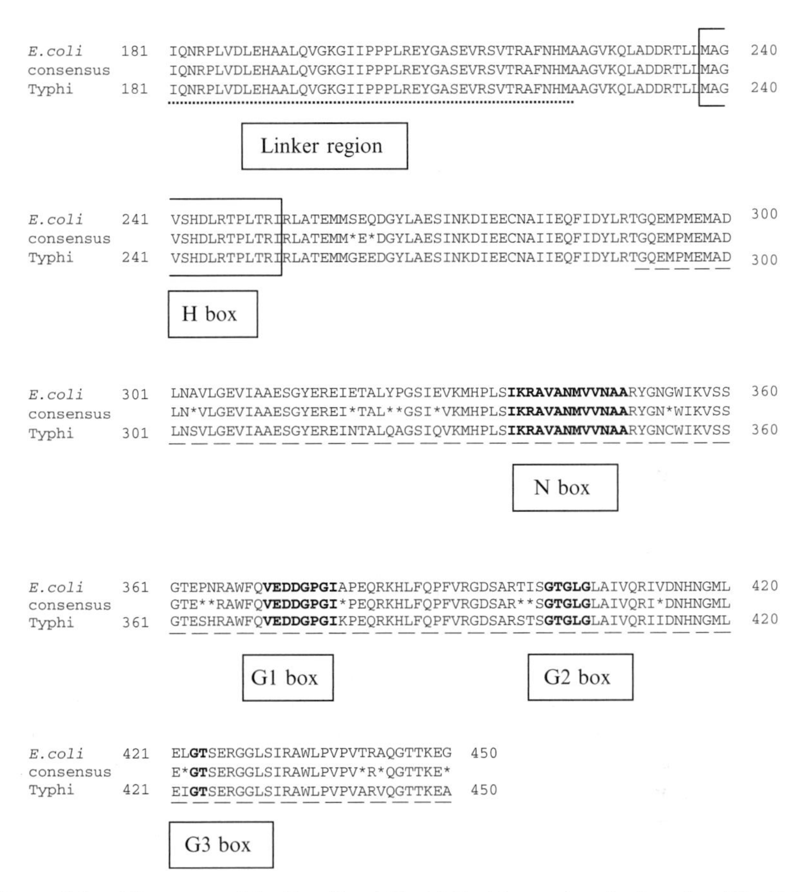

Figure 5.1 Alignment of the *E. coli* and Typhi EnvZ cytoplasmic domains. The linker region at the amino-terminal domain is indicated with dots and dashes and the B subdomain at the carboxy-terminal domain is indicated with dashed lines. Consensus boxes (H, N, G1, G2, and G3) at the A and B subdomains are indicated. The asterisks indicate the amino acid residues that differ between Typhi EnvZ and *E. coli* EnvZ, some of them were studied.

domain, known as N, G1, G2, and G3 boxes but flanking these boxes. To determine the amino acid residues affecting Typhi EnvZ autophosphorylation, single and multiple mutations were introduced to replace the amino acids present in Typhi for those present in *E. coli* EnvZ. The plasmids carrying the mutant EnvZ derivatives were generated by inverse PCR. A total of eight EnvZ Typhi derivatives were constructed containing the following changes and fused to the MBP: K379A, C354G, C354G/K379A, Q324Y/A325P, S364P/H365N, Q324Y/A325P/K379A, S364P/H365N/K379A, and Q324Y/A325P/S364P/H365N.

3. Purification of the Recombinant MBP Proteins by FPLC

The ÄKTAprime (APS) is a compact automated liquid chromatography system that was designed for standard separation applications and simple purification of proteins at the laboratory scale. It includes components for measuring UV, pH, and conductivity. The APS uses prefilled columns for different separation, desalting and purification protocols, such as purification of His- and GST-tagged proteins and poly- and monoclonal antibodies. In this case, we used a 0.7×10 cm glass column (Sigma cat. no. C3669) that was filled manually with 4 ml of amylose resin, a composite amylose/agarose bead, supplied preswollen in 20% ethanol (New England Biolabs) and washed with 50 ml of column buffer (20 m*M* Tris–HCl, pH 7.4, 0.2 *M* NaCl, 1 m*M* EDTA, 10 m*M* β-mercaptoethanol) using the APS "Manual run" method.

E. coli M15 (pREP4), harboring each of the MBP-EnvZ fusions, was grown in 100 ml of LB broth supplemented with kanamycin (30 μg/ml) and ampicillin (100 μg/ml) at 37 °C and 250 rpm. When the culture reached an optical density at 600 nm of 0.5, protein expression was induced with 1 m*M* IPTG and incubated for 4 additional hours. Bacteria were centrifuged at 12,000 rpm at 4 °C and resuspended in 5 ml of column buffer. The cell suspension was placed in an ice-water bath and sonicated using short bursts until clarification (i.e., 4–5 cycles of 30 s burst/30 s resting to avoid overheating). The sonicated extract was centrifuged for 20 min at 12,000 rpm/4 °C to eliminate cell debris, and filtrated through a 0.22 μ*M* membrane (Millipore). In contrast to the procedure described in the manual method (Riggs, 2000), the sample was not diluted with column buffer.

We introduced the file program described in Table 5.1 into the APS "Program method" section. Briefly, the whole crude extract (5 ml) was injected directly into the column with a flow rate of 200 μl/min and a pressure of 0.3 MPa (Fig. 5.2A). After extensive washing with 300 ml of column buffer, at a flow rate of 2 ml/min, UV densitometry showed that there were no more contaminating proteins eluting from the column. The MBP fusion protein was eluted with solution B (same composition as column buffer plus 10 m*M* maltose) with a flow rate of 1 ml/min and collected in 1 ml fractions. Optical density at 280 nm and the percent of solution B (Fig. 5.2A) were registered and plotted by a graph printer connected in-line. Most of the protein was obtained between fractions 5 and 15. Fractions were stored at −70 °C.

4. SDS–PAGE Analysis of Recombinant Expressed MBP-EnvZc Protein

Fractions with a high UV lecture were resolved by 10% SDS–PAGE, stained with Coomassie blue and photographed in an Alpha Imager acquisition system (AlphaInnotech) (Fig. 5.2B).

Table 5.1 Program file for purification of MBP fusion proteins in the AKTA prime

Volume (ml)	% of solution B	Flow	Volume fraction (ml)	Valve B	Injection valve	Peak collection	Autozero	Event mark	Edition *T/V*	Comment
0	100	40	0	1	W	No	No	No	0	Priming B
25	100	40	0	1	W	No	No	No	25	End of priming B
25.1	0	40	0	1	W	No	No	No	25.1	Priming A1
60	0	1	0	1	L	No	No	No	60	End of equilibration
110	0	0.2	0	1	I	No	Yes	Yes	110	Sample application
115	0	0.2	0	1	L	No	No	No	115	End of injection
130	0	2	0	1	L	No	No	No	130	Wash
430	0	1	0	1	L	No	Yes	Yes	430	End of wash
430.1	100	1	1	1	L	No	No	No	430.1	Elution
470	100	1	1	1	L	No	No	No	470	End of elution/ wash
470.1	0	40	0	1	W	No	No	No	470.1	End of wash/ priming A1
495	0	1	0	1	L	No	No	No	495	End of priming/ equilibration
550	0	1	0	1	L	No	No	No	550	Reequilibration End of method

W, wash; L, load; I, injection.

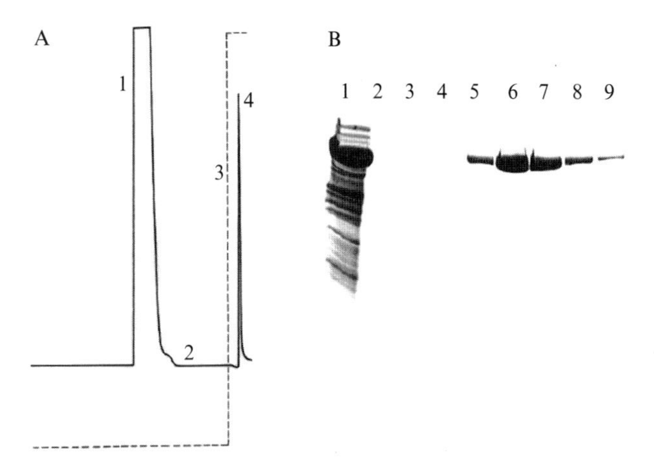

Figure 5.2 Representative plot of the MBP-EnvZ fusion purification. (A) ÄKTA-prime automatically generates a graph showing two selected parameters: $OD_{280\ nm}$ (solid line) and percent of solution B (dashed line). The sample was injected into the column (1) and, after washing with column buffer (2), no UV lecture was detected. Elution was upon addition of solution B, containing maltose (3); and samples were then collected (4). In general, the MBP fusion protein eluted between fractions 5 and 15. (B) SDS–polyacrylamide gel stained with Coomassie showing the purification process. Lanes: (1) 10 μl of crude extract, (2–9) 20 μl of fractions 5–12.

5. Results

Molecular cloning and construction of MBP-EnvZ was performed by conventional molecular biology techniques as described in a previous communication (Oropeza and Calva, 2009). Next, a novel protocol for purifying this and other MBP fusion proteins with an automated FPLC system was implemented as described above (Table 5.1). In the protocol described here, no dilution of the sample is needed, as all the extract is injected directly to the amylose column. As shown in Fig. 5.2A, all the purification procedure is monitored by UV readings and percent of the eluting solution B, containing maltose. When all the contaminating proteins were washed from the column, solution B was applied (Fig. 5.2A). In general, the MBP fusion elutes between fractions 5 and 15. These fractions were subjected to 10% SDS–PAGE. As shown in Fig. 5.2B, we observed only one band corresponding to the expected molecular weight of the MBP-EnvZc (70 kDa) with almost a 95% of purity. With this procedure, there was no need for further purification of the protein through a second chromatography column. The total yield with this procedure was between 2 and 6 mg of protein.

6. Discussion

The reported method is both less time-consuming and less expensive than the manual procedure with lab-made columns, for the following reasons: (1) the amylose volume used for packing the column is less than half the amount normally used; (2) given the pressure applied, less contaminants were observed when compared to the typical method; (3) by using the ÄKTAprime, the protein being purified can be directly estimated by optical densitometry; (4) with some adjustments to the file program, this method can be scaled for obtaining higher protein yields; (5) any commercial column with the appropriate pressure resistance could be used, after manually packing it; (6) once the column has been filled manually with the amylose resine, and for recycling purposes, it can be washed in a semiautomatic way, running the "Manual run" protocol on the ÄKTAprime and switching the A1 line among the different buffers (i.e., water, column buffer and SDS 1%). Moreover, the program designed for the ÄKTAprime could be adapted for any FPLC system, for the purification of MBP fusion proteins.

Acknowledgments

We thank all members of José Luis Puente and E. C. laboratories by their encouragement during the implementation of the procedure described here. This work was supported by grants to José Luis Puente, E. C. and R. O. from the Universidad Nacional Autónoma de México (DGAPA IN230398, IN229001, IN223603, IN201407, and IN205107) and from the CONACyT (25187-N, 37738-N, 45115-Q, 60227, and 82383).

References

Bergstrom, L. C., *et al.* (1998). Hierarchical and co-operative binding of OmpR to a fusion construct containing the ompC and ompF upstream regulatory sequences of *Escherichia coli*. *Genes Cells* **3,** 777–788.

Bernardini, M. L., *et al.* (1990). The two-component regulatory system ompR-envZ controls the virulence of *Shigella flexneri*. *J. Bacteriol.* **172,** 6274–6281.

Cai, S. J., and Inouye, M. (2003). Spontaneous subunit exchange and biochemical evidence for trans-autophosphorylation in a dimer of *Escherichia coli* histidine kinase (EnvZ). *J. Mol. Biol.* **329,** 495–503.

Calva, E., and Oropeza, R. (2006). Two-component signal transduction systems, environmental signals, and virulence. *Microb. Ecol.* **51,** 166–176.

Comeau, D. E., *et al.* (1985). Primary characterization of the protein products of the *Escherichia coli* ompB locus: Structure and regulation of synthesis of the OmpR and EnvZ proteins. *J. Bacteriol.* **164,** 578–584.

De la Cruz, M. A., *et al.* (2007). LeuO antagonizes H-NS and StpA-dependent repression in *Salmonella enterica* ompS1. *Mol. Microbiol.* **66,** 727–743.

De la Cruz, M. A., *et al.* (2009). The DNA static curvature has a role in the regulation of the ompS1 porin gene in *Salmonella enterica* serovar Typhi. *Microbiology* **155,** 2127–2136.
Delgado, J., *et al.* (1993). Identification of a phosphorylation site and functional analysis of conserved aspartic acid residues of OmpR, a transcriptional activator for ompF and ompC in *Escherichia coli. Mol. Microbiol.* **10,** 1037–1047.
di Guan, C., *et al.* (1988). Vectors that facilitate the expression and purification of foreign peptides in *Escherichia coli* by fusion to maltose-binding protein. *Gene* **67,** 21–30.
Dorrell, N., *et al.* (1998). Construction and characterisation of a *Yersinia enterocolitica* O:8 ompR mutant. *FEMS Microbiol. Lett.* **165,** 145–151.
Fernandez-Mora, M., *et al.* (1995). Isolation and characterization of ompS1, a novel *Salmonella typhi* outer membrane protein-encoding gene. *Gene* **158,** 67–72.
Fernandez-Mora, M., *et al.* (2004). OmpR and LeuO positively regulate the *Salmonella enterica* serovar Typhi ompS2 porin gene. *J. Bacteriol.* **186,** 2909–2920.
Flores-Valdez, M. A., *et al.* (2003). Negative osmoregulation of the Salmonella ompS1 porin gene independently of OmpR in an hns background. *J. Bacteriol.* **185,** 6497–6506.
Forst, S., *et al.* (1990). In vivo phosphorylation of OmpR, the transcription activator of the ompF and ompC genes in *Escherichia coli. J. Bacteriol.* **172,** 3473–3477.
Garmendia, J., *et al.* (2003). The roles of SsrA-SsrB and OmpR-EnvZ in the regulation of genes encoding the Salmonella typhimurium SPI-2 type III secretion system. *Microbiology* **149,** 2385–2396.
Hsing, W., *et al.* (1998). Mutations that alter the kinase and phosphatase activities of the two-component sensor EnvZ. *J. Bacteriol.* **180,** 4538–4546.
Huang, K. J., *et al.* (1997). Phosphorylation stimulates the cooperative DNA-binding properties of the transcription factor OmpR. *Proc. Natl. Acad. Sci. USA* **94,** 2828–2832.
Ibarra, J. A., *et al.* (2003). Identification of the DNA binding sites of PerA, the transcriptional activator of the bfp and per operons in enteropathogenic *Escherichia coli. J. Bacteriol.* **185,** 2835–2847.
Kenney, L. J., *et al.* (1995). Phosphorylation-dependent conformational changes in OmpR, an osmoregulatory DNA-binding protein of *Escherichia coli. Proc. Natl. Acad. Sci. USA* **92,** 8866–8870.
Khorchid, A., *et al.* (2005). Structural characterization of *Escherichia coli* sensor histidine kinase EnvZ: The periplasmic C-terminal core domain is critical for homodimerization. *Biochem. J.* **385,** 255–264.
Lan, C. Y., and Igo, M. M. (1998). Differential expression of the OmpF and OmpC porin proteins in *Escherichia coli* K-12 depends upon the level of active OmpR. *J. Bacteriol.* **180,** 171–174.
Martinez-Flores, I., *et al.* (1999). The ompB operon partially determines differential expression of OmpC in *Salmonella typhi* and *Escherichia coli. J. Bacteriol.* **181,** 556–562.
Martinez-Hackert, E., *et al.* (1996). Crystallization, X-ray studies, and site-directed cysteine mutagenesis of the DNA-binding domain of OmpR. *Protein Sci.* **5,** 1429–1433.
Mizuno, T., and Mizushima, S. (1990). Signal transduction and gene regulation through the phosphorylation of two regulatory components: The molecular basis for the osmotic regulation of the porin genes. *Mol. Microbiol.* **4,** 1077–1082.
Nikaido, H. (1994). Maltose transport system of *Escherichia coli*: An ABC-type transporter. *FEBS Lett.* **346,** 55–58.
Norioka, S., *et al.* (1986). Interaction of a transcriptional activator, OmpR, with reciprocally osmoregulated genes, ompF and ompC, of *Escherichia coli. J. Biol. Chem.* **261,** 17113–17119.
Oropeza, R., and Calva, E. (2009). The cysteine 354 and 277 residues of *Salmonella enterica* serovar Typhi EnvZ are determinants of autophosphorylation and OmpR phosphorylation. *FEMS Microbiol. Lett.* **292,** 282–290.

Oropeza, R., *et al.* (1999). Negative and positive regulation of the non-osmoregulated ompS1 porin gene in *Salmonella typhi*: A novel regulatory mechanism that involves OmpR. *Mol. Microbiol.* **32,** 243–252.

Perez-Martin, J., *et al.* (1997). Design of a solubilization pathway for recombinant polypeptides in vivo through processing of a bi-protein with a viral protease. *Protein Eng.* **10,** 725–730.

Puente, J. L., *et al.* (1987). Isolation of an ompC-like outer membrane protein gene from *Salmonella typhi. Gene* **61,** 75–83.

Puente, J. L., *et al.* (1991). Expression of *Salmonella typhi* and *Escherichia coli* OmpC is influenced differently by medium osmolarity; dependence on *Escherichia coli* OmpR. *Mol. Microbiol.* **5,** 1205–1210.

Qin, L., *et al.* (2003). Cysteine-scanning analysis of the dimerization domain of EnvZ, an osmosensing histidine kinase. *J. Bacteriol.* **185,** 3429–3435.

Riggs, P. (2000). Expression and purification of recombinant proteins by fusion to maltose-binding protein. *Mol. Biotechnol.* **15,** 51–63.

Russo, F. D., and Silhavy, T. J. (1991). EnvZ controls the concentration of phosphorylated OmpR to mediate osmoregulation of the porin genes. *J. Mol. Biol.* **222,** 567–580.

Schwan, W. R. (2009). Survival of uropathogenic *Escherichia coli* in the murine urinary tract is dependent on OmpR. *Microbiology* **155,** 1832–1839.

Slauch, J. M., *et al.* (1988). EnvZ functions through OmpR to control porin gene expression in *Escherichia coli* K-12. *J. Bacteriol.* **170,** 439–441.

Sutcliffe, J. A., and Foulds, J. D. (1983). Expression and regulation of protein K, an *Escherichia coli* K1 porin, in *Escherichia coli* K-12. *J. Cell Biochem.* **23,** 71–79.

Tanaka, T., *et al.* (1998). NMR structure of the histidine kinase domain of the *E. coli* osmosensor EnvZ. *Nature* **396,** 88–92.

Taylor, R. K., *et al.* (1983). Isolation and characterization of mutations altering expression of the major outer membrane porin proteins using the local anaesthetic procaine. *J. Mol. Biol.* **166,** 273–282.

Tomomori, C., *et al.* (1999). Solution structure of the homodimeric core domain of *Escherichia coli* histidine kinase EnvZ. *Nat. Struct. Biol.* **6,** 729–734.

CHAPTER SIX

Measurement of Response Regulator Autodephosphorylation Rates Spanning Six Orders of Magnitude

Robert B. Bourret,* Stephanie A. Thomas,* Stephani C. Page,† Rachel L. Creager-Allen,† Aaron M. Moore,*,[1] *and* Ruth E. Silversmith*

Contents

* Departments of Microbiology and Immunology, University of North Carolina, Chapel Hill, North Carolina, USA
† Departments of Biochemistry and Biophysics, University of North Carolina, Chapel Hill, North Carolina, USA
[1] Current address: Department of Biophysics and Biophysical Chemistry, Johns Hopkins University, Baltimore, Maryland, USA

Methods in Enzymology, Volume 471
ISSN 0076-6879, DOI: 10.1016/S0076-6879(10)71006-5

Abstract

Two-component regulatory systems, comprising sensor kinase and response regulator proteins, carry out signal transduction in prokaryotic and eukaryotic microorganisms, as well as plants. Response regulators act as phosphorylation-mediated switches, turning on and off cellular responses to environmental stimuli. Self-catalyzed dephosphorylation is an important determinant of the duration of the response regulator activated state. Reported response regulator autodephosphorylation rates vary over almost a million-fold range, consistent with control of biological processes that occur on widely different timescales. We describe general considerations for the design and execution of *in vitro* assays to measure the autodephosphorylation rates of purified response regulator proteins, as well as specific methods that utilize loss of ^{32}P, changes in fluorescence, or release of inorganic phosphate. The advantages and disadvantages of different methods are discussed, including suitability for different timescales. In addition to outlining established methods, an assay modification is proposed to measure fast autodephosphorylation rates with radioactivity, and optimization of the fluorescence/pH jump method is described.

1. Overview of Response Regulator Autodephosphorylation

The prototypical two-component regulatory system consists of two proteins, a sensor kinase and a response regulator (reviewed in Gao and Stock, 2009). Detection of stimuli by the sensor kinase modulates autophosphorylation of a conserved His residue with ATP. Environmental information is thus encoded in the form of phosphoryl groups. Phosphoryl groups are transferred from the sensor kinase to a conserved Asp residue in the response regulator. Response regulators typically consist of a receiver domain and an output domain. Phosphorylation of the receiver domain is associated with a conformational change that controls the activity of the output domain. Thus, response regulators act as phosphorylation-mediated switches to control the response of the cell to changing conditions. Response regulators can acquire phosphoryl groups via either phosphotransfer from sensor kinases or self-catalyzed phosphorylation using small molecule phosphodonors. Response regulators can lose phosphoryl groups either through a self-catalyzed reaction or with the assistance of other proteins.

The fraction of the response regulator population that is phosphorylated at any given time is determined by the net effect of opposing phosphorylation and dephosphorylation reactions. The kinetics of these biochemical reactions must be at least as fast as the *in vivo* timescales of the corresponding biological processes in order to successfully synchronize cellular response with environmental stimulus. Although all receiver domains appear to share a considerable degree of structural similarity and catalyze the same chemical reactions, published measurements of phosphoryl group stability vary dramatically between different response regulators (Goudreau *et al.*, 1998; Thomas *et al.*, 2008). The phosphoryl group half-lives for *Rhodobacter sphaeroides* CheY6 and *Myxococcus xanthus* RedF are about 2 s and 400 h, respectively (Jagadeesan *et al.*, 2009; Porter and Armitage, 2002), which establishes a lower bound for the range of response regulator autodephosphorylation rates of at least 700,000-fold. So far, two variable active site residues are known to influence the reaction rate by a factor of at least 100 (Pazy *et al.*, 2009; Thomas *et al.*, 2008). Additional factors to account for the other four orders of magnitude in reaction rate range remain to be discovered.

Although the action of an auxiliary protein with dephosphorylation-stimulating ability is often important for control of the fraction of response regulator phosphorylated *in vivo*, autodephosphorylation kinetics are physiologically relevant for at least three reasons. First, for some response regulators (e.g., CheB), autodephosphorylation is the only dephosphorylating activity (i.e., auxiliary proteins do not participate). Second, auxiliary proteins with the ability to stimulate response regulator dephosphorylation appear to do so by stimulating the intrinsic autodephosphorylation mechanism, rather than catalyzing a different reaction (Pioszak and Ninfa, 2004; Zhao *et al.*, 2002; Zhu and Inouye, 2002). Finally, the one to two order of magnitude of stimulation of dephosphorylation rate by auxiliary proteins (Keener and Kustu, 1988; Porter *et al.*, 2008; Silversmith *et al.*, 2008; Zhu and Inouye, 2002) appears to be relatively small with respect to the overall six order of magnitude range of autodephosphorylation rates.

A wide variety of experimental techniques have been developed to measure response regulator autodephosphorylation rates *in vitro*. In this chapter, we describe the rationales underlying various methods and the considerations that influence the choice of assay or guide modifications to suit particular applications. Detailed protocols are included for some procedures.

2. Purification of Response Regulator Proteins

To measure autodephosphorylation rate, it is first necessary to purify the response regulator protein. Although some receiver domains are part of membrane bound hybrid sensor kinases, virtually all response regulators are cytoplasmic proteins lacking transmembrane segments (Ulrich *et al.*, 2005) and therefore are expected to be soluble. Customized methods have been developed to purify many specific response regulators. However, even when an effective purification scheme based on classical methods exists, amino acid substitutions introduced into mutant derivatives can alter partition at various steps and thus complicate purification. In such cases, or if there is no established method to purify the response regulator to be assayed, then a cleavable affinity tag approach may be desirable. Affinity tags offer several advantages over custom purification methods. In particular, the method depends on the tag, not the protein to which the tag is attached, so the same method can be used successfully for many different proteins or for many different derivatives of the same protein. Purification is generally rapid and there are many commercially available choices of affinity tags and cloning vectors.

An obvious disadvantage of affinity tags is the possibility that the presence of the tag may alter the properties of the protein to which it is attached. This concern can be diminished by use of a tag that can be proteolytically removed following protein purification. However, many schemes leave behind a few foreign amino acids after cleavage. For response regulators, attaching the affinity tag to the amino-terminus of the protein, which is on the opposite side of the receiver domain from the active site where catalysis of phosphorylation and dephosphorylation reactions occurs, can minimize the impact of alterations to the protein. Amino terminal fusion to even large fluorescent proteins does not appear to interfere with response regulator function (Gao *et al.*, 2008).

We have used the pET28a(+) vector from Novagen to clone response regulator genes. An *Nde*I restriction site that contains the ATG start codon allows in frame fusion to vector sequences encoding an N-terminal thrombin-cleavable His tag. We have used this system to successfully purify wild type and mutant versions of CheY, PhoB, and Spo0F. We use standard manufacturer's protocols for overexpression of the His-tagged proteins in BL21(DE3) cells (Novagen) and affinity chromatography on Ni^{2+}-NTA agarose (Qiagen). After elution of the His-tagged response regulator from the Ni^{2+}-NTA agarose beads using 150 m*M* imidazole, the protein yield is estimated by UV absorbance. The pooled response regulator is then incubated with 1 unit thrombin (Novagen) per mg protein overnight at room temperature. The next day, completion of the thrombin cleavage reaction is confirmed by gel electrophoresis, and the sample is concentrated and gel filtered (Superose 12 or Superdex 75, both GE Healthcare) to separate the response regulator from the cleaved tag and the thrombin. Fractions containing response regulator are pooled, concentrated, divided into aliquots, and stored at -80 °C.

3. General Considerations for Autodephosphorylation Assays

To measure the rate of dephosphorylation, it is first necessary to generate phosphorylated response regulator. Potential phosphodonors are either a sensor kinase phosphorylated with ATP or a small molecule such as acetyl phosphate, carbamyl phosphate, phosphoramidate, or monophosphoimidazole (Lukat *et al.*, 1992; Silversmith *et al.*, 1997). The absolute extent of phosphorylation (i.e., the fraction of the response regulator population that is phosphorylated) is not particularly critical, but must be sufficient to give a signal that is detectable by the assay method and change significantly over the time of the assay. Response regulator phosphorylation and dephosphorylation, whether self-catalyzed or with the assistance of auxiliary proteins, are absolutely dependent on the presence of a divalent cation (Lukat *et al.*, 1990, 1992). Mg^{2+} is most commonly used by the response regulators that have been characterized, but other metal ions (particularly Mn^{2+}) often support phosphorylation and dephosphorylation reactions to varying degrees. Binding affinity for metal ions varies widely among the few response regulators that have been measured, and also differs substantially between cations (Feher *et al.*, 1995; Guillet *et al.*, 2002; Lukat *et al.*, 1990; Needham *et al.*, 1993; Zundel *et al.*, 1998). For example, reported K_d's for Mg^{2+} range between 0.5 and 50 m*M*. Because metal binding affinity may not be known for a response regulator of interest, preliminary experiments varying cation concentration and/or identity may be useful to optimize signal strength before measuring dephosphorylation.

To accurately measure the rate of dephosphorylation, it is preferable to directly measure dephosphorylation, rather than an output that reflects a combination of both phosphorylation and dephosphorylation reactions. One strategy is to prevent further phosphorylation, so that only dephosphorylation occurs during the assay. In this case, dephosphorylation is an exponential process with a fixed fraction of the population undergoing the reaction per unit time. Therefore, a graph of the logarithm of phosphorylated response regulator versus time should yield a straight line. Another strategy is to set up assay conditions so that phosphorylation and dephosphorylation are operating at steady state under conditions where dephosphorylation is rate limiting. In this circumstance, assay output should be linear rather than exponential with time. In either case, deviation from the expected time dependence or evidence of biphasic kinetics would merit further investigation to ascertain whether the observation reflects a flawed assay or an interesting phenomenon.

For accurate and reproducible rate measurements, temperature should be regulated. The autodephosphorylation rates of the DrrA (Goudreau *et al.*, 1998), NtrC (Pioszak and Ninfa, 2004; Weiss and Magasanik, 1988), and PhoB (Fisher *et al.*, 1996; McCleary, 1996) response regulators, as well as the hydrolysis rate of the similar acyl phosphate linkage in acetyl phosphate (Goudreau *et al.*, 1998), are all about five times faster at 37 °C than at 25 °C. Standard thermodynamic calculations based on the observed temperature dependences predict that the rate of response regulator autodephosphorylation should be about twice as fast at 25 °C as at 20 °C, which could lead to imprecision in measurements made at "room temperature."

The enormous range of response regulator autodephosphorylation rates imposes some technical challenges. Although the autodephosphorylation reaction depends on protein structure and therefore is halted by denaturation (Weiss and Magasanik, 1988), spontaneous hydrolysis of the aspartyl phosphate bond continues. Half-lives of phosphoryl groups in denatured response regulators have been reported in the range of ~8–17 h at ~25 °C (Goudreau *et al.*, 1998; Thomas *et al.*, 2008), similar to the stability of the model acyl phosphate compound acetyl phosphate (Koshland, 1952). Thus, assays to measure particularly slow autodephosphorylation rates must correct for the contribution of spontaneous hydrolysis to phosphoryl group loss. To measure autodephosphorylation rates that are slower than spontaneous hydrolysis, samples can first be denatured to halt autodephosphorylation and then frozen to halt spontaneous hydrolysis prior to analysis. In extreme cases, and depending on experimental design, it may be necessary to correct for ^{32}P decay in assays that span a significant fraction of radioisotope half-life (Goudreau *et al.*, 1998). Finally, different reaction timescales may require different assay methods, as noted in the assay descriptions that follow.

Sometimes it is informative or convenient to assay the biochemical activities of individual domains taken from a multidomain protein. In addition to the receiver domain that catalyzes phosphorylation and dephosphorylation reactions, most response regulators also contain one or more output domains (Galperin, 2006). There are known examples in which the output domain substantially affects the autophosphorylation rate of the receiver domain (Ames *et al.*, 1999; Friedland *et al.*, 2007). Data reported in Section 5.2 suggest that the autodephosphorylation rate of an isolated receiver domain may differ modestly from that of the intact parent response regulator. Pending thorough investigation, the autodephosphorylation rates of a receiver domain and intact parent response regulator cannot be assumed to be the same.

4. Assay of Autodephosphorylation by Loss of ^{32}P

Measuring response regulator autodephosphorylation rate by loss of ^{32}P-labeled phosphoryl groups offers multiple technical advantages over other assay strategies. This design allows for direct observation of the autodephosphorylation reaction, as well as confirmation that phosphotransfer from a partner sensor kinase has ceased. The radiolabel provides a very sensitive detection method and [γ-^{32}P]ATP is commercially available. Furthermore, the method is generally applicable to virtually any response regulator.

The central disadvantage of this assay strategy is that it typically requires purification of the partner sensor kinase, most of which are transmembrane proteins (Ulrich *et al.*, 2005). Therefore, it is often necessary to identify and purify a soluble cytoplasmic fragment of the sensor kinase that retains autophosphorylation function. Although there are many precedents for success, obtaining the necessary sensor kinase fragment is sometimes non-trivial. Other disadvantage of using radioisotopes are the associated safety hazards and waste disposal concerns.

4.1. Basic method

The basic method is applicable to sensor kinases that transfer phosphoryl groups rapidly to their partner response regulators and do not exhibit phosphatase activity toward the response regulator. The sensor kinase is first incubated with [γ-^{32}P]ATP to generate [^{32}P]sensor kinase. Further sensor kinase autophosphorylation is prevented by separation from ATP, which also greatly reduces the amount of radioactivity to be handled in subsequent steps. We employ ammonium sulfate precipitation followed by gel filtration to separate sensor kinase from ATP as outlined below. Another separation strategy is to remove the sensor kinase from the reaction mixture using an affinity tag (Comolli *et al.*, 2002). Alternatively, adding an excess of unlabeled ATP in a pulse-chase format (Keener and Kustu, 1988) prevents further incorporation of additional radiolabel into sensor kinase, but does not diminish the amount of radioactivity to be handled.

4.1.1. Procedure for preparation of *E. coli* [^{32}P]CheA or *B. subtilis* [^{32}P]KinA sensor kinases

1. Order [γ-^{32}P]ATP at 6000 Ci/mmol, 10 mCi/ml.
2. Prepare Sephadex G-100 column:
   a. We use a 1 cm × 25 cm column.
   b. Swell Sephadex G-100 resin with TEG buffer (50 m*M* Tris, pH 7.5, 0.5 m*M* EDTA, 10% (v/v) glycerol) and pour slurry into column. Leave 2 in. of headspace at the top.
   c. Wash column with 200 ml of TEG buffer.
3. Prepare nonradioactive portion of reaction:
   a. Make 10 m*M* nonradioactive ATP in 50 m*M* Tris, pH 8.0.
   b. Make a solution containing 170 μg sensor kinase, 35 m*M* KCl, 3.5 m*M* $MgCl_2$, and 35 m*M* Tris, pH 8.0 in a total volume of 94 μl.
4. Labeling reaction:
   a. Thaw [γ-^{32}P]ATP at room temperature for 1 h.
   b. Mix 60 μl of [γ-^{32}P]ATP with 6 μl nonradioactive 10 m*M* ATP in a 1.5-ml microfuge tube.

c. Add the 94 μl of kinase solution to the microfuge tube. Mix and let incubate at room temperature for 1 h.
d. Add ~45 mg of ammonium sulfate and mix well by tapping with finger on side of tube. Be sure solution turns cloudy.
e. Incubate on ice for 20 min.
f. Centrifuge in microcentrifuge for 10 min at maximum speed.
g. Decant supernatant into ^{32}P waste.
h. Resuspend white pellet in 200 μl of TEG buffer.

5. Separate labeled sensor kinase from ATP on G-100 column:
a. Drain buffer head to gel surface.
b. Add the 200 μl of labeled sensor kinase to column and let run into gel.
c. Add 400 μl TEG buffer and let run into the gel.
d. Add 1 ml of TEG buffer to top of gel, connect to buffer reservoir, and let flow by gravity.
e. Use Geiger counter to monitor locations of sensor kinase and ATP. The sensor kinase will move through the column faster than ATP.
f. Collect 400 μl fractions.
g. Count 5 μl of each fraction in scintillation counter.
h. Freeze fractions at -20 °C until use.
i. Assuming no loss of protein during procedure, use proportion of radioactivity in each fraction to calculate the mass and concentration of sensor kinase in each fraction.

For the dephosphorylation assay itself, a molar excess of response regulator is mixed with the [^{32}P]sensor kinase, which results in one round of synchronous phosphotransfer of virtually all the phosphoryl groups from the sensor kinase to the response regulator. At subsequent times, only autodephosphorylation occurs. Samples are removed from the reaction at suitable time points, denatured to halt autodephosphorylation, and the reaction products are separated by SDS–PAGE. Because phospho-His and phospho-Asp bonds are relatively labile, samples are not boiled prior to gel electrophoresis and the gel is not acid fixed following electrophoresis (Hess *et al.*, 1991). Gel conditions should be chosen to separate the sensor kinase, response regulator, and inorganic phosphate (P_i).

4.1.2. Procedure for autodephosphorylation assay of *E. coli* [^{32}P]CheY or *B. subtilis* [^{32}P]Spo0F response regulators

1. Prepare a 90-μl reaction, which is enough for seven samples (one of sensor kinase alone and six time points for response regulator autodephosphorylation reaction).
a. In 1.5 ml microfuge tube, make a solution of 100 m*M* Tris, pH 7.5 and $MgCl_2$ (10 m*M* for CheY or 20 m*M* for Spo0F) in a volume of 90 μl minus the volume needed to add 28 pmol of [^{32}P]sensor kinase.

   b. Add 10 μl of 2× SDS sample buffer (125 m*M* Tris, pH 6.8, 4% (w/v) SDS, 20% (v/v) glycerol, 10% (v/v) beta-mercaptoethanol, 0.2 mg/ml bromophenol blue) to each of seven 1.5 ml microfuge tubes. Label one tube "kinase" and the others "0" through "5."
   c. Add the appropriate volume of thawed [^{32}P]sensor kinase to the reaction tube to include 28 pmol and bring the volume to 90 μl. Mix.
   d. Transfer 10 μl of reaction tube contents to the "kinase" tube of sample buffer.
2. Reaction:
   a. Put the six remaining tubes with sample buffer into a Lucite block. Pipetters should be set to needed volumes and have tips.
   b. Have a sheet of paper with planned sampling times for tubes 0–5 and a place to record actual times. The time points should equally span the length of time needed for most or all of the radiolabel to be lost from the response regulator. Preliminary experiments may be needed to guide the choice of time points.
   c. Start timer.
   d. As fast possible, add 400 pmol CheY or 4 nmol Spo0F in a volume of less than 10 μl to the reaction tube and pipet up and down to mix. Discard tip.
   e. As quickly as possible, remove a 10-μl sample to sample buffer tube "0" and mix by pipeting up and down. This is the "time 0" sample. Record time from timer.
   f. Get ready for the next sample by placing a new tip on the pipetter. When desired time approaches, remove 10 μl from the reaction tube into the pipet tip, where the reaction continues. When the desired time is on the timer, add sample to tube 1 and mix with pipetter. Record the exact time the sample was put into sample buffer.
   g. Repeat with remaining time points "2" through "5."
3. SDS–PAGE:
   a. Set up gel system before performing reaction. After last time point has been taken, run gel as soon as possible.
   b. Do not boil samples prior to electrophoresis.
   c. Load samples into every other lane of the gel so radioactivity for each time point is easily distinguished.
   d. Run dye front to near end of gel. Retaining [^{32}P]P_i on the gel can be informative in terms of following phosphoryl group flow from sensor kinase through response regulator to P_i.
   e. Dry gel. Do not stain or fix with methanol/acetic acid.
4. Phosphorimage processing:
   a. Expose dried gel to phosphorimager screen overnight.
   b. On phosphorimager, collect "counts" for each protein band. Be sure the area surrounding each protein band is the same size. Make one rectangle on a portion of the gel without radiolabel to serve as a background reference value in calculations.

   c. Subtract background counts (blank rectangle on gel) from each time point.
   d. Normalize the corrected counts at each time point to the corrected counts at time 0, that is, the value of time 0 becomes 100%.
5. Calculations:
   a. Calculate actual time points as accurately as possible from record of experiment. These may be slightly different from planned time points.
   b. Plot percentage of time 0 counts versus time.
   c. Fit data to exponential curve using graphing software. The exponent is the rate constant.
6. Correction for spontaneous dephosphorylation:
   a. If the span of the time course is longer than about 1.5 h, then it is probably necessary to correct for the loss of phosphoryl groups due to spontaneous dephosphorylation in the denatured samples awaiting SDS–PAGE.
   b. First, determine the spontaneous dephosphorylation rate of the denatured response regulator as described in Thomas *et al.* (2008).
   c. Add the spontaneous dephosphorylation rate to the experimentally observed dephosphorylation rate to get the actual autodephosphorylation rate.

4.2. Modification for phosphatase activity or slow phosphotransfer

In some circumstances, the basic method described in Section 4.1 is not suitable. If the sensor kinase has phosphatase activity toward the response regulator, then loss of ^{32}P from the response regulator would be due to both sensor kinase-stimulated dephosphorylation and autodephosphorylation. Similarly, if phosphotransfer from sensor kinase to response regulator is slow (e.g., because a nonmatched sensor kinase or a mutant response regulator are used) and therefore continues over a significant portion of the assay, then the concentration of phosphorylated response regulator would depend on the rate of phosphotransfer reaction in addition to that of the autodephosphorylation reaction. In either case, sensor kinase and response regulator are first incubated together in the presence of [γ-^{32}P] ATP to ensure adequate response regulator phosphorylation. Separation of the response regulator from the sensor kinase then allows observation of the autodephosphorylation reaction alone, if reaction kinetics are slower than the separation method.

The preferred separation approach is to remove the response regulator from the reaction, because this leaves the [γ-^{32}P]ATP behind. Response regulator removal has been reported using a His-tag (Perego, 2001) or a maltose binding protein-tag (Sheeler *et al.*, 2005). Other affinity tags should also achieve separation. Affinity tags generally allow rapid separation by addition of suitable beads to the reaction, followed by brief centrifugation, removal of the supernatant, and washing and resuspension of the beads.

A second class of methods is to remove the sensor kinase from the reaction, in which case the response regulator does not need to be tagged. Sensor kinase removal has been accomplished using sensor kinase covalently coupled to beads (Wang and Matsumura, 1996), a His-tagged sensor kinase (Appleby and Bourret, 1999), a glutathione-*S*-transferase tagged sensor kinase (Janiak-Spens *et al.*, 2000), or even anti-sensor kinase antibodies (Weinstein *et al.*, 1993). The primary disadvantage of removing the sensor kinase from the reaction is that the response regulator is left in the same mixture as the [γ-^{32}P]ATP. The vast majority of the radioactivity is associated with small molecules rather than protein, so the amount of radioactivity that must be handled is much greater than when the response regulator is removed.

Traditional chromatographic methods have also been employed to separate the response regulator from the sensor kinase and [γ-^{32}P]ATP. Reported methods include gel filtration (Dahl *et al.*, 1992), ion exchange (Goudreau *et al.*, 1998), and hydroxyapatite (Zapf *et al.*, 1998). However, affinity tag technology offers the advantages of speed and general applicability.

Finally, if it is desirable to halt autodephosphorylation during separation of sensor kinase and response regulator, EDTA can be added to chelate the essential Mg^{2+}. Later addition of excess divalent cation restarts the reaction (Hess *et al.*, 1988).

4.3. Modification for fast autodephosphorylation

For response regulators with very fast autodephosphorylation rates (i.e., phosphoryl group half-life <10 s), it can be challenging or impossible using the basic method to measure the amount of [^{32}P]response regulator at multiple time points after phosphorylation is complete and before the radiolabel is entirely lost from the protein. If phosphotransfer from the partner sensor kinase is fast with respect to response regulator autodephosphorylation (e.g., from CheA-P to CheB) and the sensor kinase does not exhibit phosphatase activity toward the response regulator, then in principle the response regulator could be mixed with a large molar excess of phosphorylated sensor kinase that had been purified away from [γ-^{32}P]ATP. Because the autodephosphorylation reaction would be rate limiting, the rate of decrease of the phosphorylated sensor kinase could be measured over multiple reaction cycles and would be equal to the response regulator autodephosphorylation rate. We are not aware of a case where this strategy has been used, although we believe it to be potentially superior to the methods described in Sections 7.1 and 7.2 because fewer reactions are involved.

4.4. Modification for small molecule phosphodonors

Sensor kinases are much more efficient phosphodonors for response regulators than small molecules. However, if a partner sensor kinase is not available, then small molecule phosphodonors can be used. Unfortunately, radiolabeled versions of acetyl phosphate and phosphoramidate are not commercially available. Both a chemical method (McCleary and Stock, 1994) and an enzymatic method (Quon *et al.*, 1996) to synthesize [^{32}P]acetyl phosphate have been developed. However, acetyl phosphate does not act as a phosphodonor for all response regulators (Lukat *et al.*, 1992). Due to its chemical similarity to the phospho-His of sensor kinases, phosphoramidate generally appears to be a more effective phosphodonor for response regulators than acetyl phosphate. A method to synthesize [^{32}P]phosphoramidate is available (Buckler and Stock, 2000). Note that [^{32}P]phosphoramidate can nonspecifically donate phosphoryl groups to His-tags on proteins (Buckler and Stock, 2000); however, this modification contributes stable background noise that generally should not interfere with determination of autodephosphorylation rate.

To use small molecules as phosphodonors for the purpose of measuring dephosphorylation rates, it will be necessary to empirically establish suitable incubation conditions (concentration, time, etc.) of response regulator with phosphodonor. Success depends on the relative magnitudes of the autophosphorylation and autodephosphorylation rates. Once a useful fraction of the response regulator population is phosphorylated, reaction conditions are changed to specifically observe autodephosphorylation. One option is a pulse-chase method (Cavicchioli *et al.*, 1995). A large molar excess of unlabeled phosphodonor is added to the reaction. Autophosphorylation continues, but essentially no additional radiolabel is incorporated into protein. A second strategy is to stop autophosphorylation by separating response regulator from phosphodonor. This can be achieved by extensive washes during ultrafiltration with a membrane that retains the response regulator but allows the phosphodonor to pass through (Jagadeesan *et al.*, 2009). Because the protein concentration in phosphorylation reactions is typically μM whereas phosphodonor concentration is typically mM, it is critical that washing to remove phosphodonor is thorough. An alternate separation method would be to pull affinity-tagged response regulator out of the incubation mixture with the appropriate beads, analogous to the descriptions in Section 4.2.

5. Assay of Autodephosphorylation by Fluorescence

A subset of response regulators (e.g., *E. coli* CheY and PhoB) contain a Trp residue near the phosphorylation site whose fluorescence changes upon phosphorylation (Lukat *et al.*, 1992; McCleary, 1996). The active site Trp residue serves as a remarkably useful probe of phosphorylation status. Although only applicable to those response regulators with suitable fluorophores, optical methods can easily measure fast reaction kinetics and automated collection of data from many timepoints is simple. Avoiding the practical challenges of handling the radioisotopes used for the assays described in Section 4 is another benefit. Fluoresence assays of response regulator phosphorylation status typically avoid the complication of competing Trp residues from sensor kinases by utilizing small molecule phosphodonors instead. The tradeoff for not needing to procure the sensor kinase is that some small molecule phosphodonors are not commercially available and must be synthesized (see Section 6.1). Alternatively, an artificial fluorophore can be introduced into the active site to allow assay of response regulators without a suitable Trp residue or in the presence of other Trp-containing proteins (Stewart and VanBruggen, 2004). However, suitable control experiments are needed to determine whether adding the fluorophore affects reaction kinetics.

5.1. Kinetic theory

Incubation of a response regulator with a small molecule phosphodonor results in the following scheme of autophosphorylation and autodephosphorylation reactions (Lukat *et al.*, 1992):

$$\mathrm{RR} + \text{phosphodonor} \xrightarrow{K_s} \mathrm{RR} \cdot \text{phosphodonor} \xrightarrow{k_{\mathrm{phos}}} \mathrm{RR} - \mathrm{P}$$
$$+ \text{leaving group} \xrightarrow{k_{\mathrm{dephos}}} \mathrm{RR} + \mathrm{P_i}$$

The rate of approach to a new steady state following a change in phosphodonor concentration (k_{obs}) is given by the following equation (Mayover *et al.*, 1999):

$$k_{\mathrm{obs}} = (k_{\mathrm{phos}})[\text{phosphodonor}]/(K_s + [\text{phosphodonor}]) + k_{\mathrm{dephos}} \quad (6.1)$$

At high concentrations of phosphodonor ([phosphodonor] $\gg K_s$), one might expect the autophosphorylation reaction rate to saturate, in which case the expression for k_{obs} would simplify to:

$$k_{obs} = k_{phos} + k_{dephos} \tag{6.2}$$

However, for the CheY response regulator, when care is taken to maintain constant ionic strength, the autophosphorylation reaction does not saturate (Da Re *et al.*, 1999). In other words, $K_s \gg$ [phosphodonor] and the expression for k_{obs} simplifies to:

$$k_{obs} = (k_{phos}/K_s)[\text{phosphodonor}] + k_{dephos} \tag{6.3}$$

Note that the rate of autophosphorylation is linearly dependent on phosphodonor concentration. To the best of our knowledge, the appropriate experiments to assess saturability of autophosphorylation at constant ionic strength have only been reported for CheY and it is unknown whether Eq. (6.3) applies to other response regulators.

The nitrogen atoms of the N–P bonds in phosphoramidate and monophosphoimidazole must be protonated to act as efficient leaving groups when the N–P bond is broken during the response regulator autophosphorylation reaction. For example, nonprotonated phosphoramidate has the structure $H_2N–PO_3^{2-}$ and protonated phosphoramidate is $H_3N^+–PO_3^{2-}$. Therefore, the reaction rate with these phosphodonors is pH dependent and decreases substantially at alkaline pH (Silversmith *et al.*, 1997). In contrast, the rate of CheY autodephosphorylation is pH independent over a broad range (Silversmith *et al.*, 1997). The pH jump method (Mayover *et al.*, 1999) exploits this kinetic asymmetry by first incubating a response regulator with phosphoramidate at neutral pH, so that autophosphorylation and autodephosphorylation occur simultaneously. The pH is then abruptly raised to an alkaline value, dramatically slowing the autophosphorylation reaction, but leaving autodephosphorylation unperturbed. If reaction conditions are judiciously chosen, then the subsequent autodephosphorylation reaction can be monitored in isolation.

5.2. Fluorescence/pH jump method

Here, we outline general procedures we have developed to use the pH jump to measure the autodephosphorylation rates of response regulators with various properties. The strategy of the pH jump method is to diminish the phosphorylation term of Eqs. (6.1)–(6.3) to such an extent that it becomes negligible in comparison to the dephosphorylation term and hence $k_{obs} = k_{dephos}$. Autophosphorylation rate depends on phosphodonor concentration. Therefore, it is desirable to use a phosphodonor concentration that is large enough to result in phosphorylation of most response regulator molecules at neutral pH but too small to support significant phosphorylation at alkaline pH. Because different response regulators (or even mutant versions of CheY) have different values of k_{phos}/K_s and k_{dephos}, the appropriate phosphodonor concentration must be empirically determined for each response regulator to be assayed.

The first step is to determine $K_{1/2}$, the phosphodonor concentration at which half the response regulator is phosphorylated. After measuring the Trp fluorescence of the response regulator in the absence of phosphodonor, a small aliquot of phosphoramidate from a concentrated stock solution is added so as to result in little overall volume change. The autophosphorylation and autodephosphorylation reactions will come to a steady state, with some proportion of the population phosphorylated, and the new equilibrium fluorescence value is recorded. Then another aliquot of phosphoramidate is added and the process repeated. Eventually, the fluorescence will not change further because virtually all of the response regulator molecules are phosphorylated. Plotting the fractional change in fluorescence versus total phosphoramidate concentration (after correcting both quantities for volume changes during the experiment) yields a saturation curve that is analogous in form to first order binding between a receptor and ligand, and can be mathematically analyzed with a binding equation to deduce $K_{1/2}$. For response regulators that autophosphorylate slowly, sequential phosphoramidate additions as described above are impractical. Instead, different concentrations of phosphoramidate can be added to separate reactions in parallel to acquire analogous data.

To make the pH jump measurement of fast autodephosphorylation reactions, one syringe of a stopped-flow apparatus is filled with a solution containing 20 μM response regulator, phosphoramidate (concentration chosen as described below), 5 mM Tris, pH 7.5, and 10 mM $MgCl_2$. The other syringe contains the high pH buffer (150 mM CAPS, pH 10.5). Upon mixing, the concentration of reactive phosphoramidate is reduced substantially due to the change in pH and the rate of approach to a new stable fluorescence value is measured. The rate constant (k_{obs}) is determined by fitting the data to an equation that describes a single exponential process.

For slow autodephosphorylation reactions, a stopped-flow device is not necessary and the pH jump can be accomplished by addition of high pH buffer to a cuvette. Because gradual formation of a Mg·CAPS precipitate (Yu *et al.*, 1997) can interfere with fluorescence measurements, we instead add sodium carbonate, pH 10.2 as the high pH buffer. For routine experiments, the final buffer concentration is 75 mM. However, if a high concentration of phosphodonor is used, then the concentration of high pH buffer must be chosen to exceed the phosphodonor concentration in order to appropriately raise the pH.

A plot of k_{obs} versus phosphoramidate concentration for several different response regulators (Fig. 6.1A) has two noteworthy features. First, Eq. (6.3) implies that as [phosphodonor] $\rightarrow 0$, $k_{obs} \rightarrow k_{dephos}$. As expected, different response regulators have different values of k_{obs} at low phosphoramidate

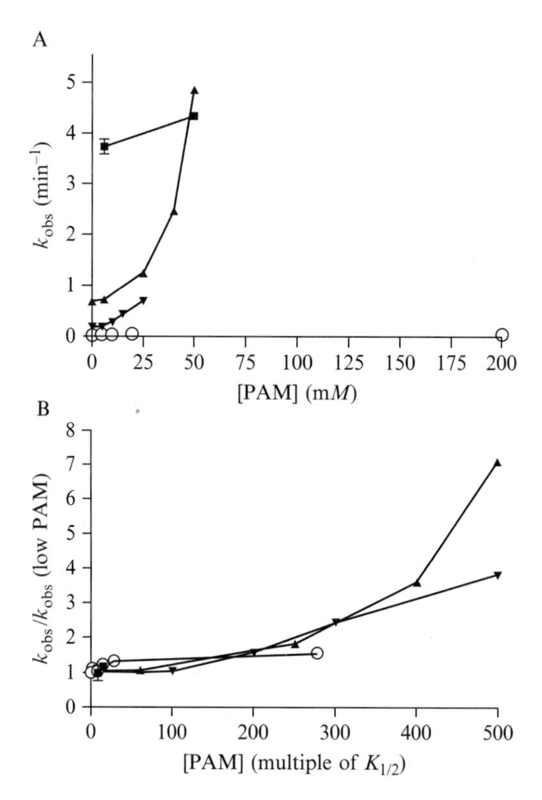

Figure 6.1 Relationship of k_{obs} to phosphodonor concentration for various response regulators. (A) k_{obs} plotted against phosphoramidate concentration. Because these were exploratory experiments conducted to identify concentrations where $k_{obs} = k_{dephos}$, most points represent a single measurement and no attempt was made to maintain constant ionic strength. Squares, CheY wild type; triangles, CheY 89EH; inverted triangles, CheY 14FE 59NM 89ER; open circles, PhoB receiver domain. (B) k_{obs} values were normalized by dividing by the value of k_{obs} obtained at the lowest phosphoramidate concentration assayed for each response regulator (CheY wild type, 3.7 min^{-1}; CheY 89EH, 0.68 min^{-1}; CheY 14FE 59NM 89ER, 0.18 min^{-1}; PhoB receiver domain, 0.022 min^{-1}). Phosphodonor concentrations were normalized to the experimentally determined $K_{1/2}$ value for each response regulator (CheY wild type, 3 mM; CheY 89EH, 0.1 mM; CheY 14FE 59NM 89ER, 0.05 mM; PhoB receiver domain, 0.7 mM). Symbols as in panel A.

concentration, corresponding to different values of k_{dephos}. Second, as [phosphodonor] increases, the magnitude of the (k_{phos}/K_s)[phosphodonor] term in Eq. (6.3) increases, even though alkaline pH dramatically slows the autophosphorylation reaction. Different response regulators exhibit different dependences of k_{obs} on phosphoramidate concentration because the relative magnitudes of (k_{phos}/K_s) and k_{dephos} differ between response regulators. At sufficiently high [phosphodonor], k_{obs} is dominated by the (k_{phos}/K_s)

[phosphodonor] term in spite of the reduction in autophosphorylation rate at high pH. Thus, accurate determination of k_{dephos} by the pH jump method requires an appropriate choice of phosphoramidate concentration, which may be different for each response regulator.

A simple calculation shows that $K_{1/2} = k_{dephos}/(k_{phos}/K_s)$. For response regulators that phosphorylate quickly and/or dephosphorylate slowly, a relatively low concentration of phosphodonor is required to achieve steady state phosphorylation of half the population. Conversely, response regulators that phosphorylate slowly and/or dephosphorylate rapidly require a relatively high concentration of phosphodonor to phosphorylate half the population at steady state. $K_{1/2}$ can also serve as a measure of the relative magnitudes of the autophosphorylation and autodephosphorylation terms in Eq. (6.3). Therefore, when k_{obs} is replotted against phosphoramidate concentration expressed in multiples of the characteristic $K_{1/2}$ for each response regulator instead of as absolute concentration, different response regulators exhibit similar behaviors (Fig. 6.1B). As a result of this analysis, we typically use a phosphoramidate concentration of less than $10 \times K_{1/2}$ in pH jump assays to measure k_{dephos}.

Measurement of k_{dephos} for several CheY mutants by the pH jump method yields values that are very similar to those determined by loss of ^{32}P (Table 6.1). The values determined by pH jump are consistently 15–60% greater than those determined by ^{32}P loss, a difference that can plausibly be attributed to the difference between 25 °C and room

Table 6.1 Comparison of CheY autodephosphorylation rate constants measured by two methods[a]

	k_{dephos} (min^{-1})			
		Assayed by fluorescence		
Response regulator[b]	Assayed by loss of ^{32}P[c]	Untagged[d]	Cleaved	Uncleaved
CheY wild type	2.5 ± 0.3	3.9 ± 0.2	4.0 ± 0.4	4.1 ± 0.3
CheY89EH	0.55 ± 0.04	0.78 ± 0.07	0.72 ± 0.00	0.85 ± 0.03
CheY89EY	0.26 ± 0.01	0.30 ± 0.03	ND[e]	ND
CheY14FE 59NM 89ER	0.11 ± 0.00	0.17 ± 0.04	ND	ND

[a] ^{32}P assay conducted at room temperature. Fluorescence assay conducted at 25 °C. Values are means ± S.D. from at least two experiments.
[b] All tested response regulators were variants of *E. coli* CheY.
[c] Data from Thomas *et al.* (2008).
[d] Untagged: no modification to N-terminus; Cleaved: three amino acids (GSH) added to N-terminus; Uncleaved: 20 amino acids (MGSSHHHHHHSSGLVPRGSH) added to N-terminus.
[e] Not determined.

temperature in the two assay methods. Thus, the pH jump is a valid method to ascertain k_{dephos} for response regulators. Furthermore, k_{dephos} values are indistinguishable between CheY derivatives that differ only in their amino termini (20 amino acid His-tag, three amino acids remaining after thrombin cleavage, or no tag) (Table 6.1). Therefore, the amino-terminal modifications introduced by the pET28a(+) vector do not affect the CheY autodephosphorylation rate.

To the best of our knowledge, autodephosphorylation rates measured by the pH jump assay have been reported only for CheY derivatives with k_{dephos} values of $\sim 3\ \text{min}^{-1}$ (Mayover *et al.*, 1999; Schuster *et al.*, 2000; Stewart and VanBruggen, 2004). Using the assay modifications for slow reactions described in this section, we measured the >100 times slower autodephosphorylation rate of the PhoB receiver domain ($k_{dephos} = 0.023 \pm 0.000\ \text{min}^{-1}$ at 25 °C). We are not aware that the autodephosphorylation rate constant for the PhoB receiver domain alone has previously been reported. However, a k_{dephos} value of $0.0085 \pm 0.0002\ \text{min}^{-1}$ at 25 °C for full-length PhoB was determined by loss of ^{32}P (Fisher *et al.*, 1996), a difference of about threefold from the receiver domain alone.

6. Assay of Autodephosphorylation by P_i Release

Response regulator autodephosphorylation rate can also be determined by measuring the steady state rate of release of P_i (Silversmith *et al.*, 2001). The enzyme purine nucleoside phosphorylase (PNP) is used to very rapidly convert P_i and 2-amino-6-mercapto-7-methylpurine riboside (MESG) into ribose-1-phosphate and 2-amino-6-mercapto-7-methylpurine, a chromogenic product that can be detected spectroscopically (Webb, 1992). The P_i release assay method offers several advantages. The EnzChek Phosphate Assay Kit containing PNP and MESG is commercially available from Invitrogen. The strategy does not require a sensor kinase, a phosphorylation-sensitive fluorophore in the response regulator, or radioisotopes. To the best of our knowledge, the P_i release method has only been used with CheY, but it may be applicable to any response regulator. A disadvantage is that synthesis of the small molecule phosphodonor monophosphoimidazole (MPI) is necessary because MPI is not commercially available. Also, response regulator phosphorylation and dephosphorylation occur simultaneously during the assay, so care must be taken to ensure that dephosphorylation is rate-limiting.

The rationale for the P_i release method is as follows. The response regulator is incubated with MPI and autophosphorylation ensues. MPI is used because it can be prepared synthetically without a high background of contaminating P_i. In addition, MPI is likely to serve as a good phosphodonor for all response regulators due to its structural similarity to phosphohistidine. Under conditions where autophosphorylation is much faster than autodephosphorylation (i.e., autodephosphorylation is rate limiting), the rate of P_i release is equal to the autodephosphorylation rate. The rate of autophosphorylation can be made to be faster than the rate of autodephosphorylation simply by increasing MPI concentration. CheY autophosphorylation exhibits the peculiar feature of nonsaturable kinetics, that is, the reaction rate is linearly dependent on phosphodonor concentration (recall Eq. (6.3)) (Da Re *et al.*, 1999). To the best of our knowledge, the saturability of autophosphorylation rate has not been determined for any response regulator other than CheY. Given the similarity of structural features and active site geometries conserved across receiver domains, it would be surprising if other response regulators did not also exhibit nonsaturable autophosphorylation kinetics.

6.1. Synthesis of phosphoramidate and monophosphoimidazole

We synthesize the potassium salt of phosphoramidate according to a published procedure (Sheridan *et al.*, 1972). We synthesize the calcium salt of MPI according to the following protocol, based on a published method (Rathlev and Rosenberg, 1956):

1. Mix 4.0 g potassium phosphoramidate, 1.06 g of imidazole, and 40 ml of water and adjust to pH 7.2 with HCl. Incubate overnight at room temperature.
2. Dissolve 4.56 g $CaCl_2 \cdot 2H_2O$ in 3–5 ml water and add the calcium solution dropwise to the incubated reaction while stirring. This should result in formation of a copious precipitate.
3. Vacuum filter the sample to remove the precipitate and save the filtrate, which contains both diphosphoimidazole (DPI) and MPI.
4. Add 2× 1 ml of 95% (v/v) ethanol with gentle swirling after each addition. Localized, transient crystal formation should be apparent at each addition.
5. Remove 0.5 ml of the filtrate and mix with 0.5 ml 95% (v/v) ethanol. Grind with a mortar and pestle and then add back to the filtrate. Repeat this process (two to four times) until stable crystallization is observed. These crystals are mainly DPI. Allow sample to sit for 15 min at room temperature and vacuum filter.
6. Take the filtrate (which contains the MPI) and add 2 volumes of 95% (v/v) ethanol. Allow to sit at room temperature for 2 h. MPI crystals should be evident. Vacuum filter to isolate the crystals. Wash the crystals with 65% (v/v) ethanol, followed by 95% (v/v) ethanol, and finally diethyl ether. Dry and store desiccated at −20 °C. Yield: 1.2 g.

6.2. Steady state P_i release assay method

The P_i release assay has been adapted for two experimental formats. The 96-well plate format, which derives rates from a single time point, is convenient when dephosphorylation is measured under a variety of conditions, for example, in the presence of different concentrations of a phosphatase (Silversmith *et al.*, 2008). However, the cuvette-based assay (Silversmith *et al.*, 2001), which continuously measures P_i release as a function of time, is most appropriate for measurement of autodephosphorylation, described as follows:

1. Place 450 μl of a solution containing 20 m*M* HEPES, pH 7.0, 20 m*M* $MgCl_2$, MPI (4 m*M* for CheY), and 200 μM MESG (from Enzchek kit) in a cuvette (10 $\times$ 2 mm) and begin continuous measurement of the absorbance at 360 nm. The absorbance should be steady at about 0.4 absorbance units due to the MESG.
2. Add 5 μl (0.5 unit) PNP (from Enzchek kit) and mix. The absorbance will increase by about 0.2–0.4 units due to contaminating P_i in the MPI preparation and then level out.
3. Add 4.5 μM CheY, mix, and continuously follow the absorbance. It is essential that the protein preparation added be free of P_i. After a brief lag time, the rate of increase in absorbance becomes linear. Measure the slope and convert to $[P_i]$/s using an extinction coefficient of 0.0091 μM^{-1} cm^{-1}. The rate is proportional to the response regulator concentration, so the concentration of response regulator can be adjusted as necessary to get an easily measureable rate. Divide the rate by the concentration of response regulator to calculate the first order rate constant for autodephosphorylation, k_{dephos}.
4. The MPI concentration in step 1 may need to be adjusted for different response regulators. To ensure that the P_i release rate is measured under conditions where autodephosphorylation is rate-limiting, the experimenter should confirm that the P_i release rate is unchanged when the measurement is repeated at a higher MPI concentration.

7. Assay of Autodephosphorylation from Systems of Reactions

Several methods to untangle the contribution of response regulator autodephosphorylation rate to the overall kinetics of multiple simultaneous phosphorylation and dephosphorylation reactions have been described. Although these methods are suboptimal due to their indirect nature, the assays are briefly described here because of their applicability to response regulators with fast autodephosphorylation rates.

7.1. Coupled ATPase assay

For sensor kinases that do not have phosphatase activity toward their response regulator, response regulator autodephosphorylation rate can be derived from reactions in which ATP, sensor kinase, and response regulator are incubated together. The net result of sensor kinase autophosphorylation (SK + ATP $\rightarrow$ SK-P + ADP), phosphotransfer from sensor kinase to response regulator (SK-P + RR $\rightarrow$ SK + RR-P), and response regulator autodephosphorylation (RR-P + H_2O $\rightarrow$ RR + P_i) reactions is hydrolysis of ATP (ATP + H_2O $\rightarrow$ ADP + P_i). The rate of ADP production in turn can be measured through two additional coupled enzymatic reactions that consume ADP (ADP + phosphoenol pyruvate $\rightarrow$ ATP + pyruvate) and NADH (NADH + H^+ + pyruvate $\rightarrow$ NAD^+ + lactate) in equimolar ratios, with NADH loss monitored spectroscopically (Norby, 1988). At a fixed concentration of response regulator, the rate of ATP hydrolysis is measured as a function of increasing concentrations of sensor kinase. At a high molar ratio of sensor kinase to response regulator, ATPase activity saturates (i.e., response regulator autodephosphorylation becomes rate limiting) and is equivalent to the autodephosphorylation rate. The coupled ATPase assay method is suitable for measurement of a wide range of autodephosphorylation rates and has been used to determine k_{dephos} for the rapidly dephosphorylating response regulators CheY (Lukat *et al.*, 1991) and CheB (Stewart, 1993). However, the indirect nature of the assay and the need to model three simultaneous reactions in order to deduce k_{dephos} are significant disadvantages of this approach. The method proposed in Section 4.3 is likely to be superior for determination of fast autodephosphorylation rates.

7.2. Steady state phosphorylation monitored by ^{32}P

To estimate autodephosphorylation rates of the eight *R. sphaeroides* CheY and CheB response regulators, the response regulators were incubated individually with [γ-^{32}P]ATP and a CheA sensor kinase (Porter and Armitage, 2002). The concentrations of phosphorylated sensor kinase and response regulator were then determined as a function of time following separation by SDS–PAGE. When both phosphorylated proteins reached a steady state concentration, the rate of sensor kinase autophosphorylation was equal to the rate of response regulator autodephosphorylation. By separately characterizing sensor kinase autophosphorylation rate constants, it was then possible to calculate k_{dephos}.

7.3. Stopped flow fluorescence measurement of approach to equilibrium

Equation (6.3) describing the rate of approach to a new equilibrium between response regulator autophosphorylation and autodephosphorylation reactions following change in small molecule phosphodonor concentration was explained in Section 5.1:

$$k_{obs} = (k_{phos}/K_s)[\text{phosphodonor}] + k_{dephos}$$

To measure autodephosphorylation, k_{obs} is first determined at multiple phosphodonor concentrations, taking care to keep ionic strength constant at each phosphodonor concentration tested. Because Eq. (6.3) has the form $y = mx + b$, a plot of k_{obs} versus phosphodonor concentration is linear and gives k_{dephos} as the y-intercept (Schuster *et al.*, 2001). This method is a rapid way to obtain an estimate of k_{dephos} and does not require either radioactivity or a sensor kinase. A fluorimeter with stopped-flow capability is required. The main limitation is that the method can only be used with response regulators that contain a phosphorylation sensitive fluorophore.

8. Future Prospects

Application of Phos-tag™ technology to quantify aspartate phosphorylation in response regulators without using radioactivity or intrinsic fluorophores was recently described (Barbieri and Stock, 2008). In principle, Phos-tag™ reagents could be utilized to measure autodephosphorylation rates. Other advances in the design of autodephosphorylation assays may arise from a more complete understanding of the autophosphorylation properties of response regulators other than CheY.

Many of the assays described to measure autodephosphorylation can be modified to quantify response regulator dephosphorylation stimulated by auxiliary proteins. Developing a more thorough understanding of the relationship between self-catalyzed and auxiliary protein-mediated dephosphorylation is a potentially informative use of the assays described in this chapter. Another active area of investigation is to identify the mechanisms that result in the remarkably broad range of response regulator autodephosphorylation rates.

If the basis for control of response regulator autodephosphorylation rate can be deciphered, then a number of potential practical applications may become possible. For example, the timescales of signal transduction processes associated with uncharacterized response regulators might be predicted based on amino acid sequence alone. A set of response regulators with the same function but different on/off switching kinetics could be engineered for use in synthetic biology circuits. Mutant response regulators with altered phosphoryl group stabilities could be created for *in vivo* analysis of specific two-component regulatory systems.

ACKNOWLEDGMENTS

The His-tagged clone of *cheY* was constructed by Kimberly Coggan. The His-tagged clone of *spo0F* was constructed by Pat McLaughlin in John Cavanagh's lab at North Carolina State University.

Investigation of response regulator autodephosphorylation in our laboratory is funded by National Institutes of Health grant GM050860. The content is solely the responsibility of the authors and does not necessarily represent the official views of the National Institute of General Medical Sciences or the National Institutes of Health.

REFERENCES

Ames, S. K., Frankema, N., and Kenney, L. J. (1999). C-terminal DNA binding stimulates N-terminal phosphorylation of the outer membrane protein regulator OmpR from *Escherichia coli*. *Proc. Natl. Acad. Sci. USA* **96,** 11792–11797.

Appleby, J. L., and Bourret, R. B. (1999). Activation of CheY mutant D57N by phosphorylation at an alternative site, Ser-56. *Mol. Microbiol.* **34,** 915–925.

Barbieri, C. M., and Stock, A. M. (2008). Universally applicable methods for monitoring response regulator aspartate phosphorylation both in vitro and in vivo using Phos-tag-based reagents. *Anal. Biochem.* **376,** 73–82.

Buckler, D. R., and Stock, A. M. (2000). Synthesis of [^{32}P]phosphoramidate for use as a low molecular weight phosphodonor reagent. *Anal. Biochem.* **283,** 222–227.

Cavicchioli, R., Schroder, I., Constanti, M., and Gunsalus, R. P. (1995). The NarX and NarQ sensor-transmitter proteins of *Escherichia coli* each require two conserved histidines for nitrate-dependent signal transduction to NarL. *J. Bacteriol.* **177,** 2416–2424.

Comolli, J. C., Carl, A. J., Hall, C., and Donohue, T. (2002). Transcriptional activation of the *Rhodobacter sphaeroides* cytochrome c_2 gene P2 promoter by the response regulator PrrA. *J. Bacteriol.* **184,** 390–399.

Dahl, M. K., Msadek, T., Kunst, F., and Rapoport, G. (1992). The phosphorylation state of the DegU response regulator acts as a molecular switch allowing either degradative enzyme synthesis or expression of genetic competence in *Bacillus subtilis*. *J. Biol. Chem.* **267,** 14509–14514.

Da Re, S. S., Deville-Bonne, D., Tolstykh, T., Veron, M., and Stock, J. B. (1999). Kinetics of CheY phosphorylation by small molecule phosphodonors. *FEBS Lett.* **457,** 323–326.

Feher, V. A., Zapf, J. W., Hoch, J. A., Dahlquist, F. W., Whiteley, J. M., and Cavanagh, J. (1995). ^{1}H, ^{15} N, and ^{13}C backbone chemical shift assignments, secondary structure, and magnesium-binding characteristics of the *Bacillus subtilis* response regulator, Spo0F, determined by heteronuclear high-resolution NMR. *Protein Sci.* **4,** 1801–1814.

Fisher, S. L., Kim, S. K., Wanner, B. L., and Walsh, C. T. (1996). Kinetic comparison of the specificity of the vancomycin resistance kinase VanS for two response regulators, VanR and PhoB. *Biochemistry* **35,** 4732–4740.

Friedland, N., Mack, T. R., Yu, M., Hung, L. W., Terwilliger, T. C., Waldo, G. S., and Stock, A. M. (2007). Domain orientation in the inactive response regulator *Mycobacterium tuberculosis* MtrA provides a barrier to activation. *Biochemistry* **46,** 6733–6743.

Galperin, M. Y. (2006). Structural classification of bacterial response regulators: Diversity of output domains and domain combinations. *J. Bacteriol.* **188,** 4169–4182.

Gao, R., and Stock, A. M. (2009). Biological insights from structures of two-component proteins. *Annu. Rev. Microbiol.* **63,** 133–154.

Gao, R., Tao, Y., and Stock, A. M. (2008). System-level mapping of *Escherichia coli* response regulator dimerization with FRET hybrids. *Mol. Microbiol.* **69,** 1358–1372.

Goudreau, P. N., Lee, P. J., and Stock, A. M. (1998). Stabilization of the phospho-aspartyl residue in a two-component signal transduction system in *Thermotoga maritima. Biochemistry* **37,** 14575–14584.

Guillet, V., Ohta, N., Cabantous, S., Newton, A., and Samama, J. P. (2002). Crystallographic and biochemical studies of DivK reveal novel features of an essential response regulator in *Caulobacter crescentus. J. Biol. Chem.* **277,** 42003–42010.

Hess, J. F., Bourret, R. B., Oosawa, K., Matsumura, P., and Simon, M. I. (1988). Protein phosphorylation and bacterial chemotaxis. *Cold Spring Harb. Symp. Quant. Biol.* **53,** 41–48.

Hess, J. F., Bourret, R. B., and Simon, M. I. (1991). Phosphorylation assays for proteins of the two-component regulatory system controlling chemotaxis in *Escherichia coli. Methods Enzymol.* **200,** 188–204.

Jagadeesan, S., Mann, P., Schink, C. W., and Higgs, P. I. (2009). A novel "four-component" two-component signal transduction mechanism regulates developmental progression in *Myxococcus xanthus. J. Biol. Chem.* **284,** 21435–21445.

Janiak-Spens, F., Sparling, D. P., and West, A. H. (2000). Novel role for an HPt domain in stabilizing the phosphorylated state of a response regulator domain. *J. Bacteriol.* **182,** 6673–6678.

Keener, J., and Kustu, S. (1988). Protein kinase and phosphoprotein phosphatase activities of nitrogen regulatory proteins NTRB and NTRC of enteric bacteria: Roles of the conserved amino-terminal domain of NTRC. *Proc. Natl. Acad. Sci. USA* **85,** 4976–4980.

Koshland, D. E. J. (1952). Effect of catalysts on the hydrolysis of acetyl phosphate. Nucleophilic displacement mechanisms in enzymatic reactions. *J. Am. Chem. Soc.* **74,** 2286–2292.

Lukat, G. S., Stock, A. M., and Stock, J. B. (1990). Divalent metal ion binding to the CheY protein and its significance to phosphotransfer in bacterial chemotaxis. *Biochemistry* **29,** 5436–5442.

Lukat, G. S., Lee, B. H., Mottonen, J. M., Stock, A. M., and Stock, J. B. (1991). Roles of the highly conserved aspartate and lysine residues in the response regulator of bacterial chemotaxis. *J. Biol. Chem.* **266,** 8348–8354.

Lukat, G. S., McCleary, W. R., Stock, A. M., and Stock, J. B. (1992). Phosphorylation of bacterial response regulator proteins by low molecular weight phosphodonors. *Proc. Natl. Acad. Sci. USA* **89,** 718–722.

Mayover, T. L., Halkides, C. J., and Stewart, R. C. (1999). Kinetic characterization of CheY phosphorylation reactions: Comparison of P-CheA and small-molecule phosphodonors. *Biochemistry* **38,** 2259–2271.

McCleary, W. R. (1996). The activation of PhoB by acetylphosphate. *Mol. Microbiol.* **20,** 1155–1163.

McCleary, W. R., and Stock, J. B. (1994). Acetyl phosphate and the activation of two-component response regulators. *J. Biol. Chem.* **269,** 31567–31572.

Needham, J. V., Chen, T. Y., and Falke, J. J. (1993). Novel ion specificity of a carboxylate cluster Mg(II) binding site: Strong charge selectivity and weak size selectivity. *Biochemistry* **32,** 3363–3367.

Norby, J. G. (1988). Coupled assay of Na^+, K^+-ATPase activity. *Methods Enzymol.* **156,** 116–119.

Pazy, Y., Wollish, A. C., Thomas, S. A., Miller, P. J., Collins, E. J., Bourret, R. B., and Silversmith, R. E. (2009). Matching biochemical reaction kinetics to the timescales of life: Structural determinants that influence the autodephosphorylation rate of response regulator proteins. *J. Mol. Biol.* **392,** 1205–1220.

Perego, M. (2001). A new family of aspartyl phosphate phosphatases targeting the sporulation transcription factor Spo0A of *Bacillus subtilis*. *Mol. Microbiol.* **42,** 133–143.

Pioszak, A. A., and Ninfa, A. J. (2004). Mutations altering the N-terminal receiver domain of NRI (NtrC) that prevent dephosphorylation by the NRII-PII complex in *Escherichia coli*. *J. Bacteriol.* **186,** 5730–5740.

Porter, S. L., and Armitage, J. P. (2002). Phosphotransfer in *Rhodobacter sphaeroides* chemotaxis. *J. Mol. Biol.* **324,** 35–45.

Porter, S. L., Roberts, M. A., Manning, C. S., and Armitage, J. P. (2008). A bifunctional kinase-phosphatase in bacterial chemotaxis. *Proc. Natl. Acad. Sci. USA* **105,** 18531–18536.

Quon, K. C., Marczynski, G. T., and Shapiro, L. (1996). Cell cycle control by an essential bacterial two-component signal transduction protein. *Cell* **84,** 83–93.

Rathlev, T., and Rosenberg, T. (1956). Non-enzymatic formation and rupture of phosphorous to nitrogen linkages in phosphoramido derivatives. *Arch. Biochem. Biophys.* **65,** 319–339.

Schuster, M., Zhao, R., Bourret, R. B., and Collins, E. J. (2000). Correlated switch binding and signaling in bacterial chemotaxis. *J. Biol. Chem.* **275,** 19752–19758.

Schuster, M., Silversmith, R. E., and Bourret, R. B. (2001). Conformational coupling in the chemotaxis response regulator CheY. *Proc. Natl. Acad. Sci. USA* **98,** 6003–6008.

Sheeler, N. L., MacMillan, S. V., and Nodwell, J. R. (2005). Biochemical activities of the *absA* two-component system of *Streptomyces coelicolor*. *J. Bacteriol.* **187,** 687–696.

Sheridan, R. C., McCullough, J. F., and Wakefield, Z. T. (1972). Phosphoramidic acid and its salts. *Inorgan. Synth.* **13,** 23–26.

Silversmith, R. E., Appleby, J. L., and Bourret, R. B. (1997). Catalytic mechanism of phosphorylation and dephosphorylation of CheY: Kinetic characterization of imidazole phosphates as phosphodonors and the role of acid catalysis. *Biochemistry* **36,** 14965–14974.

Silversmith, R. E., Smith, J. G., Guanga, G. P., Les, J. T., and Bourret, R. B. (2001). Alteration of a nonconserved active site residue in the chemotaxis response regulator CheY affects phosphorylation and interaction with CheZ. *J. Biol. Chem.* **276,** 18478–18484.

Silversmith, R. E., Levin, M. D., Schilling, E., and Bourret, R. B. (2008). Kinetic characterization of catalysis by the chemotaxis phosphatase CheZ: Modulation of activity by the phosphorylated CheY substrate. *J. Biol. Chem.* **283,** 756–765.

Stewart, R. C. (1993). Activating and inhibitory mutations in the regulatory domain of CheB, the methylesterase in bacterial chemotaxis. *J. Biol. Chem.* **268,** 1921–1930.

Stewart, R. C., and VanBruggen, R. (2004). Phosphorylation and binding interactions of CheY studied by use of Badan-labeled protein. *Biochemistry* **43,** 8766–8777.

Thomas, S. A., Brewster, J. A., and Bourret, R. B. (2008). Two variable active site residues modulate response regulator phosphoryl group stability. *Mol. Microbiol.* **69,** 453–465.

Ulrich, L. E., Koonin, E. V., and Zhulin, I. B. (2005). One-component systems dominate signal transduction in prokaryotes. *Trends Microbiol.* **13,** 52–56.

Wang, H., and Matsumura, P. (1996). Characterization of the $CheA_s$/CheZ complex: A specific interaction resulting in enhanced dephosphorylating activity on CheY-phosphate. *Mol. Microbiol.* **19,** 695–703.

Webb, M. R. (1992). A continuous spectrophotometric assay for inorganic phosphate and for measuring phosphate release kinetics in biological systems. *Proc. Natl. Acad. Sci. USA* **89,** 4884–4887.

Weinstein, M., Lois, A. F., Ditta, G. S., and Helinski, D. R. (1993). Mutants of the two-component regulatory protein FixJ of *Rhizobium meliloti* that have increased activity at the *nifA* promoter. *Gene* **134,** 145–152.

Weiss, V., and Magasanik, B. (1988). Phosphorylation of nitrogen regulator I (NRI) of *Escherichia coli*. *Proc. Natl. Acad. Sci. USA* **85,** 8919–8923.

Yu, Q., Kandegedara, A., Xu, Y., and Rorabacher, D. B. (1997). Avoiding interferences from Good's buffers: A contiguous series of noncomplexing tertiary amine buffers covering the entire range of pH 3–11. *Anal. Biochem.* **253,** 50–56.

Zapf, J., Madhusudan, M., Grimshaw, C. E., Hoch, J. A., Varughese, K. I., and Whiteley, J. M. (1998). A source of response regulator autophosphatase activity: The critical role of a residue adjacent to the Spo0F autophosphorylation active site. *Biochemistry* **37,** 7725–7732.

Zhao, R., Collins, E. J., Bourret, R. B., and Silversmith, R. E. (2002). Structure and catalytic mechanism of the *E. coli* chemotaxis phosphatase CheZ. *Nat. Struct. Biol.* **9,** 570–575.

Zhu, Y., and Inouye, M. (2002). The role of the G2 box, a conserved motif in the histidine kinase superfamily, in modulating the function of EnvZ. *Mol. Microbiol.* **45,** 653–663.

Zundel, C. J., Capener, D. C., and McCleary, W. R. (1998). Analysis of the conserved acidic residues in the regulatory domain of PhoB. *FEBS Lett.* **441,** 242–246.

CHAPTER SEVEN

Transmembrane Receptor Chimeras to Probe HAMP Domain Function

Jürgen U. Linder *and* Joachim E. Schultz

Contents

Abstract

HAMP domains are the central signal converters in bacterial chemotaxis receptors and chemosensory histidine kinases. They link the signal input modules in these proteins, that is, the ligand-binding domains, to the output modules, for example, the histidine kinase domain. A similar architecture is present in the adenylyl cyclase (AC) Rv3645 from *Mycobacterium tuberculosis*, where a HAMP domain is positioned between the N-terminal membrane anchor and the C-terminal catalytic domain. Because the activity of the catalytic domain responds to alterations in the HAMP domain, a method has been developed which uses the catalytic domain of Rv3645 as a reporter to probe the HAMP domain function of diverse bacterial proteins. A strategy for construction of chimeras between a variety of HAMP domains and the catalytic domain of the AC Rv3645 is described. The enzymes are overexpressed in *Escherichia coli* and purified by Ni^{2+}-affinity chromatography. AC activity of the chimeras is determined by a radiotracer method published earlier in the series. Results of the mutagenesis of the HAMP domain from the Af1503 protein of *Archeoglobus fulgidus* are shown as an example for the successful application of the method.

Pharmaceutical Institute, University of Tübingen, Tübingen, Germany

Methods in Enzymology, Volume 471
ISSN 0076-6879, DOI: 10.1016/S0076-6879(10)71007-7

1. Introduction

Motile bacteria respond to changes in their environment by moving toward specific attractants and away from specific repellents (Bourret *et al.*, 1991). This is accomplished either by regulating chemotactic migrations or by adjustment of expression of peculiar genes for use of appropriate nutrients and for degradation and export of untoward substances (Falke and Hazelbauer, 2001). A large segment of the signal transduction systems which coordinate these reactions operates via histidine kinases and uses sensor proteins anchored in the cytoplasma membrane (Zhu and Inouye, 2004). Most of these sensor proteins operate as dimers and possess a characteristic tripartite architecture comprising a signal input domain, a signal converter, and an output domain. Typically, the input domain consists of a membrane anchored periplasmic module which binds compounds in a receptor–ligand interaction. The signal converter consists of two short amphipathic helices arranged in a parallel coiled-coil which dimerizes to form a parallel four helix bundle (Hulko *et al.*, 2006). This universal signal converter module was termed HAMP domain, an acronym derived from the four classes of proteins in which this domain was initially recognized (*h*istidine kinases, *m*ethyl accepting chemotaxis proteins, *a*denylyl cyclases, *p*hosphatases) (Anantharaman *et al.*, 2006; Aravind and Ponting, 1999). Two major prototypes of output domains exist. In chemotaxis sensory proteins the output domain consists of a rod-like, antiparallel coiled-coil which has no intrinsic enzyme activity. It controls the activity of an associated histidine kinase which in turn affects the beating direction of the flagellar motor (Falke and Hazelbauer, 2001). In sensor proteins of two-component systems the output domain as the first component has intrinsic histidine kinase activity and abuts a histidine phosphotransfer domain. The phosphate is transferred to an aspartate residue of the receiver domain of the second component which is a transcription factor and regulates expression of genes under control of the sensory system (Inouye *et al.*, 2003).

The importance of the HAMP domain as a signal converter is emphasized by numerous, in part exhaustive studies which show that point mutations can abrogate or alter the response to chemical stimuli *in vivo* (Ames *et al.*, 2008; Zhou *et al.*, 2009). Most of these studies dealt with the chemotaxis proteins Tsr, the receptor for serine, and Tar, the receptor for aspartate. However, it has been extraordinarily difficult to derive satisfactory structural and functional models of HAMP domains from these mutational data. This was partly due to the fact that most assays used a whole cell approach because it was technically impossible to study HAMP-mediated signal conversion with purified sensors units. A possibility to overcome this problem uses bacterial adenylyl cyclases (ACs) which have a HAMP domain inserted between the membrane anchor and the catalytic domain. Thus, these ACs also have a tripartite domain composition and consist of a membrane anchor, potentially as an as-yet unrecognized input domain, a HAMP signal converter domain, and an AC catalytic domain as an output module. Operability requires dimerization like in chemotaxis receptors. In case of the AC Rv3645 from *Mycobacterium tuberculosis* it has been demonstrated that a cytosolic fragment consisting of the HAMP and the catalytic AC domains is two orders of magnitude more active than the catalytic domain alone (Linder *et al.*, 2004). Furthermore, mutations in the HAMP domain profoundly affected AC activity of the ensemble (Linder *et al.*, 2004). This established a crucial role of the HAMP domain for Rv3645 AC activity and opened the door to devise a novel *in vitro* system to probe the mechanism of signal conversion by HAMP domains. Measurements of AC activity of respective chimeras *in vitro* may help to obtain data for quantitative structure–function relationships.

The principles of construction of chimeric proteins containing the Rv3645 AC catalytic domain fused to alien HAMP domains and further seminal results are presented below. A suitable AC assay based on radiotracers has been described earlier in this series (Johnson and Salomon, 1991).

2. Design of Chimeras of HAMP Domains Fused to the AC Rv3645 Catalytic Domain

The generation of HAMP–AC fusion proteins requires the precise determination of boundaries of the HAMP domain of interest and the selection of a suitable crossover site between HAMP and AC catalytic domains. Although the HAMP domains are quite diverged in sequence usually the alignment to the HAMP domain of the AC Rv3645 is easily accomplished because the amphipathic helices show the typical spacing of hydrophobic residues at each first (a) and fourth (d) position of every heptad of amino acid residues (designated a–g) that corresponds to two full helical turns of the coiled-coil. A sample alignment is shown in Fig. 7.1.

For construction of chimeras the C-terminal residue of the HAMP domain is the fifth amino acid downstream of the last conserved hydrophobic position in the second amphipathic helix (Fig. 7.1). This residue is fused to the catalytic domain of Rv3645 comprising residues 331–549 of the enzyme (Fig. 7.1). The N-terminal residue of the HAMP domain should be the fifth amino acid upstream of a highly conserved proline in the first amphipathic helix (shaded red in Fig. 7.1). This strategy has been successfully employed for probing of the HAMP domain of the Af1503 gene from *Archeoglobus fulgidus* and others (Hulko *et al.*, 2006).

```
                 AS1              Connector       AS2
          d  a d   a d                        a d  a d  a         Cyclase
                                                                  Catalytic domain
Rv3645 278MSIADPLRQLRWALSEVQR GNYNAHMQIYDA- SELGLLQAGFNDMVRELSERQR 331 LRDLFGRYVGEDVARR
Af1503 278STITRPIIELSNTADKIAE GNLEAEVPHQNRA DEIGILAKSIERLRRSLKVAME
Tsr    216ASLVAPMNRLIDSIRHIAG GDLVKPIEVDGS- NEMGQLAESLRHMQGELMRTVG
Tar    216RMLLTPLAKIIAHIREIAG GNLANTLTIDGR- SEMDDLAGSVSHMQRSLTDTVT
NarX   176ARLLQPWRQLLAMASAVSH RDFTQRANI-SGR NEMAMLGTALNNMSAELAESYA
NarQ   174HQVVAPLNQLVTASQRIEH GQFDSPPLDTNLP NELGLLAKTFNQMSSELHKLYR
```

Figure 7.1 Alignment of bacterial HAMP domains to the AC Rv3645 from *M. tuberculosis*. The heptad periodicity of amino acids a–g is indicated by highlighting residues a and d (gray) which form the hydrophobic core in the coiled-coil of HAMP domains. AS1 and AS2 denote borders of the α-helices of the HAMP domains as deduced from the available NMR structure (Hulko *et al.*, 2006). A highly conserved proline residue in AS1 is shaded black. Rv3645, AC from *Mycobacterium tuberculosis* gi 81668722; Af1503, gene product Af1503 from *Archeoglobus fulgidus*, gi 11499098. *E. coli* chemoreceptor proteins: Tsr, gi 2506837; Tar, gi 89111064; NarX, gi 42101; NarQ, gi 581139.

3. Choice of Vector and Cloning Strategy

The plasmid *pQE30* (Qiagen) is a suitable vector for expression of HAMP–AC chimeras. Cloning into *pQE30* requires use of a *lac*-repressor overexpressing strain such as *Escherichia coli* XL1blue-MRF'. The expression cassette is cloned between the *Bam*HI and *Hin*dIII sites of the multiple cloning site (Fig. 7.2). This results in addition of an N-terminal hexahistidine-tag (MRGSH_6GS). In case of the Af1503 HAMP we employed MRGSH_6GSHM as N-terminal tag merely for a more convenient cloning from an existing PCR fragment.

To facilitate easy fusion of HAMP domain and AC we engineered a *Bgl*II site near the 5′-end of the AC Rv3645 catalytic domain (Fig. 7.2). Thus, any HAMP domain of interest can be amplified with a sense primer containing an in-frame 5′-*Bam*HI site and an antisense primer carrying a 5′-*Bgl*II site and a few bases of Rv3645 as an adaptor, that is, AGATCTCG-CAA as indicated in Fig. 7.2. The resulting PCR fragment is ligated in random orientation between the *Bam*HI site of *pQE30* and the engineered *Bgl*II site. Plasmids containing the HAMP domain cassette in the sense orientation are identified via restriction digest with *Bam*HI and *Bgl*II. Thus, the cloning strategy allows fast generation of large sets of HAMP–AC chimeras via a single cloning step.

4. Expression and Purification of the Chimeras

For expression the *pQE30* plasmid containing the HAMP–AC cassette is transformed into *E. coli* BL21(DE3)[pRep4] cells. These cells are prepared by transformation of the commercially available BL21(DE3) strain

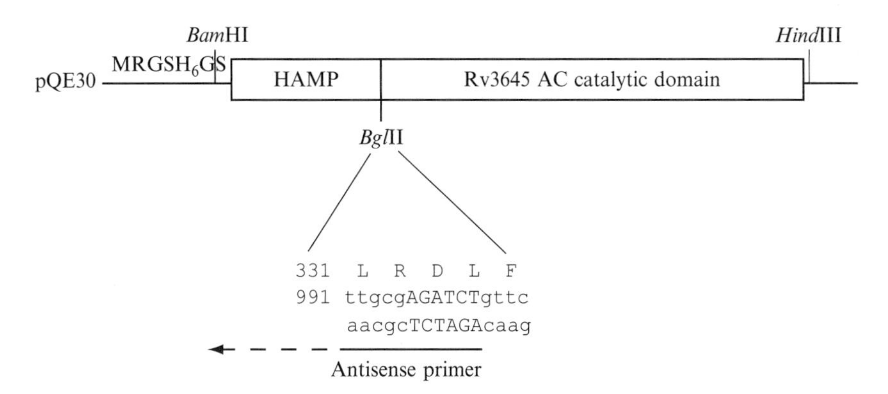

Figure 7.2 General design scheme for bacterial HAMP–AC chimeras in the expression vector *pQE30*. The primer sequence is a proposal for generating a 3′-*Bgl*II restriction site. The 5′-*Bam*H1 site is part of the N-terminal hexahistidine-tag.

(Invitrogen) with the plasmid *pRep4* (Qiagen) for overexpression of the *lac*-repressor protein.

Ten milliliters of Luria-Bertani medium containing 100 μg/ml ampicillin and 25 μg/ml kanamycin are inoculated with the expression strain and grown overnight at 37 °C. This culture is diluted into 400 ml of LB with antibiotics and grown at 37 °C to an OD_{600} of 0.3. The temperature is lowered to 30 °C and at an OD_{600} of 0.4 isopropyl-β-D-galactopyranoside (IPTG) is added at a final concentration of 1 m*M* to induce production of recombinant protein. After 4–6 h at 30 °C cells are collected by centrifugation for 15 min (4 °C, 5000×*g*). Cells are suspended in 50 ml wash buffer (50 m*M* Tris–HCl, 1 m*M* EDTA, pH 8.0) and pelleted for 15 min (0 °C, 6000×*g*). The supernatant is discarded and the pellets are shock-frozen in liquid nitrogen and stored at −80 °C.

All subsequent purification steps are carried out at 0–4 °C. Cells are suspended in 20 ml lysis buffer (50 m*M* Tris–HCl (pH 8.0), 2 m*M* 3-thioglycerol) and disrupted by two passages through a French Press at 1200 psi. Particulate material is removed by centrifugation for 30 min at a minimum of 31,000×*g*. The supernatant is transferred into a 50 ml conical tube and 1 ml of 5 *M* NaCl (final concentration 250 m*M*), 0.3 ml of 1 *M* imidazole, pH 8.0 (final concentration 15 m*M*), and 0.1 ml of 1 *M* $MgCl_2$ (final concentration 5 m*M*) are added. 0.25 ml of Ni^{2+}-nitrilotriacetic acid agarose slurry (Qiagen) is then mixed into the solution and the tube is shaken horizontally at 100 Hz for 2–3 h. Use of a rotator should be avoided because the mechanical stress caused by inversion of the tube tends to result in precipitation of the HAMP–AC fusion protein. The Ni^{2+}-NTA resin is collected by centrifugation at 1000×*g*, suspended in 5 ml buffer A (lysis buffer with 250 m*M* NaCl, 15 m*M* imidazole, 5 m*M* $MgCl_2$) and transferred onto a microfilter column. The resin is washed with another 5 ml of buffer A and with 5 ml of buffer B (lysis buffer with 15 m*M* imidazole, 5 m*M* $MgCl_2$). The HAMP–AC chimera is eluted over a period of several minutes with 0.6 ml buffer C (37.5 m*M* Tris–HCl (pH 8.0), 250 m*M* imidazole, 2 m*M* $MgCl_2$, 1.5 m*M* 3-thioglycerol). After addition of 0.4 ml glycerol the purified protein is stored at −20 °C. The enzyme is ready for use in activity assays.

5. AC Assay

The AC reaction is performed for 10 min at 37 °C in a volume of 100 μl with ATP as a substrate and manganese as a cofactor. For an analysis of the kinetic parameters of the chimeric enzymes prepare the following solutions:

1. *10× substrate* at various concentrations: 0.2–20 m*M* ATP, pH 7.5; 2.5 MBq/ml [α-^{32}P] ATP
2. *2× reaction cocktail*: 100 m*M* Tris–HCl, pH 8.0; 6 m*M* $MnCl_2$; 4 m*M* cAMP; 3 kBq/ml [2,8-^{3}H] cAMP; 44% glycerol
3. *Stop solution*: 1.5% SDS

5.1. Procedure

1. In 1 ml Eppendorf tubes dilute enzyme with water to 40 μl so that the conversion rate of ATP to cAMP will be least 0.2% and at most 10%. Therefore, the amount of enzyme protein for testing has to be determined empirically. Use the same amount of protein for all reactions which are used for kinetic analysis of a chimera.
2. Add 50 μl of *2× reaction cocktail* to each tube.
3. Start each reaction by addition of 10 μl *10× substrate solution*, vortex briefly and incubate at 37 °C for 10 min.
4. Stop by addition of 150 μl *stop solution*.
5. Separate cAMP from ATP as described (Johnson and Salomon, 1991) and determine the amount of cAMP by liquid scintillation counting (^{32}P-channel) and correct for yield (^{3}H-channel).
6. Calculate specific activities as nmol cAMP formed per mg protein and per min reaction time (nmol cAMP/mg/min).

6. Example of Applications of the Method and Results

The structure of the HAMP domain of the protein Af1503 from the thermophile *A. fulgidus* was successfully solved by NMR (Hulko *et al.*, 2006). Optimal growth of the archaebacterium occurs at 83 °C. Therefore,

Table 7.1 Mutation of Ala291 in the Af1503 HAMP–AC fusion protein tunes catalytic activity (Hulko *et al.*, 2006)

Amino acid at position 14 of Af1503 HAMP	Side chain volume (Å^3)	V_{max} (nmol cAMP/mg/min)
Gly	0	2060 ± 130
Ala (wild type)	28.5	1300 ± 100
Cys	48.4	1730 ± 100
Val	79.9	490 ± 10
Leu	106.6	320 ± 10
Ile	106.6	280 ± 10

the Af1503 HAMP domain is considered to be frozen in the ground state at room temperature. The interaction of helices in the coiled-coil structure depicts a noncanonical stacking of hydrophobic side chains, termed "x-da packing." Furthermore, it was predicted that the packing seen in the Af1503 HAMP domain is dependent on the small side chain volume of alanine 291. Therefore, it was hypothesized that replacement of alanine 291 (Hulko *et al.*, 2006), the fourteenth residue in the domain, by larger hydrophobic residues will alter the ground state of the Af1503 HAMP domain, force it to rotate and shift the conformational equilibrium toward an activated state. Because Af1503 does not contain an intrinsic output domain, this hypothesis was tested by fusion of Af1503 HAMP to the Rv3645 catalytic domain. Maximal AC velocities of the chimeras correlated well with the side chain volume of the residue at position 291 with the exception of cysteine (Table 7.1). A large volume of the side chain resulted in low activity, and *vice versa*. Therefore, the ground state of the HAMP domain corresponds to a cyclase on state, and the activated state switches the cyclase off, that is, switching between the two states can be enforced by mutation of a single amino acid residue. In conclusion, the HAMP–AC fusion system proved to be an excellent method to probe this HAMP domain, although the biochemical and physiological functions and the signaling pathway of Af1503 are completely unknown.

7. Concluding Remarks

The concept of HAMP domain mediated signaling in chemotaxis receptors and receptor histidine kinases has only evolved in recent years from the analysis of Williams and Stewart (1999) and a bioinformatics study of Galperin *et al.* (2001). Diverse methods using whole cell approaches have been used to gain insight into structure–function relationships of HAMP domains. In contrast the *in vitro* method based on a mycobacterial AC catalytic module described here allows assessment of HAMP signaling states in a single purified protein with robust measurement. We envisage that this novel HAMP probing system will enforce the understanding of HAMP domains from diverse sources be it chemotaxis receptors or histidine kinases. Meanwhile it has already successfully been extended. First, it is possible to employ other bacterial class III ACs as reporters such as a HAMP-containing AC from the cyanobacterium *Arthrospira platensis*. Second, a variety of other HAMP domains has been successfully used in this system such as HAMP domains from the chemoreceptor proteins Tsr and Tar and from the cyanobacterial AC. Finally, the complete signaling cascade of a serine or aspartate chemoreceptor can be placed in front of the Rv3645 AC and serine or aspartate signaling can be studied *in vitro* using an AC as a reporter enzyme. This opens new venues to biochemically study signal transduction through bacterial chemotaxis receptors.

ACKNOWLEDGMENT

Supported by the Deutsche Forschungsgemeinschaft (SFB766).

REFERENCES

Ames, P., Zhou, Q., and Parkinson, J. S. (2008). Mutational analysis of the connector segment in the HAMP domain of Tsr, the *Escherichia coli* serine chemoreceptor. *J. Bacteriol.* **190,** 6676–6685.

Anantharaman, V., Balaji, S., and Aravind, L. (2006). The signaling helix: A common functional theme in diverse signaling proteins. *Biol. Direct* **1,** 25.

Aravind, L., and Ponting, C. P. (1999). The cytoplasmic helical linker domain of receptor histidine kinase and methyl-accepting proteins is common to many prokaryotic signalling proteins. *FEMS Microbiol. Lett.* **176,** 111–116.

Bourret, R. B., Borkovich, K. A., and Simon, M. I. (1991). Signal transduction pathways involving protein phosphorylation in prokaryotes. *Annu. Rev. Biochem.* **60,** 401–441.

Falke, J. J., and Hazelbauer, G. L. (2001). Transmembrane signaling in bacterial chemoreceptors. *Trends Biochem. Sci.* **26,** 257–265.

Galperin, M. Y., Nikolskaya, A. N., and Koonin, E. V. (2001). Novel domains of the prokaryotic two-component signal transduction systems. *FEMS Microbiol. Lett.* **203,** 11–21.

Hulko, M., Berndt, F., Gruber, M., Linder, J. U., Truffault, V., Schultz, A., Martin, J., Schultz, J. E., Lupas, A. N., and Coles, M. (2006). The HAMP domain structure implies helix rotation in transmembrane signaling. *Cell* **126,** 929–940.

Inouye, M., Dutta, R., and Zhu, Y. (2003). Regulation of porins in *Escherichia coli* by the osmosensing histidine kinase/phosphatase EnvZ. *In* "Histidine Kinases in Signal Transduction," (M. Inouye, R. Dutta, M. Inouye, and R. Dutta, eds.), pp. 25–46. Academic Press, Inc., San Diego, CA.

Johnson, R. A., and Salomon, Y. (1991). Assay of adenylyl cyclase catalytic activity. *Methods Enzymol.* **195,** 3–21.

Linder, J. U., Hammer, A., and Schultz, J. E. (2004). The effect of HAMP domains on class IIIb adenylyl cyclases from *Mycobacterium tuberculosis*. *Eur. J. Biochem.* **271,** 2446–2451.

Williams, S. B., and Stewart, V. (1999). Functional similarities among two-component sensors and methyl-accepting chemotaxis proteins suggest a role for linker region amphipathic helices in transmembrane signal transduction. *Mol. Microbiol.* **33,** 1093–1102.

Zhou, Q., Ames, P., and Parkinson, J. S. (2009). Mutational analyses of HAMP helices suggest a dynamic bundle model of input-output signalling in chemoreceptors. *Mol. Microbiol.* **73,** 801–814.

Zhu, Y., and Inouye, M. (2004). The HAMP linker in histidine kinase dimeric receptors is critical for symmetric transmembrane signal transduction. *J. Biol. Chem.* **279,** 48152–48158.

CHAPTER EIGHT

Light-Activated Bacterial LOV-Domain Histidine Kinases

Tong-Seung Tseng,* Marcus A. Frederickson,† Winslow R. Briggs,* *and* Roberto A. Bogomolni†

Contents

Abstract

Bacteria rely on two-component signaling systems in their adaptive responses to environmental changes. Typically, the two-component system consists of a sensory histidine kinase that signals by transferring a phosphoryl group to a secondary response regulator that ultimately relays the signal to the cell. Some of these sensors use PAS (Per–Arnt–Sin) domains. A new member of the PAS super family is the LOV (light, oxygen, voltage) domain, a 10-kDa flavoprotein that functions as a light-sensory module in plant, algal, fungal, and bacterial blue-light receptors. Putative LOV domains have been identified in the genomes of many higher and lower eukaryotes, plants, eubacteria, archaebacteria, and particularly in genes coding for histidine kinases (LOV-histidine kinases, LOV-HKs) of plant and animal pathogenic bacteria, including *Brucella.* We describe here biochemical, photochemical, and biophysical methodology to purify these enzymes and to characterize their light-activation process.

* Department of Plant Biology, Carnegie Institution for Science, Stanford, California, USA
† Department of Chemistry and Biochemistry, University of California, Santa Cruz, California, USA

Methods in Enzymology, Volume 471

ISSN 0076-6879, DOI: 10.1016/S0076-6879(10)71008-9

1. Introduction

The LOV domain (light, oxygen, voltage) was first identified in the plant blue-light receptor, phototropin 1 (phot1) (Christie *et al.*, 1998, 1999; Huala *et al.*, 1997). Phototropins contain two LOV domains (LOV1 and LOV2), and LOV2 from phot1 is the best characterized (Christie *et al.*, 1998). Phototropin LOV domains bind flavin mononucleotide (FMN) as chromophore. Light absorption by the FMN results in the formation of a cysteinyl adduct of the cysteine residue with the C(4a) carbon of the flavin moiety, through formation of a C–S covalent bond. The cysteinyl-adduct formation causes a protein conformational change that activates the downstream ser/thr kinase, leading to autophosphorylation. In the dark, the flavin–cysteinyl adduct gradually breaks down, completing a photocycle (Christie, 2007; Salomon *et al.*, 2000; Swartz *et al.*, 2001). Figure 8.1 shows the typical photochemical cycle of LOV proteins.

LOV-domain-containing proteins are now found to be widespread among prokaryotes and eukaryotes (Briggs, 2007; Losi and Gaertner, 2008). The LOV domain functions as light sensor and conveys the signal to protein domains with diverse functions (Losi and Gaertner, 2008). Whereas, in higher plants, blue-light sensing in the LOV domain of phototropin activates the ser/thr kinase domain, several bacterial LOV-domain-containing histidine

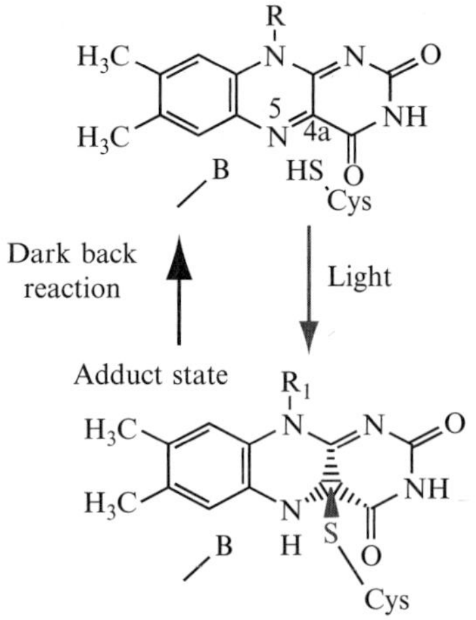

Figure 8.1 Photochemical cycle of LOV-domain proteins. Light induces the formation of a covalent adduct between the thiol sulfur and the carbon C4a of the flavin. The adduct has a broad absorption band around 390 nm. In most LOV-HKs the adduct breaks spontaneously in time ranging from seconds to hours depending on the particular system, completing a photocycle. The dark back reaction is base-catalyzed by a putative base B.

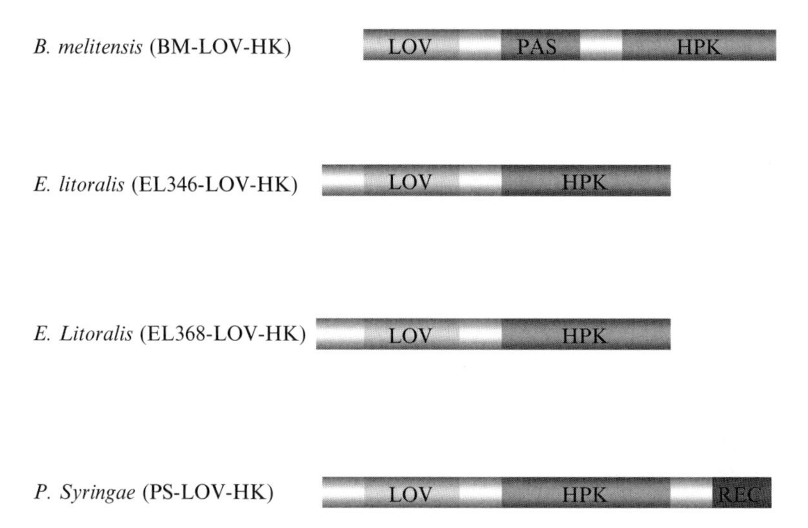

Figure 8.2 Domain alignment of four LOV histidine kinase proteins (LOV-HPKs). The bacterial species and kinase nomenclature are listed on the left of their relative domain map. *B. melitensis* gene includes an intervening PAS domain between the sensory LOV domain and the output kinase domain. *P. syringae* gene codes for a receiver domain (REC), which represents an intramolecular response-regulator domain (RR). HPK is an acronym for histidine-protein-kinase. Reproduced from Swartz *et al.* (2007), with permission from Science.

kinases (LOV-HK) (Fig. 8.2) have been identified and characterized (Purcell *et al.*, 2007; Swartz *et al.*, 2007). These LOV-HKs may function as the sensors in the canonical two-component system, inducing physiological and/or biochemical responses of the microorganisms to light signals. In *Brucella* the Brucella-LOV-HK (B-LOV-HK) mediates a large light-activated increase in virulence in a microphage virulence assay system (Swartz *et al.*, 2007). *Pseudomonas syringae* LOV-HK (PS-LOV-HK) (Swartz *et al.*, 2007) contains a putative response-regulator domain (PS-LOV-HK-RR). The expressed PS-LOV-HK can phosphorylate the RR domain (Cao *et al.*, 2008) in a light-dependent manner *in vitro*.

2. Description of Method

2.1. Cloning and purification of LOV-HKs

Brucella sp. LOV-histidine kinase (B-LOV-HK, accession number BMEII0679), *Pseudomonas syringae* LOV-histidine kinase (PS-LOV-HK, accession number NP_792694), and *Erythrobacter literalis* LOV-histidine kinases (EL368-LOV-HK, accession number ZP_00377191; EL346-LOV-HK, accession number ZP_00376813) were PCR amplified and directionally cloned into the pET/100D TOPO plasmid following the manufacturer's instructions. The proteins were expressed in *E. coli* BL21(DE3)pLysE. The *Brucella* genus is divided in six species according to their host specificity: *B. melitensis*, *B. suis*, *B. abortus*, *B. canis*, *B. ovis*, and *B. neotomae*. We have carried out most of our biochemical work on *Brucella mellitensis* LOV-histidine kinase (BM-LOV-HK), and the infection work on *Brucella abortus*. Although there are minor differences in their DNA sequences their LOV-domain histidine kinases have identical protein sequences and presumably structures. In this chapter, we will refer to the *Brucella*-LOV-HKs in general as B-LOV-HK but we will indicate the species in some experiments.

Cell cultures, protein induction, and protein purification were carried out under dim red light (wavelengths longer than 600 nm). Although this precaution may not be necessary for LOV-HKs that undergo rapid photocycles and recover in the seconds-to-minutes time scale, it is crucial in the preparation of *Brucella*-LOV-HK because it is permanently activated by near-UV/blue light. Transformed competent cells were grown in Luria Broth (LB) containing ampicillin. A 25-ml starter overnight culture was grown under strict dark conditions at 37 °C and 225 rpm (8–12 h). One-liter cultures were inoculated with the overnight starter and grown at 37 °C and 225 rpm shaking for approximately 1.5–3 h, until the culture growth reached early log phase. At this point, temperature was reduced to ambient (about 20 °C), and protein expression was induced by the addition of IPTG (isopropyl β-D-1-thiogalactopyranoside). Protein expression was continued for 4–6 h depending on growth. Protein expression and growth were halted by placing the growth flasks on ice prior to centrifugation. Cell pellets were harvested by centrifugation. Pellets were stored frozen at -20 °C until further processing. Thawed cells were ruptured under 1500–2000 lbs/in^2 using a French Press. After centrifugation for 22 min at 19 K (Sorval SS34 rotor, Dupont), the expressed LOV-HKs were affinity purified under dim red light, using TALON metal affinity resin (BD Bioscience Clontech), following the manufacturer's instructions. Polyhistidine-labeled protein bound to the resin was eluted with 200 m*M* imidizole buffer. Eluted fractions were characterized by their UV–vis absorption spectra, their photochemical activity and by SDS–PAGE (Phast Gel protocol, Pharmacia, Sweden). Imidazol acts as a catalytic base in the spontaneous breakage of the flavin–cysteinyl adduct and has a pronounced effect on its thermal decay kinetics (Alexandre *et al.*, 2007). This affects the concentration of activated kinase that can be accumulated in the photostationary state under constant illumination. To remove the imidazol and to standardize the enzymatic assay conditions, the desired fractions were desalted by size-exclusion chromatography, or had the buffer exchanged, either by dialysis or by repeated centrifugations using spin concentrators, with best results coming from desalting columns (Sephadex G-25 resin, GE-Healthcare, Life Sciences). Fractions were either freeze-dried or stored in low ionic-strength solution.

2.2. Assesment of LOV-kinase purity

The quality of the preparations was assessed by *SDS–PAGE*. We typically used the Phast Gel system programmed for resolving polyacrylamide-gradient Phast gels (GE Healthcare, Life Sciences) of 8–24%. Samples were a 1:1 mixture of protein fraction to dye (Coomassie Blue) with sodium dodecyl sulphate mixed via microcentrifugation and were boiled for 5 min. Gels were stained with filtered Coomassie Blue, and microwaved for 10–20 s. A blue Plus2 Pre-stained Standard (Invitrogen) molecular weight ladder was used to estimate protein sizes.

2.3. Spectrophotometric determination of LOV-HK concentration

The typical absorption spectrum of the LOV-HKs in the near-UV/visible range shows two major bands with maxima around 370 and 450 nm and with pronounced vibronic structure (see Fig. 8.3A). This vibronic feature results from strong interactions of the flavin with the protein in the binding pocket. This feature is lost in the spectrum of free FMN in aqueous solution. We use a molar extinction coefficient of 12,500 M^{-1} cm^{-1} (Salomon *et al.*, 2000) for the 450 nm peak to determine the LOV-HK concentration. We have observed a loss of vibronic structure as samples are diluted below the micromolar range, suggesting the presence of free FMN in solution. This change indicates that the noncovalently bound flavin is in binding equilibria with its apoprotein with dissociation constants in the micromolar range.

Because some of the absorbance at 450 nm originates from free FMN it is necessary to determine the fraction of intact (and photochemically active) chromoprotein. The fluorescence of free FMN is collisionally quenched by approximately 200 m*M* KI whereas the bound FMN is protected from quenching in the binding pocket. We rely on this differential effect of KI on the FMN fluorescence to estimate semiquantitatively the fractions of bound and free chromophore. Typically, we measure the fluorescence excitation spectra in the 300–500 nm range (fluorescence emission at 530 nm) in the absence and in the presence of approximately 200 m*M* KI (added to the solution in solid form). Because free FMN has about twice the fluorescence yield of LOV-domain bound FMN (Swartz *et al.*, 2001), the fraction of FMN bound is given by the ratio of twice the remaining fluorescence intensity to the total fluorescence intensity before addition of KI. Photoactivity of LOV kinases depends not only on proper flavin chromophore binding but also depends on the structural integrity of the binding pocket and obviously on the presence of the reactive cysteine (mutants in this cysteine show proper absorbance but are devoid of photochemical activity) (Salomon *et al.*, 2000; Swartz *et al.*, 2001). Photochemically active LOV-kinase is assayed by measuring light-induced absorbance changes indicative of adduct formation.

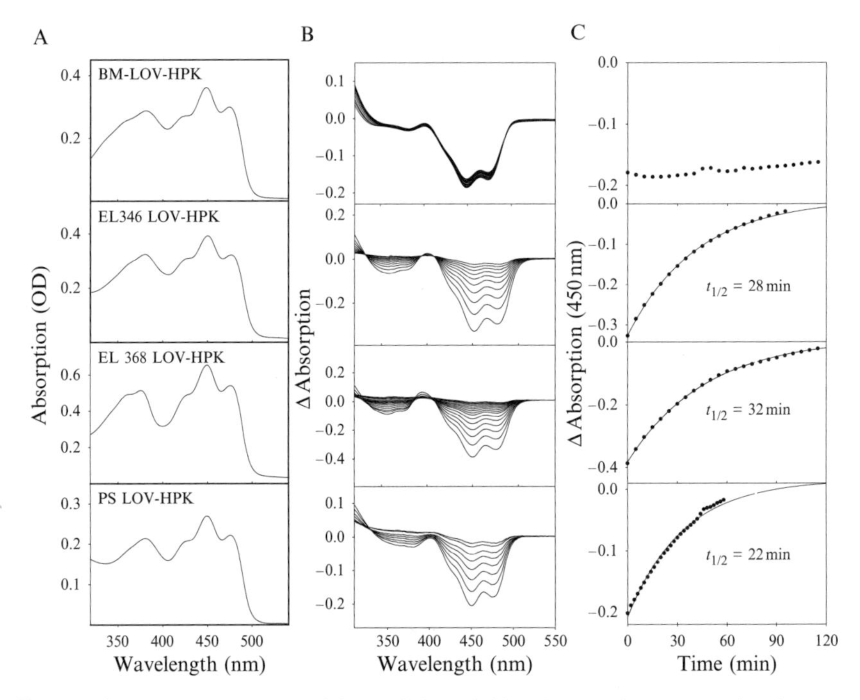

Figure 8.3 Photochemistry and thermal decay of four bacterial LOV-histidine kinases. (A) UV–vis absorbance spectra of the four sensory kinases, BM-LOV-HPK, EL346-LOV-HPK, EL368-LOV-HPK, and PS-LOV-HPK. (B) Difference spectra showing the spontaneous dark recovery of the four different photobleached LOV domains in each kinase. Note the Brucella-LOV-HK exhibits a lack of absorbance recovery. (C) Half-lives for the dark recovery of the ground state after blue-light excitation. Reproduced from Swartz *et al.* (2007), with permission from Science.

2.4. Kinase photochemical-activity assay

We induced photochemical bleaching of the BM-LOV-HK, EL368-LOV-HK, EL346-LOV-HK, and PS-LOV-HK with a camera strobe flash (full white light or broadband blue light obtained by passing the flash of white light through a Corning 5-58 blue glass filter, 1 ms, 100 mJ pulse). Absorbance changes were followed as a function of time after the flash in a Hewlett-Packard 8452A diode array spectrophotometer. Typical acquisition time for a single spectrum is about 1 s. Loss of absorption in the 450 nm band with some increase in the UV-A region (Fig. 8.3B) are characteristic of formation of a flavin–cysteinyl adduct (Christie, 2007; Salomon *et al.*, 2000). The approximate difference molar extinction coefficient at 450 nm for the four LOV-HKs described is 12,000 M^{-1} cm^{-1} (Swartz *et al.*, 2001). The stability of the cysteinyl adduct over time varies among the different LOV-HKs (Fig. 8.3C).

Unlike other LOV-HKs, the BM-LOV-HK adduct state is extremely stable and does not decay significantly in 2 h (Fig. 8.3B and C). The three other LOV-HKs do complete a photocycle, decaying thermally from the covalent adduct state back to the ground state while in the dark. EL346-LOV-HK, EL368-LOV-HK, and PS-LOV-HK decay from the adduct to the ground state with half-lifes ($t_{1/2}$) of 28, 32, and 22 min respectively, at room temperature (Fig. 8.3C).

2.5. Safe-light conditions for sample manipulation

Because of the sensitivity of the molecular system and the permanent photoconversion to the covalent adduct state, active B-LOV-HK accumulates over time if exposed to light wavelengths below 500 nm. Therefore, sample preparation and experimental data acquisition should be carried out entirely under dim red-light conditions. Spectroscopic data acquisition requires illumination within the absorption band of the flavoprotein. In this case, the intensity of light probe beams for measuring absorbance or fluorescence should be minimized. Typically, we limit the unavoidable photoconversion to less than 1%.

2.6. Kinase enzymatic-activity assay

Autophosphorylation reactions were carried out under dim red light with heterologously expressed affinity-purified LOV-HKs from *E. coli*. Each reaction contained 12.5 μg of LOV-HK protein in a final volume of 10 μl of phosphorylation buffer (50 m*M* Tris–HCl, pH 7.5, 50 m*M* KCl, 5 m*M* $MgCl_2$). Before light treatment, 1 μCi of [Υ-^{32}P] ATP (111 TBq mmol^{-1}, Perkin-Elmer), diluted 10× with cold 10 μM ATP, was spotted onto the side wall of the reaction tubes above the reaction mixture. Samples were then irradiated for 1 min with white light (fluence rate 2000 μmol m^{-2} s^{-1}; light source was the light output of a Kodak carousel 4400 slide projector, Kodak Co., Rochester, NY); a dark-control sample was mocked treated with light. After the light treatment, both light and dark samples were centrifuged for a few seconds at low speed to move the [Υ-^{32}P] ATP down into the reaction mixture. The reactions were then incubated in the dark at room temperature for 4 min, and stopped by adding SDS–PAGE loading buffer, containing 12.5 m*M* Tris–HCl (pH 6.8), 2% glycerol, 0.4% sodium dodecyl sulfate (SDS), 143 m*M* 2-mercaptoethanol, and 0.01% bromophenol blue. SDS–PAGE was carried out with 12.5% SDS gels. Gels were then dried and exposed to either X-ray film (Kodak, Rochester, NY) or to a phosphoimager (Typhoon Trio, GE).

We tested the amount of the [Υ-^{32}P] ATP required to detect any possible light enhancement of kinase activity. For all of the expressed

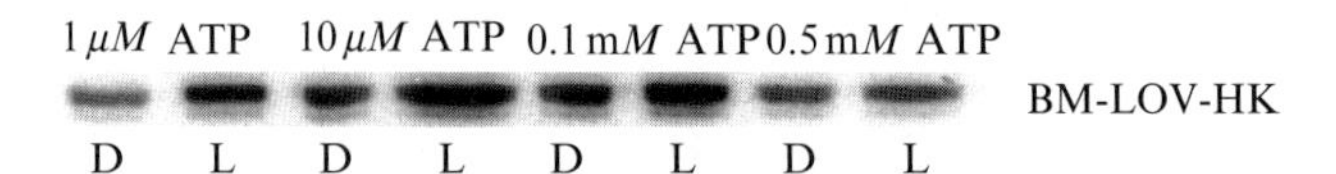

Figure 8.4 Light enhancement of BM-LOV-HK kinase activity in different concentration of cold ATP. BM-LOV-HK (Brucella-LOV-histidine kinase). D, dark control; L, white-light treated. One millicurie [ϒ-32P] ATP throughout.

LOV-HKs, 1 or 2 μCi of 10-fold-diluted [ϒ-^{32}P] ATP gave consistent light-enhanced autophosphorylation. We also tested the effect of different concentrations of cold ATP to determine how they affect the light-enhanced autophosphorylation. B-LOV-HK showed light-enhanced autophosphorylation from 1 μM ATP to 0.1 mM ATP tested, and the incorporation of [ϒ-^{32}P] ATP decreased at 0.5 mM ATP (Fig. 8.4). However, the PS-LOV-HK only exhibited a strong light/dark difference with 10 μM cold ATP. Therefore, all further reactions were carried out with 1 μCi of [ϒ-^{32}P] ATP diluted 10-fold with 10 μM ATP.

To confirm the presence of the phosphoramidite (phosphohistidine) phosphorylation product we conducted successive alkaline and acid washes of phosphorylated samples. The assay testing the alkali versus acid stability of the phosphate bond (phosophoramidite or phosphoester) indicates that in all four proteins, a phosphohistidine bond is formed upon kinase activation. Resistance to alkaline hydrolysis is a chemical property of the phosphoramidite bond (Klumpp and Krieglstein, 2002) (Fig. 8.5).

3. Concluding Remarks and Future Perspectives

Most histidine kinases are largely class I HK as classified by Bilwes *et al.* (1999). They are membrane proteins with a wide variety of extracellular sensors. However, the cytoplasmic portions have a catalytic domain with conserved regions that have been designated H, N, G1, and G2 boxes (Parkinson and Kofoid, 1992). For reference, Marina *et al.* (2005) have provided an elegant and detailed analysis of the entire cytoplasmic portion of CheA from *Thermotoga maritiae*. The HK domain from the B-LOV-HKs belongs to the new HWE family of two-component sensor kinases. This family differs from other histidine kinases in the absence of a recognizable F box and in the presence of several unique conserved amino acids, including a histidine in the N box and a WXE sequence in the G1 box (Karniol and Vierstra, 2004).

The bacterial LOV-HKs described here are the first studied and much work remains to characterize their enzymatic properties, structure and

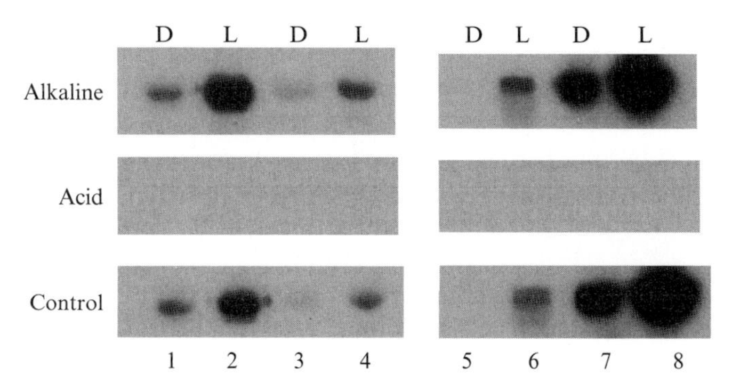

Figure 8.5 Light-activated LOV-HKs phosphorylate a histidine residue. Protein samples fractionated on polyacrylamide gels were transferred to a polyvinylidene fluoride (PDVF) membrane (Immobilon-P; Millipore, Billerica, MA). The membranes were treated with 1 *M* KOH (alkaline) at 55 °C for 2 h and autoradiographed, then treated with 6 *M* HCl (acid) at 55 °C for 2 h and autoradiographed again. The control was a duplicate membrane that had not been treated (D, L, dark, light, respectively). Lanes 1 and 2, BM-LOV-HK; lanes 3 and 4, PS-LOV-HK; lanes 5 and 6, EL346-LOV-HK; lanes 7 and 8, EL368-LOV-HK. Reproduced from Swartz *et al.* (2007), with permission from Science.

function fully. Neither the aggregation state nor the cellular localization of the LOV-kinases *in vivo* is known, although all of the recombinant proteins studied are soluble proteins. They likely are members of two-component light sensing signaling systems in bacteria but their cognate RRs need to be identified and their biological role established.

ACKNOWLEDGMENTS

This work was supported by research grants from the National Science Foundation DBM-0843617 (WRW) and DBM-0843617 (RAB).

REFERENCES

Alexandre, M., Arents, J., Van Grondelle, R., Hellingwerf, K., and Kennis, J. (2007). A base-catalyzed mechanism for dark state recovery in the *Avena sativa* phototropin-1 LOV2 domain. *Biochemistry* **46,** 3129–3137.

Bilwes, A. M., Alex, L. A., Crane, B. R., and Simon, M. I. (1999). Structure of CheA, a signal-transducing histidine kinase. *Cell* **96,** 131–141.

Briggs, W. R. (2007). The LOV domain: A chromophore module servicing multiple photoreceptors. *J. Biomed. Sci.* **14,** 499–504.

Cao, Z., Buttani, V., Losi, A., and Gärtner, W. (2008). A blue light inducible two-component signal transduction system in the plant pathogen *Pseudomonas syringae* pv. tomato. *Biophys. J.* **94,** 897–905.

Christie, J. M., Reymond, P., Powell, G. K., Bernasconi, P., Raibekas, A. A., Liscum, E., and Briggs, W. R. (1998). *Arabidopsis* NPH1: A flavoprotein with the properties of a photoreceptor for phototropism. *Science* **282,** 1698–1701.

Christie, J. M., Salomon, M., Nozue, K., Wada, M., and Briggs, W. R. (1999). LOV (light, oxygen, or voltage) domains of the blue-light photoreceptor phototropin (nph1): Binding sites for the chromophore flavin mononucleotide. *Proc. Natl. Acad. Sci. USA* **96,** 8779–8783.

Huala, E., Oeller, P. W., Liscum, E., Han, I.-S., Larsen, E., and Briggs, W. R. (1997). *Arabidopsis* NPH1: A protein kinase with a putative redox-sensing domain. *Science* **278,** 2120–2123.

Karniol, B., and Vierstra, R. D. (2004). The HWE histidine kinases, a new family of bacterial two-component sensor kinases with potentially diverse roles in environmental signaling. *J. Bactertiol.* **186,** 445–453.

Klumpp, P. S., and Krieglstein, J. (2002). Phosphorylation and dephosphorylation of histidine residues in proteins. *Eur. J. Biochem.* **269,** 1067–1071.

Losi, A., and Gaertner, W. (2008). Bacterial bilin- and flavin-binding photoreceptors. *Photochem. Photobiol. Sci.* **7,** 1168–1178.

Marina, A., Waldburger, C. D., and Hendrickson, W. A. (2005). Structure of the entire cytoplasmic portion of a sensor histidine-kinase protein. *EMBO J.* **24,** 4247–4259.

Parkinson, J. S., and Kofoid, E. C. (1992). Communication modules in bacterial signaling proteins. *Ann. Rev. Genet.* **26,** 71–112.

Purcell, E. B., Siegal-Gaskins, D., Rawling, D. C., Fiebig, A., and Crosson, S. (2007). A photosensory two-component system regulates bacterial cell attachment. *Proc. Natl. Acad. Sci. USA* **104,** 18241.

Salomon, M., Christie, J. M., Kneib, E., Lempert, U., and Briggs, W. R. (2000). Photochemical and mutational analysis of the FMN-binding domains of the plant blue light receptor, phototropin. *Biochemistry* **39,** 9401–9410.

Swartz, T. E., Corchnoy, S. B., Christie, J. M., Lewis, J. W., Szundi, I., Briggs, W. R., and Bogomolni, R. A. (2001). The photocycle of a flavin-binding domain of the blue light photoreceptor phototropin. *J. Biol. Chem.* **276,** 36493–36500.

Swartz, T. E., Tseng, T. S., Frederickson, M. A., Paris, G., Comerci, D. J., Rajashekara, G., Kim, J. G., Mudgett, M. B., Splitter, G. A., Ugalde, R. A., Goldbaum, F. A., Briggs, W. R., and Bogomolni, R. A. (2007). Blue-light-activated histidine kinases: two-component sensors in bacteria. *Science* **317,** 1090–1093.

CHAPTER NINE

Characterization of Bacteriophytochromes from Photosynthetic Bacteria: Histidine Kinase Signaling Triggered by Light and Redox Sensing

Eric Giraud,* Jérôme Lavergne,†,‡,§ *and* André Verméglio†,‡,§

Contents

* Laboratoire des Symbioses Tropicales et Méditerranéennes, IRD, CIRAD, AGRO-M, INRA, UM2, TA A-82/J, Campus de Baillarguet, Montpellier Cedex 5, France
† CEA, DSV, IBEB, Laboratoire de Bioénergétique Cellulaire, Saint-Paul-lez-Durance, France
‡ CNRS, UMR 6191, Biologie Végétale et Microbiologie Environnementales, Saint-Paul-lez-Durance, France
§ Aix-Marseille Université, Saint-Paul-lez-Durance, France

Methods in Enzymology, Volume 471
ISSN 0076-6879, DOI: 10.1016/S0076-6879(10)71009-0

Abstract

Bacteria detect environmental changes, thanks to two-component signal-transduction systems, composed, in general, of a sensor coupled to a histidine kinase and a DNA binding response regulator. Anoxygenic photosynthetic bacteria like *Rhodopseudomonas (Rps.) palustris*, possess a highly versatile metabolism and can grow via photosynthesis using light energy or via respiration through oxygen consumption. For photosynthetic bacteria, detecting changes in light quality or quantity, or in oxygen concentration, is therefore of prime importance for adjusting their metabolism for optimal development. A central role is played by bacteriophytochromes for light detection and initiation of regulatory responses. The switch of these chromoproteins between two photointerconvertible forms is the first event in the light-regulated cascade. This chapter describes *in vitro* and *in vivo* methods that have been successfully used to investigate the bacteriophytochrome dependent light regulation pathways, in several strains of *Rps. palustris* and *Bradyrhizobium*. These approaches range from biochemical and biophysical methods to genetic techniques. Such multiple approaches are indispensable for understanding these complex light-regulated pathway. In a first step, bacteriophytochromes and associated response regulators are overexpressed in *Escherichia coli* and purified. The spectral and kinetic properties of the two photointerconvertible forms of the purified bacteriophytochromes are then determined by biophysical approaches. Original spectral and kinetic properties found in some of the bacteriophytochromes that we studied necessitated the development of new methods for computing the spectra of the pure forms and the photoconversion yields. *In vitro* biochemical approaches help to assess the histidine kinase activity of bacteriophytochromes depending on light conditions, the phosphotransfer to response regulators and their affinity to promoter DNA sequences. Finally, gene inactivation tests the importance of specific genes in photosynthesis regulation under particular light and oxygen tension growth conditions. The methods described in this chapter are not restricted to the study of the light-transduction pathways of *Rps. palustris* and *Bradyrhizobium* strains but are applicable to the understanding of any bacterial light-regulatory system.

1. Introduction

Photosynthetic bacteria have developed an extraordinary metabolic versatility. They can transform light energy into biochemically useful energy but also grow with or without oxygen, via aerobic or anaerobic respirations. To benefit from this flexibility and take advantage of available resources, these bacteria have to rapidly adapt and respond to variations in their environment. This requires, in particular, sensing changes in light and in oxygen concentration.

The most studied photoreceptors in photosynthetic bacteria are bacteriophytochromes (BphPs), the bacterial counterpart of the plant phytochromes. This photoreceptor was first identified in cyanobacteria (Hughes *et al.*, 1997) and since then its presence was revealed by genome sequencing projects in many other bacteria (Giraud and Verméglio, 2008).

BphPs are chromoproteins that switch between two photointerconvertible forms via the photoisomerization of their chromophore, a linear tetrapyrrole. In most cases, these two forms absorb red or far-red light and are designated as P_r and P_{fr}, respectively.

Amino acid sequence alignment shows that BphPs possess a protein architecture similar to those of phytochromes, composed of an N-terminal photosensory core domain (PCD) and a C-terminal module involved in signal transduction and dimerization. The PCD, constituted of subdomains PAS, GAF, and PHY, is able to attach autocatalytically the bilin chromophore to a Cys residue located at the N-terminal extremity of the protein (Karniol *et al.*, 2005; Lamparter *et al.*, 2002; Rockwell *et al.*, 2006). Almost all BphPs studied thus far contain as chromophore the linear tetrapyrrole biliverdin (BV) (Bhoo *et al.*, 2001). This chromophore is synthesized thanks to a heme oxygenase (HmuO) that opens up the heme ring. Generally, the C-terminal domain contains a histidine kinase motif whose light-induced autophosphorylation triggers a phosphotransfer to a response regulator (RR). Depending on the BphP, the most active phosphorylating state may be the P_r or the P_{fr} form. The gene encoding the RR involved in the kinase cascade is often found within the BphP operon (Rockwell *et al.*, 2006) and in the vicinity of the genes that they control.

Our research over the past decade has been devoted to elucidating the roles of BphPs in photosynthetic bacteria. For that purpose, two species, *Rhodopseudomonas (Rps.) palustris* and *Bradyrhyzobium*, have been selected for the following reasons: their complete genomes are available and these bacteria are amenable to genetic studies. *Rps. palustris* strain CGA009 is an anaerobic anoxygenic (i.e., nonoxygen-evolving) phototroph, which possesses six different BphPs and a large number of putative regulatory and signaling genes (Larimer *et al.*, 2004). *Bradyrhizobium* sp. ORS278 is an aerobic anoxygenic phototroph, symbiote of *Aeschynomene* legumes and phylogenetically related to *Rps. palustris* (Giraud and Fleischman, 2004). In contrast to an anaerobic anoxygenic phototroph like *Rps. palustris*, aerobic anoxygenic phototrophs present the characteristic of forming their photosynthetic apparatus and having an efficient cyclic electron transfer only in presence of oxygen (Yurkov and Beatty, 1998).

Our studies combine biophysical, biochemical, and molecular biology techniques. Although the techniques we used are classical as regards protein biochemistry and molecular biology, only their combination with biophysical approaches renders possible the complete description of a signal-transduction pathway from light absorption to gene expression.

The spectral properties of BphPs are much more diverse, in terms of wavelength positions and overlap of the two forms, and also in kinetic aspects than those of the plants phytochromes. The classical approach developed by Butler (1972) for determining spectral characteristics and photoconversion yields requires that the spectrum of one of the interconvertible form be directly measurable. We have developed an experimental and mathematical approach to tackle the case when both spectral forms are unknown (these theoretical topics are described at the end of the chapter, Section 9).

Expression and purification of BphPs and RRs allow the *in vitro* analysis of the light-induced autophosphorylation of BphPs, their phosphotransfer to RRs and transcription factors, and the binding of these last elements to DNA promoter regions depending on their phosphorylation status.

The signal-transduction mechanism that can be inferred from these *in vitro* methods has to be finally verified by *in vivo* characterization of phenotypes of deletion mutants of these proteins under various (light and oxygen) growth conditions.

Using this global strategy, our group has shown that a bacteriophytochrome, present in *Rps. palustris* (*Rp*BphPl) and *Bradyrhizobium* (*Br*BphPl), detects the presence of light and triggers the synthesis of the entire photosynthetic apparatus under aerobic conditions by antagonizing the repressing effect of the transcription factor PpsR2 (Giraud *et al.*, 2002, 2004; Jaubert *et al.*, 2004). In addition, *Rp*BphPl regulates the synthesis of the alpha-ketoglutarate dehydrogenase enzymatic complex, a key enzyme of the Krebs cycle (Kojadinovic *et al.*, 2008). This dual regulation at the transcriptional level favors the photosynthetic activity over the respiratory activity, that is, the most favorable bioenergetic process when light is available. This allows a very finely controlled adaptation of the bacterium to environmental conditions. In addition to this BphP, other BphPs control the synthesis of peripheral light-harvesting (LH) complexes in *Rps. palustris* and *Bradyrhizobium*. In *Rps. palustris*, *Rp*BphP2 and *Rp*BphP3 act in tandem to control the synthesis of LH4 complex, characterized by a single absorption band in the near infrared at 800 nm (Giraud *et al.*, 2005). Another bacteriophytochrome, *Rp*BphP4, acts either as a light-sensitive kinase (chromo-*Rp*BphP4) or a redox-sensitive kinase (achromo-*Rp*BphP4) depending upon the strain of *Rps. palustris*. Both chromo- and achromo-*Rp*BphP4s regulate the synthesis of the peripheral light-harvesting (LH2) complexes via a two-component regulatory system upon light or redox conditions, respectively (Vuillet *et al.*, 2007).

A very similar regulatory pathway for the synthesis of peripheral LH complexes has been demonstrated in *Bradyrhizobium* BTAil involving *Br*BphP3.BTAil, a bacteriophytochrome homologous to chromo-*Rp*BphP4 of *Rps. palustris* (Jaubert *et al.*, 2008). Altogether, these results give the molecular basis of the regulation of the photosynthetic apparatus synthesis by light and oxygen in two photosynthetic bacteria.

2. Cloning BphP in Expression Vector

Thanks to the autocatalytic binding of the chromophore on apo-bacteriophytochromes, functional recombinant BphPs are easily obtained by overexpression in *Escherichia coli*. A commonly used strategy is to clone both the *BphP* (apoprotein) and *hmuO* (heme oxygenase) genes in an expression vector. Generally, this results in a high expression of an active recombinant BphP, which can be easily purified by affinity chromatography and used for biophysical, biochemical, and structural studies.

The pBAD/HisB expression vector (Invitrogen) which contains the inducible *araB* promoter has been successfully used for all the different BphPs from *Bradyrhizobium* sp. and *Rps. palustris* strains that we studied and has permitted to obtain more than 10 mg/l of reconstituted holo-BphP. However, other vectors such as pQE12, pET21, pET28 have also been used with success by other groups and gave similar yield (Chung *et al.*, 2007; Karniol and Vierstra, 2003; Lamparter *et al.*, 2002). In the case of the pBAD/HisB vector, a 38-AA tag is located at the N-ter of the recombinant protein. This tag contains, in addition to the polyhistidine region, an Xpress epitope and an enterokinase recognition cleavage site that could be used for immunodetecting the protein or removing the tag after purification, respectively.

The entire *BphP* gene is amplified by polymerase chain reaction (PCR) using the Platinium *Taq* DNA polymerase high fidelity (Invitrogen) which possesses a proof reading activity. Appropriate restriction sites are added at the 5′ end of each primer to permit the cloning in frame of the *BphP* gene in the pBAD/HisB vector and the subsequent downstream insertion of the *hmuO* gene from *Bradyrhizobium* ORS278. Other *hmuO* genes from *Synechocystis* sp. PCC6803 (Gambetta and Lagarias, 2001), *Deinococcus radiodurans* (Bhoo *et al.*, 2001), or *Rps. palustris* (unpublished data) are also perfectly functional in *E. coli*. The *hmuO* gene from *Bradyrhizobium* ORS278 was amplified using the following pairs of primers (the synthetic ribosome binding site is in bold and possible restriction sites which could differ depending on the cloning strategy are underlined): *hmuO.XbaI.f 5′-GACTCTAGAATC***AGGAGG***TGCAAGACATGAGCACGCTG-CAGAATTTG-3′* and *hmuO.EcoRI.r 5′-GTCGTCGAATTCACTCAT-TGATCGGCAGCCATCCAG-3′*. The final construct is checked by sequencing and transferred into *E. coli* LMG194 (Invitrogen). A glycerol stock at 25% (v/v) of the recombinant *E. coli* strains is stored at −80 °C.

3. Overexpression and Purification of BphPs

3.1. Culture media and buffers

Luria-Bertani (LB) broth: 10 g tryptone, 5 g yeast extract, and 10 g NaCl in 1 l of H_2O.
The medium is supplemented with ampicillin 50 μg/ml.
Lysis buffer: 50 m*M* sodium phosphate buffer, pH 8.0, 300 m*M* NaCl, and lysosyme 2 mg/ml.
Buffer A (Wash buffer): 50 m*M* sodium phosphate buffer, pH 8.0, 300 m*M* NaCl.
Buffer B (Elution buffer): 50 m*M* sodium phosphate buffer, pH 8.0, 300 m*M* NaCl, and 250 m*M* imidazole.
Storage buffer: 50 m*M* Tris/HCl, pH 8.0, 150 m*M* NaCl, and glycerol 20% (v/v).

Note: The following protocol works well with all the different BphPs of *Rps. palustris* and *Bradyrhizobium* strains that we studied except for achromo. *Rp*BphP4 for which the L-arabinose induction step is carried out at 20 °C. It may be a useful starting point for designing purification strategies for bacteriophytochromes from other species.

3.2. Growth

An overnight culture of *E. coli* LMG194 containing the pBAD/His expression vector in which the *BphP* and *hmuO* genes were cloned is grown with shaking (200 rpm) at 37 °C in LB medium supplemented with 50 μg/ml ampicillin. The culture is diluted 1:100 into fresh LB medium with 50 μg/ml ampicillin (1-l in 2-l flask) and grown at 37 °C with shaking (200 rpm) until an OD_{600} of 0.6 is reached, at which time L-arabinose is added to a final concentration of 0.66 m*M*. Protein expression is allowed to proceed for 5 h, shaking at 37 °C before harvesting the cells by centrifugation (3200×*g*, 30 min, 4 °C). The cell pellet, which displays generally a green color, is washed with 40 ml of lysis buffer without lysosyme and finally stored at −80 °C until the protein purification is initiated.

3.3. Purification

For purification, the cell pellet obtained from 1 l culture is resuspended in 40 ml of lysis buffer containing lysosyme and kept on ice during 60 min. Cells are then disrupted by sonication on ice (Branson sonifier—1 min 30 s). The cell lysate containing the soluble holo-BphP is then clarified by centrifugation (6600×*g*, 30 min, 4 °C) to remove insoluble material and cell debris, followed by a filtration at 0.22 μm.

The supernatant is applied to a 5-ml HisTrap column (Amersham Histrap Hp 17-5248-02) previously equilibrated with 100 ml of buffer A. The column is first washed with 50 ml of buffer A and then with progressive amounts of buffer B (50 ml of buffer A 90% + buffer B 10%, 50 ml of buffer A 80% + buffer B 20%, 50 ml of buffer A 70% + buffer B 30%). The BphP is finally eluted with 50 ml of buffer B. Fractions displaying a green color are pooled and desalted using an Amicon ultra centrifugal filter (30 kDa) and the storage buffer. The purified protein adjusted to a final concentration of 1 mg/ml was then aliquoted and stored at −70 °C (frozen BphPs retain their activity for more than 1 year). Purified holo-BphP is typically more than 95% pure as estimated by SDS–PAGE analysis.

4. Autophosphorylation

More than two-thirds of bacteriophytochromes identified in purple bacteria contain as transmitter domain a histidine kinase motif (Giraud and Verméglio, 2008). This kinase activity can be detected by incubating the bacteriophytochrome with [γ-^{32}P]ATP and $MgCl_2$ under appropriate light or redox conditions and measuring by autoradiography or phosphorimaging the level of radioactive phosphate incorporation on SDS–PAGE gels.

To perfectly control light and temperature conditions, we modified a thermostat apparatus (Dry bath—Fisher scientific). Light emitting diodes (LEDs) were placed at the bottom of the aluminum rack receiving Eppendorf tubes (see Fig. 9.1). For this purpose, the bottom of the rack was drilled, and LEDs were glued. This system has the advantage of allowing triggering the enzymatic reaction by adding the substrate from the top of the tube while maintaining a constant illumination from the bottom. LEDs are chosen to excite preferentially one form or the other of a given BphP. Determination of the two spectral forms of a BphP is described in Section 9.

The following protocol is an adaptation of the method described by Yeh *et al.* (1997) who first demonstrated that BphPs can act as light-sensor kinases. It was found appropriate for *Rp*BphP2, *Rp*BphP3, chromo and achromo. *Rp*BphP4 from *Rps. palustris* and *Br*BphP3 from *Bradyrhizobium* BTAi1 strain. It could be used as starting method for any BphP containing an histidine kinase domain but it certainly would be necessary to optimize the [$MgCl_2$], [KCl], pH, light, and redox conditions as well as the reaction duration.

1. *Prepare the reaction buffer*: 50 mM Tris/HCl, pH 7.6; 4 mM $MgCl_2$; 100 mM KCl; 0.4 mM EDTA; 2 mM DTT; 0.04 mM ATP (not radioactive). For 100 μl of mix add 2.5 μl of [γ-^{32}P]ATP (10 mCi/ml).

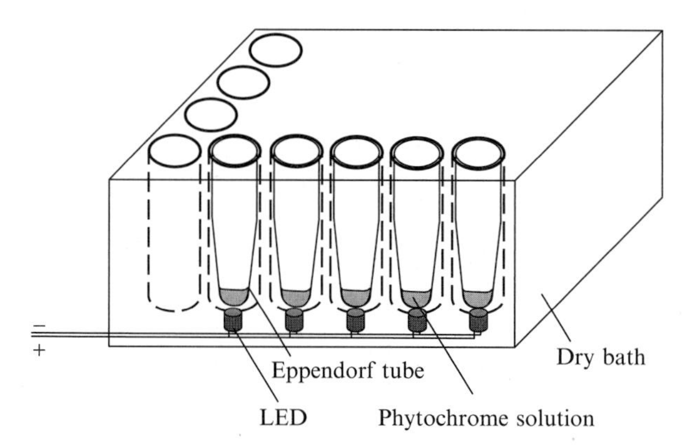

Figure 9.1 Schematic representation of the setup designed to measure autophosphorylation under controlled light and temperature conditions. Light emitting diodes (LEDs) were placed at the bottom of the thermostated aluminum rack of dry bath. This rack receives Eppendorf tubes, which could be illuminated by LEDs from the bottom. This system allows addition of the substrates from the top of the Eppendorf tube during or after a constant illumination to trigger the enzymatic reaction.

2. Distribute 10 μl of the BphP (1 mg/ml) in different Eppendorf tubes (one tube per measurement).
3. Incubate at 37 °C the tubes and irradiate them with the appropriate wavelength for 15 min. A dark cover is put over the tubes to protect the BphP from the room light (it is important to carry out all handling under dim light).
4. Start the first phosphosphorylation reaction by adding 10 μl of reaction buffer. The other reactions should be started at 30 s intervals.
5. Stop each reaction after 1–30 min (depending on the BphP) by addition of 20 μl of sample buffer (62.5 m*M* Tris/HCl, pH 6.8; 25% (v/v) glycerol; 2% (w/v) SDS; 0.01% (w/v) bromophenol blue, 5% (v/v) β-mercaptoethanol) and heating for 30 s at 55 °C.
6. The samples are then submitted to SDS–PAGE.
7. After migration, wash the gels with (40% MeOH, 7% acetic acid) during 15 min followed by two washes of 10 min with (15% MeOH).
8. Gels are dried and subjected to phosphorimage analysis or autoradiography with X-ray film.

Note: Achromo.*Rp*BphP4 acts as a redox-sensitive kinase. To study the effect of redox conditions on its kinase activity, achromo.*Rp*BpBPhP4 is first either oxidized with 1 m*M* ferricyanide or reduced with 1 m*M* DTT during 30 min. The other steps of the protocol remain unchanged except that DTT is omitted in the reaction buffer as well as β-mercaptoethanol in the sample buffer. In the cases of *Rp*BphP2 and *Rp*BphP3 autophosphorylation occurred in the dark.

5. Phosphotransfer

In some cases, the signaling pathway involves a phosphotransfer between the BphP and a RR whose gene is located in the vicinity of the *BphP*. This is the case of *Rp*BphP2, *Rp*BphP3, chromo and achromo.*Rp*. BphP4 of *Rps. palustris* strains and *Br*BphP3 from *Bradyrhizobium* BTAi1 strain (Giraud *et al.*, 2005; Jaubert *et al.*, 2008; Vuillet *et al.*, 2007). To demonstrate this, we used the following protocol:

1. The RR protein is first purified using a His tagging strategy after cloning in an expression vector. The same protocol previously described for the purification of BphPs could be used.
2. Follow steps 1–4 of the autophosphorylation protocol to permit the complete phosphorylation of the BphP (1 h is generally more than sufficient).
3. Add an equimolar concentration of RR protein.
4. Stop the reaction at 1–10 min by adding the sample buffer. The phosphotransfer can be extremely rapid and completed in less than a few seconds.
5. Proceed as described for steps 6–8 of the autophosphorylation protocol.

6. Gel Mobility Shift Assay and DNase I Footprint

In the particular case of chromo and achromo.*Rp*BphP4, the cognate RR (Rpa1489) corresponds to a transcriptional factor. The position of *RpBphP4/rpa1489* genes in the vicinity of a *pucBA.e* operon encoding the polypeptides of a light-harvesting complex was a strong indication suggesting that Rpa1489 might bind to the *pucBA.e* promoter. To demonstrate this, we used two complementary approaches: (i) a gel mobility shift assay to demonstrate the binding of Rpa1489 on the *pucBA.e* promoter and to study the effect of its phosphorylation by *RpBphP4* on its DNA binding affinity and (ii) a DNase I footprint analysis to identify precisely the DNA binding site of Rpa1489. In both cases, we used basic protocols as described in Current Protocols in Molecular Biology (Brenowitz *et al.*, 1997; Buratowski and Chodosh, 1997) that we adapted for the study of Rpa1489.

6.1. Gel mobility shift assay

The protocol can be divided into three steps: (1) preparation of a radioactively labeled probe corresponding to the promoter region of *pucBA.e*, (2) binding reaction, in which a variable amount of phosphorylated and unphosphorylated forms of Rpa1489 is bound to the DNA probe, and (3) electrophoresis of protein–DNA complexes through nondenaturating gel and analysis of the DNA binding affinity by phosphorimaging.

1. *Preparation of a radioactively labeled probe.* A 277-bp DNA region containing the *pucBA.e* promoter region of *Rps. palustris* CGA009 is obtained by PCR amplification using the primers: 5′-GTGATCGGGTCGAAGTCC-GAAAC-3′ and 5′-CAACGACGAGTCGCCTCGCGTTG-3′. For selective labeling of DNA strands, one of the primers is 5′-^{32}P-end-labeled prior to amplification. For this purpose, we use T_4 polynucleotide kinase (T_4 PNK) and the following protocol:
   (i) Dilute the primer at 10 μM in H_2O.
   (ii) Denature the primer by heating for 5 min at 95 °C and keep for 2 min in ice/EtOH.
   (iii) For the kinase reaction add 10 μl of primer, 23 μl H_2O, 5 μl T_4 PNK buffer, 2 μl T_4 PNK, 10 μl [γ-^{32}P]ATP (10 mCi/ml).
   (iv) Incubate for 60 min at 37 °C.
   (v) Stop the kinase reaction by heating for 10 min at 80 °C.
   (vi) Eliminate unincorporated [γ-^{32}P]ATP by using a spin column containing Sephadex G-50 (Amersham, ProbeQuant G-50 Microcolumns).
   (vii) The labeled primer in the aqueous phase is then ethanol-precipitated and washed with 70% ethanol.
   (viii) Dry the pellet (Vac dry) and resuspend in 10 μl H_2O. The labeled primer can then be used for the PCR reaction. After amplification, the PCR product is purified using a QIAquick PCR purification kit (QIAgen) and quantified by measurement of solution absorbance at 260 nm.
2. *DNA binding reaction.* Phosphorylated and unphosphorylated forms of Rpa1489 are obtained using the protocol previously described by incubating Rpa1489 with *Rp*BphP4 in the presence or absence of ATP, respectively. Various concentrations of both Rpa1489 forms are then added to 20 μl of reaction buffer composed of 5 fmol of ^{32}P-labeled DNA probe, 1 μg of polydIdC as competitor, 50 mM Tris–HCl (pH 8.0), 1 mM DTT, 50 mM potassium acetate, 5 μg of bovine serum albumin, glycerol 10%. The reaction mixtures are incubated at room temperature for 30 min.
3. *Electrophoresis of protein–DNA complexes.* The samples are then laid on a nondenaturing 5% Tris–glycine–EDTA-buffered polyacrylamide gel and are submitted to electrophoresis at 4 °C for 4 h at 70 V. After migration, the gels are dried and subjected to phosphorimage analysis to quantify the percentage of DNA bound in function of the concentration of phosphorylated and unphosphorylated forms of Rpa1489.

6.2. DNase I footprint analysis

The same DNA labeled probe as previously described is used. A 10-μl binding reaction mixture is first prepared containing 1 μl of DNA (100 fmol), 5 μl of H_2O and 4 μl of footprint binding buffer composed of 125 m*M* Tris–HCl (pH 8.0), 200 m*M* of NaCl, 25 m*M* of $MgCl_2$, 10 m*M* of $CaCl_2$, and 125 μg/ml bovine serum albumin. The reaction mixture is then added to a 10-μl solution containing 1 μ*M* of Rpa1489 diluted in a buffer composed of 50 m*M* Tris–HCl (pH 8.0), 150 m*M* NaCl, 1 m*M* EDTA, and 30% glycerol. After incubating the binding reaction for 30 min at room temperature, DNase I (3×10^{-4} U; Sigma) is added to the mixture and the digestion allowed to proceed for 7 min. The reaction is then stopped by addition of 180 μl of DNase I stop solution (0.6 *M* ammonium acetate, 0.5 *M* EDTA, and 0.0625 μg/ml yeast tRNA) followed by the addition of 200 μl of phenol: chloroform. The samples are vortexed and centrifuged for 5 min at 25 °C. DNA in the aqueous phase is ethanol-precipitated and washed with 70% ethanol. The pellet is then resuspended in 3 μl gel loading buffer (0.05% bromophenol blue, 0.05% xylene cyanol, EDTA 20 m*M* diluted in deionized formamide) heated for 5 min at 95 °C and subjected to electrophoresis on a 6% urea-denaturing polyacrylamide gel. A modified Maxam and Gilbert G + A chemical sequencing reaction is done at the same time for determining the location of DNase I protection (Maxam and Gilbert, 1980).

7. Gene Disruption

To determine the role of the various *BphPs* identified in *Rps. palustris* and *Bradyrhizobium* strains, we constructed isogenic mutants of each *BphP* and analyzed their photosynthetic phenotype in comparison with the wild-type (WT) strains. In all cases, these mutants were obtained by allelic exchange between a modified gene and the WT chromosomal gene. To obtain a modified *BphP* gene, we used the following general strategy: (i) amplification of a DNA region containing the *BphP* gene, (ii) cloning of the PCR product in a classical vector as pGEM-T, or pUC18, (iii) deletion of a part of the *BphP* gene by enzymatic restriction and insertion in place of the pKOK5 cassette containing a kanamycin resistance gene and a *lacZ* reporter gene (Kokotek and Lotz, 1989), and (iv) transfer of the construction in the pJQ200 suicide vector (Quandt and Hynes, 1993). The strategy is designed so that the flanking regions of the cassette which correspond to regions of the *BphP* gene are at least of 400-bp in order to promote the recombination event. The use of the pJQ200 suicide plasmid which does not replicate in *Rps. palustris* or *Bradyrhzobius* strains permits to integrate the modified *BphP* gene into the *BphP* chromosomal gene because of the homology between one of the flanking regions of the insert and the genomic DNA. Clones in which double events of recombination occurred are selected by plating on sucrose-containing medium. Cells retaining the vector backbone are lost since the *sacB* gene on the pJQ200 backbone converts sucrose to toxic levansucrose. Sucrose-resistant colonies are then screened by PCR to check that the double-crossover event has occurred.

Protocol for obtaining Rps. palustris mutants:

1. Transfer by electroporation of the pJQ200 plasmid containing the modified *BphP* gene into *E. coli* S17-1 λpir to permit its mobilization by bipartite conjugation in *Rps. palustris* strains.
2. Grow *E. coli* S17-1 containing the pJQ200 plasmid overnight at 37 °C in LB medium supplemented with kanamycin (50 μg/ml). Grow *Rps. palustris* strains for 3 days at 30 °C in Hutner medium (Clayton, 1963).
3. Harvest 2.5 ml of *E. coli* S17-1 and 10 ml of *Rps. palustris* cells by centrifugation (3200×*g*, 30 min, 4 °C).
4. Wash the pellets twice with NaCl 9 g/l.
5. Resuspend each pellet by 200 μl of NaCl 9 g/l.
6. Gently mix the two pellets and transfer on a Hutner plate. Incubate 2 days at 34 °C.
7. Resuspend the mating in 1 ml of Hutner medium, make serial dilution until 10^{-4} into NaCl 9 g/l, and spread 100 μl of each dilution onto Hutner plates containing kanamycin (50 μg/ml) and carbenicillin (50 μg/ml). The *Rps. palustris* strains are resistant to carbenicillin but not to kanamycin. The addition of carbenicillin to the medium inhibits the growth of *E. coli* cells. On the other hand, the addition of kanamycin permits to select the *Rps. palustris* recombinant cells in which the pJQ200 vector has been integrated in the chromosome.
8. Incubate plates at 30 °C for 1 week.
9. Colonies are then screened by PCR to check that the first event of recombination occurred.
10. Growth of one recombinant strain for 3 days at 30 °C in Hutner medium supplemented with kanamycin (50 μg/ml) and carbenicillin (50 μg/ml).
11. Make serial dilution until 10^{-4} into NaCl 9 g/l and spread 100 μl of each dilution onto Hutner plates containing sucrose (7%), kanamycin (50 μg/ml), and carbenicillin (50 μg/ml).
12. Incubate plates at 30 °C during 1 week.
13. Colonies are then screened by PCR to check that the second event of recombination occurred.
14. Conserved the mutant strain in glycerol (25%, v/v) at −80 °C.

Note: in the case of the *Bradyrhizobium* strains, the protocol remains the same except for the culture medium which is replaced by YM modified medium (Giraud *et al.*, 2000) supplemented with kanamycin (50 μg/ml) and nalidixic acid (35 μg/ml), and the temperature of the culture, which is 37 °C.

8. Analysis of the Photosynthetic Phenotypes of the BphP Mutants

To confirm *in vivo* the various roles of bacteriophytochromes in photosystem synthesis, WT and deletion mutants of genes of interest are grown under specific light and (or) O_2 tension. WT and mutants of *Bradyrhizobium* sp. and *Rps. palustris* were grown in a modified YM medium and in Hutner medium, respectively (Clayton, 1963; Giraud *et al.*, 2000). Appropriate antibiotics were added for growing the mutants. Bacteria were grown either in liquid medium or on Petri dishes at 37 °C for *Bradyrhizobium* and 30 °C for *Rps. palustris.*

8.1. Growth on Petri dishes

Bradyrhizobium or *Rps. palustris* cells are homogeneously inoculated on the surface of a Petri dish (usually a square 12 × 12 cm dish filled with 80 ml of the appropriate growth medium incorporated in an agar gel and sealed with double adhesive tape). Several areas of the Petri dish are subjected to continuous illumination provided by a series of LEDs of different peak wavelengths. Each LED illuminated a 3.5-cm^2 area. The wavelength peaks of the LEDs are chosen in order to span the two spectral forms of the studied bacteriophytochromes, typically between 550 and 800 nm. To obtain a correct action spectrum (clearly only of qualitative significance), irradiance has to be adjusted so that it remains below saturation even at the most efficient wavelengths. Each LED is included in the well of a Lingro-tissue culture plate (24-well). A plastic pipe, coated with silver paint, and whose extremity is covered with a piece of tracing paper, is inserted in each well. This directs and homogenizes the light. All other parts of this setup are darkened with black paint. This illumination apparatus is placed over the Petri dish and the whole system is put into a thermostated dark room. Bacteria are grown under these illumination conditions for 3 or 7 days for *Rps. palustris* and *Bradyrhizobium* respectively.

The effect of the illumination light color during growth can then be tested by several means.

1. The fluorescence from the bacteriochlorophyll pigments bound to the LH membrane complexes was used as a specific marker of the presence of the photosynthetic apparatus. The relative amounts of photosynthetic apparatus in the different illuminated areas is quantified by imaging this fluorescence ($\lambda > 800$ nm) excited by blue-green illumination over the Petri dish, using a cooled charge-coupled device (CCD) camera. The nonilluminated area of the Petri dish serves as a control.
2. The absorption spectra of intact cells could also be recorded for each illuminated area directly on the Petri dish, using a Cary

50 spectrophotometer, and compared with the nonilluminated area. Alternatively the bacteria grown in a given area could be collected using a cotton swab and resuspended in few milliliters of fresh medium before recording absorption or fluorescence spectra.

3. Insertion of a *lacZ* reporter gene in a given gene could also be used to test the effect of light on its expression. The cells are collected in the various illuminated regions and their corresponding β-galactosidase activity measured.

8.2. Illumination of round Petri dish

Some analyses require a large amount of cells. In this case, the cells are homogeneously inoculated onto a round Petri dish (diameter 9 cm). This dish is illuminated from the top by LEDs. One can use LEDs emitting at different wavelengths to observe synergetic or opposite light effects. These LEDs are fixed on the cover of another Petri dish placed on the top of a plastic pipe whose diameter is slightly larger than the diameter of the Petri dishes. This pipe is coated with a reflecting paint to funnel the light and its extremity covered by a piece of tracing paper to homogenize the illumination.

8.3. Growth in liquid medium

It is also interesting to determine the effect of light on bacterial growth and on the composition of the photosynthetic system as a function of oxygen tension. In this case, bacteria are grown in Erlenmeyer flasks of 250 ml. Illumination is provided by LEDs placed at the bottom of the flask. Pure N_2 and air were mixed using mass-flow controllers (Brooks) in order to obtain the appropriate O_2 tension ranging from 1% to 21%. The gas mixtures were flushed from the top in the gas phase of the flasks. To obtain a good equilibrium between the gas and liquid phases, we used baffled Erlenmeyer flasks. They contain 50 ml of growth medium and are shaken at 140 rpm. A side arm of the flask allows the removal of sterile aliquots through a septum during the growth period.

Nowadays, LEDs are made by a large number of manufacturers. Emission wavelengths can be found from 370 to 950 nm, the bandwidth is usually 20–30 nm and intensities up to 100 $\mu E/m^2/s$ can be obtained.

9. Photochemical Measurements

9.1. Statement of the problem

Let us assume that a phytochrome is initially present entirely in the P_r form. When submitted to a red illumination, for example at a wavelength close to the absorption peak of P_r, its absorbance decreases as some P_{fr} form is produced and, conversely the absorbance increases in the far-red region where the P_{fr} form absorbs. However, in general the P_{fr} form also absorbs in the red region, so that as soon as some P_{fr} is present, some photoconversion of P_{fr} to P_r will also occur. The system thus evolves until a steady state[1] is reached. The time course of this process is exponential; its rate and the composition of the steady state (the fractions of P_r and P_{fr}) depend on the absorption coefficients of the two forms and on their photochemical conversion quantum yields. However, unless there is a spectral region where only the P_r form absorbs, one cannot achieve a full photoconversion to the P_{fr} form, so that the absorption spectrum of this form cannot be directly measured. The accessible experimental information thus combines several unknown parameters in an intricate way: absorption spectrum of one of the two forms (or of both) and the quantum yields of both photoconversion reactions. We discuss below how this can be disentangled.

The first step aims at determining the absorption spectra of the two forms and the ratio of the quantum yields. This can be obtained from the absorption spectra of the steady-state mixture under two or three (depending on the number of unknowns) different illumination wavelengths. The absolute yields can then be determined from an additional kinetic measurement.

To our knowledge, in previous investigations of this kind the actinic illumination and the acquisition of absorption spectra were separate in time (and often in space), simply because classical spectrophotometers cannot put up with simultaneous measuring and actinic beams. This is not too much of a problem when using phytochromes with slow dark reversion rates. However, some bacteriophytochromes (e.g., *Br*BphP3.ORS278; Jaubert *et al.*, 2007) present reversion kinetics in the seconds time range and the traditional methodology would be at great difficulty in such cases. Our experimental work has benefited from the use of spectrophotometers that do allow an actinic illumination of the sample without perturbing the absorption measurement. These setups rest on the use of discrete short monochromatic flashes for the measuring beam. Due to their short duration (a few microseconds), these flashes can have a high-peak intensity while their energy is low enough to induce negligible photochemical effects on the system. On the other hand, the continuous actinic illumination does not perturb the detection, both because its instantaneous intensity is low compared with that of the flashes used as the measuring beam and because of appropriate electronic filtering. This principle has been implemented for long in the Joliot-type spectrophotometers (Joliot and Joliot, 1984; Joliot *et al.*, 1980), such as the laboratory-built machine we have in Cadarache.

[1] The steady state reached under illumination is often called "equilibrium." We prefer to restrict this term to the equilibrium occurring in the dark, whereas the situation obtained under illumination is indeed a steady state, where a metastable state is maintained at the expense of a constant flux of light to heat.

We have also been using a commercial spectrophotometer (Varian's Cary-50) which is also based on the use of a flashing detection beam.

9.2. Derivation of the unknown parameters from steady-state spectra

A general scheme for the interconversion of two phytochrome forms, A and B, is shown below:

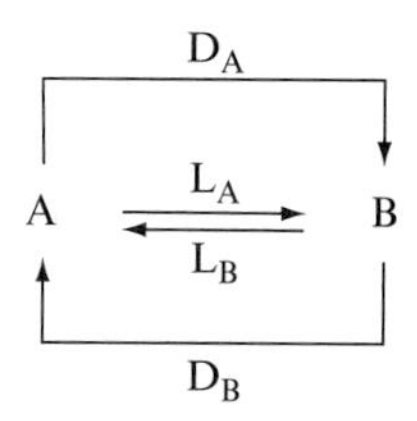

We denote D_A and D_B as the rate constants for the dark equilibrium between the two forms and L_A and L_B are the additional photochemical rate constants pertaining to some illumination conditions. This two-component scheme assumes that the more or less short-lived intermediates of the photoconversion process do not occur at a significant concentration, which implies that the illumination intensity is not too high (Kelly and Lagarias, 1985). For many (bacterio-) phytochromes, there is only one form (P_r) stable in the dark (i.e., $D_A \ll D_B$). Such was not the case, however, for the bacteriophytochrome *Br*BphP3.ORS278 from *Bradyrhizobium* described below. In this protein, the predominant form in the dark is P_o (orange form, peaking at 613 nm), but the P_r (red form, peaking at 672 nm) is also present, especially at low temperature. Another particularity of this phytochrome is that the two spectral forms overlap almost everywhere, so that it is not possible to observe significant absorption changes that are specific to only one form. These two features render inapplicable the approach developed by Butler and coworkers (Butler, 1972; Butler *et al.*, 1964; see also Johns, 1969) for computing the unknown spectra and estimating the photoconversion quantum yields. As further discussed below, in a "classical" phytochrome, the number of unknowns is reduced because the absorption spectrum of one the pure form is obtained directly, either after prolonged dark-adaptation or after a far-red preillumination in the region where P_r does not absorb; furthermore, the photochemical rate of the $P_{fr} \rightarrow P_r$ conversion is also directly obtained from the kinetics of P_{fr} destruction upon a far-red illumination. We describe below a general treatment that can handle the case of *Br*BphP3.ORS278, requiring three illumination wavelengths, or the simpler case where the spectrum of one form is directly accessible, requiring only two illumination wavelengths.

The steady-state fractions of the two forms according to the above kinetic scheme are

$$[\mathrm{A}] = \frac{D_\mathrm{B} + L_\mathrm{B}}{D_\mathrm{A} + L_\mathrm{A} + D_\mathrm{B} + L_\mathrm{B}}; \qquad [\mathrm{B}] = \frac{D_\mathrm{A} + L_\mathrm{A}}{D_\mathrm{A} + L_\mathrm{A} + D_\mathrm{B} + L_\mathrm{B}} \tag{9.1}$$

We shall assume that the illumination intensity is saturating, that is high enough to ensure that the *D*s are negligible with respect to the *L*s. We were able to satisfy this requirement by cooling the sample close to 0 °C, checking that a 50% attenuation of the actinic light caused little change. If saturation is difficult to achieve, one may attempt to extrapolate the saturated spectrum from spectra measured at different intensities. Equation (9.1) predicts a linear double reciprocal plot, for example, $1/([\mathrm{B}] - [\mathrm{B}]_\mathrm{dark})$ versus $1/I$ (with $[\mathrm{B}]_\mathrm{dark} = D_\mathrm{A}/D_\mathrm{B}$); the intercept can then be used to estimate the saturated level.

Under saturating illumination, one has

$$[\mathrm{A}] = \frac{L_\mathrm{B}}{L_\mathrm{A} + L_\mathrm{B}}; \qquad [\mathrm{B}] = \frac{L_\mathrm{A}}{L_\mathrm{A} + L_\mathrm{B}} \tag{9.2}$$

We denote the (unknown) absorption spectra of A and B as $\mathrm{S_A}(\lambda)$ and $\mathrm{S_B}(\lambda)$. Considering that in practice, one has a finite set of N wavelengths (λ_1, ..., λ_N), the spectra can be thought as vectors, $\mathbf{S}_\mathrm{A}$ and $\mathbf{S}_\mathrm{B}$, the coordinates of which are given by $\mathrm{S_A}(\lambda_i)$ and $\mathrm{S_B}(\lambda_i)$. Under a certain spectral composition of the saturating illumination (denoted as "type n," with $n = 1, \ldots, 3$), one can measure the absorption spectrum:

$$\mathbf{S}_n = [\mathrm{A}]_n \mathbf{S}_\mathrm{A} + [\mathrm{B}]_n \mathbf{S}_\mathrm{B} = \mathbf{S}_\mathrm{A} + [\mathrm{B}]_n(\mathbf{S}_\mathrm{B} - \mathbf{S}_\mathrm{A}) \tag{9.3}$$

where $[\mathrm{A}]_n$ and $[\mathrm{B}]_n$ are the steady-state fractions (we assume for simplicity that the total concentration of the protein is 1) of the two forms under illumination n. The spectra of the $\mathbf{S}_n$ family are linear combinations of the base vectors $\mathbf{S}_\mathrm{A}$ and $\mathbf{S}_\mathrm{B}$ and, conversely, one may express the unknown $\mathbf{S}_\mathrm{A}$ and $\mathbf{S}_\mathrm{B}$ as combinations of two experimental spectra (e.g., $\mathbf{S}_1$ and $\mathbf{S}_2$), that is

$$\mathbf{S}_\mathrm{A} = \mathbf{S}_1 + X_\mathrm{A}(\mathbf{S}_2 - \mathbf{S}_1); \qquad \mathbf{S}_\mathrm{B} = \mathbf{S}_1 + X_\mathrm{B}(\mathbf{S}_2 - \mathbf{S}_1) \tag{9.4}$$

where the scalar coefficients X_A and X_B are unknown.

If the illumination is monochromatic at $\lambda = \lambda_n$, the photochemical rates are the products of the rate of light absorption by the species considered and the photoconversion quantum yield ρ. Thus,

$$L_\mathrm{A} = 2.30\rho_\mathrm{A}\mathrm{S_A}(\lambda_n)I_n; \qquad L_\mathrm{B} = 2.30\rho_\mathrm{B}\mathrm{S_B}(\lambda_n)I_n \tag{9.5}$$

Consequently,

$$[\mathrm{B}]_n = \frac{\rho_\mathrm{A}\mathrm{S_A}(\lambda_n)}{\rho_\mathrm{A}\mathrm{S_A}(\lambda_n) + \rho_\mathrm{B}\mathrm{S_B}(\lambda_n)} = \frac{1}{1 + R\frac{\mathrm{S_B}(\lambda_n)}{\mathrm{S_A}(\lambda_n)}} \tag{9.6}$$

where $R \equiv \rho_B/\rho_A$. In our experiments with *Br*BphP3.ORS278, the illumination beam was monochromatic only for one of the three illumination conditions (using a He–Ne laser). In the two other cases, we used LEDs with a significant bandwidth (due to the rapid dark back-reaction, fairly high intensities are required for saturation, which could not be achieved with a lamp and interference filters with a narrow transmission band). The emission spectra of these various sources were measured and are denoted as $\mathbf{Z}_1$, $\mathbf{Z}_2$, and $\mathbf{Z}_3$. The photochemical rates now imply the overlap integrals of the emission and absorption spectra, which is the dot (scalar) product of the two associated vectors. Thus, Eq. (9.6) becomes

$$[\mathrm{B}]_n = \frac{1}{1 + R\frac{\mathbf{S}_B \cdot \mathbf{Z}_n}{\mathbf{S}_A \cdot \mathbf{Z}_n}} \tag{9.7}$$

Inserting Eqs. (9.4) and (9.7) into (9.3), one obtains

$$\mathbf{S}_n = \mathbf{S}_1 + (\mathbf{S}_2 - \mathbf{S}_1)\left[X_A + \frac{X_B - X_A}{1 + R\frac{\mathbf{S}_1\cdot\mathbf{Z}_n + X_B(\mathbf{S}_2\cdot\mathbf{Z}_n - \mathbf{S}_1\cdot\mathbf{Z}_n)}{\mathbf{S}_1\cdot\mathbf{Z}_n + X_A(\mathbf{S}_2\cdot\mathbf{Z}_n - \mathbf{S}_1\cdot\mathbf{Z}_n)}}\right] \tag{9.8}$$

In this equation one has three unknown scalars (R, X_A, and X_B) while the other quantities are experimentally determined. This equation must apply to any $\mathbf{S}_n$. In the case $\mathbf{S}_n = \mathbf{S}_1$, it reduces to

$$X_A + \frac{(X_B - X_A)}{1 + R\frac{\mathbf{S}_1\cdot\mathbf{Z}_1 + X_B(\mathbf{S}_2\cdot\mathbf{Z}_1 - \mathbf{S}_1\cdot\mathbf{Z}_1)}{\mathbf{S}_1\cdot\mathbf{Z}_1 + X_A(\mathbf{S}_2\cdot\mathbf{Z}_1 - \mathbf{S}_1\cdot\mathbf{Z}_1)}} = 0 \tag{9.9}$$

Similarly, when $\mathbf{S}_n = \mathbf{S}_2$, one has

$$X_A + \frac{(X_B - X_A)}{1 + R\frac{\mathbf{S}_1\cdot\mathbf{Z}_2 + X_B(\mathbf{S}_2\cdot\mathbf{Z}_2 - \mathbf{S}_1\cdot\mathbf{Z}_1)}{\mathbf{S}_1\cdot\mathbf{Z}_2 + X_A(\mathbf{S}_2\cdot\mathbf{Z}_2 - \mathbf{S}_1\cdot\mathbf{Z}_1)}} = 1 \tag{9.10}$$

One can use Eq. (9.9) to express R as a function of X_A and X_B:

$$R = \frac{X_B[X_A(\mathbf{S}_2\cdot\mathbf{Z}_1 - \mathbf{S}_1\cdot\mathbf{Z}_1) + \mathbf{S}_1\cdot\mathbf{Z}_1]}{X_B[X_B(\mathbf{S}_1\cdot\mathbf{Z}_1 - \mathbf{S}_2\cdot\mathbf{Z}_1) - \mathbf{S}_1\cdot\mathbf{Z}_1]} \tag{9.11}$$

This expression is inserted into Eq. (9.10), which in turn can be solved for, for example, X_B, yielding a function of the sole unknown X_A. One obtains

$$X_B = \frac{\mathbf{S}_1\cdot\mathbf{Z}_1[X_A(\mathbf{S}_1\cdot\mathbf{Z}_2 - \mathbf{S}_2\cdot\mathbf{Z}_2) - \mathbf{S}_1\cdot\mathbf{Z}_2]}{X_A[\mathbf{S}_1\cdot\mathbf{Z}_1(\mathbf{S}_1\cdot\mathbf{Z}_2 - \mathbf{S}_2\cdot\mathbf{Z}_2) + \mathbf{S}_2\cdot\mathbf{Z}_2(\mathbf{S}_2\cdot\mathbf{Z}_1 - \mathbf{S}_1\cdot\mathbf{Z}_1)] + \mathbf{S}_1\cdot\mathbf{Z}_1(\mathbf{S}_2\cdot\mathbf{Z}_2 - \mathbf{S}_1\cdot\mathbf{Z}_2)} \tag{9.12}$$

Now, if the spectrum of a pure form is known (e.g., $\mathbf{S}_A$), X_A can be directly obtained, the problem is then solved using Eqs. (9.11) and (9.12). This is the "classical" case, where the P_r spectrum can be measured in the dark-adapted state or following a far-red illumination in a region where only P_{fr} absorbs. To compute X_A, one obtains from Eq. (9.4):

$$X_A = \frac{S_A(\lambda_i) - S_1(\lambda_i)}{S_2(\lambda_i) - S_1(\lambda_i)} \quad (9.13)$$

Under the present assumptions, the right-hand member of Eq. (9.13) should give the same value at any λ_i. In practice, one obtains a more accurate value by averaging over an appropriate wavelength domain (i.e., where the amplitudes are significant and leaving out the vicinity of the isosbestic wavelength where the ratio is not defined).

When $\mathbf{S}_A$ is *not* known, a third independent equation is required for a complete solution, thus a third type of illumination $\mathbf{Z}_3$ under which the sample reaches a steady state with spectrum $\mathbf{S}_3$. In this case, Eq. (9.8) gives

$$X_A + \frac{(X_B - X_A)}{1 + R\frac{\mathbf{S}_1\cdot\mathbf{Z}_3 + X_B(\mathbf{S}_2\cdot\mathbf{Z}_3 - \mathbf{S}_1\cdot\mathbf{Z}_3)}{\mathbf{S}_1\cdot\mathbf{Z}_3 + X_A(\mathbf{S}_2\cdot\mathbf{Z}_3 - \mathbf{S}_1\cdot\mathbf{Z}_3)}} = \frac{S_3(\lambda_i) - S_1(\lambda_i)}{S_2(\lambda_i) - S_1(\lambda_i)} \equiv Y \quad (9.14)$$

Again, Eq. (9.14) applies to any λ_i chosen in the right-hand member, which should yield a constant value, denoted Y (and, again, averaging over an appropriate set of wavelengths will improve the accuracy). After performing the substitutions of the X_B and R expressions (Eqs. (9.11) and (9.12)), Eq. (9.14) can be rearranged as a ratio of second order polynomials in X_A, that is

$$\frac{N_2X_A^2 + N_1X_A}{D_2X_A^2 + D_1X_A + D_0} = Y \quad (9.15)$$

where

$$\begin{aligned}
N_1 &= \mathbf{S}_1\cdot\mathbf{Z}_1[(\mathbf{S}_1\cdot\mathbf{Z}_3)(\mathbf{S}_2\cdot\mathbf{Z}_1) - (\mathbf{S}_1\cdot\mathbf{Z}_1)(\mathbf{S}_2\cdot\mathbf{Z}_3)]\\
N_2 &= (\mathbf{S}_2\cdot\mathbf{Z}_2 - \mathbf{S}_1\cdot\mathbf{Z}_2)[(\mathbf{S}_1\cdot\mathbf{Z}_3)(\mathbf{S}_2\cdot\mathbf{Z}_1) - (\mathbf{S}_1\cdot\mathbf{Z}_1)(\mathbf{S}_2\cdot\mathbf{Z}_3)]\\
D_0 &= \mathbf{S}_1\cdot\mathbf{Z}_1[(\mathbf{S}_1\cdot\mathbf{Z}_3)(\mathbf{S}_2\cdot\mathbf{Z}_2) - (\mathbf{S}_1\cdot\mathbf{Z}_2)(\mathbf{S}_2\cdot\mathbf{Z}_3)]\\
D_1 &= (\mathbf{S}_1\cdot\mathbf{Z}_2)[(\mathbf{S}_1\cdot\mathbf{Z}_1)(\mathbf{S}_2\cdot\mathbf{Z}_3) + (\mathbf{S}_1\cdot\mathbf{Z}_3)(\mathbf{S}_2\cdot\mathbf{Z}_1) - (\mathbf{S}_2\cdot\mathbf{Z}_1)(\mathbf{S}_2\cdot\mathbf{Z}_3)]\\
&\quad +(\mathbf{S}_1\cdot\mathbf{Z}_3)(\mathbf{S}_2\cdot\mathbf{Z}_2)(\mathbf{S}_2\cdot\mathbf{Z}_1 - 2\mathbf{S}_1\cdot\mathbf{Z}_1)\\
D_2 &= [(\mathbf{S}_1\cdot\mathbf{Z}_2)(\mathbf{S}_2\cdot\mathbf{Z}_1) - (\mathbf{S}_1\cdot\mathbf{Z}_1)(\mathbf{S}_2\cdot\mathbf{Z}_2)](\mathbf{S}_2\cdot\mathbf{Z}_3 - \mathbf{S}_1\cdot\mathbf{Z}_3)
\end{aligned} \quad (9.16)$$

Equation (9.15) thus appears as a quadratic equation. Using the above notations, the solutions are

$$X_{A/B} = \frac{YD_1 - N_1 \pm \sqrt{(N_1 - YD_1)^2 - 4YD_0(YD_2 - N_2)}}{2(N_2 - YD_2)} \quad (9.17)$$

The spectra $\mathbf{S}_A$ and $\mathbf{S}_B$ are then determined using Eq. (9.4) and R is obtained from Eq. (9.11). This method was used in Jaubert *et al.* (2007), except that we were not aware that the problem could be solved analytically, so that the equation system was solved by a least square numerical procedure.

Figure 9.2 illustrates the application of the above procedure to the data obtained with BrBphP3.ORS278 (Jaubert *et al.*, 2007). In panel (A), the three steady-state spectra (dotted) are shown together with the corresponding irradiation spectra (solid black lines). The spectra computed for the P_o and P_r forms (i.e., $\mathbf{S}_A$ and $\mathbf{S}_B$) are featured in panel (B) as dashed lines. The gray lines in panel (A) show the reconstructed spectra $\mathbf{S}_n$ computed from Eqs. (9.3) and (9.7), using the parameters R, X_A, and X_B obtained from Eqs. (9.11), (9.12), and (9.17). The agreement with the data is quite good. The value of R was 3.2, implying a threefold larger efficiency for the P_r to P_o conversion compared with P_o to P_r. For comparison, in plant phytochromes, the P_r to P_{fr} transition is the most efficient, by a factor of about 2 (Kelly and Lagarias, 1985).

9.3. Absolute quantum yields

From the above procedure, one has obtained the ratio of the quantum yields $R = \rho_B/\rho_A$. The determination of the absolute values requires a kinetic measurement of the absorption change when the sample is submitted to an actinic illumination of known intensity (the photon fluence rate, in the terminology of Johns, 1969). The traditional way of carrying out such experiments consists in submitting the sample to incremental illumination periods and measuring the absorption change during "dark" intervals. The initial rate (Butler *et al.*, 1964) or the whole time course toward the steady state (Johns, 1969; Kelly and Lagarias, 1985) is then determined. As explained above, this method can only be applied to phytochromes with slow reversion kinetics, while the use of a spectrophotometer allowing the monitoring of absorption changes during the illumination is required otherwise.

We would like to outline an alternative method, in which the actinic beam is directly used for measuring the absorption change. This might seemed doomed to fail since the absorption detection requires a significant absorbance of the sample, whereas a homogeneous actinic illumination seems to be required for an exploitable measurement. Indeed, the setups described in the literature are arranged so that the sample is irradiated (e.g., at 90 °C, or see Butler *et al.* (1964) for a two-position cuvette) along a short

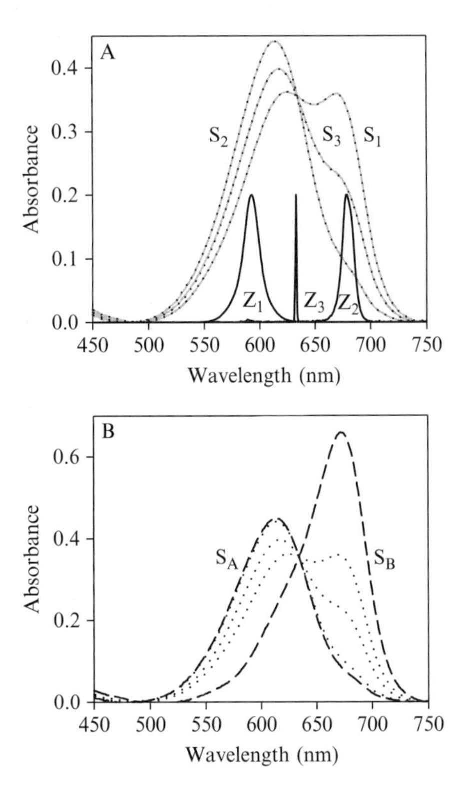

Figure 9.2 Characteristics of the two spectral forms of BrBphP3.ORS278. (A) The absorption spectra S_1, S_2, and S_3 (dotted) were measured at steady state under saturating illuminations of spectral compositions Z_1, Z_2, and Z_3 (solid lines), respectively (data from Jaubert *et al.*, 2007). The gray lines are the simulations of the S_1, S_2, and S_3 spectra, using Eqs. (9.3) and (9.6) with the pure forms spectra shown in panel (B) and the ratio of quantum yields $R = \rho_B/\rho_A \approx 3.2$. The measurement of the S_n spectra was done with a Cary-50 spectrophotometer: as explained in the text, this apparatus allows acquiring absorption spectra in the presence of continuous illumination. The actinic beam was delivered from the top of the cuvette. The Z spectra were digitized from manufacturer data (Z_1 and Z_2) or modeled as a narrow Gaussian around 632.8 nm in the case of the He–Ne laser (Z_3). The cuvette was thermostated at 5 °C, which slowed the dark reverse reaction and facilitated the attainment of saturation. (B) The spectra of the pure forms S_A (or P_o, for orange) and S_B (or P_r, for red) calculated as described in the text are shown (dashed lines), together with the spectra S_1, S_2, and S_3 (dots).

optical path, while the absorption is measured along a long optical path. On closer examination, however, it turns out that the initial rate of increase of the transmitted intensity through an absorbing sample does provide a simple way of measuring the absolute quantum yields.

We denote as $I(x)$ the beam intensity at optical depth x, and $I(L)$ the intensity transmitted through the whole optical path L. $C_A(x, t)$ and $C_B(x, t)$ are the local concentrations of the phytochrome forms and ε_A

and ε_B their extinction coefficients at the illumination wavelength (although this is not a requirement we assume a monochromatic beam for simplicity). The intensity change across a slab of width dx located at x is (omitting the t variable for brevity)

$$\mathrm{d}I(x) = -2.30[C_A(x)\varepsilon_A + C_B(x)\varepsilon_B]I(x)\,\mathrm{d}x \tag{9.18}$$

Thus,

$$\frac{\mathrm{d}\ln(I(x))}{\mathrm{d}x} = -2.30[C_A(x)\varepsilon_A + C_B(x)\varepsilon_B] \tag{9.19}$$

The time derivative of Eq. (9.19) is

$$\frac{\mathrm{d}}{\mathrm{d}t}\left(\frac{\mathrm{d}\ln(I(x))}{\mathrm{d}x}\right) = -2.30\left(\frac{\mathrm{d}C_A(x)}{\mathrm{d}t}\varepsilon_A + \frac{\mathrm{d}C_B(x)}{\mathrm{d}t}\varepsilon_B\right) = -2.30\frac{\mathrm{d}C_A(x)}{\mathrm{d}t}(\varepsilon_A - \varepsilon_B) \tag{9.20}$$

The rightmost expression takes into accounts that the rate of formation of B equals the rate of destruction of A. Denoting as ρ_A and ρ_B the photoconversion quantum yields, the rate of change of the concentrations is

$$\frac{\mathrm{d}C_A(x)}{\mathrm{d}t} = -\frac{\mathrm{d}C_B(x)}{\mathrm{d}t} = -2.30[\rho_A C_A(x)\varepsilon_A - \rho_B C_B(x)\varepsilon_B]I(x) \tag{9.21}$$

We assume that initially, the phytochrome is homogeneously distributed and entirely in the form A. Thus, at time 0, $C_A(x) = C$ (a constant at any x) and $C_B(x) = 0$. We are looking for a first order approximation of the initial value of d$I(L)/$dt. At time 0 and as long as the concentrations have not changed to a significant extent, Eq. (9.19) reduces to

$$\mathrm{d}\ln(I(x)) \approx -2.30C\varepsilon_A\,\mathrm{d}x \tag{9.22}$$

so that

$$I(x)_{t=0} = I(0)\exp(-2.30C\varepsilon_A x) \tag{9.23}$$

The initial transmitted intensity is thus

$$I(L)_{t=0} = I(0)\exp(-2.30C\varepsilon_A L) \tag{9.24}$$

Similarly, near time 0, Eq. (9.21) gives

$$\frac{\mathrm{d}C_A(x)}{\mathrm{d}t} \approx -2.30\rho_A C\varepsilon_A I(x) \tag{9.25}$$

Inserting Eq. (9.25) into Eq. (9.20), one obtains

$$\frac{d}{dt}\left(\frac{d\ln(I(x))}{dx}\right) = 2.30^2 \rho_A C\varepsilon_A(\varepsilon_A - \varepsilon_B)I(x) \tag{9.26}$$

We now integrate this expression with respect to x over the optical path. The left-hand expression gives

$$\begin{aligned}\int_0^L \frac{d}{dt}\left(\frac{d\ln(I(x))}{dx}\right)dx &= \frac{d}{dt}\left(\int_0^L d\ln(I(x))\right) \\ &= \frac{d}{dt}\left(\ln\frac{I(L)}{I(0)}\right) = \frac{d\ln(I(L))}{dt} \\ &= \frac{1}{I(L)}\frac{dI(L)}{dt}\end{aligned} \tag{9.27}$$

And for the right-hand member of Eq. (9.26) one obtains

$$2.30^2 \rho_A C\varepsilon_A(\varepsilon_A - \varepsilon_B)\int_0^L I(x)\,dx = 2.30^2 \rho_A C\varepsilon_A(\varepsilon_A - \varepsilon_B)I(0)\frac{1 - e^{-2.3C\varepsilon_A L}}{2.3C\varepsilon_A} \tag{9.28}$$

(where Eq. (9.23) was used). Equating expressions (9.27) and (9.28):

$$\frac{1}{I(L)}\frac{dI(L)}{dt} = 2.30\rho_A(\varepsilon_A - \varepsilon_B)I(0)(1 - e^{-2.3C\varepsilon_A L}) = 2.30\rho_A(\varepsilon_A - \varepsilon_B)(I(0) - I(L)) \tag{9.29}$$

where it is implied that one considers a time domain close to $t = 0$ so that $I(L)$ remains close to its initial value. Equation (9.29) is the expression that we sought for. Using the sample transmittance $T \equiv I(L)/I(0)$, one may also write:

$$\frac{dI(L)}{dt} = 2.30\rho_A(\varepsilon_A - \varepsilon_B)I(0)^2 T(1 - T) \tag{9.30}$$

These expressions require the knowledge of the absolute incident intensity, the sample transmittance and the differential extinction coefficient of the two forms. Then a measurement of the rate of increase of the transmitted intensity allows obtaining the absolute yield ρ_A. The maximum sensitivity is obtained for $T = 0.5$, thus a sample absorbance of 0.3. This method does not require an optically thin sample, or stirring. On the other hand, lateral homogeneity of the light beam is important, because of the intensity square dependence of the initial slope.

REFERENCES

Bhoo, S. H., Davis, S. J., Walker, J., Karniol, B., and Vierstra, R. D. (2001). Bacteriophytochromes are photochromic histidine kinases using a biliverdin chromophore. *Nature* **414,** 776–779.

Brenowitz, M., Senear, D. F., and Kingston, R. E. (1997). Dnase I footprint analysis of protein-DNA binding. *In* "Current Protocols in Molecular Biology," (F. Ausubel, R. Brent, R. Kingston, D. Moore, J. Seidman, J. Smith, and K. Struhl, eds.), pp. 12.4.1–12.4.16. Wiley, New York.

Buratowski, S., and Chodosh, L. A. (1997). Mobility shift DNA-binding assay using gel electrophoresis. *In* "Current Protocols in Molecular Biology," (F. Ausubel, R. Brent, R. Kingston, D. Moore, J. Seidman, J. Smith, and K. Struhl, eds.), pp. 12.2.1–12.2.11. Wiley, New York.

Butler, W. L. (1972). Photochemical properties of phytochrome *in vitro*. *In* "Phytochrome," (K. Mitrakos and W. Shropshire, eds.), pp. 185–192. Academic Press, London.

Butler, W. L., Hendricks, S. B., and Siegelman, H. W. (1964). Action spectra of phytochrome *in vitro*. *Photochem. Photobiol.* **3,** 521–528.

Chung, Y. H., Masuda, S., and Bauer, C. E. (2007). Purification and reconstitution of PYP-phytochrome with biliverdin and 4-hydroxycinnamic acid. *Methods Enzymol.* **422,** 184–189.

Clayton, R. K. (1963). Absorption spectra of photosynthetic bacteria and their chlorophylls. *In* "Bacterial Photosynthesis," (H. Gest, A. San Pietro, and L. P. Vernon, eds.), pp. 495–500. Antioch Press, Yellow Springs, OH.

Gambetta, G. A., and Lagarias, J. C. (2001). Genetic engineering of phytochrome biosynthesis in bacteria. *Proc. Natl. Acad. Sci. USA* **98,** 10566–10571.

Giraud, E., and Fleischman, D. (2004). Nitrogen-fixing symbiosis between photosynthetic bacteria and legumes. *Photosynth. Res.* **82,** 115–130.

Giraud, E., and Verméglio, A. (2008). Bacteriophytochromes in anoxygenic photosynthetic bacteria. *Photosynth. Res.* **97,** 141–153.

Giraud, E., Hannibal, L., Fardoux, J., Verméglio, A., and Dreyfus, B. (2000). Effect of *Bradyrhizobium* photosynthesis on stem nodulation of *Aeschynomene sensitiva*. *Proc. Natl. Acad. Sci. USA* **97,** 14795–14800.

Giraud, E., Fardoux, J., Fourrier, N., Hannibal, L., Genty, B., Bouyer, P., Dreyfus, B., and Verméglio, A. (2002). Bacteriophytochrome controls photosystem synthesis in anoxygenic bacteria. *Nature* **417,** 202–205.

Giraud, E., Zappa, S., Jaubert, M., Hannibal, L., Fardoux, J., Adriano, J.-M., Bouyer, P., Genty, B., Pignol, D., and Verméglio, A. (2004). Bacteriophytochrome and regulation of the synthesis of the photosynthetic apparatus in *Rhodopseudomonas palustris*: Pitfalls of using laboratory strains. *Photochem. Photobiol. Sci.* **3,** 587–591.

Giraud, E., Zappa, S., Vuillet, L., Adriano, J.-M., Hannibal, L., Fardoux, J., Berthomieu, C., Bouyer, P., Pignol, D., and Verméglio, A. (2005). A new type of bacteriophytochrome acts in tandem with a classical bacteriophytochrome to control the antennae synthesis in *Rhodopseudomonas palustris*. *J. Biol. Chem.* **280,** 32389–32397.

Hughes, J., Lamparter, T., Mittmann, F., Hartmann, E., Gärtner, W., Wilde, A., and Börner, T. (1997). A prokaryotic phytochrome. *Nature* **386,** 663.

Jaubert, M., Zappa, S., Fardoux, J., Adriano, J.-M., Hannibal, L., Elsen, S., Lavergne, J., Verméglio, A., Giraud, E., and Pignol, D. (2004). Light and redox control of photosynthesis gene expression in *Bradyrhizobium*: Dual roles of two PpsR. *J. Biol. Chem.* **279,** 44407–44416.

Jaubert, M., Lavergne, J., Fardoux, J., Hannibal, L., Vuillet, L., Adriano, J.-M., Bouyer, P., Pignol, D., Giraud, E., and Verméglio, A. (2007). A singular bacteriophytochrome acquired by lateral gene transfer. *J. Biol. Chem.* **282,** 7320–7328.

Jaubert, M., Vuillet, L., Hannibal, L., Adriano, J.-M., Fardoux, J., Bouyer, P., Bonaldi, K., Fleischman, D., Giraud, E., and Verméglio, A. (2008). Control of peripheral light-harvesting complex synthesis by a bacteriophytochrome in the aerobic photosynthetic bacterium *Bradyrhizobium* strain BTAi1. *J. Bacteriol.* **190,** 5824–5831.

Johns, H. E. (1969). Photochemical reactions in nucleic acids. *Methods Enzymol.* **16,** 253–316.

Joliot, P., and Joliot, A. (1984). Electron transfer between the two photosystems. I. Flash excitation under oxidizing conditions. *Biochim. Biophys. Acta* **765,** 210–218.

Joliot, P., Béal, D., and Frilley, B. (1980). Une nouvelle méthode spectrophotométrique destinée à l'étude des réactions photosynthétiques. *J. Chim. Phys.* **77,** 209–216.

Karniol, B., and Vierstra, R. D. (2003). The pair of bacteriophytochromes from *Agrobacterium tumefaciens* are histidine kinases with opposing photobiological properties. *Proc. Natl. Acad. Sci. USA* **100,** 2807–2812.

Karniol, B., Wagner, J. R., Walker, J. M., and Vierstra, R. D. (2005). Phylogenetic analysis of the phytochrome superfamily reveals distinct microbial subfamilies of photoreceptors. *Biochem. J.* **392,** 103–116.

Kelly, J. M., and Lagarias, J. C. (1985). Photochemistry of 124-Kilodalton *Avena* phytochrome under constant illumination *in vitro*. *Biochemistry* **24,** 6003–6010.

Kojadinovic, M., Laugraud, A., Vuillet, L., Fardoux, J., Hannibal, L., Adriano, J.-M., Bouyer, P., Giraud, E., and Verméglio, A. (2008). Dual role for a bacteriophytochrome in the bioenergetic control of *Rhodopsdeudomonas palustris*: Enhancement of photosystem synthesis and limitation of respiration. *Biochim. Biophys. Acta* **1777,** 163–172.

Kokotek, W., and Lotz, W. (1989). Construction of a *lacZ*-kanamycine-resistance cassette, useful for site-directed mutagenesis and as a promoter probe. *Gene* **84,** 467–471.

Lamparter, T., Michael, N., Mittmann, F., and Esteban, B. (2002). Phytochrome from *Agrobacterium tumefaciens* has unusual spectral properties and reveals an N-terminal chromophore attachment site. *Proc. Natl. Acad. Sci. USA* **99,** 11628–11633.

Larimer, F. W., Chain, P., Hauser, L., Lamerdin, J., Malfatti, S., Do, L., Land, M. L., Pelletier, D. A., Beatty, J. T., Lang, A. S., Tabita, F. R., Gibson, J. L., *et al.* (2004). Complete genome sequence of the metabolically versatile photosynthetic bacterium *Rhodopseudomonas palustris*. *Nat. Biotechnol.* **22,** 55–61.

Maxam, A. M., and Gilbert, W. (1980). Sequencing and labeled DNA with base-specific chemical cleavages. *Methods Enzymol.* **65,** 499–560.

Quandt, J., and Hynes, M. F. (1993). Versatile suicide vectors which allow direct selection for gene replacement in gram-negative bacteria. *Gene* **127,** 15–21.

Rockwell, N. C., Su, Y. S., and Lagarias, J. C. (2006). Phytochrome structure and signaling mechanisms. *Annu. Rev. Plant. Biol.* **57,** 837–858.

Vuillet, L., Kojadinovic, M., Zappa, S., Jaubert, M., Adriano, J. M., Fardoux, J., Hannibal, L., Pignol, D., Verméglio, A., and Giraud, E. (2007). Evolution of a bacteriophytochrome from light to redox sensor. *EMBO J.* **6,** 3322–3331.

Yeh, K. C., Wu, S. H., Murphy, J. T., and Lagarias, J. C. (1997). A cyanobacterial phytochrome two-component light sensory system. *Science* **277,** 1505–1508.

Yurkov, V., and Beatty, J. T. (1998). Aerobic anoxygenic phototrophic bacteria. *Microbiol. Mol. Biol. Rev.* **62,** 695–724.

CHAPTER TEN

Biophysical Assays for Protein Interactions in the Wsp Sensory System and Biofilm Formation

Nabanita De,[1] Marcos V. A. S. Navarro, Qi Wang, Petya V. Krasteva, *and* Holger Sondermann

Contents

Abstract

Many signal transduction and regulatory events are mediated by a change in oligomeric state upon posttranslational modification or ligand binding. Hence, the characterization of proteins and protein complexes with respect to their size and shape is crucial for elucidating the molecular mechanisms that control their activities. Commonly used methods for the determination of molecular weights of biological polymers such as standard size-exclusion chromatography or

Department of Molecular Medicine, College of Veterinary Medicine, Cornell University, Ithaca, New York, USA
[1] Current address: Department of Molecular Biology, The Scripps Research Institute, La Jolla, California, USA

Methods in Enzymology, Volume 471
ISSN 0076-6879, DOI: 10.1016/S0076-6879(10)71010-7

analytical ultracentrifugation have been applied successfully but have some limitations. Static multiangle light scattering presents an attractive alternative approach for absolute molecular weight measurements in solution. We review the biophysical principles, advantages, and pitfalls of some popular methods for determining the quaternary structure of proteins, using the response regulator diguanylate cyclase WspR from *Pseudomonas* and FimX, a protein involved in *Pseudomonas aeruginosa* twitching motility, as examples.

1. Introduction

As it is the case for many signaling proteins, diguanylate cyclases and phosphodiesterases that control cellular levels of the bacterial second messenger c-di-GMP function as multimers and are often regulated by a change in oligomeric state upon posttranslational modification (e.g., phosphorylation) or ligand binding (Barends *et al.*, 2009; Chan *et al.*, 2004; De *et al.*, 2008, 2009; Wassmann *et al.*, 2007). Similarly, proteins governing bacterial chemotaxis or transcription undergo changes in oligomeric state and engage in distinct complexes, often triggered by reversible phosphorylation as part of two-component signaling systems (Gao and Stock, 2009). These examples highlight the prevalence of changes in oligomerization as means to establish different functional states of a protein, and underline the importance of studying quaternary structure and complex formation.

We provide a general overview of popular approaches (size-exclusion chromatography, SEC; analytical ultracentrifugation, AUC; small-angle X-ray scattering, SAXS) that we have used to study response regulator proteins from *Pseudomonas* with an emphasis on their respective advantages and possible experimental limitations. More recently, innovative detector development and methodological advances promoted static multiangle light scattering (MALS) to a versatile technique with many beneficial characteristics for the determination of absolute molecular weight of macromolecules in solution. Key differences will be illustrated based on two cases: For FimX, a protein involved in twitching motility and biofilm formation in *Pseudomonas aeruginosa*, we identified a dimerization module in solution that is different from that observed in crystal lattices formed by the phosphodiesterase-like domain. Modeling of SAXS data yielded a low-resolution shape reconstruction of the full-length protein (Navarro *et al.*, 2009). In the case of the diguanylate cyclase response regulator WspR, a potent switch in bacterial biofilm formation, studying the oligomeric state in combination with structural data revealed an intricate switching behavior between distinct oligomeric and activation states (De *et al.*, 2008, 2009; Hickman *et al.*, 2005). Static light scattering-based experiments were crucial to reveal the quaternary structure of WspR and to establish the functional role of distinct oligomeric species (De *et al.*, 2008).

2. Analyses of the Shape and Molecular Weight of Proteins and Protein Complexes

In this section, we provide an incomplete overview of common techniques that have been used successfully to study the size and shape of proteins. We mainly concentrate on solution approaches for native proteins that do not require protein engineering or modification such as labeling of the sample.

2.1. Size-exclusion chromatography

Given the availability of basic chromatography systems in many labs, analytical SEC (or gel filtration) is often the first method to be used for studying the quaternary structure of proteins in solution. Gel filtration columns separate molecules by their hydrodynamic or Stokes radius (Lathe and Ruthven, 1955, 1956; Porath and Flodin, 1959). The method relies on the principle that molecules of different size elute at different rates while passing through a porous stationary phase, usually a dextran, agarose, silica, or polyacrylamide matrix. The molecules in the sample partition between the mobile phase and the pores of the stationary phase, with smaller molecules penetrating further into the matrix than larger ones. This differential diffusion results in a longer retention time on the column for smaller molecules. Molecules or complexes that are too large to enter the pores of the matrix migrate with the mobile phase front and define the void volume of the column.

The apparent partition coefficient K_{av} for a given molecule can be calculated as

$$K_{av} = \frac{V_e - V_0}{V_t - V_0} \tag{10.1}$$

where V_e is the elution volume of the molecule, V_0 is the column void volume, and V_t is the total column bed volume. The elution time of a molecule with a size too big to diffuse into the stationary phase determines V_0 (usually $\sim 1/3$ of the column volume).

The elution volume of a molecule is linearly proportional to the logarithm of its hydrodynamic radius. For globular proteins, this proportionality can be extended to molecular weight. Based on the elution volume of several globular standard proteins with known molecular weights, a linear calibration curve can be derived by plotting their apparent partition coefficient K_{av} versus the logarithm of their molecular weight. This curve can be used to estimate the molecular weight (or size) of a protein based on its elution volume. Molecular weight determinations based on this calibration method are only accurate for globular molecules with a similar relationship between molecular weight and size as that of the standard proteins. In addition, associating systems may elute aberrantly if molecular interactions occur on a fast time-scale. By using a concentration series, a shift in elution peak position is indicative of an oligomeric system in fast exchange. Alternatively, aberrant elution from the SEC column could be indicative of unspecific interactions of the solute with the column matrix.

The method allows for large flexibility with regard to choice of chromatography matrix and mobile phase. Also, samples within a wide range of concentrations can be analyzed, only being restricted by the detection limit of the UV detector and the loading capacity of the gel filtration column.

2.2. Analytical ultracentrifugation

A versatile alternative to SEC and an approach to determine the absolute molecular weight of macromolecules, their quaternary structure, and thermodynamic properties is AUC (Hansen *et al.*, 1994; Hensley, 1996; Howlett *et al.*, 2006; Lebowitz *et al.*, 2002; Minton, 2000; Schachman and Edelstein, 1966; Schachman *et al.*, 1962). Two main methods can be distinguished: Sedimentation velocity and sedimentation equilibrium.

Similar to SEC, sedimentation velocity experiments rely on volumetric measurements that report on the hydrodynamic properties of macromolecules, providing their distribution of sedimentation (and diffusion) coefficients as they move through a centrifugal field. The sedimentation coefficient s is defined as the ratio of radial sedimentation velocity u of the macromolecule to its centrifugal acceleration (or the strength of the applied centrifugal field) $\omega^2 r$. This relationship is described by the Svedberg equation:

$$s = \frac{u}{\omega^2 r} = \frac{M(1 - \bar{v}\rho)}{N_A f} \tag{10.2}$$

where ω is the angular velocity, r is the distance from the center of rotation, M is the molar mass, ρ is the density of the solvent, $\bar{v}$ is the partial-specific volume, N_A is Avogadro's number, and f is the frictional coefficient. Based on this equation, information about the molecular mass of the macromolecule and its shape and size (related to the frictional coefficient f) in solution can be readily obtained from the measured data. Equation (10.2) can be rewritten as follows by applying the Einstein–Smoluchowski relation:

$$M = \frac{sRT}{D(1 - \bar{v}\rho)} \tag{10.3}$$

where R is the gas constant, T is the absolute temperature, and D is the diffusion coefficient. While $\bar{v}$ and ρ can be calculated based on the protein sequence and buffer composition, respectively, s and D are obtained from fitting the experimental data.

In contrast, sedimentation equilibrium experiments take advantage of thermodynamic measurements that are shape-independent and report on the steady-state of the system at a point where the gravitational force is opposed by diffusion. In addition to the molecular weight of particles, this approach yields equilibrium-binding constants for self-associating system or heterocomplex formation.

Experiments are carried out using an XL-I analytical ultracentrifuge (Beckman Coulter). Detection of macromolecules is achieved by recording absorbance or by using interference optics. The former depends on the molar extinction coefficient of the macromolecules, which sets the limits for sample concentration. Typically, protein samples at low micromolar concentration (and within one order of magnitude) can be analyzed. To study the oligomeric state at higher protein concentrations the wavelength at which data is collected can be chosen farther away from the absorbance peak wavelength.

2.3. SEC-coupled static multiangle light scattering

Static MALS relies on the fact that the intensity of scattered light produced by a macromolecule is proportional to its molecular weight. In SEC-coupled MALS, the intensity of the scattered laser light (also known as Rayleigh scattering), collected simultaneously at different angles, is measured as the protein elutes from a gel filtration column. A second detector, known as differential refractive index detector, accurately monitors protein concentration by measuring a change in refractive indices between the eluate and solvent. Protein concentration and scattering intensity as a function of angle can be used to determine the molar mass of a scattering particle, for example, a protein or protein complex, on an absolute scale (Wen *et al.*, 1996; Wyatt, 1997). The relationship between the intensity of the scattered light at a certain angle and the molecular weight is described by the Zimm equation (Debye, 1947; Zimm, 1948):

$$\frac{Kc}{R(\theta)} = \frac{1}{M_{\mathrm{w}}P(\theta)} + 2A_2c \tag{10.4}$$

where c is the solute concentration, $R(\theta)$ is the Rayleigh ratio of the solution at scattering angle θ and concentration c and is directly proportional to the light scattering intensity, M_w is the weight-averaged molecular mass, $P(\theta)$ is a descriptor for the angular dependence of the scattered light, and A_2 is the second virial coefficient. K is the constant $4\pi^2 n^2 (dn/dc)^2/(\lambda^4 N_A)$, where n is the refractive index of the solvent, dn/dc is the refractive index increment of the solution, λ is the wavelength of the light source, and N_A is Avogadro's number. Double extrapolation to zero angle and zero concentration from multiangle measurements provides the molecular weight and second virial coefficient for the solute particles. For experiments at a single concentration, as it is the case for SEC-coupled MALS approaches, only the molecular weight can be determined. The second virial coefficient term is negligible due to the low solute concentration in SEC peaks. While MALS experiments can yield the radius of gyration (R_g) of larger proteins (with an $R_g > 10$ nm), scattering from smaller proteins (with an $R_g < 10$ nm) shows little angular dependence, and Eq. (10.4) can be simplified to

$$\frac{Kc}{R(\theta)} = \frac{1}{M_w} \tag{10.5}$$

The measurements are independent of the shape of a protein and its retention time in SEC. In addition, polydispersity indices can be calculated that provide an estimate for the mass distribution in the sample (Mogridge, 2004; Sondermann *et al.*, 2005; Wyatt, 1997). While fractionation via SEC in-line with MALS yields accurate measurements for various species that may be present in the sample, experiments can be also carried out in batch mode. Batch mode measurements at multiple protein concentrations provide the second virial coefficient, a measure of nonspecific interactions, as well as stoichiometry and equilibrium constants for reversible, specific interactions (Attri and Minton, 2005; Kameyama and Minton, 2006; Valente *et al.*, 2005).

Similar to stand-alone SEC experiments, SEC-coupled MALS provides flexibility regarding the choice of mobile phase. Buffers may contain ligands such as nucleotides or phosphomimetic compounds that can stabilize certain conformations or molecular assemblies. In addition, the method is amenable to transmembrane or peripheral membrane proteins since samples can be analyzed in the presence of detergent or lipids. The new generation of detectors have a large dynamic range that allows for the analysis of protein samples at concentrations ranging from $\sim$1 μg/ml to $\sim$20 mg/ml.

The experimental approach is amenable to interacting systems, self-associating macromolecules or heterocomplexes. At slow exchange, distinct oligomeric species may elute as separate peaks in SEC, and their molecular weights can be determined (e.g., in the case of nucleotide-free WspR;

see below) (De *et al.*, 2008). If the peaks are not well separated by SEC due to close retention times, careful analysis of the molecular weight distribution across a peak may reveal the presence of multiple species (Margarit *et al.*, 2003). Considering a protein in fast exchange, the position of the peak will be concentration dependent. The average molecular weight determined for a system with fast exchange rates will be an intermediate between the molecular weight of a fully associated complex and that of the associating units in isolation, with values closer to the higher oligomeric species with increasing protein concentration (Pirruccello *et al.*, 2006; Zoltowski and Crane, 2008). A concentration series can be used to estimate an apparent dissociation constant. In these cases, peak shapes often deviate from a Gaussian distribution, and the molecular weight distribution across the peak may show a trend, usually with higher molecular weights at the leading edge of the peak and lower molecular weights as the protein gets diluted on the chromatography column.

A typical instrument setup and sample data are illustrated in Fig. 10.1. Experiments are carried out using a KW-803 SEC column (Shodex) connected to an isocratic HPLC pump/in-line degasser combination (1100 Series, Agilent). It is our experience that HPLC columns and pumps provide superior flow properties and stability compared to FPLC systems yet both can be used in MALS-coupled experiments. The column is connected in-line to a 3-angle or 18-angle light scattering detector (mini-DAWN TREOS or DAWN HELEOS II, Wyatt Technology) and a refractive index detector (Optilab rEX, Wyatt Technology). Optionally, the system can be combined with UV absorbance detection to measure protein extinction coefficients, and quasielastic light scattering for in-line size determination (see below). Recently, field flow fractionation has been introduced as a fractionation technique that can replace SEC (Luo *et al.*, 2006).

2.4. Small-angle X-ray scattering

SAXS relies on the interaction of X-rays with orbital electrons, and presents another classical volumetric method for the determination of geometric descriptors and molecular weight of particles in solution (Koch *et al.*, 2003; Putnam *et al.*, 2007). Similar in principle to static laser light scattering, information including the radius of gyration, maximum diameter, shape, and the oligomeric state of a particle can be extracted from the scattering data.

In a monodispersed, noninteracting system, the scattering is linearly proportional to the number of particles in the sample. After normalization to particle concentration, the scattering intensity corresponds to that of a single particle. Specifically, the absolute scattering intensity at zero-angle (I_0, in units of cm^{-1}) is proportional to the square of the number of electrons in the particle, and related to the macromolecule's partial-specific

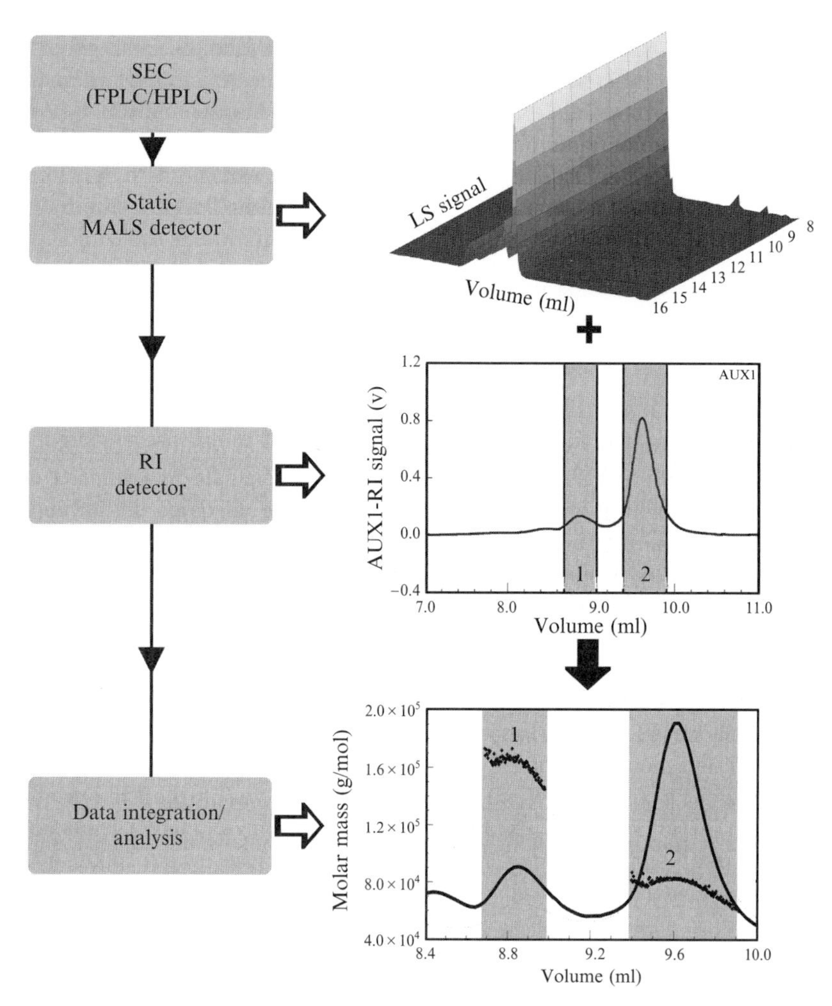

Figure 10.1 SEC-coupled MALS. A typical MALS setup with in-line SEC is shown. Upon fractionation on a gel filtration column, the eluate enters first a static multiangle laser light detector, then a differential refractive index detector (left column). Light scattering (LS) and refractive index (RI) signals are collected (in units of volts), and produce absolute molecular weight measurements (in units of g/mol) (right column). In this case, the sample contained two species that were separated by SEC. Each peak is analyzed separately (peak areas 1 and 2) in the same experiment. The molecular weight distribution across a single peak provides a measure of the polydispersity index.

volume ($\bar{v}$). The partial-specific volume varies very little among different proteins, and can be assumed as constant (Mylonas and Svergun, 2007). Therefore, the solute scattering intensity normalized to its concentration c (I_0/c, in units of $\mathrm{mg}^{-1}\ \mathrm{cm}^{-1}$ ml) is linearly related to the molecular weight of the solute:

$$\frac{I_0}{c} = F \times \left(\frac{\sum_j m_j}{M}\right)^2 \times N_A \times M = k \times M \tag{10.6}$$

where F is the density factor ($F = 1 - \bar{v}\rho$), $\sum_j m_j/M$ is the constant ratio of the number of electrons to the molecular weight of the particle. I_0 can be determined from data fitting to the linear region of the Guinier plot (ln $I(Q)$ versus Q^2) at low angles or to the autocorrelation function (also known as distance distribution or $P(r)$ function). The linear constant k is determined by using standard protein samples such as lysozyme or bovine serum albumin with known molecular weights. The molecular weight (M) of a protein is calculated as follows:

$$M^{\text{sample}} = M^{\text{standard}} \times \frac{I_0^{\text{sample}} \times c^{\text{standard}}}{I_0^{\text{standard}} \times c^{\text{sample}}} \tag{10.7}$$

Inaccuracies in determining protein concentrations and intensity fluctuations of the incident beam during data acquisition present known sources of error. For an associating system, the calculated molecular weight reflects an average molar mass of the system, and such an analysis is often used to determine the oligomerization states of proteins. Rigid body modeling using high-resolution structures determined by X-ray crystallography or NMR has been used successfully to derive structural information of proteins and protein complexes (Petoukhov and Svergun, 2005; Svergun *et al.*, 1995). In addition, modeling of SAXS data can be used to generate low-resolution *ab initio* shape reconstructions of macromolecules in solution (Svergun *et al.*, 2001; Volkov and Svergun, 2003). These approaches work most reliably for monodispersed samples, and conformational heterogeneity and/or higher order oligomerization may interfere with model accuracy.

While access to an X-ray source is a requirement for SAXS experiments, data acquisition is fast and requires only little protein. Typically, sample concentrations range from 0.5 to 10 mg/ml. Depending on the sample cell and setup (batch vs. flow-cell measurements), 10–200 μl of protein solution is required per sample and protein concentration. Data acquisition times vary with beam intensity but usually range from seconds to minutes. Dose–response experiments are used to optimize exposure times. Investigation of Guinier plots yields information regarding suitable radiation dose at which radiation damage is minimal but sufficient signal-to-noise ratio is obtained. Investigation of Kratky plots yields information regarding the folded state of the samples and the degree of conformational flexibility. Due to the high sensitivity of the measurements, background correction requires an accurate buffer control, preferably by using buffer samples collected as flow-through during protein concentration via ultrafiltration devices or from SEC.

Molecular weight determination, shape, reconstruction and oligomerization analysis by SAXS is also applicable to high-throughput formats (Hura *et al.*, 2009).

2.5. Other methods for the biophysical characterization of proteins in solution

Dynamic light scattering (DLS), also known as quasielastic light scattering or photon correlation spectroscopy, is based on the measurement of time-dependent, short-term fluctuation of scattered light by molecules under Brownian motion (Brown, 1993; Fujime, 1972; Nobbmann *et al.*, 2007; Pecora, 1972; Schurr, 1977). Analysis of the intensity fluctuation yields a distribution of diffusion coefficients of the particles in solution and their hydrodynamic radius is calculated through the Stokes–Einstein equation. In this way, DLS presents a fast method for assessing the homogeneity of macromolecular samples and it has been widely applied as a tool for buffer screening and for predicting and monitoring protein crystal growth (George and Wilson, 1994; Wilson, 2003). However, in contrast to static light scattering where absolute molecular weights can be extracted from measurements of the intensity of the scattered light, the molecular weights of scattering macromolecules in a DLS experiment can only be estimated based on comparisons of their calculated R_h with standard curves calibrated for different molecular shape models.

Isothermal titration calorimetry (ITC) is used to determine the change in enthalpy upon molecular interactions that form or break, as well as the binding affinities and changes in entropy (Ladbury and Chowdhry, 1996; Velazquez-Campoy *et al.*, 2004). Unlike surface plasmon resonance (SPR)-based approaches, which require the selective immobilization of one of the interacting species, ITC is a solution measurement in which a ligand is titrated into a solution containing the receptor. The instrument measures the amount of heat absorbed or released upon molecular association as a function of power required to maintain a constant temperature between a reference cell and the reaction chamber. Similarly, complex dissociation is accessible by titrating small amounts of a concentrated complex sample into matching buffer (Wassmann *et al.*, 2007).

Methods described so far do not require the introduction of protein modification such as fluorescent or spin labels. Experimental approaches to analyze protein complex formation in a quantitative manner using fluorescent probes include FRET-based techniques or fluorescence anisotropy measurements (Heyduk *et al.*, 1996; Jares-Erijman and Jovin, 2003; Sondermann *et al.*, 2004). While the latter requires the labeling of a single protein, FRET measurements rely on the energy transfer between two fluorophores. Such approaches provide flexibility in the choice of buffers and protein concentration. Also, the option to work at steady-state or in a time-resolved mode yields both equilibrium-binding constants as well as kinetic details of the molecular interaction.

Extensive studies on the bacterial chemotaxis pathway have established disulfide mapping as a versatile approach to analyze protein structure and oligomerization (Bass *et al.*, 2007; Falke and Koshland, 1987; Falke *et al.*, 1988). The method relies on cross-linking of cysteine side chains in close proximity that potentially traps distinct conformations and/or complexes covalently, which can be analyzed by native or SDS–polyacrylamide gel electrophoresis or by functional assays. Similar to the fluorescence-based methods, disulfide mapping studies require protein engineering in which cysteines are introduced at specific sites that are likely to come close enough to form a covalent bond upon chemical oxidation. Native cysteines may interfere with the analysis and may have to be mutated to a nonreactive residue. Although not a requirement, the analysis may be facilitated by knowledge about the three-dimensional structure of a protein and its oligomeric state.

Another elegant approach for studying protein interaction that has been applied to the bacterial chemotaxis system is pulsed electron spin resonance spectroscopy (Bhatnagar *et al.*, 2007; Park *et al.*, 2006). The measurements report the distance between spin labels that have been covalently attached to cysteine residues (Borbat and Freed, 2007). As for disulfide mapping, cysteine residues have to be introduced into the protein by site-directed mutagenesis, while undesired native cysteines have to be mutated to non-reactive residues. Unlike fluorescence labels that are usually too bulky to react with buried side chains that are solvent inaccessible, spin labels are small enough to penetrate into the folded core of a protein and all cysteines in a protein sequence have to be considered as potential sites for modification. For measuring conformational changes within a protein, two identical spin labels are introduced. Depending on the geometry and distances, homooligomerization may only require a single spin label per protein whereas heterocomplex formation relies on the labeling with a single spin probe in both binding partners.

In the following section, we will briefly discuss the experimental considerations for choosing a particular method, pointing out some of the limitation of the approaches introduced above.

3. Experimental Considerations

Many approaches have been established that utilize different biophysical properties of macromolecules for determining their molecular weight and shape in solution. More traditional methods for the determination of the oligomeric state and molar mass of proteins and their complexes are standard SEC or AUC. These methods have been demonstrated to be versatile and effective, but some limitations exist that should be considered when choosing the appropriate approach.

In SEC studies, it is not possible to distinguish between conformational changes affecting the hydrodynamic radius and elution properties of a molecule and/or a change in stoichiometry, an obstacle for the reliable calibration of this method (Oliva *et al.*, 2001). Standardization of SEC by using globular, monomeric proteins with known molecular weight produces an apparent molecular weight and as such only a rough estimate that can be misleading. Yet, if the quaternary structure of a protein or complex is known from other approaches, SEC can yield low-resolution shape information. Another drawback of SEC-based approaches comes from the fact that samples get diluted while passing through the gel filtration column (typically between 10 and 20-fold), and measurements may be taken at nonequilibrium conditions.

AUC is a potent method for determining the absolute molecular weight of protein species at chemical equilibrium, and for determining the dissociation constant of oligomeric assemblies. It is a true solution approach, not requiring any separation matrix (as in SEC) or surface immoblilization (as in SPR spectroscopy) that could potentially bias the results due to unspecific interactions. Molecular weight determination by sedimentation velocity AUC for associating systems in which the complex has a different shape than its components presents a nontrivial case since frictional ratios of the species may be different. In such a case, distinct peaks may be analyzed separately. A typical AUC experiment requires centrifugation times of multiple hours up to days. Many of the proteins and their biologically relevant oligomers may not be stable in isolation for such long data acquisition periods due to the short-lived, transient nature of some posttranslational modification (e.g., the phosphorylation in two-component signaling systems) or due to inherent dynamic properties of the proteins.

SEC-coupled MALS presents a versatile approach combining some of the advantages of standard SEC and AUC, yielding accurate, shape-independent absolute molecular weight measurements in solution with short data acquisition times. For many applications, it can replace the more traditional techniques, or provide complementary data. The large dynamic range and buffer flexibility provides a platform for measurements across a large chemical space and environment (such as in lipid mixtures or detergent). Similar to AUC, matrix interactions are negligible since measurements are taken downstream of SEC and are not affected by elution time, and SEC functions only as a fractionation step. At the same time, since it is coupled to SEC, shape information can be obtained from an experiment by calibration of the SEC column with globular proteins with known molecular weights. Polydispersity indices calculated from MALS data provide an estimate for the mass distribution in the sample, a critical parameter describing the homogeneity of a sample that can be informative for protein crystallization and functional assays.

In DLS, an alternative light scattering method to MALS, data decomposition of the measured fluctuations from complex systems into the individual time-dependent correlation functions is difficult. Also, given the time-dependence of DLS measurements, this approach is more suitable for batch measurements, and less so for the coupled assays with in-line SEC fractionation. Yet, when combined with MALS, online DLS is a viable approach to obtain preliminary structural information. Fast data acquisition times and instruments in plate reader format for batch measurements provide a tool for quick evaluation of buffer conditions and buffer optimization, using sample polydispersity as a read-out.

Molecular weight determination by SAXS yields accurate, shape-independent measurements for monodispersed samples. Samples should resist radiation-damage over the course of an experiment, yet producing strong scattering for a good signal-to-noise ratio. In addition, SAXS data can be used to model low-resolution structural models of proteins in solution, a unique feature of this approach. Limitations are the requirement of homogeneous protein samples that should be free of protein aggregation. Also, although sophisticated data analysis tools exist for dealing with conformational mixtures or solutions containing different oligomers (Bernado *et al.*, 2007, 2008; Krukenberg *et al.*, 2008; Wang *et al.*, 2008a), not all states may be accessible at the experimental conditions.

In addition to method-specific advantages and limitations, sample properties need to be considered as well. If structural information is available and suggests a nonglobular shape of the protein or complex, SEC may not be a reliable method for the determination of molecular weight but could yield shape information in solution based on the elution volume and profile. Posttranslational modifications in bacterial signaling proteins are often rather short-lived and labile. In the case of response regulator proteins such as CheY, phosphorylation is counteracted by an autophosphatase activity encoded in the phosphorylated domain itself (Hess *et al.*, 1987, 1988). Often, phosphomimetic reagents such as beryllium fluoride are used but their affinity for the active site is rather low and requires a high concentration of the compounds to be present (Yan *et al.*, 1999). Also, distinct states and conformations of purified proteins may be less stable outside of the cell, potentially posing solubility issues over prolonged incubation and measurement times.

For associating systems, proteins that form homo- or heterocomplexes, it may be important to consider the time scale at which molecular interactions take place (relative to the measurement time), as it may affect the results and analysis. These scenarios point out some aspects that need to be considered when choosing the appropriate approach. As exemplified by the case studies introduced below, a combination of various biophysical methods, especially if they report on distinct molecular properties and are complementary, is often desirable for the unambiguous structural characterization of proteins and protein complexes.

4. Case Studies

4.1. FimX

FimX (PA4959) is a multidomain protein that is involved in twitching motility and biofilm formation (Fig. 10.2A) (Huang *et al.*, 2003; Kazmierczak *et al.*, 2006). It contains an N-terminal receiver–PAS domain unit and a C-terminal diguanylate cyclase–phosphodiesterase domain module. The sequence of the active sites of the latter domains is degenerate and the protein lacks significant activity towards c-di-GMP (Navarro *et al.*, 2009; Rao *et al.*, 2008). Our structural and biochemical studies support a model in which FimX is catalytically inactive but retained c-di-GMP-binding properties via the phosphodiesterase domain (Navarro *et al.*, 2009). High-affinity binding of c-di-GMP to the phosphodiesterase domain was established by using ITC. In order to elucidate the tertiary and quaternary structure of FimX, we employed a combination of approaches described above.

Initially, we determined the crystal structure of the phosphodiesterase domain of FimX in its c-di-GMP-bound and apo-state, and in the context of the diguanylate cyclase–phosphodiesterase dual-domain module. In some crystal lattices such as the one formed by the dual-domain module, the EAL domain forms a cage-like dimer in which the nucleotide-binding sites are sequestered into the inside of the assembly, unavailable for and incompatible with c-di-GMP binding (Fig. 10.2A) (Navarro *et al.*, 2009). Several biophysical approaches that report on the molecular weight and shape of proteins were used to determine whether a similar dimer exists in solution, or whether crystal-packing contacts maintained the oligomer. Based on a comparison of AUC, SAXS, and MALS data that agreed with each other, we showed that the diguanylate cyclase–phosphodiesterase module of FimX is a monomer in solution although its phosphodiesterase domains formed such tight dimers in crystals (Fig. 10.2B). Calibrated SEC experiments reported a significantly larger molecular weight, indicating a somewhat elongated shape of the molecule. Calculations of the frictional ratio based on sedimentation velocity AUC data and of the internal distance distribution function based on SAXS data supported such an observation (Fig. 10.2). The full-length protein and fragments containing the N-terminal domains formed dimers even when the phosphodiesterase domain was deleted suggesting that the receiver and/or PAS domains mediate oligomerization (Navarro *et al.*, 2009).

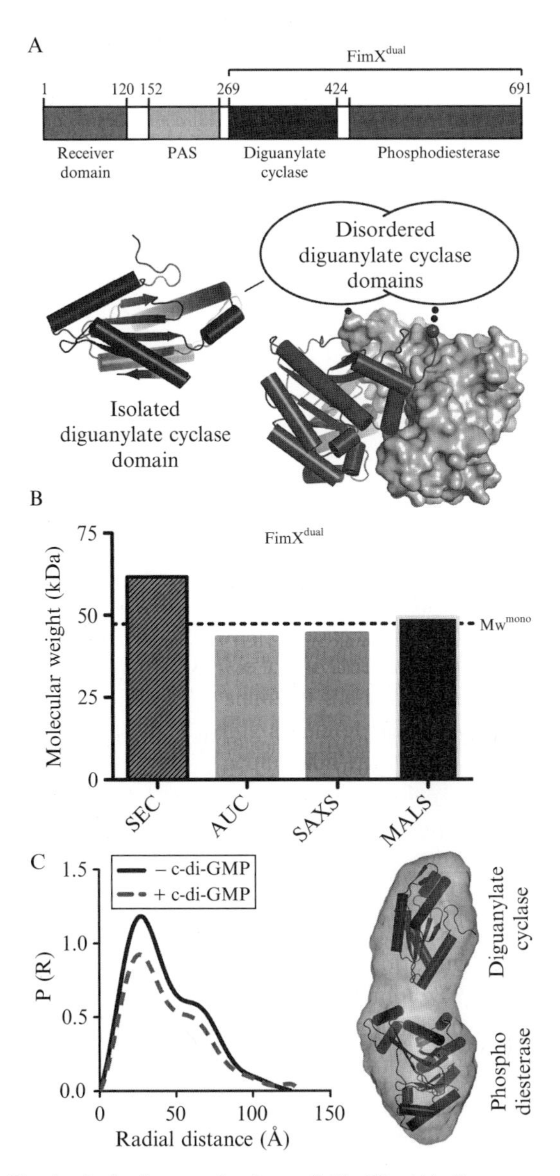

Figure 10.2 Biophysical characterization of FimX. (A) Structure of FimX. The domain organization of FimX from *P. aeruginosa* is shown (top panel). Structures of the isolated diguanylate cyclase domain (PDB code: 3HVA; left) and the diguanylate cyclase–phosphodiesterase module (FimX^{dual}; PDB code: 3HVB; right) determined by X-ray crystallography are shown (bottom panel) (Navarro *et al.*, 2009). FimX^{dual} forms a dimer in the crystal lattice via the phosphodiesterase domains. One dimer half is shown in cartoon presentation, the other is shown as surface presentation. The c-di-GMP-binding site is buried in the center of the dimer. The diguanylate cyclase domains appeared disordered in the FimX^{dual} crystals. (B) Molecular weight determination for

In addition to molecular weight and shape determination by SAXS, we combined SAXS data and available crystal structures of the isolated domains to model a low-resolution envelope for the monomeric fragments and dimeric, full-length FimX (Navarro *et al.*, 2009). The distance distribution function and modeling of the SAXS data for the diguanylate cyclase–phosphodiesterase module is shown in Fig. 10.2C, demonstrating the potential of this approach for reconstructing accurate low-resolution models (Hura *et al.*, 2009; Koch *et al.*, 2003; Putnam *et al.*, 2007).

In summary, the absolute molecular weight of FimX was accurately determined by AUC, MALS, and SAXS. While SEC produced an aberrant molecular weight, highlighting the limitations for the calibration of this method, it served as a diagnostic tool to assess protein shape. The homogeneity of the samples with regard to conformation and stoichiometry and their resistance to radiation damage allowed us to model SAXS data providing low-resolution structural models with more detailed shape information. Based on the structural data, we proposed molecular mechanisms that may explain how c-di-GMP binding to the phosphodiesterase domain of FimX could affect bacterial biofilm formation and motility.

4.2. WspR

In this study, we described an inhibitory feedback loop of the response regulator diguanylate cyclase WspR, a potent cellular switch that controls biofilm formation in *Pseudomonas*, based on its full-length crystal structure and biochemical analyses (Fig. 10.3A) (De *et al.*, 2008). WspR contains an N-terminal receiver domain and a C-terminal diguanylate cyclase domain

the diguanylate cyclase–phosphodiesterase fragment of FimX. In light of the dimeric state of $FimX^{dual}$ in the protein crystal, its solution state was characterized biophysically in order to determine its quaternary structure. Three methods (AUC, SAXS, and MALS) indicated that $FimX^{dual}$ is monomeric in solution (De and Sondermann, unpublished data; Navarro *et al.*, 2009). SEC overestimated the molecular weight, which could be indicative of a nonglobular shape of the protein or unspecific column interactions. Based on the frictional ratio obtained from AUC and SAXS data shown in (C), the aberrant behavior in SEC could be attributed to an elongated shape of $FimX^{dual}$ (Navarro *et al.*, 2009). (C) Modeling of solution SAXS data of the diguanylate cyclase–phosphodiesterase fragment of FimX. Distance distribution functions calculated from SAXS data of $FimX^{dual}$ in the nucleotide-free (solid line) and c-di-GMP-bound state (dashed line) are shown after averaging and subtraction of solvent background scattering (left panel). The bimodal shape of the curves and maximal radial distance indicate an extended structure containing at least two folded domains. Modeling of the SAXS data yielded a low-resolution shape reconstruction of $FimX^{dual}$ (right panel). In the approach, the position of spherical scattering units representing residues in a protein are adjusted by simulated annealing with the scattering curve or distance distribution function as the target (Koch *et al.*, 2003; Putnam *et al.*, 2007).

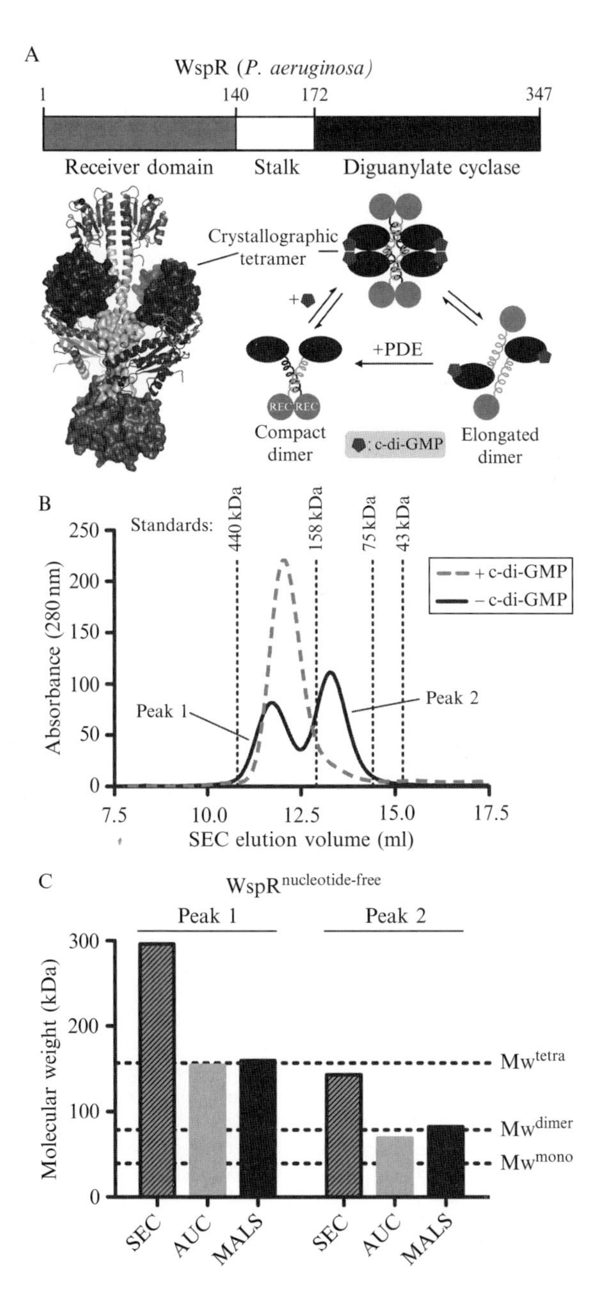

Figure 10.3 Biophysical characterization of WspR. (A) Structure of WspR. The domain organization of WspR from *P. aeruginosa* is shown (top panel). In the crystal lattice, WspR forms dimers of dimers yielding a tetrameric assembly (PDB code: 3BRE; De *et al.*, 2008). In the structure, one dimer is shown as cartoon presentation, the second one is shown as surface presentation (left panel). The model for the

(Hickman *et al.*, 2005). From the structural studies, we learned that the two domains are connected via a long helical stalk, an extension of the terminal helix of the receiver domain (Fig. 10.3A). The product of the diguanylate cyclase reaction, c-di-GMP, was bound to the inhibitory site of WspR, as has been observed in other diguanylate cyclase structures (Chan *et al.*, 2004; De *et al.*, 2008; Wassmann *et al.*, 2007). Examination of the crystal lattice indicated a WspR tetramer as the biological unit. Such an assembly is unique to WspR and differs from structures determined for a related enzyme, PleD from *Caulobacter crescentus* (Chan *et al.*, 2004; De *et al.*, 2008; Wassmann *et al.*, 2007).

In solution, the activation and product inhibition cycle of WspR involved the switching between distinct states that were sensitive to enzyme and c-di-GMP concentration, and could be distinguished by SEC (De *et al.*, 2008). To correlate the crystallographic model with the functional states, biophysical characterization by SEC-coupled MALS was chosen for several reasons. It was clear from the crystal structure that some of the states were likely to be nonglobular, deeming molecular weight determination by SEC inaccurate. While SAXS data was successfully used for the structural characterization of FimX, WspR samples existed in equilibrium between different states at concentrations producing good signal-to-noise ratio, complicating the analysis (De and Sondermann, unpublished data). Similarly, although some samples could be analyzed by AUC, others did not withstand the long data acquisition times. For example, prolonged incubation with beryllium fluoride, a compound that mimics receiver domain phosphorylation, led to unspecific aggregation of the protein (De and Sondermann, unpublished data). Hence, the analysis of the quaternary structure of these proteins required short experimentation times for the determination of the true molecular weight in solution. Also, the presence

oligomeric switching of WspR was derived from structural, biophysical, and functional assays (right panel) (De *et al.*, 2008). (B) SEC analysis of WspR in its nucleotide-free and c-di-GMP-bound form. Analyses of WspR by SEC revealed distinct, c-di-GMP-dependent elution profiles for WspR. The nucleotide-bound protein purified from *E. coli* eluted in a single peak, while the nucleotide-free protein produced a bimodal elution profile. Both peaks were distinct in the elution peak position from that of the c-di-GMP-bound WspR. Calibration using globular standard proteins with known molecular weight indicated that WspR (with monomeric molecular weight of 39.2 kDa) forms higher order oligomers. (C) Molecular weight determination for nucleotide-free WspR. The molecular weight of nucleotide-free WspR was determined by three independent methods (SEC, AUC, and MALS). The dashed lines indicate the theoretical molecular weight of monomeric, dimeric, and tetrameric assemblies. While AUC and MALS analyses agreed, indicating that nucleotide-free WspR forms dimers and tetramers in solution, the SEC-based approach overestimated the molecular weight significantly (De and Sondermann, unpublished data; De *et al.*, 2008). Given the available structural information, a nonglobular shape is likely to be responsible for this discrepancy.

of protein-bound c-di-GMP in some samples increased the molar extinction coefficient of WspR significantly, affecting protein detection by UV absorbance.

In contrast, SEC-coupled MALS provided many advantages. Molecular weight determination by MALS is shape-independent yielding the absolute molecular weight of the solute. Since different oligomers could be distinguished in SEC (Fig. 10.3B), MALS could be used to assign the molecular weights to distinct elution peaks in the same experiment, providing a protein-specific calibration. Some of the homotypic WspR interactions appeared to have fast exchange rates, which could be estimated from MALS experiments (De and Sondermann, unpublished data). Finally, short data acquisition times allowed characterization of several relevant oligomeric states of WspR (De *et al.*, 2008).

By using SEC-coupled MALS, we could show that nucleotide-bound WspR purified as a tetramer–dimer mixture. At lower concentrations, the protein was predominantly dimeric. Based on the SEC component of the analysis, its elution volume was significantly smaller than that of a globular protein of similar molecular weight (Fig. 10.3B). Such aberrant behavior in SEC suggested either a nonglobular shape of WspR and/or higher order oligomerization that affected its interactions with the chromatography matrix. Since samples are subject to significant dilution during gel filtration, complexes may fall apart on the column. At higher protein concentrations, a molecular weight between a WspR dimer and a tetramer was determined, and the non-Gaussian peak shape and molecular weight distribution across the peak suggested indeed a tetramer–dimer equilibrium under these conditions (De and Sondermann, unpublished data). The results were also consistent with a tetramer forming at high protein concentration used in crystallization trials.

Treatment of the wild-type protein with a c-di-GMP-specific phosphodiesterase or mutation of the inhibitory c-di-GMP-binding site on WspR yielded nucleotide-free protein that was highly active with regard to c-di-GMP production (De *et al.*, 2008). This species was analyzed in a similar manner (Fig. 10.3B and C). The nucleotide-free protein eluted from the SEC column in a bimodal distribution with one peak eluting earlier, and the second peak eluting later than c-di-GMP-inhibited WspR (Fig. 10.3B) (De *et al.*, 2008). According to MALS measurements, the first peak corresponded to a tetrameric species, whereas the second peak contained dimeric protein (Fig. 10.3C) (De *et al.*, 2008). The results suggested two characteristics for the nucleotide-free protein: (1) The dimer detected in this sample was distinct from that formed by c-di-GMP-bound WspR. (2) The dimer–tetramer equilibrium of the nucleotide-free protein was in slower exchange than the nucleotide-bound protein since two distinct species could be detected in the sample with a clear separation of the elution peaks. In contrast, nucleotide-bound WspR showed a more continuous shift in molecular weight that was dependent on protein concentration (De and Sondermann, unpublished data).

Finally, we could show that the tetrameric species functioned as a scaffold, bridging the two distinct dimeric assemblies (De *et al.*, 2008). It assembled from the more compact WspR dimer, and—in the presence of c-di-GMP—partially fell apart into the more elongated, dimeric species. As seen with FimX, determination of the apparent molecular weight based on the SEC elution volume would have been misleading, and retrospectively was indicative of an elongated shape of the tetramer and one of the dimer species.

In summary, WspR appears to be an unusual response regulator that formed dimers even in the absence of phosphorylation of its receiver domain. This is in contrast to canonical receiver domain-containing proteins that often require phosphorylation to switch between monomers and dimers (Gao and Stock, 2009). The distinct oligomerization of WspR could in part be attributed to the helical stalk motif that connects the receiver and diguanylate cyclase domains, and contributes to both dimerization and tetramerization (De *et al.*, 2008). Site-directed mutagenesis of a hydrophobic residue that contributes to coiled-coil formation of two stalks in the more compact WspR dimer affected oligomerization, rendering the protein predominantly monomeric (De *et al.*, 2008). Enzymatic assays established a correlation between distinct oligomeric species and diguanylate cyclase activity, describing a molecular mechanism for the c-di-GMP-mediated feedback inhibition of WspR.

5. Concluding Remarks

We are often faced with the challenge to probe interactions observed in crystal lattices (De *et al.*, 2008; Navarro *et al.*, 2009). Some techniques provide only apparent molecular weight determination such as SEC but can be useful tools to assess the shape of proteins in solution if the molecular weight has been determined by using shape-independent methods. An integrated approach employing multiple techniques, which report on different properties of proteins and protein complexes, can yield a detailed structural characterization (Mao *et al.*, 2005; Wang *et al.*, 2008a,b, 2009). Application of redundant or complementary approaches also increases the confidence in molecular weight (and shape) determination (De *et al.*, 2008; Navarro *et al.*, 2009). In addition to technical differences between various methods, practical considerations concerning protein stability, sample homogeneity, and kinetics weigh in on the choice of an appropriate approach for the biophysical analysis of a particular system.

ACKNOWLEDGMENTS

We are grateful to Michelle Chen for helpful discussions and critical reading of the manuscript. This work was supported by the NIH under award 1R01GM081373 (H. S.) and a PEW Scholar award in Biomedical Sciences (H. S.).

REFERENCES

Attri, A. K., and Minton, A. P. (2005). New methods for measuring macromolecular interactions in solution via static light scattering: Basic methodology and application to nonassociating and self-associating proteins. *Anal. Biochem.* **337,** 103–110.

Barends, T. R., Hartmann, E., Griese, J. J., Beitlich, T., Kirienko, N. V., Ryjenkov, D. A., Reinstein, J., Shoeman, R. L., Gomelsky, M., and Schlichting, I. (2009). Structure and mechanism of a bacterial light-regulated cyclic nucleotide phosphodiesterase. *Nature* **459,** 1015–1018.

Bass, R. B., Butler, S. L., Chervitz, S. A., Gloor, S. L., and Falke, J. J. (2007). Use of site-directed cysteine and disulfide chemistry to probe protein structure and dynamics: Applications to soluble and transmembrane receptors of bacterial chemotaxis. *Methods Enzymol.* **423,** 25–51.

Bernado, P., Mylonas, E., Petoukhov, M. V., Blackledge, M., and Svergun, D. I. (2007). Structural characterization of flexible proteins using small-angle X-ray scattering. *J. Am. Chem. Soc.* **129,** 5656–5664.

Bernado, P., Perez, Y., Svergun, D. I., and Pons, M. (2008). Structural characterization of the active and inactive states of Src kinase in solution by small-angle X-ray scattering. *J. Mol. Biol.* **376,** 492–505.

Bhatnagar, J., Freed, J. H., and Crane, B. R. (2007). Rigid body refinement of protein complexes with long-range distance restraints from pulsed dipolar ESR. *Methods Enzymol.* **423,** 117–133.

Borbat, P. P., and Freed, J. H. (2007). Measuring distances by pulsed dipolar ESR spectroscopy: Spin-labeled histidine kinases. *Methods Enzymol.* **423,** 52–116.

Brown, W. (1993). Dynamic Light Scattering: The Method and Some Applications. Clarendon Press, Oxford University Press, Oxford, New York.

Chan, C., Paul, R., Samoray, D., Amiot, N. C., Giese, B., Jenal, U., and Schirmer, T. (2004). Structural basis of activity and allosteric control of diguanylate cyclase. *Proc. Natl. Acad. Sci. USA* **101,** 17084–17089.

De, N., Pirruccello, M., Krasteva, P. V., Bae, N., Raghavan, R. V., and Sondermann, H. (2008). Phosphorylation-independent regulation of the diguanylate cyclase WspR. *PLoS Biol.* **6,** e67.

De, N., Navarro, M. V. A. S., Raghavan, R. V., and Sondermann, H. (2009). Determinants for the activation and autoinhibition of the diguanylate cyclase response regulator WspR. *J. Mol. Biol.* **393,** 619–633.

Debye, P. (1947). Molecular-weight determination by light scattering. *J. Phys. Colloid. Chem.* **51,** 18–32.

Falke, J. J., and Koshland, D. E. Jr. (1987). Global flexibility in a sensory receptor: A site-directed cross-linking approach. *Science* **237,** 1596–1600.

Falke, J. J., Dernburg, A. F., Sternberg, D. A., Zalkin, N., Milligan, D. L., and Koshland, D. E. Jr. (1988). Structure of a bacterial sensory receptor. A site-directed sulfhydryl study. *J. Biol. Chem.* **263,** 14850–14858.

Fujime, S. (1972). Quasi-elastic scattering of laser light. A new tool for the dynamic study of biological macromolecules. *Adv. Biophys.* **3,** 1–43.

Gao, R., and Stock, A. M. (2009). Biological insights from structures of two-component proteins. *Annu. Rev. Microbiol.* (Epub ahead of print). *Annu. Rev. Microbiol.* **63,** 133–154.
George, A., and Wilson, W. W. (1994). Predicting protein crystallization from a dilute solution property. *Acta Crystallogr. D Biol. Crystallogr.* **50,** 361–365.
Hansen, J. C., Lebowitz, J., and Demeler, B. (1994). Analytical ultracentrifugation of complex macromolecular systems. *Biochemistry* **33,** 13155–13163.
Hensley, P. (1996). Defining the structure and stability of macromolecular assemblies in solution: The re-emergence of analytical ultracentrifugation as a practical tool. *Structure* **4,** 367–373.
Hess, J. F., Oosawa, K., Matsumura, P., and Simon, M. I. (1987). Protein phosphorylation is involved in bacterial chemotaxis. *Proc. Natl. Acad. Sci. USA* **84,** 7609–7613.
Hess, J. F., Oosawa, K., Kaplan, N., and Simon, M. I. (1988). Phosphorylation of three proteins in the signaling pathway of bacterial chemotaxis. *Cell* **53,** 79–87.
Heyduk, T., Ma, Y., Tang, H., and Ebright, R. H. (1996). Fluorescence anisotropy: Rapid, quantitative assay for protein-DNA and protein-protein interaction. *Methods Enzymol.* **274,** 492–503.
Hickman, J. W., Tifrea, D. F., and Harwood, C. S. (2005). A chemosensory system that regulates biofilm formation through modulation of cyclic diguanylate levels. *Proc. Natl. Acad. Sci. USA* **102,** 14422–14427.
Howlett, G. J., Minton, A. P., and Rivas, G. (2006). Analytical ultracentrifugation for the study of protein association and assembly. *Curr. Opin. Chem. Biol.* **10,** 430–436.
Huang, B., Whitchurch, C. B., and Mattick, J. S. (2003). FimX, a multidomain protein connecting environmental signals to twitching motility in *Pseudomonas aeruginosa. J. Bacteriol.* **185,** 7068–7076.
Hura, G. L., Menon, A. L., Hammel, M., Rambo, R. P., Poole Ii, F. L., Tsutakawa, S. E., Jenney, F. E. Jr., Classen, S., Frankel, K. A., Hopkins, R. C., Yang, S. J., Scott, J. W., *et al.* (2009). Robust, high-throughput solution structural analyses by small angle X-ray scattering (SAXS). *Nat. Methods* (Epub ahead of print). *Nat. Methods* **6,** 606–612.
Jares-Erijman, E. A., and Jovin, T. M. (2003). FRET imaging. *Nat. Biotechnol.* **21,** 1387–1395.
Kameyama, K., and Minton, A. P. (2006). Rapid quantitative characterization of protein interactions by composition gradient static light scattering. *Biophys. J.* **90,** 2164–2169.
Kazmierczak, B. I., Lebron, M. B., and Murray, T. S. (2006). Analysis of FimX, a phosphodiesterase that governs twitching motility in *Pseudomonas aeruginosa. Mol. Microbiol.* **60,** 1026–1043.
Koch, M. H., Vachette, P., and Svergun, D. I. (2003). Small-angle scattering: A view on the properties, structures and structural changes of biological macromolecules in solution. *Q. Rev. Biophys.* **36,** 147–227.
Krukenberg, K. A., Forster, F., Rice, L. M., Sali, A., and Agard, D. A. (2008). Multiple conformations of *E. coli* Hsp90 in solution: Insights into the conformational dynamics of Hsp90. *Structure* **16,** 755–765.
Ladbury, J. E., and Chowdhry, B. Z. (1996). Sensing the heat: The application of isothermal titration calorimetry to thermodynamic studies of biomolecular interactions. *Chem. Biol.* **3,** 791–801.
Lathe, G. H., and Ruthven, C. R. (1955). The separation of substances on the basis of their molecular weights, using columns of starch and water. *Biochem. J.* **60,** xxxiv.
Lathe, G. H., and Ruthven, C. R. (1956). The separation of substances and estimation of their relative molecular sizes by the use of columns of starch in water. *Biochem. J.* **62,** 665–674.
Lebowitz, J., Lewis, M. S., and Schuck, P. (2002). Modern analytical ultracentrifugation in protein science: A tutorial review. *Protein Sci.* **11,** 2067–2079.

Luo, J., Leeman, M., Ballagi, A., Elfwing, A., Su, Z., Janson, J. C., and Wahlund, K. G. (2006). Size characterization of green fluorescent protein inclusion bodies in *E. coli* using asymmetrical flow field-flow fractionation-multi-angle light scattering. *J. Chromatogr. A* **1120,** 158–164.

Mao, X., Ren, Z., Parker, G. N., Sondermann, H., Pastorello, M. A., Wang, W., McMurray, J. S., Demeler, B., Darnell, J. E. Jr., and Chen, X. (2005). Structural bases of unphosphorylated STAT1 association and receptor binding. *Mol. Cell* **17,** 761–771.

Margarit, S. M., Sondermann, H., Hall, B. E., Nagar, B., Hoelz, A., Pirruccello, M., Bar-Sagi, D., and Kuriyan, J. (2003). Structural evidence for feedback activation by Ras.GTP of the Ras-specific nucleotide exchange factor SOS. *Cell* **112,** 685–695.

Minton, A. P. (2000). Quantitative characterization of reversible macromolecular associations via sedimentation equilibrium: An introduction. *Exp. Mol. Med.* **32,** 1–5.

Mogridge, J. (2004). Using light scattering to determine the stoichiometry of protein complexes. *Methods Mol. Biol.* **261,** 113–118.

Mylonas, E., and Svergun, D. I. (2007). Accuracy of molecular mass determination of proteins in solution by small-angle X-ray scattering. *J. Appl. Crystallogr.* **40,** s245–s249.

Navarro, M. V. A. S., De, N., Bae, N., Wang, Q., and Sondermann, H. (2009). Structural analysis of the GGDEF-EAL domain-containing c-Di-GMP receptor FimX. *Structure* **17,** 1–13.

Nobbmann, U., Connah, M., Fish, B., Varley, P., Gee, C., Mulot, S., Chen, J., Zhou, L., Lu, Y., Shen, F., Yi, J., and Harding, S. E. (2007). Dynamic light scattering as a relative tool for assessing the molecular integrity and stability of monoclonal antibodies. *Biotechnol. Genet. Eng. Rev.* **24,** 117–128.

Oliva, A., Llabres, M., and Farina, J. B. (2001). Comparative study of protein molecular weights by size-exclusion chromatography and laser-light scattering. *J. Pharm. Biomed. Anal.* **25,** 833–841.

Park, S. Y., Borbat, P. P., Gonzalez-Bonet, G., Bhatnagar, J., Pollard, A. M., Freed, J. H., Bilwes, A. M., and Crane, B. R. (2006). Reconstruction of the chemotaxis receptor-kinase assembly. *Nat. Struct. Mol. Biol.* **13,** 400–407.

Pecora, R. (1972). Quasi-elastic light scattering from macromolecules. *Annu. Rev. Biophys. Bioeng.* **1,** 257–276.

Petoukhov, M. V., and Svergun, D. I. (2005). Global rigid body modeling of macromolecular complexes against small-angle scattering data. *Biophys. J.* **89,** 1237–1250.

Pirruccello, M., Sondermann, H., Pelton, J. G., Pellicena, P., Hoelz, A., Chernoff, J., Wemmer, D. E., and Kuriyan, J. (2006). A dimeric kinase assembly underlying autophosphorylation in the p21 activated kinases. *J. Mol. Biol.* **361,** 312–326.

Porath, J., and Flodin, P. (1959). Gel filtration: A method for desalting and group separation. *Nature* **183,** 1657–1659.

Putnam, C. D., Hammel, M., Hura, G. L., and Tainer, J. A. (2007). X-ray solution scattering (SAXS) combined with crystallography and computation: Defining accurate macromolecular structures, conformations and assemblies in solution. *Q. Rev. Biophys.* **40,** 191–285.

Rao, F., Yang, Y., Qi, Y., and Liang, Z. X. (2008). Catalytic mechanism of cyclic di-GMP-specific phosphodiesterase: A study of the EAL domain-containing RocR from *Pseudomonas aeruginosa*. *J. Bacteriol.* **190,** 3622–3631.

Schachman, H. K., and Edelstein, S. J. (1966). Ultracentrifuge studies with absorption optics. IV. Molecular weight determinations at the microgram level. *Biochemistry* **5,** 2681–2705.

Schachman, H. K., Gropper, L., Hanlon, S., and Putney, F. (1962). Ultracentrifuge studies with absorption optics. II. Incorporation of a monochromator and its application to the study of proteins and interacting systems. *Arch. Biochem. Biophys.* **99,** 175–190.

Schurr, J. M. (1977). Dynamic light scattering of biopolymers and biocolloids. *CRC Crit. Rev. Biochem.* **4,** 371–431.

Sondermann, H., Soisson, S. M., Boykevisch, S., Yang, S. S., Bar-Sagi, D., and Kuriyan, J. (2004). Structural analysis of autoinhibition in the Ras activator Son of Sevenless. *Cell* **119,** 393–405.

Sondermann, H., Zhao, C., and Bar-Sagi, D. (2005). Analysis of Ras:RasGEF interactions by phage display and static multi-angle light scattering. *Methods* **37,** 197–202.

Svergun, D., Barberato, C., and Koch, M. H. J. (1995). CRYSOL—A program to evaluate X-ray solution scattering of biological macromolecules from atomic coordinates. *J. Appl. Crystallogr.* **28,** 768–773.

Svergun, D. I., Petoukhov, M. V., and Koch, M. H. (2001). Determination of domain structure of proteins from x-ray solution scattering. *Biophys. J.* **80,** 2946–2953.

Valente, J. J., Verma, K. S., Manning, M. C., Wilson, W. W., and Henry, C. S. (2005). Second virial coefficient studies of cosolvent-induced protein self-interaction. *Biophys. J.* **89,** 4211–4218.

Velazquez-Campoy, A., Leavitt, S. A., and Freire, E. (2004). Characterization of protein-protein interactions by isothermal titration calorimetry. *Methods Mol. Biol.* **261,** 35–54.

Volkov, V. V., and Svergun, D. I. (2003). Uniqueness of ab initio shape determination in small-angle scattering. *J. Appl. Crystallogr.* **36,** 860–864.

Wang, Q., Kaan, H. Y., Hooda, R. N., Goh, S. L., and Sondermann, H. (2008a). Structure and plasticity of endophilin and sorting nexin 9. *Structure* **16,** 1574–1587.

Wang, Q., Shui, B., Kotlikoff, M. I., and Sondermann, H. (2008b). Structural basis for calcium sensing by GCaMP2. *Structure* **16,** 1817–1827.

Wang, Q., Navarro, M. V., Peng, G., Molinelli, E., Lin Goh, S., Judson, B. L., Rajashankar, K. R., and Sondermann, H. (2009). Molecular mechanism of membrane constriction and tubulation mediated by the F-BAR protein pacsin/syndapin. *Proc. Natl. Acad. Sci. USA* (Epub ahead of print). *Proc. Natl. Acad. Sci. USA* **106,** 12700–12705.

Wassmann, P., Chan, C., Paul, R., Beck, A., Heerklotz, H., Jenal, U., and Schirmer, T. (2007). Structure of BeF3- -modified response regulator PleD: Implications for diguanylate cyclase activation, catalysis, and feedback inhibition. *Structure* **15,** 915–927.

Wen, J., Arakawa, T., and Philo, J. S. (1996). Size-exclusion chromatography with on-line light-scattering, absorbance, and refractive index detectors for studying proteins and their interactions. *Anal. Biochem.* **240,** 155–166.

Wilson, W. W. (2003). Light scattering as a diagnostic for protein crystal growth—A practical approach. *J. Struct. Biol.* **142,** 56–65.

Wyatt, P. J. (1997). Multiangle light scattering: The basic tool for macromolecular characterization. *Instrum. Sci. Technol. Instrum. Sci. Technol.* **25,** 1–18.

Yan, D., Cho, H. S., Hastings, C. A., Igo, M. M., Lee, S. Y., Pelton, J. G., Stewart, V., Wemmer, D. E., and Kustu, S. (1999). Beryllofluoride mimics phosphorylation of NtrC and other bacterial response regulators. *Proc. Natl. Acad. Sci. USA* **96,** 14789–14794.

Zimm, B. H. (1948). The scattering of light and the radial distribution function of high polymer solutions. *J. Chem. Phys.* **16,** 1093–1099.

Zoltowski, B. D., and Crane, B. R. (2008). Light activation of the LOV protein vivid generates a rapidly exchanging dimer. *Biochemistry* **47,** 7012–7019.

CHAPTER ELEVEN

High-Throughput Screening of Bacterial Protein Localization

John N. Werner *and* Zemer Gitai

Contents

Department of Molecular Biology, Princeton University, Princeton, New Jersey, USA

Methods in Enzymology, Volume 471
ISSN 0076-6879, DOI: 10.1016/S0076-6879(10)71011-9

Abstract

The ever-increasing number of sequenced genomes and subsequent sequence-based analysis has provided tremendous insight into cellular processes; however, the ability to experimentally manipulate this genomic information in the laboratory requires the development of new high-throughput methods. To translate this genomic information into information on protein function, molecular and cell biological techniques are required. One strategy to gain insight into protein function is to observe where each specific protein is subcellularly localized. We have developed a pipeline of methods that allows rapid, efficient, and scalable gene cloning, imaging, and image analysis. This work focuses on a high-throughput screen of the *Caulobacter crescentus* proteome to identify proteins with unique subcellular localization patterns. The cloning, imaging, and image analysis techniques described here are applicable to any organism of interest.

1. Introduction

It has become clear that subcellular protein localization is as essential for bacterial cells as it is for their eukaryotic counterparts (Gitai, 2005; Thanbichler and Shapiro, 2008). Since bacteria do not contain membrane-bound organelles to sequester specific enzymatic functions from one another, it is essential that they tightly coordinate the structural, enzymatic, and regulatory activities required for viability. Recent studies have established that the function of many proteins depends on their spatial and temporal localization: division proteins go to the division plane, polar development proteins go the pole, DNA replication proteins go to the replisome, etc. Thus, we are now in a position to use protein localization as a systematic method for characterizing protein function. With this understanding and the emergence of new molecular and cell biological methods, protein localization studies can be performed on a proteomic scale.

The development of advanced methods for observing specific subcellular protein localizations by high-resolution imaging of fluorescent protein fusions has coincided with an explosion of genomic resources and techniques. While protein localization studies have traditionally been performed on small subsets of individual proteins, genome-scale analysis of protein localization has the potential to identify functions for uncharacterized proteins and enhance understanding of previously studied proteins, identify potential protein interactions, and propose localization mechanisms. Several groups have performed large-scale efforts that have cataloged the localizations of most of the proteins of *Saccharomyces cerevisiae*, *Schizosaccharomyces pombe*, and *Escherichia coli* (Huh *et al.*, 2003; Kitagawa *et al.*, 2005; Matsuyama *et al.*, 2006); however, the laborious nature of generating, imaging, and scoring these libraries has limited their reanalysis under different conditions. In this work, we developed a pipeline of high-throughput cloning and imaging methods that allowed rapid, efficient, and scalable analysis of protein localization.

The pipeline of methods we describe here can be applied to a wide range of species for a variety of experimental applications. Here, we will focus on our initial application of the pipeline to a high-throughput protein localization screen in the gram-negative aquatic bacterium *Caulobacter crescentus* (Werner *et al.*, 2009). *Caulobacter* cells represent excellent models for studying bacterial cell biology due to their asymmetric polarity that can be readily visualized by the presence of a stalk that protrudes from one cell pole. This polarity provides cellular landmarks for specifically defining protein localization. Also, *Caulobacter* has well-studied cell shape, cell division, and cell-cycle regulated proteins known to have specific subcellular localization patterns that serve as positive controls for our protein localization screen (Ausmees *et al.*, 2003; Collier and Shapiro, 2007; Gitai *et al.*, 2004; Thanbichler and Shapiro, 2006). These characteristics along with the relative ease of manipulating *Caulobacter* made this an attractive subject for this study.

2. Pipeline Overview

The effort to generate a library of fluorescent protein fusions in *C. crescentus* involved five steps that we will detail in the sections below: (1) construction of a *Caulobacter* ORFeome library, (2) creating expression vectors by mobilization of the open reading frames (ORFs) into *Caulobacter* mCherry fusion destination vectors, (3) transferring of the expression vectors into *Caulobacter*, (4) imaging the resulting fusion strains, and (5) analysis of the imaging data.

For ease and modularity we adopted the Gateway (Invitrogen) system of recombinational cloning (Walhout *et al.*, 2000) in order to construct the expression clones. In the Gateway system, each ORF of interest is first cloned into a "donor" vector to create an "entry" vector (BP reaction). This library of entry vectors, also known as an ORFeome (Brasch *et al.*, 2004), is then mobilized into any "destination" vector to create an "expression" vector (LR reaction) for specific applications. For this study, our destination vectors created an inducible fluorescent fusion protein for each *Caulobacter* ORF. The Gateway system greatly reduces the time and effort required for traditional cut and paste cloning methods, and we were able to implement an "*in vivo* LR" method that significantly reduced the labor and expense of mobilizing the ORFs into destination vectors with similar cloning efficiency to traditional LR reactions. Once the expression vectors were generated in *E. coli* via *in vivo* LRs, we developed a high-throughput conjugation strategy to transfer them into *Caulobacter*, thereby generating the final desired fusion strains.

Upon completion of strain construction, thousands of *Caulobacter* strains containing fluorescent protein fusions needed to be imaged. This quantity of imaging required its own strategy as there were no previously developed methods for high-throughput imaging at the resolution necessary to visualize protein localization in bacteria. Traditional methods of imaging only a few samples on a slide would not be an efficient use of time for the number of strains that needed to be imaged, and available high-throughput imaging methods could not provide the resolution necessary for this study. Therefore, we designed a "pedestal slide" system that enabled the imaging of 48 samples on a single slide. Coupled with a robotic stage and semiautomated image acquisition script, we were able to decrease the time and effort to a minimum for this large-scale project.

Besides strain construction and imaging, an undertaking of this magnitude required a means to organize, store, validate, and analyze thousands of strains and images. A web site was created that organized all of the information regarding strain and primer locations. Scoring of each image was performed by individuals with the assistance of specially designed image viewing and analysis programs. Upon scoring of these images, the final steps to our complete library were validating the localization patterns by reimaging and verifying gene identity.

In this work, we developed several new methods and adapted others to create a novel integrated cloning and imaging pipeline. This pipeline of methods is applicable to many projects and may be adapted to any organism. Here we describe the generation of a library of localized proteins along with resources that have already been useful in other studies. The following sections will provide a more detailed description of the methods used in creating this localization library and describe the results we obtained.

3. Construction of a *Caulobacter* ORFeome

3.1. Rationale of entry vector cloning

The goal of this work was to create a library of proteins with specific subcellular localizations. In the process of attaining this goal we built a resource with which future high-throughput screens can be performed by creating a *Caulobacter* ORFeome. This ORFeome contains entry vectors that each possesses a single *Caulobacter* ORF. The modular nature of the Gateway system allows these ORFs to be transferred to the destination vector created for this study and any destination vector created in the future. Because of the relative ease of transfer from the entry vector to the destination vector, the initial step of entry vector construction was one of the most critical steps in the cloning process.

Cloning efficiency using the Gateway system is greatly enhanced by selection features in the "Gateway cassette" contained in the donor and destination vectors (Walhout *et al.*, 2000). The Gateway cassette is approximately 1.7 kb and is flanked by the phage lambda-derived *attP* and *attR* sequences in the donor and destination vectors, respectively. Within the *attP* and *attR* borders are the cytotoxic *ccdB* gene and a chloramphenicol resistance (Cam^R) marker (Walhout *et al.*, 2000). When a successful BP or LR reaction occurs, the *ccdB* gene and Cam^R marker are recombined out of the vector. If the recombination reaction is unsuccessful the CcdB protein will kill the CcdB-sensitive recipient cell, ensuring that only reactions resulting in inserted DNA into the donor or destination vectors will be transformed into the recipient strain. This can be verified by checking for chloramphenicol sensitivity. Recipient strains will be chloramphenicol sensitive if the resistance marker has been removed by a successful recombination reaction. This selection method greatly enhances the likelihood of transformed colonies containing the desired entry vector or expression vector and provides a simple method for validation.

3.2. Procedure for entry vector cloning

To create a *Caulobacter* ORFeome we started by PCR amplifying every ORF in the genome. A total of 3763 primers (Operon Biotechnologies) were designed to PCR amplify every protein encoding ORF using chromosomal DNA from the wild-type *Caulobacter* strain CB15 (Nierman *et al.*, 2001) as the template. Each primer was designed with a specific sequence added to the 5′ end of the "forward" (5′ end of the ORF) and "reverse" (3′ end of the ORF) primer that allowed specific recombination into the donor vector pDONR223 (Rual *et al.*, 2004). An ATG start codon was also included in the Forward primer that replaced the endogenous start codon. The 3′ end of each primer contained 15–20 bases of homology to the ORF of interest. The donor vector-specific sequence for each primer is shown below (Rual *et al.*, 2004):

Forward—5′ GGGGACAACTTTGTACAAAAAAGTTGGC**ATG** 3′
Reverse—5′ GGGGACAACTTTGTACAAGAAAGTTGGG 3′

These sequences contain the *attB* sites that recombine with *attP* sites (BP reaction) on the donor vector. All PCR reactions were performed in a 96-well format with KOD polymerase (Novagen) using the manufacturer's instructions. Initial verification of a successful PCR reaction was done by examining the length of the PCR product on E-gels (Invitrogen).

After confirming the size of each PCR product, BP reactions were performed using the Gateway protocols and reagents provided by

Invitrogen. The BP reaction, consisting of the recombination of the *attB* sites on the PCR product with the *attP* sites on pDONR223, was performed using BP clonase II and its specific protocol (Invitrogen). Upon completion of the BP reaction, samples were transformed into competent DH5α *E. coli* cells. After this validation method, sequencing (Agencourt) of the 5′ and 3′ end of each ORF in its entry vector was performed using M13 primers. Quality trimmed sequences were analyzed using BLAST analysis. This analysis confirmed that the correct ORF was present in its full length with no detectable point mutations.

3.3. Results

All cloning and imaging results and analysis, along with comprehensive tables and figures, are present in our earlier publication (Werner *et al.*, 2009). In order to create the *Caulobacter* ORFeome, we attempted PCR amplification of all 3763 ORFs for which we had PCR primers designed. Several rounds of PCR with slightly altered conditions were performed in order to meet the appropriate PCR conditions for as many ORFs as possible. Combining the results from all of the PCR attempts, we recovered PCR products for 3744 ORFs (99.5% of all ORFs); however, only 3184 PCR products were the correct length when observed on a gel (85.0% of all ORFs). BP reactions were performed with these PCR products and sequencing of the entry vector insert was performed to verify the identity, size, and sequence integrity of the ORF. Upon analysis of the sequencing results, we had obtained 2786 entry vectors containing a unique correct-size ORF. This represented 74.0% of the ORFs for which we designed primers. This set of clones represents version 1.0 of the *Caulobacter* ORFeome. Future work may attempt to increase this number of successfully cloned entry vectors for even better coverage of the *Caulobacter* genome.

4. Construction of the Fluorescently Tagged Protein Library

4.1. Overview

While the construction of the *Caulobacter* ORFeome generally followed manufacturers procedures, many of the subsequent methods were developed or adapted for this project (but applicable to many others). In the process of constructing inducible fluorescent fusion *Caulobacter* strains, we designed and created unique destination vectors, implemented new high-throughput methods for performing LR reactions (*in vivo* LR), and utilized new high-throughput conjugation methods for moving vectors from *E. coli* to *Caulobacter* cells. To build our library of fluorescently tagged proteins each ORF needed to be moved into a vector that would create a fluorescent protein fusion. We created our desired expression vectors by moving ORFs from entry vectors to destination vectors containing the mCherry fluorescent protein (Shaner *et al.*, 2004) via *in vivo* LR reactions. The resulting constructs also allowed us to control the expression of the fluorescent fusions by using the xylose-inducible *xylX* promoter (Meisenzahl *et al.*, 1997). The use of an inducible promoter allowed us to minimize potential toxic effects of protein overexpression and enables future studies on how protein localization responds to a range of protein concentrations. The expression vectors were constructed in *E. coli* and then transferred into *Caulobacter* for examination of protein localization.

4.2. Designing xylose-inducible mCherry fusion destination vectors

The first step in creating the fluorescent fusion library was to design and construct destination vectors. Our goal was to create fluorescent fusions to both the N- and C-terminus of each protein. Initially, we started by constructing the destination vector, gXRC (*G*ateway, *X*ylose-inducible, *R*ed fluorescent protein, *C*-terminal fusion), that would create an mCherry fusion to the C-terminus of each protein. This was followed by the construction of the N-terminal fusion destination vector gXRN. We chose the mCherry fluorescent protein because of its ability to fold and fluoresce in all cellular compartments (cytoplasm, periplasm, inner membrane, and outer membrane). We also included the *xylX* promoter so that the resulting fusion proteins would be under the control of a xylose-inducible promoter. In the absence of xylose no fluorescence is visible in cells; however, in the presence of xylose there is enough expression to visualize the localization of a fluorescent protein in *Caulobacter.*

The gXRC destination vector was created from the Kanamycin resistant (Kan^R) pXGFP4 (gift of M.R. Alley) which has an origin of replication (oriV) that allows replication in *E. coli* but not in *Caulobacter.* For *Caulobacter* to survive on media containing Kan the vector must integrate into the chromosome. pXGFP4 contains approximately 2.3 kilobases of the *Caulobacter* chromosome upstream and containing the *xylX* promoter which greatly increases the probability that the expression vector will integrate into this specific region when moved into *Caulobacter.* We generated gXRC by digesting pXGFP4 with *Nde*I and *Asp*718 and ligating with a similarly digested Gateway cassette PCR product amplified from pTGW (*Drosophila* Genomics Resource Center). The green fluorescent protein (GFP) in pXGFP4 was removed using *Not*I and *Asp*718 and replaced with the similarly digested mCherry PCR product amplified from pmCherry (Clontech). Thus, this cloning effort creates a vector with the Gateway cassette flanked on its 5′ end with the xylose promoter and its 3′ end with *mCherry* which creates the xylose-inducible C-terminal fusion.

Once gXRC was constructed and validated, gXRN was generated. The vector pXGFP4-C1 (gift of M. R. Alley), which creates N-terminal GFP fusions, was used to create this destination vector. pXGFP4-C1 was digested with *Bgl*II and *Asp*718 and ligated with a similarly digested PCR product amplified from pTGW. The GFP was removed from pXGFP4-C1 by digesting with *Bgl*II and *Nde*I and replaced with the appropriately digested mCherry PCR product amplified from pmCherry. This creates a vector with *mCherry* flanked by the xylose promoter and the Gateway cassette creating an N-terminal fusion vector. Successful construction of gXRC and gXRN allowed us to proceed to the next step of performing LR reactions with the *Caulobacter* ORFeome.

4.3. Procedure for the *in vivo* LR reaction

With the completion of the construction of gXRC and gXRN, expression vectors were made by performing LR reactions. The LR reaction recombines the *attL* sequences created on the entry vector, as a result of the BP reaction, with *attR* sequences in the destination vectors. As with the BP reaction, Invitrogen offers a commercial enzyme to facilitate this recombination step. However, this purified enzyme, while cost effective on a small scale, is a large expense when used at the scale of cloning the *Caulobacter* genome. Also, performing the traditional LR reaction *in vitro* required purifying destination vectors and transforming the LR reaction product into competent cells. These steps would be very labor intensive to perform at this scale. Therefore, we implemented a method to perform this LR reaction *in vivo* without the need for purified recombinase enzymes. We have named this approach the "*in vivo* LR" reaction.

The *in vivo* LR method builds upon work studying the *xis* and *int* genes that mediate the LR recombination reaction (Platt *et al.*, 2000), as well as earlier low-throughput methods to perform Gateway reactions in living cells (Schroeder *et al.*, 2005). The *in vivo* LR requires three features besides the entry clone and destination vectors needed for a traditional LR reaction. First, an *E. coli* strain containing the pXINT129 plasmid (Platt *et al.*, 2000) that expresses the *xis* and *int* genes under the control of an isopropyl β-D-thiogalactopyranoside (IPTG)-inducible promoter was necessary to provide the recombinase enzymes for the LR reaction. Second, destination vectors must contain an origin of transfer (oriT) that allows the vector to be transferred from one organism to another via conjugation. Third, a conjugation helper strain (LS980) was used to facilitate conjugation between multiple strains, allowing the oriT-containing destination vector to be transferred from one strain to another.

The *in vivo* LR reaction is a series of recombination and conjugation events that transfers an ORF from the entry vector to the destination vector. The first step in the process was to transform isolated Spectinomycin/Streptomycin resistant ($Spec^R/Strep^R$) entry vectors into a chemically competent *E. coli* strain containing pXINT129 using a traditional heat shock transformation protocol. Four microliters of isolated entry vectors were added to each well of a 96-well plate containing 50 μl of chemically competent pXINT129 *E. coli* cells in each well. Each plate was heat shocked in a 42 °C water bath for 60 s. Immediately following the heat shock, plates were placed on ice and 100 μl LB broth (Silhavy *et al.*, 1984) was added to each well. Plates were then incubated at 37 °C for 1 h. Fifty microliters of each transformation was added to 750 μl LB broth containing Spectinomycin (Calbiochem), Streptomycin (Fisher Bioreagents), and Kanamycin (Agri-bio, Inc.) to isolate *E. coli* strains containing the pXINT129 (Kan^R) plasmid and the entry vector and grown overnight.

To perform the *in vivo* LR, the entry vector/pXINT129 containing strains were combined with an *E. coli* strain containing gXRC (Kan^R), a conjugation helper *E. coli* strain (Chloramphenicol resistant; Cam^R), and a *ccdB*-sensitive, Rifampin resistant ($CcdB^S$ Rif^R) recipient DH5α *E. coli* strain at a 1:1:1:1 ratio (50 μl each). This mixture of cultures was spotted with a 48-pin pinning tool (frogger, DanKar Corporation) on LB agar plates containing 1 mM IPTG (Ambion, Inc.) in order to express the recombinase enzymes. After overnight growth, the spots containing the *E. coli* strain mixtures were transferred with a frogger to LB broth containing Rifampin (Fisher Bioreagents) and Kanamycin to select for expression vectors that have been transferred to the recipient strain.

While we have not studied the specific course of events that occurs in the *in vivo* LR step, we can describe the likely path gXRC follows to become an expression vector. gXRC contains an *oriT* so it can be transferred from one strain to another via conjugation. It is likely gXRC is first transferred to the strain containing the entry vector and pXINT129. The Xis and Int proteins expressed from pXINT129 mediate the recombination of the ORF from the entry vector into gXRC. The recombined gXRC that now contains the gene of interest in place of the *ccdB* gene is then transferred again via conjugation from this strain to the recipient Rif^R DH5α strain. This desired strain is purified from the other *E. coli* strains by growing in LB broth containing Rifampin and Kanamycin. All *E. coli* strains but the recipient strain containing the expression vector are killed by one or both of these antibiotics or the CcdB protein resulting from the unrecombined gXRC. It is probable that this is an inefficient process at the cellular level but we only need a small fraction to work for the procedure to be successful.

4.4. Protocol for fast and efficient transfer of expression vectors from *E. coli* to *Caulobacter*

The final step in constructing the strains to perform the large-scale localization screen was to transfer the expression vectors from *E. coli* to *Caulobacter*. Again, we performed this in a high-throughput manner using conjugation. In 96-well plates, we combined cultures of the *E. coli* strains containing expression vectors, the conjugation helper strain, and the wild-type *Caulobacter* strain CB15N at a 1:1:1 ratio (50 μl each). This mixture was spotted on PYE plates (Ely, 1991) with a 48-pin frogger to allow conjugation to occur between the *E. coli* and *Caulobacter* strains. The spots were incubated at 30 °C for 48 h and then streaked on PYE plates containing Nalidixic acid (Nal, Fisher Bioreagents) and Kanamycin. Kanamycin selected for the expression vector while the Nalidixic acid selected against *E. coli*. Only after completing this step did we discover that DH5α cells have resistance to Nalidixic acid which explained much of the difficulty we had obtaining pure cultures of *Caulobacter* without *E. coli* contamination. For future studies we will use an *E. coli* recipient strain that is sensitive to Nalidixic acid or use a different selection to isolate *Caulobacter* from *E. coli*. To verify that the Kan^R and Nal^R *Caulobacter* strain we obtained by purification actually possessed an expression vector containing an insert, we patched two colonies from each strain on PYE plates containing either Kanamycin or Kanamycin and Chloramphenicol (Fisher Bioreagents). Strains that were Kan^R but Cam^S were deemed to have a successfully cloned expression vector due to the removal of the Cam^R Gateway cassette.

The generation of multiple sets of strains along the path to creating the final *Caulobacter* strains required careful handling and storage. At each step along the cloning and conjugation pipeline *E. coli* and *Caulobacter* strains were stored at −80 °C. For future use, we stored the *E. coli* strains containing entry vectors, *E. coli* strains containing pXINT129 and the entry vectors, *E. coli* strains containing expression vectors, and the *Caulobacter* strains containing expression vectors. All strains were stored in 96-well plates; *E. coli* were stored in LB broth containing 20% glycerol and *Caulobacter* were stored in PYE broth containing 20% glycerol.

4.5. Results

For this step of the pipeline we created a system that allowed us to perform LR reactions without purified recombinase enzymes or isolated entry vectors and destination vectors. This system involved mixing four cultures together and letting the bacteria perform the work. We also developed a high-throughput way of transferring expression vectors from *E. coli* to *Caulobacter* without the need to isolate vectors and perform transformations. Once these methods were optimized, the entire process (from the transformation of entry vectors into the pXINT129-containing *E. coli* cells, to the final step of patching to confirm the isolation of the desired *Caulobacter* strains) could be performed in less than 2 weeks.

5. Imaging the *Caulobacter* Protein Localization Library

5.1. Overview

Once strain construction was complete we were ready to image the *Caulobacter* protein localization library. To visualize the *Caulobacter* cell borders and the fluorescent protein it was necessary to use phase contrast and fluorescent imaging. Traditional methods of examining protein localization in bacterial strains were not sufficient for the throughput necessary to image this volume of strains. Air immersion objectives could be used for high-throughput work but did not provide the resolution necessary to observe protein localization. Therefore, we developed a method utilizing a robotic stage and custom-made pedestal slides that allowed us to image strains in high-throughput at maximal diffraction-limited resolution using oil immersion objectives.

5.2. Protocol for high-throughput imaging of fluorescent fusion proteins in *Caulobacter*

To prepare this volume of strains for imaging we needed to optimize cell growth and induction conditions. All strains were grown overnight in a 96-well format containing PYE broth with Kanamycin. Four hours prior to imaging, cultures were diluted 1:15 in PYE broth with Kanamycin to allow the strains to recover from stationary phase. Two hours prior to imaging, cultures were induced with 0.03% xylose. This level of xylose was sufficient to observe protein localization in control strains but low enough to reduce toxicity effects caused by overexpression.

High-throughput high-resolution imaging was performed by developing a 48-pad "pedestal slide" and by writing a script so that the microscope automatically performs many of the repetitive imaging tasks. The 48-pad pedestal slide was developed by designing a stainless steel mold as shown in Fig. 11.1A and using custom-made coverglass slips of the same length and width dimensions as the mold. To create the pedestal slide (Fig. 11.1B), one coverglass was placed on a lab bench with two molds placed on top flush with each other and the bottom coverglass. Molten 1% agarose was poured over the molds and coverglass filling all of the holes. While the agarose was still in liquid form, a second coverglass was placed on top of the molds expelling the excess agarose from the top of the molds. The agarose was then allowed to solidify in the mold for approximately 5 min. Once the

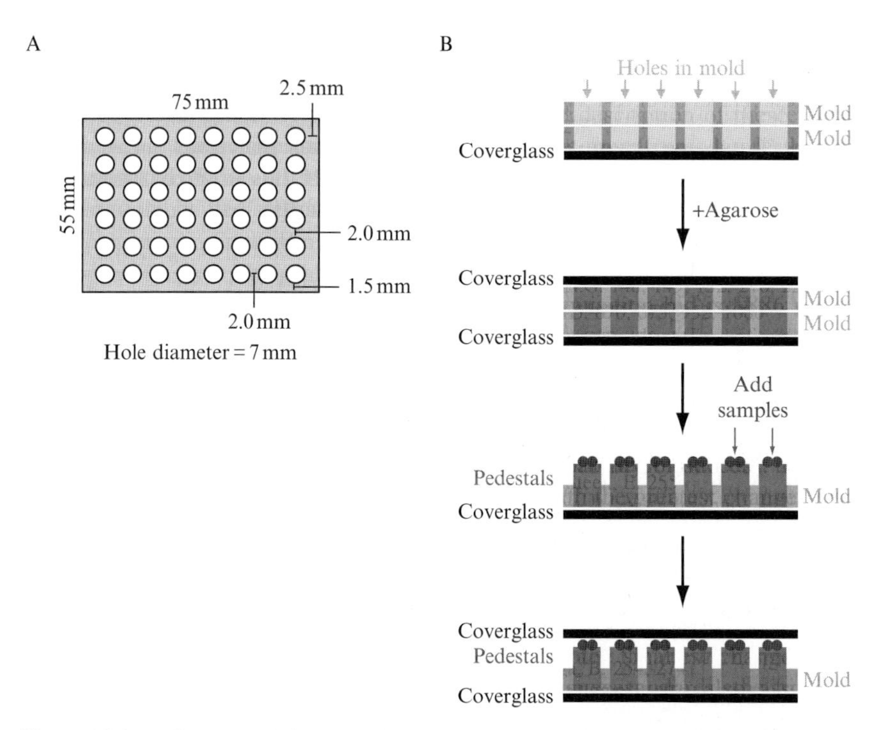

Figure 11.1 Schematic of the 48-well pedestal slide mold and pedestal slide assembly. (A) Measurements of mold are 75 mm long by 55 mm wide by 1 mm thick. The diameter of each hole is 7 mm. The width of the long border, short border, and width between holes is 1.5, 2.5, and 2.0 mm, respectively. (B) Flow chart of pedestal slide assembly as described in the text. A side view of a slice through the assembly sandwich is shown with the stainless steel molds in brown, empty holes in the mold in light gray, coverglass in black, agarose pedestals in dark gray, and culture samples in red. (See Color Insert.)

agarose solidified, the sandwich of coverglasses and molds was flipped over so the bottom coverglass was on top and this coverglass was removed by carefully sliding it off the sandwich. The top mold was then pried off, exposing 48 agarose pads 1 mm in height. These pads were allowed to dry for 10–15 min prior to the application of culture samples.

Imaging of the fluorescent fusion library in *Caulobacter* was ready to be performed after the preparation of the pedestal slides. Culture growth and induction was timed to coincide with the preparation of the pedestal slides. Culture samples (3 μl) were pipetted on the 48-pedestal slides using a multichannel pipettor and fields of approximately 50–200 cells were chosen for imaging. All imaging was performed with a Nikon90i epifluorescent microscope equipped with a 100× 1.4 NA objective (Nikon), Rolera XR cooled CCD camera (QImaging), and NIS Elements software (Nikon). A script automated the tasks of taking phase and fluorescent images, naming and saving each image, and moving the objective to the next pedestal on the slide. After optimization, a single 48-pedestal slide could be imaged in as little as 15–20 min.

5.3. Results

Without the development of these high-throughput methods for imaging, this project would not have been feasible. Over 6400 strains (gXRC and gXRN expression vectors) were imaged at least once and most were imaged multiple times. It would have taken months to image that volume of strains using traditional imaging methods and would have been prohibitory to any future efforts to reimage this library. With these methods, three or four 96-well plates were comfortably imaged per day, which allowed the completion of this effort in merely a few weeks. We currently estimate that the entire library of C- and N-terminal fusions could be reimaged by a single person in less than 1 month.

6. Image Scoring and Analysis

6.1. Overview

The next step, after imaging all of the C- and N-terminal fusion proteins, was to examine the images to identify proteins that localized to discrete nonuniform patterns in *Caulobacter*. While image analysis software is constantly improving and certain aspects of localization analysis are immensely aided by quantitation, we found that the best initial analysis method was to simply examine each image by eye. Since the images were collected using a script that automated standardized file naming, we used a MATLAB (Matworks) script to automate opening the image files, aid in adjusting contrast levels, provide a standardized menu of scoring options, and save the scoring results in an easily accessible format. The use of these scripts allowed us to efficiently view each image, assign a localization pattern, and easily access and manipulate the scoring information. This scoring data allowed us to readily pare down our set of fluorescent protein fusions to a library of proteins with specific localization patterns.

In order to collect meaningful data from thousands of images of bacterial cells we needed to define what constituted a "localized" protein. For our study, we defined a localized protein as having a specific subcellular non-diffuse fluorescence pattern. The most abundant and easily recognized pattern was diffuse fluorescence defined as a confluent localization across the entire cell body. It is possible that we missed some cells with a subtle localization pattern that we were unable to detect by eye, but as a first pass we were satisfied to categorize these proteins as diffuse. Another set of proteins localized to the periphery of the cell creating a fluorescent cell outline suggesting that these proteins may reside in an extracytoplasmic compartment; however, these proteins did not show any other specific localization pattern. We chose to categorize both the diffuse and peripherally located proteins as "not localized" because both patterns appeared diffuse even if confined to a specific compartment of the cell.

While cells containing diffuse and peripherally localized proteins were the most common and easiest to identify, there were also strains that exhibited obvious localization patterns that were not difficult to discern. Many proteins localized as a tight focus in the cell. Interestingly, there could be one or multiple foci localized in several locations in the cell. A single focus could be localized at a pole or at a central location in the cell. One of the benefits of using *Caulobacter* as a model organism is the ability to differentiate the stalked and swarmer poles. Although we did not utilize this feature of *Caulobacter* for this study, future work will define polar localization more specifically. Some cells displayed more than one focus in the cell. Cells containing foci at both poles were termed "bipolar" while cells with foci at a pole and in the central region of the cell were called "Middle and Pole." Other cells displayed a single focus at different locations in the cell. Cells that consistently showed a single focus at either a pole or a central location were defined as "Middle or Pole" while cells that displayed a focus at an inconsistent location throughout the cell were called "Variable Focus."

Some fusions also localized in patterns other than a focus. Several proteins localized as a band across the middle of the cell while other proteins localized as a line of varying lengths along the long axis of the cell. Also, a previously unidentified localization category was proteins that were localized exclusively in the stalk. A final category was created that encompassed other obviously nonuniform yet less precise localization patterns. The "patchy/spotty" category encompassed proteins that may localize as helices or other structures. Patchy referred to light and dark patches in the cell that were not as defined as foci or bands while spotty referred to more than two foci in the cell often at less reproducible locations. The patchy/spotty category was the largest of the localized proteins but the most difficult to define.

6.2. Protocol for image scoring and validation

In order to score each image for the localization patterns described above we needed a program to assist collecting and organizing the scoring data. Besides localization patterns we could also describe problems with the image that may have prevented choosing a localization pattern such as too little fluorescence, a blurry picture, or *E. coli* contamination. All of this information was saved to a Microsoft Excel spreadsheet file that allowed us to easily summarize and manipulate the localization scoring data.

In order to eliminate some of the scoring bias that would be present by having a single individual score all of the pictures, several lab members were involved in the scoring process. Initially, two individuals scored all of the images and created a library of "possibly localized" proteins. Once this more manageable set of images was created, seven members of the lab independently scored each image to decide if each protein was "localized" or "not localized." Finally, if five of the seven individuals deemed a protein to be localized, a specific pattern, as described above, was assigned to its localization. The "Localization Scorer" program is freely available upon request.

Once the set of localized proteins was collected it was necessary to validate the identities of the fusion proteins in each *Caulobacter* strain. Sequencing of PCR products amplified with primers from inside the *xylX* promoter and *mCherry* gene for gXRC and from within the *mCherry* gene to a region downstream from the Gateway cassette in gXRN verified the expected ORF was present in the expression vector. Only the *Caulobacter* strains in the localized set were sequence verified. Proteins that were both deemed localized by the consensus and were sequence verified constituted our final library of localized *Caulobacter* proteins.

The creation of a large set of localized proteins allowed us to perform statistical and quantitative analyses to examine general localization trends and to analyze the accuracy and distributions of localization patterns. For example, we statistically compared our set of localized proteins with their gene ontology (GO) functional classifications (Harris *et al.*, 2004) to identify cellular activities that are more or less likely to function in precise cellular locations. To quantitate the localization patterns themselves, we used a software suite termed PSICIC (Guberman *et al.*, 2008). PSICIC allowed us to perform a precise examination of protein localization and fluorescence intensity using interpolated contours to achieve subpixel resolution and generate an internal coordinate system for each cell. This allowed us to directly compare localization data between multiple cells regardless of cell geometries.

6.3. Results

After scoring and validating the localization library we were able to analyze the abundance of information we collected. Of the potential 5572 fusions in gXRC and gXRN, we identified a total of 352 localized fusion proteins (6.3% of all fusions) of which 187 localized as C-terminal fusions and 165 localized as N-terminal fusion proteins. Of these 352 fusions, 63 proteins localized with both a C- or N-terminal fusion, which gave a total of 289 unique proteins with localization patterns (Werner *et al.*, 2009). These results testify to the value of the high-throughput methods we developed.

Without the ability to create both C- and N-terminal fusion libraries we would have missed roughly one-third of the localized proteins.

After identifying whether the fusion proteins were localized, they were categorized into the specific classes described above. Of the 63 proteins that localized as both a C- or N-terminal fusion, 58 produced the same localization pattern giving us a "high-confidence" set of localized proteins. Another means to validate the localized proteins was to compare our set to proteins that have previously been reported as localized as fluorescent fusions. We identified 29 proteins whose genes were present in our ORFeome that localized as fluorescent protein fusions in previous studies. The localization patterns of 24 of the 29 proteins (83%) were recapitulated in our study indicating a relatively low rate of false-negative localizations. It is possible that the five localizations that did not match could be due to expression levels or the fluorescent protein used.

The statistical analysis correlating GO classifications to localized proteins provided insight into the cellular location of specific processes. This analysis found that proteins involved in small molecule metabolism were underrepresented in our localized protein library. This suggests that rapid diffusion of small molecules throughout the cell may not require precisely localized biosynthetic machinery. As expected, proteins involved in cellular processes such as cell division and chromosome partitioning, motility, and DNA replication/recombination/repair were overrepresented in the library of localized proteins along with less obvious processes such as signal transduction, intracellular trafficking and secretion, and membrane and cell wall biogenesis (Werner *et al.*, 2009). These overenriched GO classes suggest that processes involved in spatial and temporal regulation inside the cell require localized proteins. This is consistent with the finding that genes whose transcription is cell-cycle regulated (Laub *et al.*, 2000) are overrepresented in our set of localized proteins.

The use of PSICIC to precisely analyze protein localization provided intriguing information on general localization trends in *Caulobacter*. For analysis using PSICIC we excluded proteins classified as having a "patchy/spotty" localization pattern to reduce complexity. By examining the distribution of the mean position of each fluorescent fusion we found that there was an enrichment of localized proteins at the cell pole and near the middle of the cell. The standard deviation of the peak intensity position of these proteins was low suggesting that these localization patterns are highly reproducible. Interestingly, there were relatively few proteins that localized to the region between the pole and midcell. Proteins that did localize to this region had higher standard deviations of the peak intensity position, suggesting their localization patterns were more variable. These data suggest that from the cell pole to quartercell protein, localization becomes less accurate, while accuracy increases again between quarter- and midcell (Werner *et al.*, 2009). PSICIC also allowed us to examine how tightly a protein is localized in a single cell by examining the distance between the peak fluorescence intensity and its half-maximal value. These data found that polar proteins localized more tightly than proteins localized to the midcell (Werner *et al.*, 2009). The combination of the findings that proteins are more likely to localize to the poles or midcell but proteins localized to the pole are more tightly localized suggests that there may be fundamentally different mechanisms involved in localizing these proteins.

An organized and accessible database was created to allow other individuals to access all of the information produced in this study. We designed a web site that allows public access to the list of proteins represented in our library and images of *Caulobacter* strains with localized proteins. Access to the list and images is found at the Gitai lab web site: www.molbio1.princeton.edu/labs/gitai/.

7. Conclusion

Methods and resources created in this study proved valuable for providing interesting data for this large-scale localization project but also provided the means to perform future high-throughput experimentation. In this study, we created a *Caulobacter* ORFeome that can be used to easily clone any ORF into any destination vector. To do this in a high-throughput manner, a protocol for performing the Gateway LR reaction was developed that reduces the effort and expense of traditional methods. Also, a technique to perform high-throughput high-resolution microscopy was developed that allows the reimaging of the current library or imaging of other libraries quickly and in a highly organized way. As useful as our library of localized proteins is for studying protein function and finding protein localization mechanisms, the techniques and resources we generated are just as valuable.

Generating our library of localized *Caulobacter* proteins identified new aspects of protein localization. Prior to this study there were no proteins known to specifically localize to the stalk. Also, there was only one protein, crescentin (Ausmees *et al.*, 2003), shown to localize as a line while our study has identified two others. Computational analysis using PSICIC identified regions of the cell that lack as many localized proteins as other regions when examining only the proteins that localized in patterns other than patchy/spotty. Correlating protein localization to GO classifications discovered trends that were previously suspected, such as that motility and cell division proteins are more likely to be localized, but also found that other groups such as secretion and cell wall/membrane biogenesis proteins are also more likely to be localized. The creation of this library has provided a resource for future studies to be performed on the scale of an individual protein or at a proteomic scale of looking for localization trends or mechanisms.

Looking ahead only at the future projects that can be performed with the localization library provides just a small glimpse of the usefulness of this resource. While the feats of performing large-scale localization projects in yeast and *E. coli* were tremendous achievements, the difficulty associated with reimaging these libraries under various conditions without huge resource expenditures limited their subsequent utility. While the construction and imaging of our initial library took months to complete, subsequent imaging of the localized set of proteins can be performed in a single day. Future studies may include screening the localized set of proteins under various nutrient conditions or examining this library in the presence or absence of toxic compounds or antibiotics.

Creation of the *Caulobacter* ORFeome coupled with the *in vivo* LR reaction and high-throughput microscopy has provided a resource for future studies. Further examination of these proteins is only limited by the destination vector that can be created. Vectors are commercially available that produce tags for protein purification or pull-down assays and we are in the process of generating other destination vectors for a range of functional genomic studies. While we plan to increase the comprehensiveness of this ORFeome in the future, it has already provided a resource to perform many high-throughput studies.

This study has provided a template for manipulating genomic data at an experimental level. The ability to clone ORFs for an entire genome into a modular system such as the Gateway system has enabled researchers to perform new screens or comprehensive studies of protein function at a proteomic scale. We were able to couple the Gateway cloning system with new microscopy methods to perform high-throughput high-resolution microscopy-based screens for the first time. These methods are not confined to *Caulobacter* studies. Nearly all of the methods described in this work can be adapted to work in any bacteria or higher organism. These microscopy methods can be used to study any organism that fits under a microscope. This work has thus created a resource for studying protein localization, a more efficient and cost-effective method for high-throughput cloning, and a template for performing microscopy-based screens at the genomic level.

ACKNOWLEDGMENTS

The authors thank Eric Chen for his assistance in creating the *Caulobacter* ORFeome and imaging, Jonathan Guberman for writing the image acquisition and image scoring programs, Angela Zippilli and Joe Irgon for gene sequencing and validation, and the entire Gitai lab for assistance with image scoring. We also thank Greg Phillips (Iowa State University, Ames, IA), Michael Kahn (Washington State University, Pullman, WA), Martin Thanbichler (Max Planck Institute for Terrestrial Microbiology, Marburg, Germany), and M. R. Alley (Anacor Pharmaceuticals, Palo Alto, CA) for materials, and Denis Dupuy and the rest of Marc Vidal's lab for help in constructing the *Caulobacter* ORFeome entry library. John N. Werner is supported by a postdoctoral fellowship, Grant 1F32AI073043-01A1, from the National Institute of Allergy and Infectious Diseases and Zemer Gitai is supported by funding from Grant DE-FG02-05ER64136 from the U.S. Department of Energy Office of Science (Biological and Environmental Research) and a National Institute of Health New Innovator Award Number 1DP2OD004389-01.

REFERENCES

Ausmees, N., Kuhn, J. R., and Jacobs-Wagner, C. (2003). The bacterial cytoskeleton: An intermediate filament-like function in cell shape. *Cell* **115,** 705–713.

Brasch, M. A., Hartley, J. L., and Vidal, M. (2004). ORFeome cloning and systems biology: Standardized mass production of the parts from the parts-list. *Genome Res.* **14,** 2001–2009.

Collier, J., and Shapiro, L. (2007). Spatial complexity and control of a bacterial cell cycle. *Curr. Opin. Biotechnol.* **18,** 333–340.

Ely, B. (1991). Genetics of *Caulobacter crescentus. Methods Enzymol.* **204,** 372–384.

Gitai, Z. (2005). The new bacterial cell biology: Moving parts and subcellular architecture. *Cell* **120,** 577–586.

Gitai, Z., Dye, N., and Shapiro, L. (2004). An actin-like gene can determine cell polarity in bacteria. *Proc. Natl. Acad. Sci. USA* **101,** 8643–8648.

Guberman, J. M., Fay, A., Dworkin, J., Wingreen, N. S., and Gitai, Z. (2008). PSICIC: Noise and asymmetry in bacterial division revealed by computational image analysis at sub-pixel resolution. *PLoS Comput. Biol.* **4,** e1000233.

Harris, M. A., Clark, J., Ireland, A., Lomax, J., Ashburner, M., Foulger, R., Eilbeck, K., Lewis, S., Marshall, B., Mungall, C., Richter, J., Rubin, G. M., *et al.* (2004). The gene ontology (GO) database and informatics resource. *Nucleic Acid Res.* **32,** 258–261.

Huh, W. K., Falvo, J. V., Gerke, L. C., Carroll, A. S., Howson, R. W., Weissman, J. S., and O'Shea, E. K. (2003). Global analysis of protein localization in budding yeast. *Nature* **425,** 671–672.

Kitagawa, M., Ara, T., Arifuzzaman, M., Ioka-Nakamichi, T., Inamoto, E., Toyonaga, H., and Mori, H. (2005). Complete set of ORF clones of *Escherichia coli* ASKA library (a complete set of *E. coli* K-12 ORF archive): Unique resources for biological research. *DNA Res.* **12,** 291–299.

Laub, M. T., McAdams, H. H., Feldblyum, T., Fraser, C. M., and Shapiro, L. (2000). Global analysis of the genetic network controlling a bacterial cell cycle. *Science* **290,** 2144–2148.

Matsuyama, A., Arai, R., Yashiroda, Y., Shirai, A., Kamata, A., Sekido, S., Kobayashi, Y., Hashimoto, A., Hamamoto, M., Hiraoka, Y., Horinouchi, S., and Yoshida, M. (2006). ORFeome cloning and global analysis of protein localization in the fission yeast *Schizosaccharomyces pombe. Nat. Biotechnol.* **24,** 841–847.

Meisenzahl, A. C., Shapiro, L., and Jenal, U. (1997). Isolation and characterization of a xylose-dependent promoter from *Caulobacter crescentus. J. Bacteriol.* **179,** 592–600.

Nierman, W. C., Feldblyum, T. V., Laub, M. T., Paulsen, I. T., Nelson, K. E., Eisen, J. A., Heidelberg, J. F., Alley, M. R., Ohta, N., Maddock, J. R., Potocka, I., Nelson, W. C., *et al.* (2001). Complete genome sequence of *Caulobacter crescentus. Proc. Natl. Acad. Sci. USA* **98,** 6533.

Platt, R., Drescher, C., Park, S. K., and Phillips, G. J. (2000). Genetic system for reversible integration of DNA constructs and *lacZ* gene fusions into the *Escherichia coli* chromosome. *Plasmid* **43,** 12–23.

Rual, J. F., Hirozane-Kishikawa, T., Hao, T., Bertin, N., Li, S., Dricot, A., Li, N., Rosenberg, J., Lamesch, P., Vidalain, P. O., Clingingsmith, T. R., Hartley, J. L., *et al.*

(2004). Human ORFeome version 1.1: A platform for reverse proteomics. *Genome Res.* **14,** 2128–2135.

Schroeder, B. K., House, B. L., Mortimer, M. W., Yurgel, S. N., Maloney, S. C., Ward, K. L., and Kahn, M. L. (2005). Development of a functional genomics platform for *Sinorhizobium meliloti*: Construction of an ORFeome. *Appl. Environ. Microbiol.* **71,** 5858–5864.

Shaner, N. C., Campbell, R. E., Steinbach, P. A., Giepmans, B. N., Palmer, A. E., and Tsien, R. Y. (2004). Improved monomeric red, orange and yellow fluorescent proteins derived from *Discosoma* sp. red fluorescent protein. *Nat. Biotechnol.* **22,** 1567–1572.

Silhavy, T. J., Berman, M. L., and Enquist, L. W. (1984). Experiments with Gene Fusions. Cold Spring Harbor Laboratory Press, Cold Spring Harbor, NY.

Thanbichler, M., and Shapiro, L. (2006). MipZ, a spatial regulator coordinating chromosome segregation with cell division in *Caulobacter. Cell* **126,** 147–162.

Thanbichler, M., and Shapiro, L. (2008). Getting organized—How bacterial cells move proteins and DNA. *Nat. Rev. Microbiol.* **6,** 28–40.

Walhout, A. J., Temple, G. F., Brasch, M. A., Hartley, J. L., Lorson, M. A., van den Heuvel, S., and Vidal, M. (2000). GATEWAY recombinational cloning: Application to the cloning of large numbers of open reading frames or ORFeomes. *Methods Enzymol.* **328,** 575–592.

Werner, J. N., Chen, E. Y., Guberman, J. M., Zippilli, A. R., Irgon, J. J., and Gitai, Z. (2009). Quantitative genome-scale analysis of protein localization in an asymmetric bacterium. *Proc. Natl. Acad. Sci. USA* **106,** 7858–7863.

CHAPTER TWELVE

In Vitro and *In Vivo* Analysis of the ArcB/A Redox Signaling Pathway

Adrián F. Alvarez *and* Dimitris Georgellis

Contents

Abstract

The Arc (anoxic redox control) two-component system (TCS) is a complex signal transduction system that plays an important role in regulating energy metabolism at the level of transcription in bacteria. This system comprises the ArcB protein, a hybrid membrane-associated sensor kinase, and the ArcA protein, a typical response regulator. Under anoxic growth conditions, ArcB

Departamento de Genética Molecular, Instituto de Fisiología Celular, Universidad Nacional Autónoma de México, México D.F., México

Methods in Enzymology, Volume 471
ISSN 0076-6879, DOI: 10.1016/S0076-6879(10)71012-0

autophosphorylates and transphosphorylates ArcA via a His → Asp → His → Asp phosphorelay. Under aerobic conditions, the ArcB kinase activity is silenced by the oxidation of two cytosol-located redox-active cysteine residues that participate in intermolecular disulfide bond formation. Under such conditions, ArcB acts as a phosphatase that catalyzes the dephosphorylation of ArcA-P and thereby releasing its transcriptional regulation. This chapter describes general *in vitro* and *in vivo* assays and strategies that have been used to characterize the ArcB/A two-component signal transduction system, which could, also, be applied to most other TCS.

1. Introduction

Bacteria are equipped with a variety of mechanisms to monitor and respond to diverse environmental stimuli and elicit appropriate adaptive responses, to assure growth and survival. In prokaryotic cells, the sensing and processing of these stimuli rely largely on two-component signal transduction pathways that depend on histidine and aspartyl residues as phosphoryl-group donors and acceptors. The prototypical two-component system (TCS) comprises two protein components, a sensor histidine kinase (SK) and a response regulator (RR). A typical SK consists of a signal-sensing (or input) domain and a transmitter domain, which has a conserved kinase core with an invariant histidine residue, whereas a typical RR consists of an N-terminal receiver domain, which has an invariant aspartate residue and a C-terminal output domain. Signal reception by the SK stimulates ATP-dependent autophosphorylation at the conserved His residue in its transmitter domain. The SK-P then donates the phosphoryl group to the conserved Asp residue in the receiver domain of the cognate RR, thereby rendering it functional. Many TCS, however, are more elaborate. For instance, signal transmission may involve hybrid SKs that contain both transmitter and receiver domains, and/or additional proteins as phosphorelay components (Parkinson and Kofoid, 1992).

The Arc (anoxic redox control) TCS, which plays an important role in the complex transcriptional regulatory network that allows facultative anaerobic bacteria to sense and respond to changes in respiratory growth conditions (Georgellis *et al.*, 2001b; Jung *et al.*, 2008; Malpica *et al.*, 2006), consists of the ArcB hybrid SK and the ArcA RR (Iuchi and Lin, 1988; Iuchi *et al.*, 1990). The ArcB protein (Fig. 12.1), a membrane-associated SK, comprises three catalytic domains: a typical transmitter domain (H1) with a conserved His at position 292, a central receiver domain (D1) with a conserved Asp at position 576, and a secondary C-terminal histidine-containing phosphotransfer domain (HPt) with a conserved His at position 717 (Ishige *et al.*, 1994; Iuchi *et al.*, 1990). Moreover, in the linker region, that is the region connecting the catalytic domains with the transmembrane region, there are two redox-active cysteines

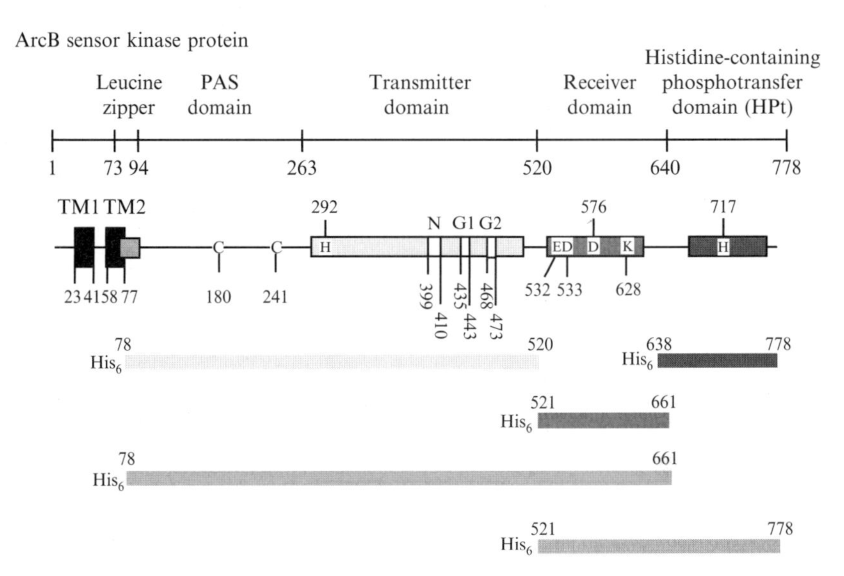

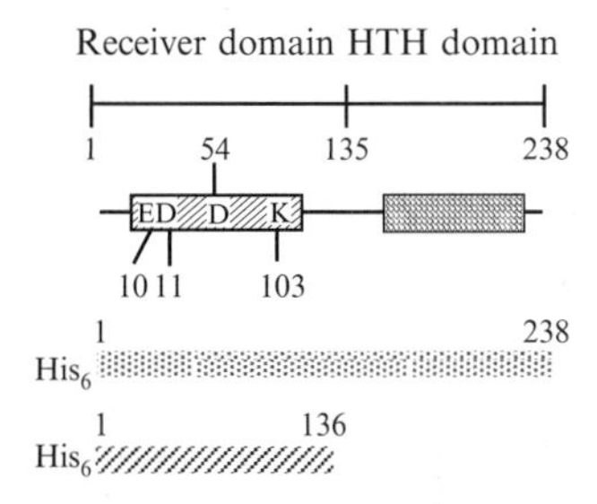

Figure 12.1 Schematic representation of ArcB and ArcA. *Top*: the ArcB sensor kinase protein. Two N-terminal transmembrane segments (TM1 and TM2) are shown. Residues 73–94 contain a putative leucine zipper motif. The transmitter domain is shown with the conserved His292 and the catalytic determinants N, G1, and G2. The receiver domain is shown with the conserved Asp576, and the histidine-containing phosphotransfer domain is shown with the conserved His717. The gray bars represent the ArcB functional modular peptides produced as His-tagged proteins. *Bottom*: the ArcA response regulator protein. The N-terminal receiver domain contains the conserved Asp54, and the C-terminal domain contains the helix-turn-helix (HTH) motif. The bars indicate the peptides from ArcA produced and purified as His-tagged proteins.

(Malpica *et al.*, 2004), a putative leucine zipper (Georgellis *et al.*, 1998), and a PAS domain (Zhulin *et al.*, 1997). ArcA (Fig. 12.1), on the other hand, is a classical RR, comprising an N-terminal receiver domain with a conserved Asp residue at position 54, and a C-terminal helix-turn-helix domain for DNA binding.

Under reducing conditions, ArcB undergoes ATP-dependent autophosphorylation, a process that is enhanced by certain anaerobic metabolites, such as D-lactate and acetate (Georgellis *et al.*, 1999; Rodriguez *et al.*, 2004), and transphosphorylates ArcA via a $His^{292} \rightarrow Asp^{576} \rightarrow His^{717} \rightarrow Asp^{54}$ phosphorelay (Georgellis *et al.*, 1997; Kwon *et al.*, 2000a). Phosphorylated ArcA represses the expression of many operons involved in respiratory metabolism, and activates a few operons encoding proteins involved in fermentative metabolism (Gunsalus and Park, 1994; Liu and De Wulf, 2004; Lynch and Lin, 1996). Under oxic growth conditions the quinone electron carriers inhibit the kinase activity of ArcB (Georgellis *et al.*, 2001a) through the oxidation of two redox-active cysteine residues (Cys^{180} and the Cys^{241}) that participate in intermolecular disulfide bond formation (Malpica *et al.*, 2004). Under such conditions, ArcB acts as a phosphatase that catalyzes the dephosphorylation of ArcA-P by a reverse $Asp^{54} \rightarrow His^{717} \rightarrow Asp^{576} \rightarrow P_i$ phosphorelay, thereby releasing its transcriptional regulation (Georgellis *et al.*, 1998; Peña-Sandoval *et al.*, 2005).

This chapter focuses on the general *in vitro* and *in vivo* strategies and methods that have been used to characterize the ArcB/A two-component signal transduction system.

2. *In Vitro* Characterization of the Arc TCS

In vitro reconstitution can be a first experimental approach in the understanding of a given system. In the case of TCS, this should involve the characterization of SK autophosphorylation, the routes of phosphoryl-group transfer(s) between the conserved His and Asp residues of the transmitter, receiver, and phosphotransfer domains, as well as the dephosphorylation of the RR-P. To this end, wild-type and functional mutant proteins, have to be cloned as full-length proteins and/or individual subdomains in a suitable vector, overexpressed, and purified. Membrane-associated SKs, like ArcB, can be overexpressed and isolated as membrane embedded proteins (in inverted membrane vesicles), or overexpressed and purified as soluble proteins lacking the hydrophobic transmembrane segments.

2.1. Dissection of the Arc components to individual subdomains

Computer-based alignment analyses of two-component proteins have revealed that the overall sequence identity between pairs of SKs or pairs of RRs is very low. However, it has been possible to identify conserved sequences characteristic for transmitter, receiver, and phosphotransfer modules. Typical transmitter modules are about 200 amino acids in length and their structural features include an H-box with an invariant histidine residue, the site of autophosphorylation, the N, D, and F boxes, and a G-box that is the site for ATP binding. Typical receiver modules are about 125 amino acids in length and their principal structural features include an acid pocket composed by two aspartate residues (or a glutamate contiguous to an aspartate) approximately at amino acid positions 10 and 11, an invariant aspartate residue, which is the site of phosphorylation, and a highly conserved lysine residue at their C-terminus (Parkinson and Kofoid, 1992). Finally, the phosphotransfer (HPt) modules are approximately 150 amino acids in length and posses an invariant histidine residue (Mizuno, 1998). Considering the above facts, ArcB was successfully dissected in the following modular peptides: $ArcB^{1-520}$ or $ArcB^{78-520}$ (H1), containing the linker region and the transmitter domain with the invariant histidine at position 292 and the ATP-binding box at position 468–473, with or without the transmembrane domain; $ArcB^{521-661}$ (D1), containing the transmitter domain with the conserved aspartate at position 576; and $ArcB^{638-778}$, containing the HPt domain with the conserved histidine at position 717 (Fig. 12.1) (Georgellis *et al.*, 1997). Once the limits of the various protein subdomains have been predicted, the respective DNA fragments are PCR-amplified and cloned into a high-copy number plasmid, such as pBluescript KS^- (Stratagene), to generate pB7 (Fig. 12.2). This plasmid is used to generate desired site-directed mutations in the ArcB-encoding DNA fragment (pB7n), which can subsequently be cloned into a suitable overexpressing vector (such as pQE30), or used to replace the *arcB* allele on the chromosome (Fig. 12.2). The pQE30 vector (Qiagen), producing His_6-tagged peptides has been successfully used to clone and overexpress soluble ArcB proteins. Also, the pBAD vectors (Guzman *et al.*, 1995) have been useful to clone and overexpress ArcB for preparation of ArcB-enriched inverted membrane vesicles. Remarkably, the individual transmitter- and receiver-containing ArcB polypeptides are able to fold into functional tertiary structures, with or without covalent linkage of the contiguous domains. Also, the N-terminal His_6-tag, derived from the pQE30, and used for peptide purification, does not seem to be functionally perturbing (Georgellis *et al.*, 1997, 1998).

2.2. Detailed methods

2.2.1. ArcB-enriched inverted vesicles

Inverted membrane vesicles provide an important *in vitro* tool to study a membrane bound SK. The main advantage of inverted vesicles is that they reproduce the overall fold of the membrane protein as found in whole cells, except that the exterior surface of the vesicles corresponds to the inner face of the cytoplasmic membrane. This arrangement allows the catalytic domains of the SK to be exposed to impermeant compounds, such as ATP, without disruption of the vesicles. The use of a mutant strain lacking the SK and transformed with a plasmid overexpressing the modified SK (or a particular subdomain) is recommended. This is in order to avoid formation of

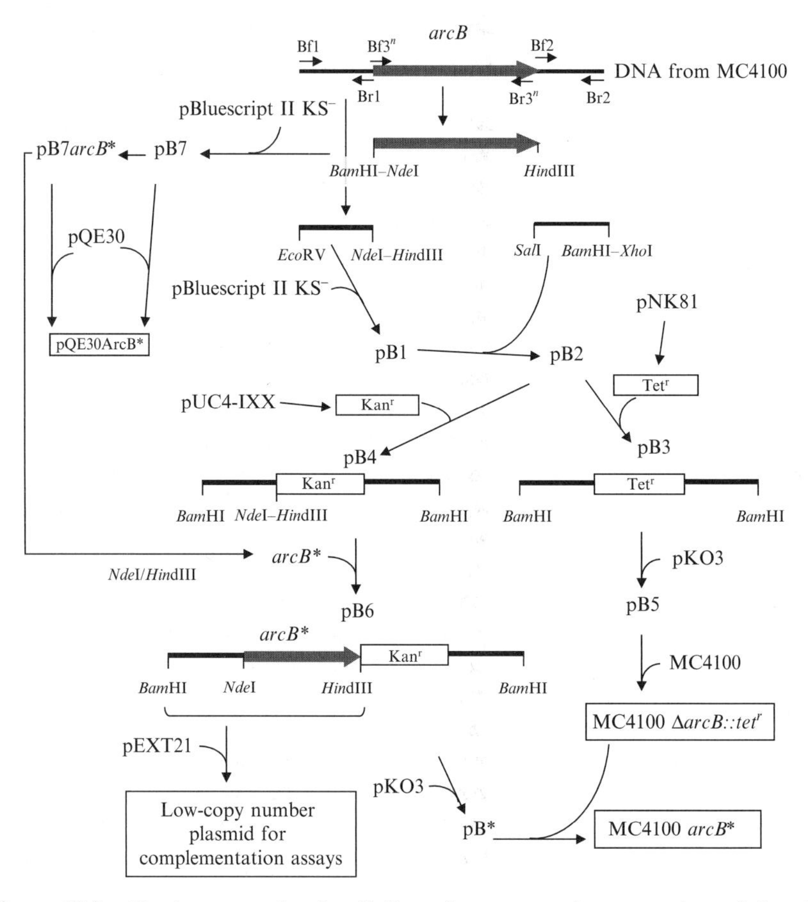

Figure 12.2 Cloning strategies for allelic replacement and construction of plasmid used for complementation assays or overproduction of different variants of ArcB.

heterodimers, which could complicate the interpretation of the results, in case that a functional mutant of the SK is to be studied. For instance, a Δ*arcB* strain, such as ECL5012 (Kwon *et al.*, 2000b), is transformed with a pBAD-based plasmid carrying $arcB^{1-778}$ under the control of the arabinose promoter (the same procedure can be used for $ArcB^{1-661}$ or $ArcB^{1-520}$). Ampicillin-resistant transformants are selected, a single colony is inoculated into 5 ml of LB medium supplemented with 100 μg/ml ampicillin, and incubated overnight with shaking at 37 °C. This culture is used as an inoculum for 250 ml of LB supplemented with 100 μg/ml ampicillin in a 2-l flask. The culture is incubated in a rotary shaker at 37 °C, and at an OD_{600} of 0.5 the expression of ArcB is induced by addition of L-arabinose to a final concentration of 0.5 m*M*. Cells are harvested 4 h after induction, by centrifugation at 6000×*g* for 10 min at 4 °C. The cell pellet can be stored at −20 °C for several months or processed immediately. The cell pellet is thawed on ice, resuspended in 6 ml of ice-cold MOPS buffer (50 m*M* K-MOPS, pH 7.0, 5 m*M* $MgSO_4$, and 100 m*M* KCl), and the cell suspension is passed through a French Press at 4000 psi. The cell lysate is incubated with 30 μg/ml DNase for 15 min at 4 °C to reduce viscocity, and centrifuged at 10,000×*g* for 15 min at 4 °C to remove cell debris. The supernatant, which contains ArcB embedded in inner membranes, is pelleted by centrifugation at 32,500×*g* for 40 min at 4 °C, the pellet is washed with ice-cold MOPS buffer, solubilized in 500 μl of MOPS buffer containing 30% glycerol, and stored at −20 °C. A small fraction (5 μl should be enough) is analyzed by SDS–PAGE. ArcB should be the major component in these membrane preparations. However, overproduction of some membrane proteins frequently results in misfolded peptides that form inclusion bodies. Therefore, the overexpression conditions of each protein must be optimized in small-scale experiments. For example, lowering the temperature of growth (between 15 and 37 °C) before and during induction, which slows down the rate of bacterial metabolism, and/or lowering the concentration of inducer, in order to avoid the overloading of the biosynthetic and folding machinery of the cell, could help to prevent the formation of inclusion bodies (Sambrook *et al.*, 1989).

2.2.2. Expression and purification of soluble forms of full-length ArcB or its subdomains

The following protocol has been successfully used to purify $ArcB^{78–778}$, $ArcB^{78–661}$, $ArcB^{78–520}$, $ArcB^{521–778}$, $ArcB^{521–661}$, $ArcB^{638–778}$, $ArcA^{1–238}$, and $ArcA^{1–136}$. The DNA fragment encoding any of the above ArcB or ArcA peptides is PCR-amplified and cloned into the pQE30 vector (Qiagen). The constructed plasmid is transformed into *Escherichia coli* M15 cells already harboring pREP4, and a single transformant colony is inoculated in 20 ml of LB medium supplemented with 100 μg/ml ampicillin and 25 μg/ml kanamycin and grown overnight. This culture is used as an inoculum for 1 l of LB supplemented with 100 μg/ml ampicillin and 25 μg/ml kanamycin, which is then incubated in a rotary shaker at 37 °C. At an OD_{600} of 0.7 the expression of the His-tagged protein is induced by addition of 1 m*M* isopropyl-β-D-thiogalactopyranoside (IPTG). Cells are harvested 5 h after induction by centrifugation at 6000×*g* for 10 min at 4 °C. The cell pellet can be stored at −20 °C for several months or processed immediately. Cell pellet is thawed on ice, resuspended in 10 ml of lysis buffer (50 m*M* sodium phosphate—pH 8.0, 300 m*M* NaCl, 10 m*M* imidazole), 1 mg/ml lysozyme is added to the suspension, and incubated on ice for 15 min. Cell suspension is passed through a French Press cell at 4000 psi, treated with 25–50 μg/ml DNase for 15 min at 4 °C, and the lysate is centrifuged at 10,000×*g* for 15 min at 4 °C to remove cell debris. The supernatant is mixed with 2 ml of a 50% Ni-NTA agarose (Qiagen) which has been previously equilibrated with lysis buffer, and incubated with gentle agitation for 60 min at 4 °C. The agarose slurry is loaded into a column, and allowed to flow by gravity. Agarose columns are washed twice with 20 ml of washing buffer (50 m*M* sodium phosphate—pH 8.0, 300 m*M* NaCl, 20 m*M* imidazole). Finally, the protein is eluted with 10 ml of elution buffer (50 m*M* sodium phosphate—pH 8.0, 300 m*M* NaCl, 300 m*M* imidazole), and 1 ml fractions are collected. Fractions are analyzed by SDS–PAGE, and Coomassie blue stained to determine integrity and purity of the purified protein. Usually, the protein elutes between fractions 2 and 4. Fractions that contain substantial amounts of ArcB are pooled and dialyzed against storage buffer (0.1 *M* sodium phosphate—pH 8.0, 1 m*M* Na_2EDTA, 150 m*M* NaCl, 10 m*M* β-mercaptoethanol, 25% glycerol) at 4 °C. After dialysis, proteins are concentrated in Centricon 10 units (Amicon) and stored at −20 °C. Protein concentrations can be estimated by using the Coomassie Plus protein assay reagent (Pierce) or the Bio-Rad protein assay kit (Bio-Rad), using bovine serum albumin as a standard.

2.3. Autophosphorylation and transphosphorylation assays

Autophosphorylation of SKs, which are homodimeric proteins, relies on ATP binding at the G-box in one monomer and transfer of the γ-P of ATP to the histidine residue of the H-box in the other monomer. Thus, when [γ-^{32}P]ATP is used, ^{32}P is incorporated into the protein as a function of time and/or protein concentration, and easily detected by autoradiography of the resolved proteins on SDS–PAGE. Also, the intensity of the signal on the autoradiogram is proportional to the amount of ^{32}P incorporated to the protein.

Transphosphorylation refers to the transfer of the phosphoryl group from a phosphorylated SK to a RR, which can be a one step transfer or via a phosphorelay, depending whether the SK involved is a typical kinase or a hybrid kinase. In the case of phosphorelay, the phosphoryl group of the phosphorylated histidine residue of the transmitter domain is sequentially transferred to the conserved aspartate residue of the receiver domain, and the conserved histidine residue of the HPt domain in the sensor kinase, and finally from the HPt domain to the conserved aspartate residue of the receiver domain in the RR. In the case of proteins with different molecular masses, the loss of ^{32}P of one protein and the respective gain of it in another protein can be easily visualized by autoradiography after electrophoretic separation of the proteins. It is important to mention that the half-life of phosphohistidines of the transmitter and phosphotransfer domains appears to be quite long, whereas the half-life of phosphoaspartates of the receiver domains can vary from seconds, as is the case of CheB-P and CheY-P (Hess *et al.*, 1988) to more than an hour, as is the case of OmpR-P and ArcA-P (Georgellis *et al.*, 1998; Igo *et al.*, 1989), and therefore the intensity of the signal on the autoradiogram represents the net phosphorylation of the protein.

2.4. Detailed methods

2.4.1. Autophosphorylation

To measure ArcB autophosphorylation, purified ArcB$^{78-520}$ (or 1 μg of ArcB$^{1-520}$-enriched inverted vesicles) is added at a concentration of 2.5–10 μM to phosphorylation buffer (33 mM HEPES at pH 7.5, 50 mM KCl, 5 mM $MgCl_2$, 1 mM dithiothreitol, 0.1 mM EDTA, and 10% glycerol). Mg^{2+} is essential for the autophosphorylation of ArcB. The reaction is initiated by the addition of [γ-^{32}P]ATP (1 μl of a 10× solution with a specific activity of 2 Ci/mmol is added; this specific activity is prepared by mixing [γ-^{32}P]ATP of higher specific activity with unlabeled ATP), and aliquots are removed at times ranging between 0.5 and 10 min. Samples are mixed with the same volume of 4× SDS sample buffer to stop the reaction, and kept on ice until the last aliquot is taken. Samples are immediately subjected to SDS–PAGE (do not boil the samples prior to electrophoresis because ArcB-P is heat labile), analyzed by autoradiography and/or phosphorimaging. In case that the reaction kinetics are too fast, slower ones can be achieved by lowering the reaction temperature (however, no ArcB phosphorylation is observed at temperatures below 15 °C).

If the ArcB$^{78-661}$ or ArcB$^{78-778}$ peptides are used in the above reaction, the autoradiographic signals represent netphosphorylation of ArcB and cannot be directly compared to the kinetics of ArcB$^{78-520}$ autophosphorylation. If a direct comparison between the autophosphorylation kinetics of ArcB$^{78-520}$ and ArcB$^{78-661}$ or ArcB$^{78-778}$ is desired, the Asp^{576} of the receiver domain should be mutated. This is because the ^{32}P from His^{292}, the site of autophosphorylation, is rapidly transferred to Asp^{576}, and the half-life of this phosphoaspartate is very short.

2.4.2. Transphosphorylation

Phosphoryl-group transfer between the transmitter, receiver, and phosphotransfer domains of a given system can be followed by assembling a phosphorylation reaction of the transmitter domain, as described above, and adding the phosphoaccepting peptide (containing a receiver or phosphotransfer domain) to the reaction mixture. As ATP is present in the reaction, it is important to carry out a control reaction in which the transmitter domain is not present, in order to demonstrate that the phosphoaccepting peptide is not able to autophosphorylate. Alternatively, the phosphorylated transmitter domain can be purified free of ATP prior to the phosphotransfer reaction. Also, two or more competing phosphoreceiving domains (of different molecular masses in order to make possible their separation of SDS–PAGE) can be simultaneously added to a reaction with a transmitter domain. In such an experiment, the preference and/or specificity of the phosphodonor peptide for a phosphor-acceptor peptide can be discerned. Below, we use the Arc subdomains as an example to describe the design, result, and interpretation of such experiments.

1. If ArcB$^{521–661}$ (harboring the receiver domain with the conserved Asp^{576}) is added to a phosphorylation reaction containing ArcB$^{78–520}$ (harboring the transmitter domain with the conserved His^{292}) and [γ-^{32}P]ATP, both ArcB$^{78–520}$-^{32}P and ArcB$^{521–661}$-^{32}P will be formed and detected after SDS–PAGE and analysis by autoradiography or phosphorimaging. This is because ArcB$^{78–520}$ autophosphorylates at His^{292} and rapidly transfers the phosphoryl group to ArcB$^{521–661}$ at Asp^{576}. Although Asp^{576}-P has a short half-life, it is stable enough to allow its detection.
2. On the other hand, in a phosphorylation reaction containing ArcB$^{78–520}$ and ArcB$^{638–778}$ (harboring the phosphotransfer domain with the conserved His^{717}), only ArcB$^{78–520}$-^{32}P will be formed and detected. This is because the phosphoryl-group transfer from H^{292} to H^{717} is not possible. Even so, considerable transphosphorylation of ArcB$^{638–778}$ at His^{717} by ArcB$^{78–520}$ can be achieved after prolonged times (1–2 h) of incubation. However, if peptide ArcB$^{521–661}$ is added to the above reaction (or if ArcB$^{78–661}$, containing both the transmitter and the receiver domain, is used instead of ArcB$^{78–520}$), rapid phosphorylation of ArcB$^{638–778}$ will occur. This result demonstrates that the phosphoryl group of H^{292}-P needs to be transferred to Asp^{576} of the receiver domain before ending to His^{717} of the phosphotransfer domain.
3. If in a reaction containing ArcB$^{78–520}$, ArcB$^{521–661}$, and ArcA$^{1–238}$ (harboring the receiver domain of ArcA with the conserved Asp^{54}) are simultaneously added, high amounts of ArcB$^{521–661}$-^{32}P, but almost no ArcA$^{1–238}$-^{32}P, will be rapidly formed and detected. This result demonstrates that the route of phosphoryl-group transfer from the transmitter domain of ArcB is via Asp^{576} of the ArcB receiver domain and not directly to ArcA.
4. Finally, if ArcA is added in a reaction containing ArcB$^{78–520}$-^{32}P and ArcB$^{638–778}$-^{32}P, ArcA-^{32}P and concomitant lose of ^{32}P from ArcB$^{638–778}$ will occur within seconds, whereas the amount of ArcB$^{78–520}$-^{32}P will remain almost unaffected. This result demonstrates that the phosphoryl group is transferred to ArcA through the HPt domain and not through the transmitter domain of ArcB.

2.5. RR-P dephosphorylation

Phospho-RR dephosphorylation could occur by spontaneous hydrolysis of the phosphoaspartyl bond, or catalyzed by either the cognate sensor kinase or another specific phosphatase. In the Arc system, it appears that ArcB is responsible for ArcA-P dephosphorylation.

To *in vitro* analyze the route of ArcA-P dephosphorylation, it is first necessary to generate and purify the phosphorylated form of ArcA. However, attempts to purify ArcA-P were unsuccessful because of its high tendency to aggregate during the purification process. This problem was bypassed by the use of $ArcA^{1-136}$, which contains the receiver domain, but not the C-terminal DNA-binding domain. This truncated ArcA protein retains its activity as a phosphorylation substrate of ArcB (Georgellis *et al.*, 1998), and can also autophosphorylate, using acetyl-P as a phosporyl-group donor.

2.6. Detailed methods

2.6.1. Phosphorylation and isolation of $ArcA^{1-136}$-P using ArcB as kinase

To phosphorylate $ArcA^{1-136}$, a 20-μl reaction mixture containing Buffer A (33 mM HEPES at pH 7.5, 50 mM KCl, 5 mM $MgCl_2$, 1 mM dithiothreitol, 0.1 mM EDTA, and 10% glycerol), $ArcA^{1-136}$ (20 μM—purified as a His_6-tagged peptide according to the above described protocol), $ArcB^{78-778}$ (2 μM) and 40 μM [γ-^{32}P]ATP (specific activity 2 Ci/mmol) is prepared. After 10 min of incubation at 25 °C, the reaction is stopped by addition of 100 μl of Buffer B (50 mM Tris–HCl at pH 7.0, 150 mM KCl, 5 mM EDTA, and 3% Triton X-100). Separation of $ArcA^{1-136}$-P (its molecular mass is estimated to 16.941 Da) from $ArcB^{78-778}$-P (estimated molecular mass 80.917 Da), ATP and P_i is achieved by ultrafiltration using a Nanosep 30 K device (Pallfiltron), which retains $ArcB^{78-778}$ but not $ArcA^{1-136}$-P, ATP and P_i. The eluate is then passed through a Nanosep 10K device (Pallfiltron), washed four times with 500 μl of Buffer B to remove ATP and P_i, and once with 500 μl of Buffer A to reestablish the reaction conditions. The retained material containing $ArcA^{1-136}$-^{32}P (essentially free of His_6-$ArcB^{78-778}$, ATP, and P_i) is recovered in 200 μl of Buffer A, aliquoted, and used in dephosphorylating assays as described below.

2.6.2. Synthesis of [^{32}P] labeled dilithium acetyl phosphate

Acetyl-^{32}P can be used as an alternative phosphoryl donor for ArcA. It can be synthesized from acetic anhydride and orthophosphate. In an Eppendorf tube are mixed 90 μl of pyridine, 30 μl of K_2HPO_4 1 M and 90 μl of $H_3{}^{32}PO_4$ (ortophosphoric acid, 10 μCi/μl), and the mixture is stirred on ice for 1 h. Then, pure acetic anhydride (7.2 μl) is added and the mixture is vortexed vigorously. After 2 min, 4 M lithium hydroxide is added to adjust the pH to 7.5. Thereafter, 1.8 ml of chilled ethanol (− 15 °C) is added slowly to the reaction mixture. After 1 h on ice, the precipitate is collected by centrifugation at 6000 × g for 15 min at 4 °C, washed twice with 900 μl of ice cold ethanol, and dried over calcium chloride for 24 h. The dry-pellet-containing tube is weighted before it is resuspended in 100 μl of deionized water and centrifuged again 15 min at 6000×g at 4 °C. Now, the supernatant contains the [^{32}P] dilithium acetyl phosphate. The remaining pellet is dried and weighted again to estimate how much dried insoluble material is left. The specific activity of the reaction product is usually about 80 mCi/mmol at the day of synthesis.

2.6.3. Autophosphorylation of ArcA$^{1-136}$ by acetyl phosphate

Acetyl-P-dependent phosphorylation of ArcA$^{1-136}$ is carried out by addition of the protein in a 50-μl reaction mixture containing Buffer A and acetyl-^{32}P (about 80 mCi/mmol). The reaction is incubated for 30 min at 25 °C, and ArcA$^{1-136}$-^{32}P is purified by ultrafiltration using a Nanosep 10K device. The retained material is washed three times with 500 μl of Buffer A, and recovered in 200 μl of the same buffer.

2.6.4. Phosphatase assays

To study whether ArcB is involved in the dephosphorylation of ArcA$^{1-136}$-P, two reactions are carried out at 25 °C in mixtures of 45 μl containing Buffer A and purified ArcA$^{1-136}$-^{32}P (1.5 μM). One of them is used to measure the spontaneous dephosphorylation of ArcA$^{1-136}$-^{32}P, whereas the other is supplemented with either of ArcB$^{78-778}$, ArcB$^{78-661}$, ArcB$^{78-520}$, ArcB$^{521-778}$, ArcB$^{521-778,\ D576A}$, or ArcB$^{520-778,\ H717Q}$. A 5-μl sample is withdrawn (corresponds to time 0), 0.15 μM ArcB$^{78-778}$, or any of the other peptides, is added to one of the reaction mixtures, and samples of 5 μl are withdrawn from each reaction at 0.5, 1, 2.5, 5, 10, 15, and 20 min. Samples are mixed with 5 μl of 4× SDS sample buffer, and kept on ice. One microliter of each sample is analyzed for P_i by thin-layer chromatography (TLC) using polyethyleneimine-cellulose plates (Aldrich), and 2 N formic acid and 0.5 M LiCl as mobile phase for 45 min. The TLC plate is dried with a hair-dryer, and radioactivity of resolved P_i is determined by autoradiography/phosphorimaging. The other 9 μl are subjected to SDS–PAGE, and analyzed by autoradiography/phosphorimaging. If the SK (or its subdomains) catalyzes the dephosphorylation of the phospho-RR, the half-life of phospho-RR should be drastically lowered in the presence of the SK, and this should also be reflected in a corresponding increase in the rate of P_i release.

The following results were obtained in the above experiment with ArcA$^{1-136}$-Pand ArcB or its subdomains: (1) peptides ArcB$^{78-520}$, ArcB$^{78-661}$, and the mutant ArcB$^{520-778,\ H717Q}$ protein did not affect the dephosphorylation of ArcA$^{1-136}$-^{32}P or the rate of P_i liberation; (2) the ArcB$^{521-778,\ D576A}$ mutant protein did not affect the rate of P_i liberation but it was able to receive the phosphoryl group from ArcA$^{1-136}$-^{32}P, as judged by the loss of ^{32}P from ArcA$^{1-136}$-^{32}P that was accounted for by the ^{32}P gained ArcB$^{521-778,\ D576A}$; and (3) the ArcB$^{78-778}$ and ArcB$^{521-778}$ peptides rapidly dephosphorylate ArcA$^{1-136}$-^{32}P resulting in the corresponding increase in the rate of P_i liberation. During the course of this reaction, there was a slight transient rise of ArcB$^{78-778}$-^{32}P and ArcB$^{521-778}$-^{32}P followed by a gradual decay. Taken together these results demonstrate that the ^{32}P is transferred from ArcA$^{1-136}$-^{32}P to the conserved His^{717} of the HPt domain and subsequently to the conserved Asp^{576} of the receiver domain before it is released as P_i.

2.7. Amplification and inhibition of ArcB autophosphorylation

D-Lactate, acetate, and pyruvate, whose intracellular levels are elevated during anoxia, enhance the level of ArcB autophosphorylation (Georgellis *et al.*, 1999; Iuchi, 1993). On the other hand, the quinone electron carriers, and the oxidant chloramine T inhibit ArcB autophosphorylation (Georgellis *et al.*, 2001a). The effect of these molecules on the activity of ArcB can be tested by direct addition of these compounds in the phosphorylation reaction, as described above, and the rate of autophosphorylation (in the case of ArcB$^{78-520}$, ArcB$^{78-661,\ Asp576Ala}$, or ArcB$^{78-778,\ Asp576Ala}$) or net phosphorylation (in the case of ArcB$^{78-661}$ or ArcB$^{78-778}$) is determined. It is important that the compound to be tested is preincubated with ArcB in the phosphorylation mixture for 10 min prior to [γ-^{32}P]ATP addition. Because quinones are very hydrophobic and insoluble in water, the soluble analog ubiquinone-0 (Sigma) can be used. Ubiquinone-0 (Q0) is dissolved as 10× solution in 50% ethanol and therefore all control reaction mixtures must be supplemented with an appropriate amount of ethanol (5%, v/v). If the concentration for half maximal inhibition is to be tested, various 10 μl phosphorylation reactions can be set up, as described above. One of the reactions serves as control, whereas increasing concentrations of ubiquinone-0 (1–1000 μM) are added to the other ones, and incubated for 10 min. Phosphorylation is initiated by addition of [γ-^{32}P]ATP, and incubated for 2 min at 20 °C. The reactions are stopped and analyzed as above (Fig. 12.3).

2.8. Redox regulation of ArcB autophosphorylation

Various observations in our laboratory provided a number of indications regarding the mechanism for quinone-dependent silencing of the ArcB kinase. First, the purified ArcB$^{78-778}$, lacking the transmembrane domain, exhibited a kinase activity that was inhibited by Q0, even when purified under denaturing conditions. Thus, no prosthetic groups necessary for the regulation of the kinase could be present in this preparation of ArcB. Also, since ArcB$^{78-778}$ lacks the transmembrane domain, a cytoplasm-located redox-active site is deduced. Second, the kinase activity of this protein was inhibited not only by Q0 but also by chloramine T, and therefore ArcB is sensitive to oxidation rather than to the allosteric binding of Q0. Third, two unique cysteine residues (Cys^{180} and Cys^{241}) are present in the linker region. Cysteine residues are uniquely suited for sensing a range of redox signals

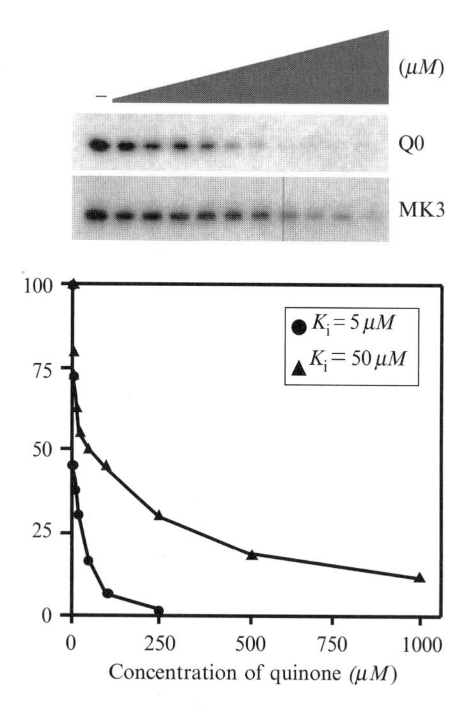

Figure 12.3 Kinase inhibion of ArcB$^{78-778}$ by ubiquinone-0 (Q0) and menaquinone (MK3). To test the half maximal inhibition, ArcB$^{78-520}$ was incubated with increasing concentrations of Q0 or MK3 (1–1000 *μM*) for 10 min. The reaction was initiated by addition of [γ-^{32}P]ATP, and after 2 min it was stopped by addition of an equal volume of 4× SDS sample buffer. Phosphorylated products were resolved on SDS–PAGE and analyzed by autoradiography (top) and phosphorimaging (bottom).

because the thiol side-chain can be oxidized to several different redox states, many of which are readily reversible.

2.9. Detailed methods

2.9.1. Blocking cysteine residues with *S*-methyl methanethiosulfonate

To find out whether the two cysteine residues participate in the mechanism that controls the activity of ArcB we took advantage of the fact that upon alkylation, cysteine residues are protected from further oxidation, and also unable to form disulfide bonds. To this end, four phosphorylation reactions with ArcB$^{78-778}$ are set up, and two of them are supplemented with 1 m*M* *S*-methyl methanethiosulfonate (MMTS) (Sigma). MMTS is a small, neutral, highly soluble molecule that reacts specifically with the free sulfhydryl groups on cysteine side chains to form $-S-S-CH_3$. The reaction mixtures are incubated for 30 min at 25 °C, and 0.250 m*M* of Q0 (or 0.100 m*M* of Chloramine T) is added to one untreated sample and one treated with MMTS reactions. After additional 10 min of incubation, the phosphorylation reactions are initiated by the addition of [γ-^{32}P]ATP, terminated by addition of equal volume of 4× SDS sample buffer, and analyzed as above.

The results obtained in this laboratory are as follows:

Reaction 1: ArcB$^{78-778}$ alone → control reaction of ArcB$^{78-778}$ phosphorylation kinetics

Reaction 2: ArcB$^{78-778}$ and Q0 → ArcB$^{78-778}$ kinase activity is strongly inhibited

Reaction 3: ArcB$^{78-778}$ and MMTS → no effect on ArcB$^{78-778}$ phosphorylation kinetics

Reaction 4: ArcB$^{78-778}$, MMTS and Q0 → no effect on ArcB$^{78-778}$ phosphorylation kinetics

Thus, alkylation *per se* does not affect the phosphorylation kinetics of ArcB. However, in contrast to the inhibition of the unprotected protein by Q0, the MMTS-blocked protein is unresponsive to Q0, indicating that cysteine oxidation is involved in ArcB regulation.

2.9.2. Tagging Cys residues with methoxy-poly(ethylene glycol) maleimide

Methoxy-poly(ethylene glycol) maleimide (MAL-PEG) (M_w = 5000) (Shearwater polymers, Inc.) can be used to covalently tag free thiols on proteins. Once MAL-PEG forms a covalent bond with a free thiol in the protein, the MAL-PEG–protein conjugate can be detected as a band shift on Coomassie blue stained gels, or by Western blot analysis if antibodies against the protein are available.

To determine whether both Cys^{180} and Cys^{241} undergo Q0-dependent oxidation, ArcB$^{78-778}$ (or the mutant ArcB$^{78-778,\ Cys180Ala}$, ArcB$^{78-778,\ Cys241Ala}$, ArcB$^{78-778,\ Cys180Ala,\ Cys241Ala}$ proteins) is added at a concentration of 10 μM to two 10 μl reactions containing 20 m*M* Tris/HCl (pH 7.0), and 1 m*M* EDTA. Q0 is added to a concentration of 0.250 m*M* to one of the reactions, incubated at 25 °C for 10 min, and 1 m*M* MAL-PEG is added to each reaction. After 60 min of incubation the reaction is terminated by addition of an equal volume of 2× SDS sample buffer and immediately subjected to SDS–PAGE. Differences in the mobility of the tagged and untagged proteins are then visualized by Commassie blue staining or by Western blot with an ArcB polyclonal antibody. Detection of MAL-PEG–protein conjugates, as indicated by a shift in the mobility of the protein, signifies free thiols. Thus, from the pattern of the detected conjugates it can be inferred whether one or both cysteine residues undergo Q0-dependent oxidation. For instance, if both cysteine residues undergo Q0-dependent oxidation, conjugates should be detected for the Q0 untreated $ArcB^{78–778}$, $ArcB^{78–778,\ Cys180Ala}$, and $ArcB^{78–778,\ Cys241Ala}$, but not for the Q0 treated proteins. The $ArcB^{78–778,\ Cys180Ala,\ Cys241Ala}$ double mutant protein serves as a control, and no MAL-PEG–protein conjugates should be detected at any condition.

2.9.3. Trapping intermolecular disulfide bonds *in vivo*

The pathway for thiol oxidation begins with the formation of sulfenic acid (–SOH), which rapidly condenses to form disulfide bonds (–S–S–). If the disulfide bond is intermolecular, the two involved monomers form a stable dimer in the absence of a reductant. To test whether the two cysteines of ArcB participate in inter- or intramolecular disulfide bond formation, a wild-type strain (MC4100) is grown aerobically (in a partially filled baffled Erlenmeyer flask) or anaerobically (in a completely filled screw capped test tube) in 5 ml of LB medium at 37 °C. At an OD_{600} of ~0.5 the cultures are treated with 1/10 volume of 100% trichloroacetic acid (TCA) to precipitate the proteins and stop further thiol/disulfide exchange. After 1 h incubation on ice, the precipitated proteins are isolated by centrifugation, and the pellets are washed with acetone to remove the TCA. The pellet is dissolved in nonreducing 4× SDS sample buffer and split into two equal portions. One portion is treated with 100 m*M* DTT, and the other receives an equivalent amount of SDS sample buffer. Finally, the proteins are separated by nonreducing SDS–PAGE, and ArcB is visualized by Western blot analysis using an ArcB polyclonal antibody. The predominant portion of ArcB in the aerobic extracts should migrate as a ~170 kDa dimer, the mobility of which is shifted to that of the monomer (~85 kDa) at the presence of DTT, indicative for intermolecular disulfide bond formation. On the other hand, as the disulfide bond is regulatory, no dimer should be detected in the anaerobic extracts.

3. *In Vivo* Characterization of the Arc TCS

Despite the relative complexity of TCSs, genetic approaches can be readily used to confirm and extend the results obtained from biochemical experiments. This is facilitated by the fact that the majority of TCSs are not essential for cell survival, at least under laboratory conditions. Also, since in many cases one or several target genes of a given TCS are known, it is possible to explore the effect of mutations on the activity of target promoters, by the use of reporter gene constructs (e.g., promoter fusions to β-galactosidase). To assay the effect of mutations implies that the chromosomal wild-type gene must be replaced by the mutant gene. Alternatively, the mutant gene is expressed from a low-copy number plasmid in a deletion mutant strain. In general, expression from a plasmid involves fewer genetic manipulations. However, the level of expression of the mutant protein must be carefully verified by Western blotting with an antibody raised against the wild-type protein, to ensure that it is similar to that of the chromosomal one.

3.1. Construction of a deletion mutant and chromosomal gene replacement

The first step in the genetic study of a given gene product is the deletion of the gene from the chromosome (assuming that it is not essential). For example, to construct a Δ*arcB* strain, the chromosomal *arcB* gene is replaced with a tetracycline resistance cassette by homologous recombination (Fig. 12.2). To this end, approximately 1000 bp flanking both ends of the *arcB* open reading frame is PCR-amplified from chromosomal DNA of a wild-type strain (MC4100). To amplify the upstream region, primers Bf1 (forward) and Br1 (reverse) are used. Bf1 has an *Eco*RV restriction site on its 5′ end and Br1 has a *Hin*dIII and an *Nde*I restriction sites on its 5′ end. The *Nde*I site includes the ATG start codon of *arcB*. The PCR-amplified product is cloned into the *Eco*RV–*Hin*dIII sites of pBluescript KS^{-} (Stratagene) to generate plasmid pB1. To amplify the downstream region, primers Bf2 and Br2 are used. Bf2 has a *Sal*I on its 5′ and Br2 has a *Xho*I and a *Bam*HI at its 5′. The PCR-amplified product is cloned into the *Sal*I–*Xho*I sites of pB1 to generate pB2. A Tet^{r} or a Kan^{r} cassette, isolated from plasmids pNK81 (Way *et al.*, 1984) and pUC4-IXX (Barany, 1985), are inserted between the two *arcB* flanking fragments in the *Cla*I site of pB2 to generate pB3 and pB4, respectively. pB3 is used to construct a Δ*arcB::tet*^{r} strain, whereas pB4 is used to construct an *arcB*-shuttle vector (see below). The *Bam*HI–*Bam*HI fragment of pB3, containing the upstream *arcB* flanking region, the Tet^{r} cassette and the downstream flanking region of *arcB*, is cloned into the unique *Bam*HI site of the suicide vector pKO3 (Link *et al.*, 1997) to generate pB5. pKO3 contains a temperature-sensitive origin of replication and the *sacB* and *cat* markers for positive and negative selection, respectively. The *sacB* gene renders *E. coli* sensitive to sucrose, thus, LB-agar plates containing sucrose can be used to select against cells containing this gene (Gay *et al.*, 1985). pB5 is transformed (by electroporation) into a wild-type MC4100 strain and the cells are allowed to recover at 30 °C for 1 h. The cell suspension is plated out on LB-agar plates containing 20 μg/ml of chloramphenicol and incubated overnight at 43 °C. The next day, five colonies are picked and inoculated into 1 ml of LB broth, serially diluted, plated on LB-agar plates containing 5% (w/v) sucrose and 12 μg/ml tetracycline, and incubated overnight at 30 °C. The third day, the plates are replica plated on LB-agar plates containing either 20 μg/ml chloramphenicol, 5% (w/v) sucrose, or 12 μg/ml tetracycline and incubated at

30 °C. Colonies that are Tet^r, Cm^s, and Suc^r are selected, and gene replacement is confirmed by PCR.

3.2. Construction of an *arcB* shuttle vector

To construct an *arcB*-shuttle vector (Fig. 12.2), the *arcB* ORF is PCR amplified with primers Bf3^n and Br3^n and chromosomal DNA as template. In these primers "*n*" symbolizes the starting and ending amino acid in the *arcB* DNA fragment to be amplified. For example, Bf3^1 and Br3^{778} should be the primers to amplify the region corresponding to the full-length ArcB, whereas Bf3^{520} and Br3^{661} should be the primers to amplify the region corresponding to the receiver domain ArcB. Bf3^n has a *Bam*HI and an *Nde*I site in its 5′ and Br3^n has a *Hin*dIII site at its 5′. The PCR product is digested with *Bam*HI and *Hin*dIII and cloned into the *Bam*HI—*Hin*dIII sites of pBluescript KS$^-$ (Stratagene), to generate pB7. This plasmid is used to generate desired site-directed mutations in the ArcB-encoding DNA fragment generating plasmid pB7arcB*. The *Nde*I–*Hin*dIII fragment of pB7arcB* is then cloned into the *Nde*I–*Hin*dIII sites of pB4 (constructed above) to generate pB6. The *Bam*HI–*Bam*HI fragment of pB6 (containing the *arcB* promoter—the *arcB* ribosome binding site—an introduced *Nde*I site that includes the initiation codon of *arcB*—the *arcB* ORF (with or without modifications) with its stop codon—a kanamycin cassette—and approximately 1 kbp downstream *arcB*) is cloned into the *Bam*HI site of pKO3 to generate pB*. This plasmid is used to replace the Tet^r cassette of the Δ*arcB::tet^r* strain by the desired modified *arcB* sequence. The protocol for the gene replacement is the same as described above, except that the selection must be done with kanamycin instead of tetracycline, and the resulting colonies should be Km^r, Tet^s, Cm^s, and Suc^r. Finally, the *arcB** sequence can be transferred to any other strain by P1 transduction. Alternatively, the *Bam*HI—*Hin*dIII fragment of pB6 (containing the *arcB* promoter—the *arcB* ribosome binding site—an introduced *Nde*I site that includes the initiation codon of *arcB*—the *arcB* ORF (with or without modifications, and a stop codon)) is cloned into any low-copy number vector (such as pXT21 or pEXT22; Dykxhoorn *et al.*, 1996), which, in turn, can be used for complementation studies.

3.3. Detection of signaling defects in the Arc TCS

Genetic dissection and reconstitution of a signal transduction pathway is suitable in the same manner as the biochemical approach. *In vivo*, SK and RR proteins participate in an ordered series of reactions whose end product influences the behavioral response and/or phenotype of the cell. Thus, a defect at any step in the sequence, blocking the pathway, could be detected by aberrant transcription of the target genes, or an altered phenotype. For example, in the case of the Arc system, null *arc* mutants or functional *arc* mutants, blocking any step in the phosphorelay (i.e., *arcB*$^{\mathrm{His292Gln}}$), are growth sensitive to dyes such as toluidine blue O (Buxton *et al.*, 1983). That is a Δ*arcB::tet*^{r} strain or a Δ*arcA::tet*^{r} will not be able to grow on tryptone–toluidine blue O-agar plates (10 g/l tryptone, 8 g/l NaCl, 0.2 mg/ml toluidine blue O, and 15 g/l of Bacto Agar). Alternatively, defects in the Arc signaling pathway could be analyzed by detecting changes in the *in vivo* level of phospho-ArcA, as indicated by the expression of target operons. For instance, the *cydAB* and the *lldPRD* operons, which are, respectively, activated and repressed by ArcA-P under anaerobic growth (Cotter *et al.*, 1997; Iuchi *et al.*, 1994), could be employed to construct a λΦ(*cydA′-lacZ*) and a λΦ (*lldP′-lacZ*) reporter carrying strain (Kwon *et al.*, 2000b). To minimize undesired errors, the use of both the activated and the repressed reporter is recommended in the analysis of all modified proteins.

3.4. Genetic analysis of the Arc TCS

The interpretation of the phenotypic effects of mutations in RRs and SKs is not always straightforward. For example, mutating the phosphoaccepting aspartate of a RR should lead to a null phenotype. However, in the case of VirG, the above mutation (Asp52Glu) in combination with a mutation of Ile77Val results in a constitutively activated RR, most probably because the mutation interferes with the proper fold of the protein in such a way that the active conformation of the protein is favored (Gao *et al.*, 2006). In the case of mutant SKs, problems could stem from two sources: the phosphatase activities that are often associated with SK, and the existence of alternative phosphodonors, that is, acetyl-P. Mutants with deletions of a given SK, grown under specific conditions, such as growth on pyruvate as sole carbon and energy source, could exhibit patterns of regulated genes similar to those of wild type. This is attributed to the elevation of the intracellular concentration of acetyl-P (McCleary and Stock, 1994), which is used by the RR to autophosphorylate, and to the RR-P phosphatase deficiency of the strain, as a consequence of the deletion of the cognate SK.

3.5. *In vivo* reconstitution of the Arc signaling pathways

The amplitude and duration of an adaptive response depend on the balance between the rates of phosphorylation and dephosphorylation of the RR. *In vitro* studies (described above) demonstrate that ArcB not only phosphorylates ArcA via an ArcB$^{\mathrm{His292}}$-P → ArcB$^{\mathrm{Asp576}}$-P → ArcB$^{\mathrm{His717}}$-P → ArcA$^{\mathrm{Asp54}}$-P phosphorelay, but also dephosphophosphorylates ArcA-P via an ArcA$^{\mathrm{Asp54}}$-P → ArcB$^{\mathrm{His717}}$-P → ArcB$^{\mathrm{Asp576}}$-P → P_i reverse pathway. To verify the *in vitro* results for the forward phosphorelay with *in vivo* experiments, the following two strategies can be used. First, the wild-type *arcB* allele of a strain harboring the *cydA′-lacZ* or the *lldP′-lacZ* reporter could be replaced with either of the *arcB*$^{\mathrm{His292Gln}}$, *arcB*$^{\mathrm{Asp576Ala}}$, or *arcB*$^{\mathrm{His717Gln}}$ mutant alleles. Alternatively, a Δ*arcB::tet*^{r} strain harboring either of the two reporters could be transformed with plasmids carrying the corresponding mutant *arcB* allele. No aerobic/anaerobic regulation of the reporter should be obtained for any of the mutant constructs. Second, the wild-type *arcB* allele of the above strain could be replaced with either of the *arcB*$^{1-520}$, or *arcB*$^{1-661}$ mutant alleles. With this replacement, no aerobic/anaerobic regulation of the reporter should become evident. However, if the *arcB*$^{1-520}$ or the *arcB*$^{1-661}$ carrying mutant strain is transformed with a plasmid carrying *arcB*$^{521-778}$ or *arcB*$^{638-778}$, respectively, wild-type aerobic/anaerobic regulation of the reporter should be obtained.

On the other hand, to verify the *in vitro* results for the reverse phosphorelay with *in vivo* experiments, the following strategy can be used. First, the *lldP′-lacZ*-harboring wild-type strain could be transformed with a plasmid carrying either of *arcB*$^{521-778}$, or *arcB*$^{520-778,\ \mathrm{Asp576Ala,\ His717Gln}}$ under the control of the *ara* promoter (Peña-Sandoval *et al.*, 2005), and transformants are grown anaerobically in the presence of 1.3 m*M* of arabinose. The expression of the *lldP′-lacZ* reporter should be completely derepressed in the case of *arcB*$^{521-778}$, whereas no derepression of *lldP′-lacZ* should be observed in the case of the *arcB*$^{520-778,\ \mathrm{Asp576Ala,\ His717Gln}}$ mutant peptide. To fully reconstitute the reverse phosphorelay, plasmids overexpressing *arcB*$^{520-778,\ \mathrm{Asp576Ala}}$ and *arcB*$^{520-778,\ \mathrm{His717Gln}}$ should also be included in the experiment.

3.6. Ablations and transplantations

Most SKs are membrane-associated proteins with two or more hydrophobic membrane-spanning sequences, generally assumed to embody the signal input domain, which upon signal sensing cause conformational changes that modulate kinase activity. Because of the modular nature of SK, two conceptually simple genetic tests could distinguish whether the transmembrane domain is important for signal sensing or not: ablation of the hydrophobic membrane-spanning sequences, and transplantation of a foreign transmembrane domain. For instance, if a truncated SK, lacking the transmembrane segments, exhibits kinase regulation would imply that it responds to a cytosol-located signal. On the other hand, if no kinase activity were present, it would imply that either the transmembrane segments are involved in signal sensing and regulation of the kinase activity or that they play an important role in the folding of the cytosolic portion of the SK. Finally, if constitutive kinase activity were observed would imply that the transmembrane segments inhibit kinase activity, and that the presence of the signal alleviates this inhibition.

For example, to test whether the membrane-spanning regions of ArcB are important for signal sensing and regulation of the kinase activity, the wild-type *arcB* allele of the reporter-harboring strains was replaced with *arcB*$^{78-778}$. Monitoring the aerobic/anaerobic expression of the *cydA′-lacZ* or the *lldP′-lacZ* reporters, it was found that liberation of ArcB from the plasma membrane results in a partially constitutive kinase activity *in vivo*, preserving only a small degree of regulation (Kwon *et al.*, 2000b; Rodriguez *et al.*, 2004). Thus, this stimulus independent response would imply that the hydrophobic membrane-spanning sequences inhibit the kinase activity in the absence of stimulus. This mode of control appears to be common in SKs because removal of their input domain leads to activation of transmitter kinase activity, at least *in vitro* (Yamamoto *et al.*, 2005). However, subsequent transplantation experiments demonstrated that the above conclusion was wrong. The transmembrane segments of ArcB neither participate in signal sensing nor in regulation of the ArcB kinase activity. Instead, they serve as an anchor to keep the protein close to the source of a silencing signal, the membrane embedded quinone electron carriers (Georgellis *et al.*, 2001a; Kwon *et al.*, 2003). Thus, extra caution is recommended when results from such experiments are interpreted.

As stated earlier, SKs function as dimers, and the transmembrane segments can play a pivotal role for proper dimerization and/or orientation of the cytosolic portions of the protein. Therefore, the length and flexibility of the linker segment might prove critically important for proper domain interactions. An example in case is the Tar-ArcB hybrids (Kwon *et al.*, 2003). A series of hybrids with different spatial relationships between the Tar anchor and the cytosolic ArcB portion (Tar$^{1-212}$-ArcB$^{78-778}$, Tar$^{1-213}$-ArcB$^{78-778}$, Tar$^{1-214}$-ArcB$^{78-778}$) were constructed and incorporated at the *arcB* locus in the chromosome (as described above) of a $\lambda\Phi$*lldP′-lacZ*-bearing host strain. Also, to avoid heterodimer formation between the wild-type Tar and Tar-ArcB hybrids, the *tar* sequence was deleted. The transmembrane domain of Tar forms a stable dimer in the cytoplasmic membrane (Russo and Koshland, 1983), and therefore the relative orientation between the two cytosolic ArcB monomers in a dimer was altered by fusing a set of Tar transmembrane domains with progressively extended C-terminal region to a constant cytosolic portion of ArcB. Since the rigid helical structure of the second transmembrane segment of Tar protrudes into the cytoplasm, each extra residue of Tar added at the Tar-ArcB junction rotates the attached ArcB kinase by approximately 100°. The reported Tar-ArcB chimeras presented three different phenotypes: a redox-regulated, a constitutively active, and a catalytically inactive ArcB kinase. This implies that the transmembrane domain of ArcB is not involved in signal sensing, and that the orientation of the cytosolic domain of one ArcB monomer relative to the other in the homodimer is of major importance.

In retrospective, and knowing that the ArcB kinase activity is regulated by the reversible oxidation of Cys^{180} and Cys^{241}, the above results can be explained as follows: the relative orientation of the kinase moieties in the redox-regulated chimera, most likely represent that of the wild-type ArcB. In this conformation it is expected that Cys^{180} and Cys^{241} are located on the same side on the structure of the monomer and that they face the ones of the other monomer. A 100° rotation of the ArcB monomers relative to each other will place the cysteines almost at opposite sides, rendering disulfide bond formation practically impossible, and therefore resulting in a constitutively active kinase.

4. Conclusions

In the foregoing sections, *in vitro* and *in vivo* techniques have been illustrated that frequently have been used in the analysis of the Arc signal transduction pathway, including phosphoryl-group transfer between the various subdomains, signal identification, and kinase regulation. The herein provided protocols should be considered as a starting point and not necessarily inclusive of all approaches that can or should be taken. Clearly, these protocols can be modified and the same results can be obtained in more than one way.

ACKNOWLEDGMENTS

We thank Georges Dreyfus, Diego Gonzalez Halphen, Bertha Gonzalez Pedrajo, and Bertha Michel for helpful discussions and for critically reading the manuscript. This work was supported by Grants 37342-N from the Consejo Nacional de Ciencia y Tecnología (CONACyT), IN221106/17 from DGAPA-PAPIIT, UNAM and CRP/MEX08-02 from the International Centre for Genetic Engineering and Biotechnology (ICGEB).

REFERENCES

Barany, F. (1985). Single-stranded hexameric linkers: A system for in-phase insertion mutagenesis and protein engineering. *Gene* **37,** 111–123.

Buxton, R. S., Drury, L. S., and Curtis, C. A. (1983). Dye sensitivity correlated with envelope protein changes in *dye* (*sfrA*) mutants of *Escherichia coli* K12 defective in the expression of the sex factor F. *J. Gen. Microbiol.* **129,** 3363–3370.

Cotter, P. A., Melville, S. B., Albrecht, J. A., and Gunsalus, R. P. (1997). Aerobic regulation of cytochrome *d* oxidase (*cydAB*) operon expression in *Escherichia coli*: Roles of Fnr and ArcA in repression and activation. *Mol. Microbiol.* **25,** 605–615.

Dykxhoorn, D. M., St Pierre, R., and Linn, T. (1996). A set of compatible *tac* promoter expression vectors. *Gene* **177,** 133–136.

Gao, R., Mukhopadhyay, A., Fang, F., and Lynn, D. G. (2006). Constitutive activation of two-component response regulators: Characterization of VirG activation in *Agrobacterium tumefaciens*. *J. Bacteriol.* **188,** 5204–5211.

Gay, P., Le Coq, D., Steinmetz, M., Berkelman, T., and Kado, C. I. (1985). Positive selection procedure for entrapment of insertion sequence elements in gram-negative bacteria. *J. Bacteriol.* **164,** 918–921.

Georgellis, D., Lynch, A. S., and Lin, E. C. (1997). *In vitro* phosphorylation study of the Arc two-component signal transduction system of *Escherichia coli*. *J. Bacteriol.* **179,** 5429–5435.

Georgellis, D., Kwon, O., De Wulf, P., and Lin, E. C. (1998). Signal decay through a reverse phosphorelay in the Arc two-component signal transduction system. *J. Biol. Chem.* **273,** 32864–32869.

Georgellis, D., Kwon, O., and Lin, E. C. (1999). Amplification of signaling activity of the Arc two-component system of *Escherichia coli* by anaerobic metabolites. An *in vitro* study with different protein modules. *J. Biol. Chem.* **274,** 35950–35954.

Georgellis, D., Kwon, O., and Lin, E. C. (2001a). Quinones as the redox signal for the Arc two-component system of bacteria. *Science* **292,** 2314–2316.

Georgellis, D., Kwon, O., Lin, E. C., Wong, S. M., and Akerley, B. J. (2001b). Redox signal transduction by the ArcB sensor kinase of *Haemophilus influenzae* lacking the PAS domain. *J. Bacteriol.* **183,** 7206–7212.

Gunsalus, R. P., and Park, S. J. (1994). Aerobic-anaerobic gene regulation in *Escherichia coli*: Control by the ArcAB and Fnr regulons. *Res. Microbiol.* **145,** 437–450.

Guzman, L. M., Belin, D., Carson, M. J., and Beckwith, J. (1995). Tight regulation, modulation, and high-level expression by vectors containing the arabinose P_{BAD} promoter. *J. Bacteriol.* **177,** 4121–4130.

Hess, J. F., Oosawa, K., Kaplan, N., and Simon, M. I. (1988). Phosphorylation of three proteins in the signaling pathway of bacterial chemotaxis. *Cell* **53,** 79–87.

Igo, M. M., Ninfa, A. J., Stock, J. B., and Silhavy, T. J. (1989). Phosphorylation and dephosphorylation of a bacterial transcriptional activator by a transmembrane receptor. *Genes Dev.* **3,** 1725–1734.

Ishige, K., Nagasawa, S., Tokishita, S., and Mizuno, T. (1994). A novel device of bacterial signal transducers. *EMBO J.* **13,** 5195–5202.

Iuchi, S. (1993). Phosphorylation/dephosphorylation of the receiver module at the conserved aspartate residue controls transphosphorylation activity of histidine kinase in sensor protein ArcB of *Escherichia coli*. *J. Biol. Chem.* **268,** 23972–23980.

Iuchi, S., and Lin, E. C. (1988). *arcA* (*dye*), a global regulatory gene in *Escherichia coli* mediating repression of enzymes in aerobic pathways. *Proc. Natl. Acad. Sci. USA* **85,** 1888–1892.

Iuchi, S., Matsuda, Z., Fujiwara, T., and Lin, E. C. (1990). The *arcB* gene of *Escherichia coli* encodes a sensor-regulator protein for anaerobic repression of the *arc* modulon. *Mol. Microbiol.* **4,** 715–727.

Iuchi, S., Aristarkhov, A., Dong, J. M., Taylor, J. S., and Lin, E. C. (1994). Effects of nitrate respiration on expression of the Arc-controlled operons encoding succinate dehydrogenase and flavin-linked L-lactate dehydrogenase. *J. Bacteriol.* **176,** 1695–1701.

Jung, W. S., Jung, Y. R., Oh, D. B., Kang, H. A., Lee, S. Y., Chavez-Canales, M., Georgellis, D., and Kwon, O. (2008). Characterization of the Arc two-component signal transduction system of the capnophilic rumen bacterium *Mannheimia succiniciproducens*. *FEMS Microbiol. Lett.* **284,** 109–119.

Kwon, O., Georgellis, D., and Lin, E. C. (2000a). Phosphorelay as the sole physiological route of signal transmission by the *arc* two-component system of *Escherichia coli*. *J. Bacteriol.* **182,** 3858–3862.

Kwon, O., Georgellis, D., Lynch, A. S., Boyd, D., and Lin, E. C. (2000b). The ArcB sensor kinase of *Escherichia coli*: Genetic exploration of the transmembrane region. *J. Bacteriol.* **182,** 2960–2966.

Kwon, O., Georgellis, D., and Lin, E. C. (2003). Rotational on-off switching of a hybrid membrane sensor kinase Tar-ArcB in *Escherichia coli*. *J. Biol. Chem.* **278,** 13192–13195.

Link, A. J., Phillips, D., and Church, G. M. (1997). Methods for generating precise deletions and insertions in the genome of wild-type *Escherichia coli*: Application to open reading frame characterization. *J. Bacteriol.* **179,** 6228–6237.

Liu, X., and De Wulf, P. (2004). Probing the ArcA-P modulon of Escherichia coli by whole genome transcriptional analysis and sequence recognition profiling. *J. Biol. Chem.* **279,** 12588–12597.

Lynch, A. S., and Lin, E. C. (1996). Transcriptional control mediated by the ArcA two-component response regulator protein of *Escherichia coli*: Characterization of DNA binding at target promoters. *J. Bacteriol.* **178,** 6238–6249.

Malpica, R., Franco, B., Rodriguez, C., Kwon, O., and Georgellis, D. (2004). Identification of a quinone-sensitive redox switch in the ArcB sensor kinase. *Proc. Natl. Acad. Sci. USA* **101,** 13318–13323.

Malpica, R., Sandoval, G. R., Rodriguez, C., Franco, B., and Georgellis, D. (2006). Signaling by the Arc two-component system provides a link between the redox state of the quinone pool and gene expression. *Antioxid. Redox Signal.* **8,** 781–795.

McCleary, W. R., and Stock, J. B. (1994). Acetyl phosphate and the activation of two-component response regulators. *J. Biol. Chem.* **269,** 31567–31572.

Mizuno, T. (1998). His-Asp phosphotransfer signal transduction. *J. Biochem.* **123,** 555–563.

Parkinson, J. S., and Kofoid, E. C. (1992). Communication modules in bacterial signaling proteins. *Annu. Rev. Genet.* **26,** 71–112.

Peña-Sandoval, G. R., Kwon, O., and Georgellis, D. (2005). Requirement of the receiver and phosphotransfer domains of ArcB for efficient dephosphorylation of phosphorylated ArcA *in vivo*. *J. Bacteriol.* **187,** 3267–3272.

Rodriguez, C., Kwon, O., and Georgellis, D. (2004). Effect of D-lactate on the physiological activity of the ArcB sensor kinase in *Escherichia coli*. *J. Bacteriol.* **186,** 2085–2090.

Russo, A. F., and Koshland, D. E. Jr. (1983). Separation of signal transduction and adaptation functions of the aspartate receptor in bacterial sensing. *Science* **220,** 1016–1020.

Sambrook, J., Fritsch, E. F., and Maniatis, T. (1989). Molecular Cloning: A Laboratory Manual. Cold Spring Harbor Laboratory, Cold Spring Harbor, NY.

Way, J. C., Davis, M. A., Morisato, D., Roberts, D. E., and Kleckner, N. (1984). New Tn*10* derivatives for transposon mutagenesis and for construction of *lacZ* operon fusions by transposition. *Gene* **32,** 369–379.

Yamamoto, K., Hirao, K., Oshima, T., Aiba, H., Utsumi, R., and Ishihama, A. (2005). Functional characterization *in vitro* of all two-component signal transduction systems from *Escherichia coli*. *J. Biol. Chem.* **280,** 1448–1456.

Zhulin, I. B., Taylor, B. L., and Dixon, R. (1997). PAS domain S-boxes in Archaea. Bacteria and sensors for oxygen and redox. *Trends Biochem. Sci.* **22,** 331–333.

CHAPTER THIRTEEN

Potassium Sensing Histidine Kinase in *Bacillus subtilis*

Daniel López, Erin A. Gontang, *and* Roberto Kolter

Contents

Abstract

The soil-dwelling organism *Bacillus subtilis* is able to form multicellular aggregates known as biofilms. It was recently reported that the process of biofilm formation is activated in response to the presence of various, structurally diverse small-molecule natural products. All of these small-molecule natural products made pores in the membrane of the bacterium, causing the leakage of potassium cations from the cytoplasm of the cell. The potassium cation leakage was sensed by the membrane histidine kinase KinC, triggering the genetic pathway to the production of the extracellular matrix that holds cells within the biofilm. This chapter presents the methodology used to characterize the leakage of cytoplasmic potassium as the signal that induces biofilm formation in *B. subtilis* via activation of KinC. Development of novel techniques to monitor activation of gene expression in microbial populations led us to discover the differentiation of a subpopulation of cells specialized to produce the matrix that holds all cells together within the biofilm. This phenomenon of cell differentiation was previously missed by conventional techniques used to monitor transcriptional gene expression.

Department of Microbiology and Molecular Genetics, Harvard Medical School, Boston, Massachusetts, USA

Methods in Enzymology, Volume 471
ISSN 0076-6879, DOI: 10.1016/S0076-6879(10)71013-2

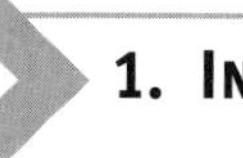

1. Introduction

Bacterial communities survive in their natural habitats due to their ability to sense environmental changes and respond accordingly. To do this, bacteria have evolved complex sensing apparatuses to monitor external fluctuations, allowing them to adjust their behavior in response to specific cues (Mascher *et al.*, 2006). These cues might come directly from the environment, such as changes in temperature or nutrient limitation, or be small-molecule natural products secreted by neighboring microorganisms.

Small-molecules produced by one microorganism can be perceived by other microorganisms cohabitating within the same ecological niche. The small-molecules can serve as signals to activate distinct pathways and trigger changes in gene expression, allowing bacteria to adapt as the environment and other members of the microbial community change. Moreover, these molecules can be produced by members of the bacterial community with the purpose of triggering a self-response, a process known as quorum sensing (Camilli and Bassler, 2006; Ng and Bassler, 2009).

How bacteria respond to the presence of small-molecules varies. Some organisms respond by differentiating, creating a subpopulation of specialized cells that express a certain pattern of genes (Kearns and Losick, 2005). The subpopulation of cells may in turn produce or respond to different signals and serve a distinct function within the community. For example, the soil bacterium *Bacillus subtilis* differentiates several subpopulations of specialized cell types in response to different environmental cues (Lopez *et al.*, 2009b). One of these specialized subpopulations produces the extracellular matrix necessary to hold cells together during the process of biofilm formation (Chai *et al.*, 2008). Differentiation of this subpopulation of matrix producers is initiated by the activation of a gene regulator termed Spo0A (Lopez *et al.*, 2009a). The active form of Spo0A (Spo0A~P) indirectly activates a regulon of genes that includes the two operons responsible for the production of extracellular matrix, the 15-gene operon epsA-O (henceforth simply eps) (Branda *et al.*, 2001, 2004), responsible for the production of the extracellular polysaccharide, and the yqxM-sipW-tasA operon (henceforth simply yqxM), responsible for the production of the matrix-associated protein TasA (Branda *et al.*, 2004, 2006).

Phosphorylation of Spo0A to trigger matrix production is driven by the phosphorylative action of two transmembrane sensor histidine kinases, KinC and KinD (Hamon and Lazazzera, 2001; Jiang *et al.*, 2000; Kobayashi *et al.*, 2008; LeDeaux *et al.*, 1995). KinD is a canonical membrane kinase with two transmembrane segments connected by a large extracellular sensor domain, presumably involved in recognizing and binding to an unknown extracellular signal. On the contrary, the membrane kinase KinC shows two transmembrane segments connected by only six amino acids, leaving no room for an extracellular sensor domain. Instead, KinC has a PAS sensor domain located in the cytoplasmic region of the kinase. PAS domains are signaling modules known to monitor changes in light, redox potential, oxygen, and small ligands in the cytoplasm of the cell (Gu *et al.*, 2000; Taylor and Zhulin, 1999).

Recently, the PAS domain of KinC was described as the domain responsible for sensing the leakage of potassium cations from the cytoplasm of *B. subtilis*, leakage induced by diverse small-molecules able to form pores in the membrane of the bacterium. The leakage of potassium cations serves as the stimulus to activate the sensor kinase and trigger the phosphorylation of Spo0A, which in turn, leads to the differentiation of a subpopulation of matrix producing cells and the formation of biofilm (Lopez *et al.*, 2009a). While the nature of the stimulus is always consistent (potassium leakage caused by the formation of pores in the membrane), the various small-molecules identified that induce matrix production via KinC differ wildly in their molecular structure. Among these molecules are the macrolide polyenes nystatin and amphotericin and the peptide antibiotics gramicidin and valinomycin, all of which are derived from soil-dwelling bacteria. However, perhaps the most interesting small-molecule described to trigger matrix production via KinC is the self-generated lipopeptide, surfactin (Arima *et al.*, 1968; Kluge *et al.*, 1988; Lopez *et al.*, 2009a). Once produced by the community of *B. subtilis*, surfactin causes the leakage of potassium by the formation of pores in the membrane (Sheppard *et al.*, 1991) and that triggers the differentiation of a subpopulation of matrix producers that secretes the extracellular matrix. Surfactin is now described as the first paracrine signaling molecule in bacteria (Lopez *et al.*, 2009c).

Recognizing the mode of action of a signaling molecule rather than its structure is a remarkable strategy that allows *B. subtilis* to sense a diverse array of signals and to respond not only to self-produced molecules but also to natural products secreted by other soil-dwelling organisms. In this chapter, we discuss the procedures and protocols used to describe the mechanism of action of the kinase KinC as a membrane kinase that senses potassium leakage from the cytoplasm of the cell. We analyze the methods and techniques used, highlighting tricks discovered over the course of the study.

2. Screen for Molecules that Stimulate KinC Sensor Kinase

To screen for small-molecule natural products capable of activating biofilm formation in *B. subtilis*, we developed a bioassay to identify one or more substances that, when added to a culture of *B. subtilis* in small concentrations, induce the formation of biofilm. Development of such an assay was remarkably difficult given the intrinsic nature of undomesticated strains of *B. subtilis* to form a biofilm under almost all conditions. The likely reason for constitutive biofilm production under most conditions is that activation of the cascade to biofilm formation is driven by the phosphorylation of the regulator Spo0A by the action of five different kinases, each one of them able to sense different stimuli, most of them unknown (Hamon and Lazazzera, 2001; LeDeaux *et al.*, 1995).

To identify the precise culture conditions under which *B. subtilis* does not produce a biofilm and Spo0A remains dephosphorylated, we cultured *B. subtilis* using different media and under various conditions. Following the comparative study, MSgg was chosen as the positive control medium because it was discovered that *B. subtilis* always makes biofilm when grown in this medium. MSgg is a defined medium composed of 5 m*M* potassium phosphate (pH 7), 100 m*M* Mops (pH 7), 2 m*M* $MgCl_2$, 700 μM $CaCl_2$, 50 μM $MnCl_2$, 50 μM $FeCl_3$, 1 μM $ZnCl_2$, 2 μM thiamine, 0.5% glycerol, 0.5% glutamate, 50 μg/ml tryptophan, and 50 g/ml phenylalanine. MSgg, with glycerol and glutamate as the carbon and nitrogen sources, respectively, is commonly used to induce the process of cell differentiation in *B. subtilis* (Branda *et al.*, 2001). When *B. subtilis* is plated on MSgg and incubated for 3 days at 30 °C, biofilm formation is observed and colonies exhibit several morphological features indicative of the differentiation of distinct cell subpopulations involved in the process of biofilm formation. For example, the production of an extracellular matrix that encases all cells within the biofilm results in the formation of wrinkles on the colony surface (Branda *et al.*, 2004). Similarly, the raising of aerial structures on the surface of the biofilm is indicative of the presence of a subpopulation of sporulating cells. Spores are localized in the apical area of these aerial structures (Branda *et al.*, 2001; Vlamakis *et al.*, 2008).

In contrast to *B. subtilis* always producing biofilm when grown in MSgg, it was found that *B. subtilis* grown in Luria-Bertani (LB) medium did not show any of the developmental characteristics observed in MSgg, even after extended incubation. *B. subtilis* incubated in LB medium, composed of 1% peptone, 0.5% yeast extract, and 0.5% NaCl, developed into colonies composed of a mass of undifferentiated cells, in which cell differentiation was rarely observed according to the occurrence of spores. LB medium proved to be an excellent base medium in which to grow *B. subtilis* to screen for small-molecules that induce biofilm formation.

To screen each small-molecule independently, *B. subtilis* was grown shaken in liquid LB medium rather than stationary on LB agar plates. Upon reaching stationary phase, 1 ml of the shaken culture was used to inoculate each well of a polystyrene 24-well plate. The experimental design allowed each small-molecule to be independently tested at many different concentrations because a different concentration of each small-molecule could be directly added to a separate well. Surprisingly, simply modifying the protocol to grow *B. subtilis* in LB liquid culture resulted in *B. subtilis* producing a weak biofilm. Without the addition of any small-molecule, in liquid LB medium *B. subtilis* produces a weak biofilm or thin floating pellicle. We hypothesize that biofilm development is caused by activation of the master regulator Spo0A~P. Because liquid shaken cultures had reached stationary phase prior to transfer to the 24-well plate, phosphorylation of Spo0A had likely already taken place, leading to low-level, background biofilm production.

To resolve the issue of producing low background levels of biofilm during the assay, a protocol was developed such that *B. subtilis* cultures were in exponential phase when tested against the library of small-molecules. We reasoned that cultures should be grown in exponential phase for prolonged time periods to eliminate any remaining Spo0A~P. To maintain cultures in exponential phase, we defined a culture protocol to grow *B. subtilis* at low optical density for almost 12 h. Following the repeated passaging of cultures, cultures did not exhibit any background pellicle formation when incubated standing and undisturbed. We used *B. subtilis* cultures in exponential phase to assay a battery of small-molecules for their ability to induce *B. subtilis* biofilm formation (detailed protocol provided below).

2.1. Protocol to screen for pellicle formation by adding small-molecules

(1) Undomesticated *B. subtilis* strain NCIB3610 was cultured overnight in LB medium at 37 °C with vigorous agitation.
(2) The overnight culture was used to inoculate new LB medium at dilution 1:100. We routinely used 30 μl of culture to inoculate a 15-ml tube containing 3 ml of liquid LB. The culture was incubated at 37 °C with vigorous agitation until reaching $OD_{600} > 0.4$.
(3) The culture at $OD_{600} > 0.4$ was used to inoculate new LB medium at dilution 1:100. The new culture was incubated at 37 °C with vigorous agitation until reaching $OD_{600} > 0.4$. The passaging was repeated three additional times, to keep cultures growing in exponential phase for a prolonged period.
(4) The final inoculation requires a scale up in volume according to the number of wells that are ultimately needed. For example, a 24-well plate requires 24 ml of LB to be inoculated with 240 μl of culture at $OD_{600} > 0.4$. When this culture has reached $OD_{600} > 0.4$, the culture is transferred to the wells (1 ml of culture per well) where a solution of the small-molecule at the desired final concentration was previously placed.

(5) 24-well plates were incubated at 30 °C with no shaking. To avoid dehydration of the wells, plates were either incubated in a covered container with wet towels at the bottom or wrapped with parafilm.

Several repetitions of each experiment were required to standardize results. Variations due to differential dehydration (positioning of the plate close to the bottom and the wet towels led to less dehydration), disturbances during the incubation (avoid incubation near shakers or platforms in motion that can create vibration), and edge effects within the plate (wells on the side of the plate may experience slight dehydration) may influence pellicle formation.

Using the pellicle formation assay, a preselected battery of small-molecules at a variety of concentrations were tested and it was observed that only compounds defined to cause potassium leakage from the cytoplasm of the cell were able to induce pellicle formation (See Table 13.1). Compounds that induced pellicle formation included the polyene nystatin, produced by *Streptomyces noursei*. Nystatin is known to create pores in the membrane that selectively allow the leakage of potassium and other related cations (Bolard, 1986). Nystatin induced pellicle formation in *B. subtilis* when added to the culture at a final concentration of 60 μM. Amphotericin, a natural product produced by *Streptomyces nodosus* and similar in structure and function to nystatin, also induced pellicle formation. The final concentration to induce pellicle formation with amphotericin was also 60 μM. Surfactin is a lipopeptide produced by *B. subtilis* itself and it too induced pellicle formation in our pellicle formation assay when added at a final concentration of 20 μM. Surfactin is known to cause pores in the membrane allowing for the selective leakage of cytoplasmic cations, generally potassium (Sheppard *et al.*, 1991). In addition to these molecules, the antimicrobials valinomycin and gramicidin, produced by some species of *Streptomyces* and *Bacillus*, respectively, induced pellicle formation. Both valinomycin and gramicidin induce potassium leakage from the cell. Particularly interesting is the case of valinomycin. Valinomycin does not make pores in the membrane but rather works as an ionophore that evacuates potassium exclusively (Marrone and Merz, 1995). The concentrations of valinomycin and gramicidin that induced pellicle formation in *B. subtilis* were less than 2 μM.

The molecules discovered to induce biofilm formation in wild-type (WT) *B. subtilis* failed to induce pellicle formation in cultures of the Δ*kinC*-deficient mutant. The mutant lacking KinC was generated by deleting the gene that encodes the kinase, *kinC*, and replacing the gene with an antibiotic resistance cassette. We used the long flanking homology PCR technique to create the DNA fragment used to replace the gene *kinC* (Wach, 1996). To create the cassette, we used primers to amplify the upstream and downstream regions of *kinC*, used the appropriate primers to amplify the resistance gene, and then performed a joining PCR to

Table 13.1 Battery of small-molecules tested to induced biofilm formation in *B. subtilis*, using the pellicle formation assay

Molecule	Function	Biofilm induction
Nystatin	Cation-selective pore formers	Yes
Amphotericin		
Gramicidin		
Surfactin	Potassium-selective pore former	
Valinomycin	Potassium-selective ionophore	No
CCCP	Proton-selective ionophore	
Polymyxin	Cell wall synthesis inhibitors or membrane disaggregators	
Bacitracin		
Filipin		
Vancomycin		
Aculeacin		
Microcystin		
Nisin	Unselective pore formers	
Iturin		
Syringomycin		
Nonactin	Ammonium-selective pore former	
Triton X-100	Detergents	
Nonidet P-40		
Tween-20		
Sodium dodecyl sulfate		
Novobiocin	Structurally related compounds	
Stigmatellin		
Antimycin		

The name of each compound tested is provided in the left column. The function of each molecule is detailed in the center column and the ability of each compound to induce pellicle formation is given in the right column. This table is credited to Lopez *et al.* (2009a).

generate the cassette. To amplify the upstream and downstream regions of *kinC*, the following primers were used: upfoward: 5′-tcttcttgtgattaacccgc-caaga-3′, upreverse: 5′-gaacaacctgcaccattgcaagaagtattttcaatttgcatcg ctccaa-3′, downforward: 5′-ttgatcctttttttataacaggaattcttcatattgaaagtgaagtgcgaaga-3′, downreverse: 5′-tgtttaagatattcttcacctgggta-3′. The resulting cassette was inserted into the genome of the domesticated strain of *B. subtilis* 168 by inducing its natural competence (Hardwood and Cutting, 1990). Constructs were subsequently transferred to the undomesticated strain NCIB3610 by SPP1 phage transduction (Yasbin and Young, 1974).

Furthermore, when surplus amounts of potassium were present in the LB medium, the small-molecules also failed to induce pellicle formation in the WT strain. Addition of extracellular potassium to levels at or near the cytoplasmic concentration of potassium apparently blocks the efflux of potassium ions from the cytoplasm when the pores are made. Remarkably, blockage in the signaling cascade leading to biofilm formation caused by excess extracellular potassium results in an overproduction of the signaling molecule surfactin, as described by Fall and coworkers (Fall *et al.*, 2006; Kinsinger *et al.*, 2003, 2005). To cause this inhibition, any given potassium salt can be dissolved in LB medium at a final concentration of 150 m*M* or more. The medium may need to be buffered to pH 7.4 depending on the potassium salt dissolved. To test that extracellular levels of potassium would inhibit biofilm formation, LB + K^+ was used as the culture medium for the final passage of *B. subtilis* (step 4 of the protocol) prior to the transfer of the culture to individual wells where the small-molecules were tested. The 24-well plates were incubated at 30 °C with no shaking over 8 h to induce biofilm formation.

3. Quantitative Analysis of the Activation of KinC

Surfactin, nystatin, and several other potassium-selective pore formers stimulate the membrane sensor kinase KinC. Stimulation of KinC in turn, leads to the differentiation of a subpopulation of cells responsible for the production of extracellular matrix, which encapsulates and holds cells together within the biofilm. The subpopulation of matrix producers expresses the *eps* operon (Branda *et al.*, 2001, 2004) and the *yqxM* operon responsible for the production of the extracellular exopolysaccharide and the matrix-associated protein TasA (Branda *et al.*, 2004, 2006), respectively.

Hence, the activation of KinC was observed by monitoring the activation of the genetic cascade that triggers the differentiation of the subpopulation of matrix producers. The activation of the expression of both the *eps* and *yqxM* operons was used as a direct readout for the activation of the cascade. Transcriptional fusions, of the promoter controlling the expression of each operon to the reporter genes *lacZ* or *yfp*, were created such that the activation of operon expression would simultaneously activate the reporters. The expression of the reporters was observed either by the production of the β-galactosidase enzyme or yellow fluorescent protein (YFP), respectively. The amount of protein produced can be quantitatively measured either by monitoring enzymatic activity of the β-galactosidase or the fluorescence emitted by the protein YFP. β-Galactosidase activity can be measured with a colorimetric assay using the substrate *ortho*-nitrophenyl-β-galactoside (ONPG). ONGP is normally colorless. However, when

β-galactosidase is present, it hydrolyzes the ONPG molecule into galactose and *ortho*-nitrophenol. The latter compound has a yellow coloration that can be measured. The YFP has the advantage that it can be detected at the single-cell level, either qualitatively by using a microscope equipped with fluorescence detection or quantitatively, by measuring fluorescent cells using flow cytometry.

To construct the appropriate transcriptional fusions, promoters from the desired genes were amplified by using PCR, excluding their natural ribosome-binding site (RBS). The DNA fragments were cloned into the plasmid pDG1661 or pKM008 (Fig. 13.1) using a multicloning site located upstream of the *lacZ* or *yfp* gene, respectively, and fused to an optimized RBS. Additionally, both vectors have a specific resistance marker downstream of the reporter gene (pDG1661 has a chloramphenicol (cm^R) marker whereas pKM008 has a spectinomycin (spc^R) resistance marker), which permits positive selection when the construct integrates into the genome. Integration is successful by double recombination. DNA fragments of the *amyE* locus flank the cassette harboring the transcriptional fusion with the resistance gene. Prior to the transfer of individual plasmids to *B. subtilis* by natural competence, each plasmid is linearized. Selection for the appropriate resistance will yield only the colonies that have integrated the transcriptional fusion into the *amyE* locus of the genome.

When quantifying gene expression with transcriptional fusions, it is important to consider that not all of the cells are expressing the genes being reported by the fusions. For instance, *eps* and *yqxM* operons are expressed in only the subpopulation of cells specialized as matrix producers. The matrix producing subpopulation represents approximately 35% of the total cells, meaning that only 35% of the signal will be detected when the whole population is analyzed. Taking accurate measurements of subpopulations becomes an issue when transcriptional fusions are coupled to the expression of enzymes such as β-galactosidase. Quantification of gene expression is given by the ratio of the enzymatic activity to the total number of cells and therefore a correct interpretation of the results may be confounded as dramatic pellicle formation would be correlated with a low increase in the expression of the *eps* and *yqxM* operons (threefold increase). Because subpopulations exist, it is recommended that gene expression be quantified using transcriptional fusions coupled to the expression of fluorescent proteins. By using fluorescent proteins as reporters, it is possible to track the specific cells that are expressing the reporter and study the activation of the pathway in this specific subpopulation. The single-cell analysis using fluorescent proteins as reporters estimated that cells, which differentiated to become matrix producers, increase their fluorescence more than 10-fold when the signal molecule is added. Monitoring of fluorescence at the single-cell level was done using flow cytometry. The flow cytometer is a cell counter coupled with a fluorescence detector. Samples are measured by

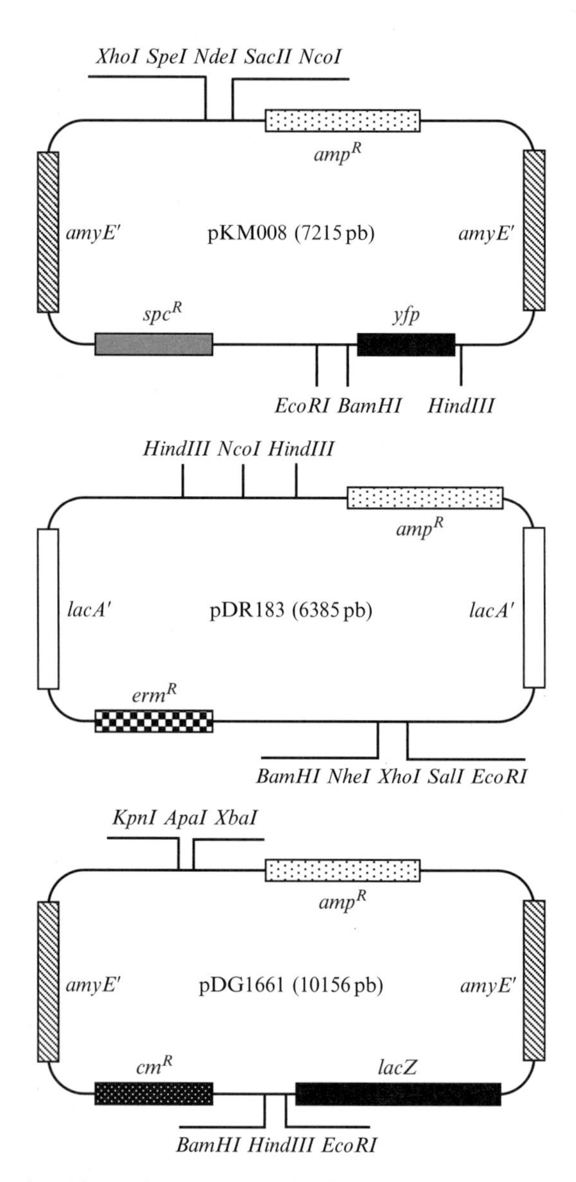

Figure 13.1 Plasmids used to integrate the distinct transcriptional fusions into the genome of *B. subtilis*. Schematic representation of the three plasmids (pKM008, pDR183, and pDG1661) used to construct the transcriptional fusions for integration into the neutral loci *amyE* (in the case of pKM008 and pDG1661) and *lacA* (in the case of pDR183), in the genome of *B. subtilis*. Selected promoters were fused to yellow fluorescent protein gene (*yfp*), in the case of pKM008 and to the β-galactosidase gene (*lacZ*), in the case of pDG1661. When required, constructions were transferred to the *lacA* locus by subcloning into the pDR183 vector. Restriction sites are indicated by the name of the endonucleases that recognize their respective sequences. Integration of these vectors into the neutral loci of the genome was performed by double recombination. The vectors carry different antibiotic resistance makers: amp^R, ampicillin resistance marker; spc^R, spectinomycin resistance marker; erm^R, erythromycin resistance marker; cm^R, chloramphenicol resistance marker. The plasmids and the genes are not to scale.

counting a predetermined number of cells. A total of 50,000 cells were measured by the flow cytometer for each sample and the number of fluorescent cells and the intensity of fluorescence, when present, were determined.

Several considerations need to be made with regard to the preparation of samples for flow cytometry. First, the flow cytometer requires a strong fluorescent signal for proper detection of the fluorescent cells. To ensure proper detection, the promoter coupled to the fluorescence protein gene must strongly induce the expression of the fluorescence protein when activated. As such, the promoter used in the construction of transcriptional fusions must be carefully selected for use in flow cytometry. Selection of the appropriate reporter led to the use of a transcriptional fusion of the *yqxM* promoter to *yfp* (using the primers PyqxMforward; 5′-tggcgaattctcagagt-taaatggtattgcttcact-3′ and PyqxMreverse: 5′-cctaagcttgtaaaacactgtaacttgatat-gacaa-3′). Transcriptional fusions to the promoter of the *eps* operon were not used because the signal obtained with the activation of the promoter was not strong enough to be detected by flow cytometry. Additionally, samples of cells prepared for flow cytometry need to be dispersed as single cells in solution. Obtaining solutions of single cells is extremely difficult when working with any type of biofilm and especially difficult with *B. subtilis*' biofilms because the cells grow in chains and are encased within the matrix. The flow cytometer counts and measures single cells. If the samples contain clumps of cells, these clumps will likely be counted by the flow cytometer as a single cell and the variable fluorescent background is likely to confuse the detector. To avoid this problem, samples must be mildly sonicated prior flow cytometry. Sonication will disperse the clumps into single cells without causing cell lysis. To avoid any alteration of the gene expression due to sonication, samples were fixed with paraformaldehyde before the cells were sonicated. Below is the detailed protocol used to prepare samples for flow cytometry.

(1) Cultures were sampled, centrifuged and washed with 1 ml of PBS buffer. Pellicles formed at the top of a well were easily disrupted by repetitive passage through a pipette tip or a needle.
(2) Washed samples were centrifuged and fixed with 4% paraformaldehyde. Pellets were dissolved in 500 μl of 4% paraformaldehyde and incubated at room temperature for exactly 7 min.

4% paraformaldehyde was prepared as follows:

10 ml stock 16% paraformaldehyde
26 ml distilled water
4 ml 10× PBS
4 μl 10 N NaOH

The final solution is filtered through a 0.22-μm filter and aliquoted.

(3) After fixation, cells were washed twice with 1 ml of PBS and resuspended in GTE buffer (50 m*M* glucose, 10 m*M* EDTA at pH 8, 20 m*M* Tris–HCl at pH 8). GTE buffer is preferred to PBS as it preserves cells under more optimal osmolar conditions.

Prior to analysis by flow cytometry, cells were subjected to mild sonication under conditions that disrupt the cells from the extracellular matrix but do not lyse cells at detectable levels (Branda *et al.*, 2006). Mild sonication is defined as two series of 12 pulses (output 5, amplitude 0.7 s). The efficiency of sonication may be confirmed by light microscopy.

In preparation for injection into the flow cytometer, cells were diluted in PBS and directly loaded into a BD LSR II flow cytometer (BD Biosciences) operating a solid-state laser at 488 nm. For each sample, at least 50,000 cells were analyzed. Data containing the fluorescent signals were collected by a 505LP and a 530/30-bp filter, and the photomultiplier voltage was set between 300 and 500 V. Data were captured using FACS Diva software (BD Biosciences) and further analyzed using FlowJo 8.5.2 software (http://www.flowjo.com) (Vlamakis *et al.*, 2008).

Flow cytometry was used to compare cultures that had been exposed to a signal molecule, such as surfactin, to cultures that had not been exposed to a signal molecule. Flow cytometry of cultures exposed to a signaling molecule during the pellicle formation assay resulted in the visualization of the subpopulation of cells differentiated as matrix producers. For the mutant lacking the membrane kinase KinC, which failed to sense the presence of surfactin in the pellicle formation assay, visualization of a subpopulation of matrix producers by flow cytometry was not observed (Fig. 13.2). No subpopulation of matrix producers was observed by flow cytometry following the addition of excess potassium in the form of KCl to the pellicle formation assay (data not shown).

4. Structural Analysis of KinC

The activation of KinC leads to biofilm formation. Thus, the activation of KinC can be monitored qualitatively using the pellicle formation assay and quantitatively by monitoring the differentiation of matrix producers using flow cytometry. To characterize the sensor kinase and determine how KinC senses potassium leakage, different alleles of the kinase KinC were constructed, each lacking different domains of the protein, in an effort to pinpoint the critical components of the kinase involved in sensing potassium leakage. Various constructs were generated and their ability to complement the mutant deficient in KinC monitored using both the qualitative and quantitative measurement techniques.

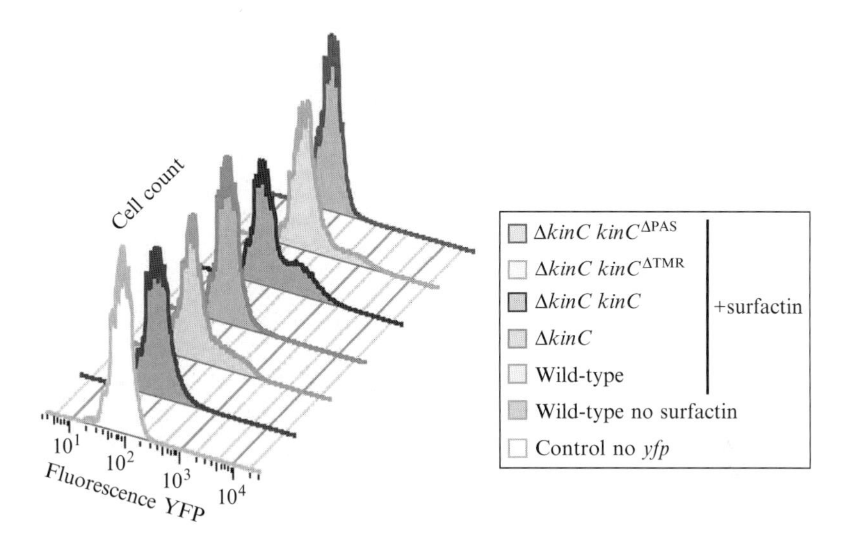

Figure 13.2 Flow cytometry analysis of cells harboring the transcriptional fusion P_{yqxM}-*yfp*. The profiles illustrate cultures of untreated cells and cells in the presence of the small-molecule inducer surfactin (20 μM). The control profile is represented by untreated cells that do not harbor any *yfp* gene and a single population of cells with very low relative fluorescence is observed. Cells harboring the P_{yqxM}-*yfp* fusion also show a single population, similar to the control, in the absence of surfactin (blue peak). When surfactin was added (orange peak), subpopulations of cells with high relative fluorescence were observed as the shoulder to the right of the primary peak. This subpopulation represents approximately 35% of the total cells. Addition of surfactin to the Δ*kinC*-defective mutant does not induce the differentiation of the subpopulation (red peak). Complementation of the Δ*kinC*-defective mutant with a WT copy of kinC resulted in the restoration of the subpopulation of cells expressing P_{yqxM}-*yfp* (purple peak). If the allele of kinC does not have the PAS–PAC sensor domain, it cannot complement the Δ*kinC*-defective mutant (dark green peak), contrasting to the allele lacking the transmembrane region (pale green peak). This figure is credited to Lopez *et al.* (2009a). (See Color Insert.)

KinC is a membrane kinase that has two transmembrane segments connected by six extracellular amino acid residues in the N-terminal region. Following the transmembrane region is an intracellular PAS–PAC domain. PAS domains have been described as able to monitor changes in light, redox potential, oxygen, small ligands, and the overall energy level of a cell (Gu *et al.*, 2000; Taylor and Zhulin, 1999). The PAC motif is proposed to contribute to the folding of the PAS domain. The C-terminal region of KinC harbors the phosphoacceptor and the ATPase domains of the kinase. While the PAS–PAC domain was likely responsible for sensing the leakage of potassium from the cytoplasm of the cell, experiments were conducted to conclusively determine whether the transmembrane domain or the PAS–PAC domain of kinC was the sensory domain that activated the kinase.

To determine whether the transmembrane domain or the PAS–PAC domain of kinC was the sensory domain responsible for activating the kinase, the Δ*kinC*-deficient mutant was complemented with three different alleles of the kinase KinC: the WT version of the protein, an allele lacking the transmembrane region, and another allele lacking the PAS–PAC sensor domain (Fig. 13.3). Each allele was fused to a cyan fluorescence protein (cfp) gene to allow their expression in *B. subtilis* to be followed by western blot. Technically, the alleles were designed by amplifying the DNA fragments independently. To do this, the following primers were employed: PASupfw: 5′-aaaagaattcgt-catgccgattgagttgag-3′, PASuprev: 5′-ctgggattccctcgccagttcagaaagctgttta-tacttc-3′, PASdwfw: 5′-ctggcgaggga-atcccag-3′, PASdwrev: 5′-ttttggatcc gtatacaaacagaagcgag-3′, TMRuprev: 5′-accagctgctgtttctcat cccattgatattttct-catatgaccacc-3′, TMRdwfw: 5′-gatgagaaacagcagctggt-3′. The DNA fragments amplified were combined by joining PCR as described by Wach *et al.*, for LFH-PCR. The WT allele of *kinC* was constructed by joining the entire structural *kinC* gene (including its own promoter) to the *cfp* gene. The allele of KinC lacking the transmembrane region was created by joining the kinC promoter with the initial six amino acids upstream of the transmembrane region, the PAS–PAC sensor domain, and the *cfp* gene. Similarly, the allele of

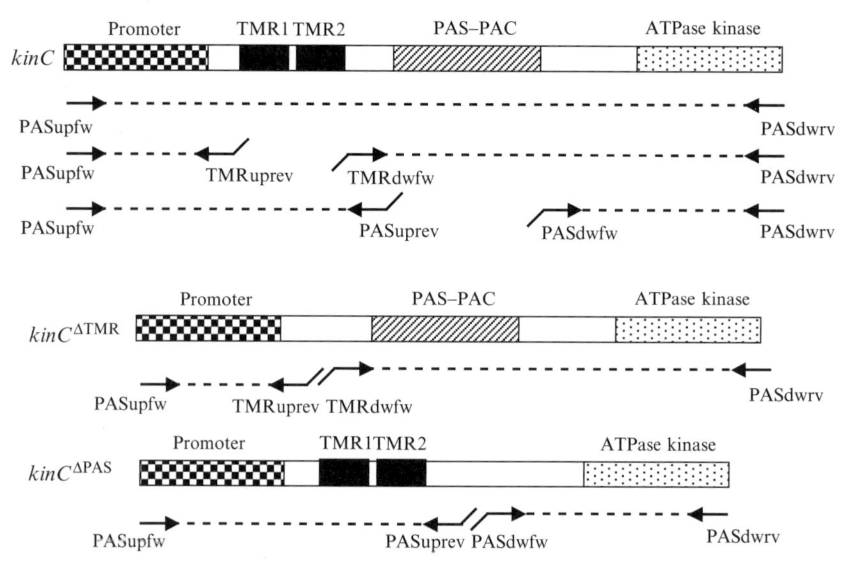

Figure 13.3 Scheme of joining PCRs used to generate the different alleles of KinC. The gene *kinC* is represented as a rectangle in which the coding regions of the domains are marked. Primers are illustrated and named. Amplified DNA fragments are represented by dashed lines. In addition to the WT allele, a subsequent joining PCR yields two modified alleles, one lacking the transmembrane region and another lacking the PAS–PAC sensor domain of the kinase KinC.

KinC lacking the PAS–PAC sensor domain was created by joining the promoter region and the transmembrane region of kinC, the phosphoacceptor and ATPase domains of the kinase, and the *cfp* gene. To successfully construct and express all three different alleles of the protein KinC, it was critical to join the DNA fragments in such a way as to preserve the same open reading frame. Preservation of the open reading frame ensured each allele would be transcribed into a single molecule of mRNA that in turn, would be translated into the desired protein (allele).

The three distinct alleles of KinC were cloned into a modified pKM008 vector (lacking *cfp*) and integrated into the genome of the Δ*kinC*-deficient strain by double recombination into the *amyE* locus, as described above. The expression of the different versions of KinC was confirmed by the detection of the CFP protein fused to each allele. To detect CFP, the conventional western blot technique was employed, using monoclonal antibodies (Promega) against CFP (monoclonal antibodies against CFP can also be used to recognize GFP and YFP proteins, since the three proteins share a common epitope). Results indicated that both the expression of the WT allele and the expression of the allele lacking the transmembrane region restored the ability of the Δ*kinC*-deficient mutant to form biofilm communities in the biofilm-inducing medium MSgg. Also, these two alleles induced the formation of pellicle in the presence of surfactin in the pellicle formation assay. The allele of KinC lacking the PAS–PAC sensor domain was neither able to induce the formation of biofilm in MSgg nor respond to the presence of surfactin in the pellicle formation assay.

To further confirm the results observed during the pellicle formation assay, quantitative measurements were made using flow cytometry. The experiment required the expression of the transcriptional fusion P_{yqxM}-*yfp* but the neutral *amyE* locus used previously to integrate the constructs was occupied in every case by the distinct kinC alleles that complemented the Δ*kinC*-deficient strain. An additional neutral locus was required to integrate the transcriptional reporter of the *yqxM* operon. For that purpose the vector pDR183 was used (Doan *et al.*, 2005). Vector pDR183 has a multicloning site next to an erythromycin-resistant marker and the cassette is flanked by DNA fragments of a second neutral *B. subtilis* locus *lacA* (See Fig. 13.1). The transcriptional fusion P_{yqxM}-*yfp* was transferred from pKM008 to pDR183 by endonuclease digestion and T4 ligase ligation. The vector was linearized and integrated into the *lacA* locus by double recombination using *B. subtilis*' natural competence. Colonies carrying the transcriptional fusion P_{yqxM}-*yfp* within the *lacA* locus were resistant to erythromycin.

The transcriptional fusion P_{yqxM}-*yfp* under the control of the distinct alleles of kinC resulted in the occurrence of a subpopulation of matrix producers only in the presence of the WT allele or the allele lacking the transmembrane region, and only when surfactin was added. The allele of

KinC lacking the PAS–PAC sensor domain neither responded to surfactin nor differentiated to create a subpopulation of matrix producers.

5. Monitoring the Signals Using Indirect Measurements

Whether the membrane kinase KinC is directly activated by the leakage of potassium cations from the cytoplasm or is indirectly activated remains to be determined. Other cellular responses associated with the leakage of potassium from the cytoplasm may be considered as stimuli, leading to the activation of the membrane kinase KinC. For instance, the efflux of potassium cations might be counteracted by the uptake of other cations, to maintain the electrophysiology of the bacterium. Thus, the PAS–PAC domain might possibly sense the uptake or increased concentration of another cation or cations, rather than the loss of potassium ions directly.

To identify and study the nature of the signal that activates KinC in detail, several specific experiments were performed. First, we attempted to determine whether the efflux of potassium acts as the signal triggering the kinase KinC. To do this, a potassium-sensitive electrode was used to measure the extracellular concentration of potassium (Katsu *et al.*, 2002; Orlov *et al.*, 2002; Yasuda *et al.*, 2003). By measuring the fluctuation in extracellular potassium following the addition of surfactin, it was possible to calculate the concentration of potassium evacuated from the cytoplasm of the cells.

Several technical problems were encountered while using the potassium-sensitive electrode. The most prominent issue was the tremendous sensitivity of the electrode for potassium cations. The sensitivity of the electrode required measurements to be made in a buffer free of potassium cations. Otherwise, the background obscured detection of any variation in the concentration of extracellular potassium when surfactin was added. However, the complete absence of potassium in the extracellular milieu represents neither natural nor laboratory conditions under which surfactin was previously tested. Therefore, working with a buffer free of potassium cations was unlikely to yield any additional information about the nature of the signal detected by KinC.

Ultimately, cytoplasmic potassium leakage could be measured by the electrode after resuspending cells in a buffer containing 10 m*M* KCl. Under these conditions, a 1.2 m*M* increase in the extracellular concentration of potassium was detected an hour after cells were treated with surfactin. Rough calculations, considering only the volume of cell mass versus volume of resuspension buffer, suggested the cytoplasmic decrease of potassium was 40 m*M* when surfactin was added. Although the experiments could be both repeated and refined, the fluctuation in cytoplasmic potassium appears to be negligible considering the concentration of potassium in the cytoplasm has been estimated to be between 350 and 650 m*M*.

Possibly, the leakage of cytoplasmic potassium may be associated with other internal effects such as pH variation or the uptake of other cations such as calcium. We monitored pH variation in the cytoplasm using pH-sensitive dyes. For instance, 5(6)-carboxynaphthofluorescein is a pH-sensitive fluorescent dye with $\lambda_{ex} > 512$ nm and $\lambda_{em} > 567$ nm under acidic or neutral conditions or $\lambda_{ex} > 598$ nm and $\lambda_{em} > 668$ nm under basic conditions (Butterfield *et al.*, 2009; Song *et al.*, 1997). While exposed to the dye, *B. subtilis* cells were observed to increase cytoplasmic acidity upon treatment with surfactin. The result suggests that potassium leakage is probably linked to an uptake of H^+. Additional experiments need to be completed to decipher the nature of the signal that stimulates the activation of the membrane kinase KinC. Thus far, we cannot discard that other factors might be involved.

6. Applications of the System Signal-Kinase

Characterization of any system which senses and recognizes a known and defined signal is a remarkable molecular tool that may permit us to control desired genetic pathways to direct the production of enzymes, antibiotics or developmental processes in *B. subtilis*, or other bacteria. In our case, the signaling molecules surfactin and nystatin activate the sensor kinase KinC and in turn, KinC triggers the phosphorylation of its cognate regulator Spo0A. The phosphorylation domain of KinC recognizes exclusively its cognate regulator Spo0A, a common feature of all kinases is to selectively recognize their cognate regulators at a molecular level. Thus, phosphorylation of any other cognate regulator by the phosphorylative action of KinC is possible by changing the phosphorylation domain, which recognizes Spo0A, to a phosphorylation domain that recognizes another regulator. It is therefore possible to create a kinase chimera, composed of the sensing domain of KinC and a phosphorylation domain that phosphorylates and activates another regulator of interest.

For instance, motility in *B. subtilis* is inhibited when the regulator DegU is activated by phosphorylation (Amati *et al.*, 2004; Kobayashi, 2007; Verhamme *et al.*, 2007). Phospshorylation of DegU is governed by the action of the kinase DegS (Dahl *et al.*, 1991). Thus, activation of DegS leads to an inhibition of motility. We controlled motility in *B. subtilis* in response to the presence of nystatin by creating a kinase chimera, which was comprised of the transmembrane region and the PAS–PAC sensor domain of KinC and the phosphoacceptor and ATPase domain of the kinase DegS

(Fig. 13.4A). The activation of the chimera required the signaling molecule nystatin. However, instead of inducing biofilm formation, nystatin activated the regulator DegU and in turn, motility was inhibited.

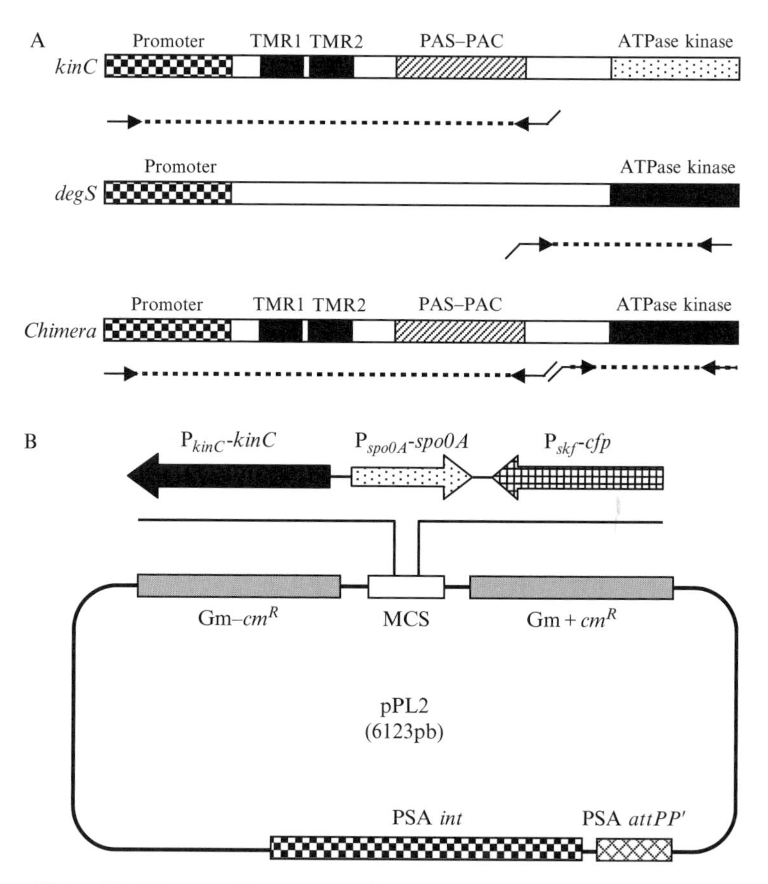

Figure 13.4 (A) Scheme of joining PCRs used to generate the kinase chimera KinC–DegS. Both genes *kinC* and *degS* are represented as rectangles in which the coding regions of the domains are marked. PCR amplification of the regions of the genes required to generate the chimera are represented as dashed lines. Primers required for amplification are illustrated as arrows. The region coding for the transmembrane domain and the PAS–PAC sensor domain of *kinC* was amplified and joined to the kinase domain of degS by the joining PCR as represented in the figure. (B) Construction of the cassette KinC-Spo0A-P_{skf}-*cfp* and integration into the genome of *Listeria monocytogenes*. The multicloning site (MCS) of the vector pPL2 was used to clone the joining PCR KinC-Spo0A-P_{skf}-*cfp*. This plasmid integrates into the genome of *L. monocytogenes* due to the PSA prophage integrase gene (PAS *int*) and the PSA prophage attachment site in $tRNA^{Arg}$ (PSA *attPP'*). Selection is given by the chloramphenicol resistance marker (cm^R). The plasmid harbors two different cm^R genes to be selectable in Gram-positive strains (Gm−cm^R) and Gram-negative strains (Gm+cm^R).

In order for the above chimera to be the exclusive phosphorylator of the regulator DegU, the strain must be defective in the native kinase DegS. Without the native DegS, the phosphorylation of DegU will be done wholly by the DegS domain of the kinase chimera and the results will not be confounded by a native DegS. The *degS* gene was deleted by LFH-PCR as explained above. To create the deletion cassette we used the primers: degSupforward: 5′-agccctacaactaccaatagtgcaa-3′, degSupreverse: 5′-caattcgccctatagtgagtcgta gcactttggaatccatctttgtttt-3′, degSdownforward: 5′-ccagcttttgt tccctttagtgagataggtcttgggacatttattatg att-3′, degSdownreverse: 5′-tcaatttcttcacggtaagtct-cct-3′. We complemented the Δ*degS*-deficient mutant with the kinase chimera by integrating the cassette in the *amyE* locus by double recombination. The KinC–DegS chimera was constructed by combining a PCR-amplified DNA fragment of the *kinC* gene harboring the natural promoter of the kinase, the transmembrane region and the PAS–PAC sensor domain, to a PCR-amplified fragment of the *degS* gene corresponding to the phosphoacceptor and the ATPase domain. The fragments were linked by joining PCR. Again, the experiment conserved the open reading frame across the resulting DNA fragment such that the expression of the kinase chimera was not hindered. The resulting DNA fragment was cloned into the modified pKM008 vector (lacking *yfp*) and integrated into the *amyE* locus of the genome of a Δ*degS*-deficient mutant, using the resistance to spectinomycin provided by the cassette.

The resulting chimera was tested for the inhibition of motility in the presence of nystatin. The assay for the inhibition of motility was prepared using the LB medium described by Kearns and coworkers, to promote motility (Kearns and Losick, 2003). The assay was performed on freshly prepared LB plates with a low agar concentration (1%). We inoculated the plates with different concentrations of nystatin. Three microliters of an overnight culture of the strain harboring the chimera were spotted onto the plate and the plates were then incubated at 30 °C for 24 h.

The process of making kinase chimeras is not always straightforward. Some kinases require specific conformational changes to permit the transference of the phosphate group to its cognate regulator. Often, the process of making chimeras changes the particular configuration of the kinase making it difficult or impossible for the activation of the chimera. It is difficult to avoid this problem. The generation of a functioning KinC/DegU chimera required experimentation with various lengths of each individual PCR-amplified fragment.

Regardless of the difficulties inherent in creating modified kinases and chimeras, the signal-kinase systems can be used to control the expression of a desired gene in any given bacterial system. It is possible to construct a transcriptional fusion of a desired gene, placing it under the control of a promoter that is recognized by a cognate regulator of the kinase. To test this principle, the gene *cfp* was expressed in *Listeria monocytogenes*, a Gram-positive pathogen that does not have the KinC-Spo0A system. We made a transcriptional fusion of the *cfp* gene to the promoter of the *skf* operon. The *skf* operon is an operon in *B. subtilis*, the transcription of which is directly activated by the regulator Spo0A~P (Hobbs, 2006). We used PCR to construct a cassette that joined *kinC* with its natural promoter, *spo0A* with its natural promoter, and the transcriptional fusion P_{skf}-*cfp*. The primers used to amplify the fragments and construct the cassette are: KinC-up: 5′-cactcttcattacgctgcttggccgcattgtcatgcc-3′, KinC-dw: 5′-gatcgagcttatgtatacac-3′, Spo0A-up: 5′-gtgtatacataagctcgatcgaaaattaccgatccaagac-3′, Spo0A-dw: 5′-aaaagtcgactacttacgattttg caggac-3′, Pskf-up: 5′-aagcagcgtaatgaagagtc-3′, Pskf-upSalI: 5′-aaaagtcgacaagcagcgta atgaagagtg-3′, and Pskf-cfp-dw: 5′-aaaacggccgtgatcgaaatagtacataatg-3′. As illustrated in Fig. 13.4, the joining PCR cassette containing *kinC*, *spo0A*, and the transcriptional fusion P_{skf}-*cfp* was made by joining the three genes ordered in divergent transcriptional frames. The use of divergent transcriptional frames in the expression of each gene eliminates false activation of the reporter fusion P_{skf}-*cfp* due to possible interference in the expression of the different promoters.

The cassette was cloned in the pPL2 (Lauer *et al.*, 2002) plasmid, using the multicloning site of the vector. The pPL2 plasmid harbors a phage attachment site of the *L. monocytogenes* chromosome, which leads to the integration of the plasmid at the 3′ end of an arginine tRNA gene ($tRNA^{Arg}$). The sequence reconstitutes a complete $tRNA^{Arg}$ gene after the integration of the plasmid. Transformation of *L. monocytogenes* was carried out using electroporation, as previously described for other studies (Lemon *et al.*, 2007). Colonies with the plasmid integrated into the genome can be positively selected by using chloramphenicol 5 μg/ml.

Cultures of *L. monocytogenes* carrying the construction were grown in TSBYE medium (TSB medium + 1% yeast extract) and dispensed in wells where the signal molecule surfactin had been previously added to a final concentration of 50 μg/ml. Activation of the recombinant KinC led to the phosphorylation of the recombinant Spo0A and thus, activation of the expression of the transcriptional fusion P_{skf}-*cfp*. Fluorescent cells were detected by fluorescence microscopy. We were unable to use flow cytometry for quantitative analysis because the flow cytometer was unable to count *L. monocytogenes* cells due to their smaller size as compared to *B. subtilis*.

7. Conclusions

We have presented in this chapter the procedures, methods, and protocols established in the laboratory to characterize the leakage of cytoplasmic potassium as a signal to induce biofilm formation. The various small-molecules identified to cause potassium leakage by making pores in the membrane of *B. subtilis* are sensed by the membrane kinase KinC and trigger the differentiation of the subpopulation of cells responsible for the production of the extracellular matrix that holds cells together in a biofilm (Lopez *et al.* 2009a). Recognizing the mode of action of a signaling molecule rather than the molecular structure allows a bacterium to sense of a large number of signals and to respond not only to self-produced molecules but also to natural products secreted by other organisms in close proximity.

Although it is clear that the PAS–PAC domain of the kinase KinC is responsible for sensing the leakage of potassium cations from the cell, the nature of the signal has yet to be elucidated. Whether the membrane kinase KinC is activated directly by the leakage of potassium cations from the cytoplasm or indirectly, due to variation of another parameter linked to potassium leakage, remains unknown.

The ability to monitor populations at the single-cell level through the use of transcriptional fluorescent reporters that could be detected by microscopy techniques or flow cytometry was critical to the success of our work. The new techniques employed permitted the characterization of the subpopulation of cells responsible for the production of the extracellular matrix (Smits *et al.*, 2005; Vlamakis *et al.*, 2008). This subpopulation represents 35% of the total cell population and showed a 10-fold induction of the genes directly responsible for matrix production. Conventional techniques frequently used to monitor transcriptional changes in gene expression, such as the β-galactosidase assay or other enzymatic activities coupled to colorimetric assays, would fail to detect this effect. This limitation resides in the fact that conventional techniques estimate the relative values of gene expression with respect to the size of the whole cell population, normally considered as a measurement of the optical density OD_{600}. Conventional techniques will systematically and drastically underestimate the induction value of gene expression when gene expression only occurs in a subpopulation of cells.

REFERENCES

Amati, G., Bisicchia, P., and Galizzi, A. (2004). DegU-P represses expression of the motility fla-che operon in *Bacillus subtilis*. *J. Bacteriol.* **186,** 6003–6014.

Arima, K., Kakinuma, A., and Tamura, G. (1968). Surfactin, a crystalline peptidelipid surfactant produced by *Bacillus subtilis*: Isolation, characterization and its inhibition of fibrin clot formation. *Biochem. Biophys. Res. Commun.* **31,** 488–494.

Bolard, J. (1986). How do the polyene macrolide antibiotics affect the cellular membrane properties? *Biochim. Biophys. Acta* **864,** 257–304.

Branda, S. S., Gonzalez-Pastor, J. E., Ben-Yehuda, S., Losick, R., and Kolter, R. (2001). Fruiting body formation by *Bacillus subtilis*. *Proc. Natl. Acad. Sci. USA* **98,** 11621–11626.

Branda, S. S., Gonzalez-Pastor, J. E., Dervyn, E., Ehrlich, S. D., Losick, R., and Kolter, R. (2004). Genes involved in formation of structured multicellular communities by *Bacillus subtilis*. *J. Bacteriol.* **186,** 3970–3979.

Branda, S. S., Chu, F., Kearns, D. B., Losick, R., and Kolter, R. (2006). A major protein component of the *Bacillus subtilis* biofilm matrix. *Mol. Microbiol.* **59,** 1229–1238.

Butterfield, S. M., Hennig, A., and Matile, S. (2009). Colorful methods to detect ion channels and pores: Intravesicular chromogenic probes that respond to pH, pM and covalent capture. *Org. Biomol. Chem.* **7,** 1784–1792.

Camilli, A., and Bassler, B. L. (2006). Bacterial small-molecule signaling pathways. *Science* **311,** 1113–1116.

Chai, Y., Chu, F., Kolter, R., and Losick, R. (2008). Bistability and biofilm formation in *Bacillus subtilis*. *Mol. Microbiol.* **67,** 254–263.

Dahl, M. K., Msadek, T., Kunst, F., and Rapoport, G. (1991). Mutational analysis of the *Bacillus subtilis* DegU regulator and its phosphorylation by the DegS protein kinase. *J. Bacteriol.* **173,** 2539–2547.

Doan, T., Marquis, K. A., and Rudner, D. Z. (2005). Subcellular localization of a sporulation membrane protein is achieved through a network of interactions along and across the septum. *Mol. Microbiol.* **55,** 1767–1781.

Fall, R., Kearns, D. B., and Nguyen, T. (2006). A defined medium to investigate sliding motility in a *Bacillus subtilis* flagella-less mutant. *BMC Microbiol.* **6,** 31.

Gu, Y. Z., Hogenesch, J. B., and Bradfield, C. A. (2000). The PAS superfamily: Sensors of environmental and developmental signals. *Annu. Rev. Pharmacol. Toxicol.* **40,** 519–561.

Hamon, M. A., and Lazazzera, B. A. (2001). The sporulation transcription factor Spo0A is required for biofilm development in *Bacillus subtilis*. *Mol. Microbiol.* **42,** 1199–1209.

Hardwood, C. R., and Cutting, S. M. (1990). Molecular Biological Methods for Bacillus. Wiley, New York.

Hobbs, E. C. (2006). Control of cannibalism in *Bacillus subtilis*. Department of Molecular and Cellular Biology. Harvard University, Cambridge.

Jiang, M., Shao, W., Perego, M., and Hoch, J. A. (2000). Multiple histidine kinases regulate entry into stationary phase and sporulation in *Bacillus subtilis*. *Mol. Microbiol.* **38,** 535–542.

Katsu, T., Nakagawa, H., and Yasuda, K. (2002). Interaction between polyamines and bacterial outer membranes as investigated with ion-selective electrodes. *Antimicrob. Agents Chemother.* **46,** 1073–1079.

Kearns, D. B., and Losick, R. (2003). Swarming motility in undomesticated *Bacillus subtilis*. *Mol. Microbiol.* **49,** 581–590.

Kearns, D. B., and Losick, R. (2005). Cell population heterogeneity during growth of *Bacillus subtilis*. *Genes Dev.* **19,** 3083–3094.

Kinsinger, R. F., Shirk, M. C., and Fall, R. (2003). Rapid surface motility in *Bacillus subtilis* is dependent on extracellular surfactin and potassium ion. *J. Bacteriol.* **185,** 5627–5631.

Kinsinger, R. F., Kearns, D. B., Hale, M., and Fall, R. (2005). Genetic requirements for potassium ion-dependent colony spreading in *Bacillus subtilis*. *J. Bacteriol.* **187,** 8462–8469.

Kluge, B., Vater, J., Salnikow, J., and Eckart, K. (1988). Studies on the biosynthesis of surfactin, a lipopeptide antibiotic from *Bacillus subtilis* ATCC 21332. *FEBS Lett.* **231,** 107–110.

Kobayashi, K. (2007). Gradual activation of the response regulator DegU controls serial expression of genes for flagellum formation and biofilm formation in *Bacillus subtilis*. *Mol. Microbiol.* **66,** 395–409.

Kobayashi, K., Kuwana, R., and Takamatsu, H. (2008). kinA mRNA is missing a stop codon in the undomesticated *Bacillus subtilis* strain ATCC 6051. *Microbiology* **154,** 54–63.

Lauer, P., Chow, M. Y., Loessner, M. J., Portnoy, D. A., and Calendar, R. (2002). Construction, characterization, and use of two *Listeria monocytogenes* site-specific phage integration vectors. *J. Bacteriol.* **184,** 4177–4186.

LeDeaux, J. R., Yu, N., and Grossman, A. D. (1995). Different roles for KinA, KinB, and KinC in the initiation of sporulation in *Bacillus subtilis*. *J. Bacteriol.* **177,** 861–863.
Lemon, K. P., Higgins, D. E., and Kolter, R. (2007). Flagellar motility is critical for Listeria monocytogenes biofilm formation. *J. Bacteriol.* **189,** 4418–4424.
Lopez, D., Fischbach, M. A., Chu, F., Losick, R., and Kolter, R. (2009a). Structurally diverse natural products that cause potassium leakage trigger multicellularity in *Bacillus subtilis*. *Proc. Natl. Acad. Sci. USA* **106,** 280–285.
Lopez, D., Vlamakis, H., and Kolter, R. (2009b). Generation of multiple cell types in *Bacillus subtilis*. *FEMS Microbiol. Rev.* **33,** 152–163.
Lopez, D., Vlamakis, H., Losick, R., and Kolter, R. (2009c). Paracrine signaling in a bacterium. *Genes Dev.* **23,** 1631–1638.
Marrone, T. J., and Merz, K. M. (1995). Molecular recognition of K^+ and Na^+ by valinomycin in methanol. *J. Am. Chem. Soc.* **117,** 779–791.
Mascher, T., Helmann, J. D., and Unden, G. (2006). Stimulus perception in bacterial signal-transducing histidine kinases. *Microbiol. Mol. Biol. Rev.* **70,** 910–938.
Ng, W. L., and Bassler, B. L. (2009). Bacterial quorum-sensing network architectures. *Annu. Rev. Genet.* **43,** 197–222.
Orlov, D. S., Nguyen, T., and Lehrer, R. I. (2002). Potassium release, a useful tool for studying antimicrobial peptides. *J. Microbiol. Methods* **49,** 325–328.
Sheppard, J. D., Jumarie, C., Cooper, D. G., and Laprade, R. (1991). Ionic channels induced by surfactin in planar lipid bilayer membranes. *Biochim. Biophys. Acta* **1064,** 13–23.
Smits, W. K., Eschevins, C. C., Susanna, K. A., Bron, S., Kuipers, O. P., and Hamoen, L. W. (2005). Stripping *Bacillus*: ComK auto-stimulation is responsible for the bistable response in competence development. *Mol. Microbiol.* **56,** 604–614.
Song, A., Parus, S., and Kopelman, R. (1997). High-performance fiber-optic pH micro-sensors for practical physiological measurements using a dual-emission sensitive dye. *Anal. Chem.* **69,** 863–867.
Taylor, B. L., and Zhulin, I. B. (1999). PAS domains: Internal sensors of oxygen, redox potential, and light. *Microbiol. Mol. Biol. Rev.* **63,** 479–506.
Verhamme, D. T., Kiley, T. B., and Stanley-Wall, N. R. (2007). DegU co-ordinates multicellular behaviour exhibited by *Bacillus subtilis*. *Mol. Microbiol.* **65,** 554–568.
Vlamakis, H., Aguilar, C., Losick, R., and Kolter, R. (2008). Control of cell fate by the formation of an architecturally complex bacterial community. *Genes Dev.* **22,** 945–953.
Wach, A. (1996). PCR-synthesis of marker cassettes with long flanking homology regions for gene disruptions in *S. cerevisiae*. *Yeast* **12,** 259–265.
Yasbin, R. E., and Young, F. E. (1974). Transduction in *Bacillus subtilis* by bacteriophage SPP1. *J. Virol.* **14,** 1343–1348.
Yasuda, K., Ohmizo, C., and Katsu, T. (2003). Potassium and tetraphenylphosphonium ion-selective electrodes for monitoring changes in the permeability of bacterial outer and cytoplasmic membranes. *J. Microbiol. Methods* **54,** 111–115.

CHAPTER FOURTEEN

Two-Component Systems and Regulation of Developmental Progression in *Myxococcus xanthus*

Bongsoo Lee,* Andreas Schramm,* Sakthimala Jagadeesan,† *and* Penelope I. Higgs*

Contents

* Department of Ecophysiology, Max Planck Institute for Terrestrial Microbiology, Marburg, Germany
† Department of Molecular Biology, Max Planck Institute for Infection Biology, Berlin, Germany

Methods in Enzymology, Volume 471
ISSN 0076-6879, DOI: 10.1016/S0076-6879(10)71014-4

Abstract

Myxococcus xanthus is a prokaryotic model system for multicellular development and cell differentiation. Two-component signal transduction genes are abundant in this organism and the majority is likely organized into complex signaling pathways. This chapter describes *in vivo* genetic and *in vitro* biochemical methods used to define signal transduction systems in *M. xanthus*. We also describe a series of phenotypic analyses utilized to define how a specific set of atypical histidine kinases (HKs) influence progression through the complex developmental program.

1. Introduction

Two-component signal (TCS) transduction systems play a major role in controlling adaptive and behavioral responses in prokaryotes by coupling of a specific signal input to a specific output response via histidyl-aspartyl (His-Asp) phosphorelay (Hoch and Silhavy, 1995). TCS family members can be identified by the presence of conserved His-Asp signal transmission domains, including (1) a dimerization/phosphoaccepting (HisKA; Dhp) domain, which accepts a phosphoryl group on an invariant histidine residue, (2) a ATPase domain (HATPase_c; CA) which mediates ATP hydrolysis to donate a phosphoryl group to the HisKA domain, and (3) a receiver (REC) domain which accepts a phosphoryl group onto an invariant aspartic acid (Stock *et al.*, 2000). In the paradigm TCS system, these domains are organized into two proteins, a histidine kinase (HK) containing a sensor domain fused to the HisKA and HATPase_c domains, and a response regulator (RR) containing the REC domain fused to an effector domain. Upon reception of an environmental signal, the HK autophosphorylates, the phosphoryl group is transferred to the REC domain of the RR, and (typically) the associated DNA-binding domain is activated to regulate transcription of target genes. Genes encoding HK and RR partners are most often expressed from a single transcript. Thus, the paradigm TCS system is a 1 HK:1 RR (1:1) paired system in which the core system players are easily identified.

However, the highly modular nature of the conserved TCS signal transmission domains allows for generation of signaling systems that are far more complex. These systems include multistep (His-Asp-His-Asp) phosphorelays (Burbulys *et al.*, 1991; Uhl and Miller, 1996), and branched pathways in which more than one HK signals to a single RR (many-to-one systems) (Jiang *et al.*, 2000; Matsubara *et al.*, 2000) or one HK signals to more than one RR (one-to-many systems) (Hess *et al.*, 1988). These more complex signaling systems are often reflected in the genomic arrangement of TCS genes, including TCS genes which are orphan (i.e., not encoded adjacent to a signaling partner) or complex (i.e., more than two TCS genes are encoded in the same locus). Although the signal systems in which orphan and complex TCS genes participate are sometimes difficult to identify, they are worth investigation since these systems can reflect the true signaling potential of histidyl-aspartyl (His-Asp) phosphorelay.

A premier model in which to investigate TCS complexity and system variability is the deltaproteobacteria species *Myxococcus xanthus*, which is remarkable for a complex life cycle in which social (group) behavior is preferred (Shimkets, 1999). *M. xanthus* has a large repertoire of TCS genes corresponding in part to the large (9.3 Mbp) genome (Goldman *et al.*, 2007; Shi *et al.*, 2008; Whitworth and Cock, 2008a,b). Depending on the exact definition criteria, there are at least 272 TCS genes encoding 21 Che-like-TCS proteins, 118 HKs, 14 HK-like proteins, and 119 RRs (Shi *et al.*, 2008; Whitworth and Cock, 2008a,b). Interestingly, the genomic organization of the non-Che-like-TCS genes indicates that only 29% are organized as simple 1:1 paired systems, with the majority expressed either as orphans (55%) or within complex clusters (19%) (Shi *et al.*, 2008). HKs considered atypical (i.e., cytoplasmic or hybrid) are overrepresented in the orphan and complex cluster gene arrangements. RRs fused to DNA-binding domains are overrepresented in the paired organization, whereas RRs containing other output domains, or even lacking an output domain, were overrepresented in complex or orphan genes. Not surprisingly, then, many atypical TCS systems have been described that regulate the complex *M. xanthus* life cycle (listed in Shi *et al.*, 2008; Whitworth and Cock, 2008a,b).

In a nutrient rich environment, *M. xanthus* favors group behavior because nutrients are most efficiently obtained cooperatively via a "wolf-pack" feeding mechanism (Rosenberg *et al.*, 1977). Cells swarming in groups can secrete sufficient quantities of antibiotics and digestive enzymes to paralyze and digest prey microorganisms or decaying organic material (Reichenbach, 1999). During periods of nutrient limitation, the cells enter an approximately 72 h developmental program which culminates in the formation of multicellular fruiting bodies filled with environmentally resistant spores (Fig. 14.1). Dissemination of spore-filled fruiting bodies to more favorable environments enables cells to reenter the vegetative cycle in groups.

During the developmental program, the initial population of cells segregates into at least three subpopulations that undergo different developmental fates. Approximately 80% of the cells are thought to undergo programed cell death (PCD) (Nariya and Inouye, 2008; Wireman and Dworkin, 1977). Approximately 15% of cells aggregate into mounds of 10^5–10^6 cells and then differentiate into spores. Finally, the remaining 5% of cells do not aggregate and remain as peripheral rods (O'Connor and Zusman, 1991). Thus, the *M. xanthus* developmental program represents a prokaryotic model system in which to investigate the sophisticated signaling

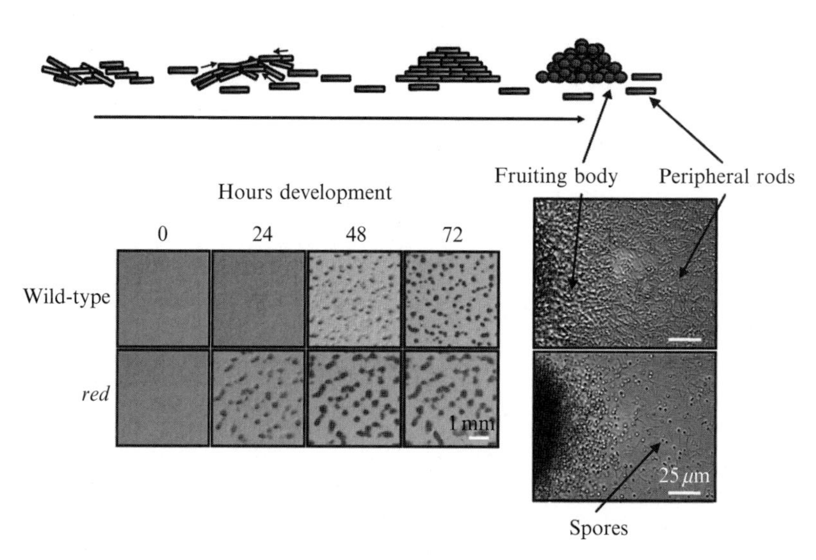

Figure 14.1 The effect of negative regulator TCS systems on the *M. xanthus* developmental program. Top: Schematic of the developmental program. Cells aggregate into mounds and differentiate into spores. Peripheral rods do not aggregate and remain outside of the fruiting bodies. Bottom left: Stereomicroscope images of 4×10^7 wild-type (strain DZ2) and DZ2 Δ(*redC-F*) (strain DZ4659) cells spotted on nutrient limited agar plates and incubated for the indicated times at 32 °C. In the wild type, cells move into aggregation centers (translucent mounds) between 24 and 48 h. Mounds darken when the cells differentiate into spores (48–72 h). *red* mutants aggregate and sporulate 24 h earlier than wild type and form disorganized fruiting bodies. Bottom right: Light microscopy images of the edge of a fruiting body surrounded by peripheral rods. Spores can be found outside of the fruiting bodies in the early developing mutants (bottom) but not in the wild type (top).

systems that are necessary to coordinate multicellular behavior and cell fate decisions.

Our interests lay in a unique group of genes (*espA*, *todK*, *espC*, *red*) that are necessary for negative regulation of progression through the developmental program. Deletions in any of these genes do not cause cells to enter the developmental program inappropriately, but aggregation and sporulation are observed earlier than in wild type (Fig. 14.1). In all cases, these mutants do not display a defect in sporulation efficiency, but fruiting bodies are disorganized and spores are observed outside of fruiting bodies. All of these genes encode atypical members of the TCS family. EspA and EspC are orphan hybrid HKs (Cho and Zusman, 1999; Lee *et al.*, 2005), TodK is an orphan cytoplasmic HK (Rasmussen and Sogaard-Andersen, 2003), and Red is a novel four component TCS signaling system (Higgs *et al.*, 2005; Jagadeesan *et al.*, 2009). We have been interested in analysis of these signal transduction proteins to identify novel signaling schemes, and to understand the role of negative regulators in *M. xanthus* development. These TCS proteins are not organized in paired arrangements, and are not associated with obvious effector (output) domains. This chapter presents a series of genetic, biochemical, and phenotypic analyses that we are using to characterize these TCS systems in *M. xanthus*.

2. Generation of In-Frame Deletions or Point Mutations in the *M. xanthus* Genome

The first step in analysis of particular two-component signal transduction systems is the generation of null mutants to assign a phenotype to the TCS system. In paired TCS systems, the phenotypes of the individual HK and RR mutations can be compared to the deletion of both genes to determine the order of the corresponding proteins in the signaling pathway (Parkinson, 1995). In complex TCS systems, in addition to deletion of each individual gene, the phenotypes associated with pairwise combinations of gene deletions can also reveal the order in which each TCS protein functions in the system. For example, in the Red four-component system (Higgs *et al.*, 2005) which consists of two RRs (RedD and RedF), an HK (RedC), and a HK-like protein (RedE), deletion of *redC*, *redF*, or of the entire system leads to early development, while deletion of *redD* or *redE* causes delayed development. Pairwise combinations of mutants indicated, for example, that the early development phenotype of the *redF* deletion was dominant (epistatic) over *redD* and *redE* deletions, indicating that RedF is the output to the system (Jagadeesan *et al.*, 2009).

To fully understand the signal flow in a given TCS system, point mutants in the conserved phosphorylation sites should also be generated. For HKs, the invariant histidine should be substituted to determine whether phosphorylation is required *in vivo*. For the hybrid HK EspC, deletion of the gene causes an early development phenotype. However, mutation of the conserved histidine has no phenotype, although autophosphorylation analyses indicated this residue can be phosphorylated *in vitro* (unpublished data). Some HKs are bifunctional and have both kinase and phosphatase activity on a cognate receiver domain (Stock *et al.*, 2000). Depending on the residue substituted for the histidine, either kinase activity alone or kinase plus phosphatase activity could be altered. For receiver domains, substitution of the conserved aspartic acid with asparagine or alanine residues can determine whether phosphorylation is important for activity. Furthermore, certain receiver domains can be rendered constitutively active by substitution of the invariant aspartic acid with glutamic acid (Stock *et al.*, 2000). Finally, deletions of entire domains (i.e., sensing, receiver, output), can also be used to reveal signal transduction mechanisms.

M. xanthus is readily genetically amenable, but generation of mutants is relatively time-consuming because colonies arising from single cells need at least 5 days of growth. Common plasmids do not stably replicate in *M. xanthus*, and are therefore suitable for generating gene interruptions by single crossover homologous recombination with an approximately 500 bp internal gene fragment inserted in a standard cloning vector. Genetic material is easily introduced into *M. xanthus* via electroporation (see protocol below), and both kanamycin (100 μg ml^{-1}) and oxytetracycline (10 μg ml^{-1}) are widely used for selection.

We prefer to generate markerless in-frame deletions to avoid polar effects on downstream genes, and so that multiple mutations can be generated in the same background. In-frame deletions are generated by a two-step selection/counter selection method modified from the Inouye laboratory (Ueki *et al.*, 1996) based on resistance to kanamycin (kan^R), and toxicity in the presence of galactose and the *galK* gene, respectively (Fig. 14.2 and protocol below). Generation of mutants by this method requires at least 2 weeks.

Efficient site-specific recombination is also possible using vectors derived from pSWU30 (Wu and Kaiser, 1996) carrying the Mx8 phage *attP* site and integrase gene which mediate recombination into the genomic *attB* site (Magrini *et al.*, 1999). These vectors are useful for genetic complementation experiments.

We generate point mutations (and domain deletions) in the endogenous locus using a similar strategy to that used for in-frame deletions (see Fig. 14.2 and protocol below). Although time-consuming, the resulting mutant proteins are then more likely to be produced at levels and in complexes most similar to the wild type. Alternatively, mutant genes can be tested for complementation when expressed from the phage attachment (*attB*) locus in a background in which the wild-type gene is deleted. Regardless, it is imperative to demonstrate that the mutant proteins accumulate to the same levels as wild type before valid conclusions from the phenotypic analyses can be drawn.

2.1. Plasmid construction

An in-frame deletion is constructed so that (at least) the first 6 codons of the gene are fused in-frame to the last 6 codons. Two PCR fragments are generated corresponding to approximately 500 bp upstream (UP) and 500 bp downstream (DOWN) of the first and last 6 codons of the gene to be deleted, respectively (Fig. 14.2 inset, step 1). Primers (A and B) for the UP PCR fragment are designed such that resulting PCR fragment contains at the 3′ end, 9 bp (3 codons) of sequence corresponding to the 5′ end of the DOWN PCR fragment (Fig. 14.2, "HS1"). Correspondingly, primers (C and D) for the DOWN PCR fragment are designed such that the resulting

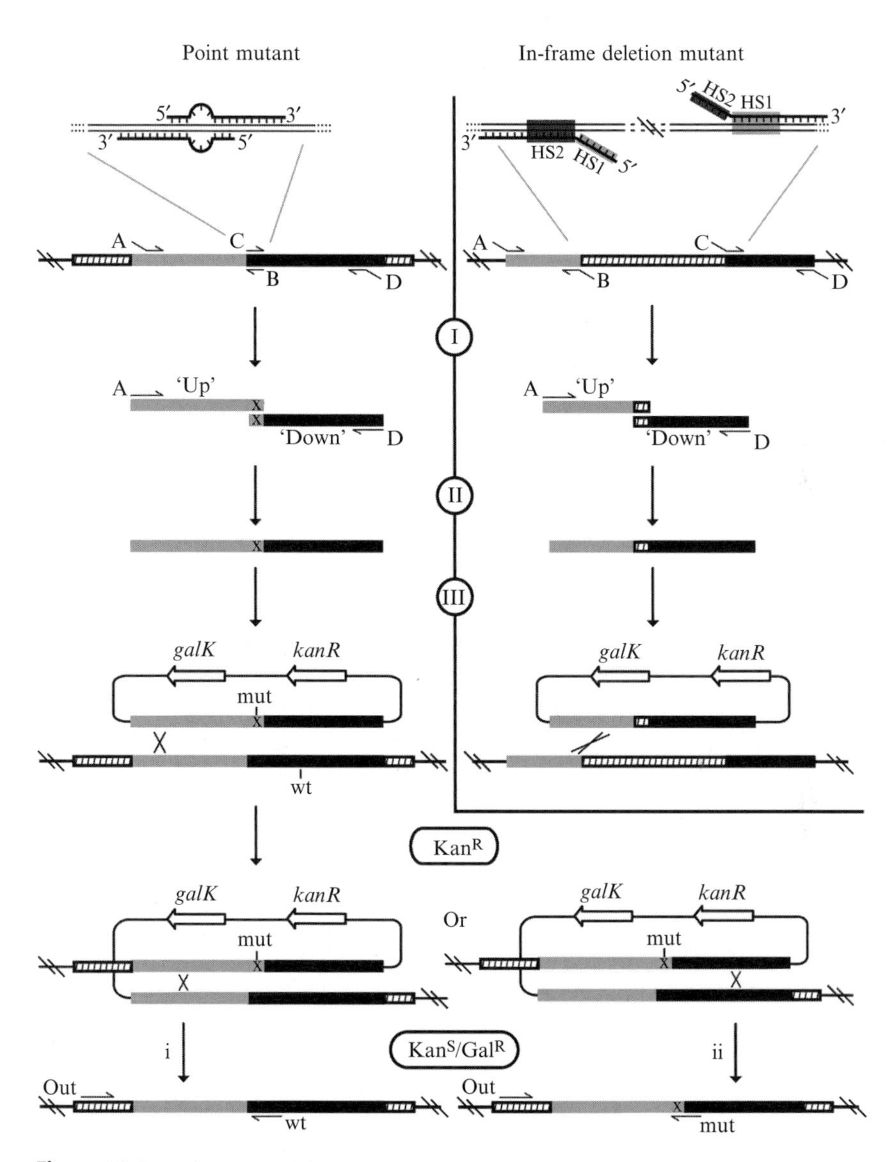

Figure 14.2 Schematic of the strategy used to generate point mutations (left) and in-frame deletions (right) in the *M. xanthus* genome via a homologous recombination and a positive/negative selection strategy. See text for details. The thick lines correspond to the *M. xanthus* genome including regions upstream (UP, in gray) or downstream (DOWN, in black) of the desired mutation (X) in the gene (hatched) to be mutated or deleted. The scheme of plasmid excision for deletion mutants (not shown) is the same as for the point mutation (shown). Kan^R, kanamycin resistant; Gal^R, growth in the presence of galactose.

PCR fragment contains at the 5′ end, 9 bp (3 codons) of overlapping sequence to the 3′ end of the UP PCR product (Fig. 14.2, "HS2"). The total overlap between the UP and DOWN fragments finally corresponds to 6 codons (18 bp; HS2 and HS1). The two PCR fragments are ultimately fused using primers A and D, each containing a unique restriction site for subsequent cloning (Fig. 14.2 inset, step 2).

Point mutations are generated by substituting the wild-type codon with a codon specific for the desired amino acid substitution. Bearing in mind that *M. xanthus* is 68% GC rich, the substituted codon should correspond to host codon preference. Two PCR fragments are generated corresponding to approximately 500 bp upstream (UP) and 500 bp downstream (DOWN) of the mutated codon (Fig. 14.2, left panel step 1). Following the scheme outlined above, primers B and C will contain, at the respective 5′ ends, 3 codons of complementary sequence in which the mutated codon is in the center of the complementary sequence. As above, the two PCR fragments are ultimately fused using primers A and D (Fig. 14.2, step 2).

(1) Amplify the 500 bp UP fragment using the primers A and B and the DOWN fragment using primers C and D as follows (Fig. 14.2, step 1):

PCR protocol	**PCR program**
Approximately 100 ng genomic DNA	95 °C 3 min
	25 steps of:
0.25 μl 50 μM A or C primer	95 °C 30 s
0.25 μl 50 μM B or D primer	62 °C 15 s
12.5 μl FailSafe™ 2× PreMixJ*,	68 °C 2 min kb^{-1} then,
0.25 μl Pfx (Strategene)	68 °C 5 min
sdH_2O to 25 μl final volume	4 °C hold
*Epicentre®	

Note: PreMixJ is designed for PCR with GC rich templates and works very well for *M. xanthus*. Alternatively, PCR in the presence 5% DMSO or 10% glycerol may be successful. Although primers will have a high T_m (often >75 °C), generally 62 °C works well as an annealing temperature in the presence of Buffer J. It is crucial to use a high fidelity polymerase.

(2) Purify the PCR products via a commercial column if single PCR bands are observed or by gel excision if there are contaminating bands. Perform PCR as above except genomic DNA is substituted with 40 ng each UP and DOWN PCR fragment and primers A and D are used (Fig. 14.2, step 2). Gel purify the resulting ~1000 bp fused PCR fragment.
(3) Clone the PCR product into the pBJ114 vector (Julien *et al.*, 2000) using the restriction enzymes compatible to the restriction sites

incorporated into the PCR product (Fig. 14.2, step 3). pBJ114 is compatible with blue/white screening.

Note: Overlap PCR often results in PCR generated sequence errors; it is crucial to sequence at least three clones to confirm sequences are error-free and that the desired in-frame deletion or point mutation was generated.

2.2. Mutant generation in *M. xanthus*

To generate mutants, the plasmid is introduced by electroporation into *M. xanthus* cells, and recombination through either the UP (Fig. 14.2, step 4) or DOWN (Fig. 14.2, not shown) fragment is selected by resistance to kanamycin. Resulting colonies are confirmed for plasmid integration at the expected region by colony PCR. PCR is performed using a plasmid-specific and a genome-specific primer. To select for cells in which the plasmid has excised via homologous recombination, cells are plated in the presence of galactose. Resulting kanamycin sensitive, galactose resistant colonies are screened by colony PCR to discriminate between colonies in which plasmid excision has generated the wild type versus the desired mutation (Fig. 14.2, step 1 versus 2, respectively). *M. xanthus* cells can accumulate random mutations during prolonged manipulation. To be sure any observed phenotype results from the desired deletion/point mutation, the phenotype of at least three independent mutant colonies should be compared to wild type.

Preparation of electrocompetent *M. xanthus*

(1) Using a sterile wooden stick, inoculate a scrape of cells from a fresh plate into 10 ml of CYE growth medium [0.1% (w/v) Bacto^{TM} Casitone, 0.5% (w/v) yeast extract, 10 m*M* morpholinepropanesulfonic acid (MOPS) (pH 7.6), 8 m*M* magnesium sulfate ($MgSO_4$)] in a sterile 100 ml flat-bottomed flask. Incubate the broth culture at 32 °C with shaking at 250 rpm in the dark to obtain an absorbance at 550 nm (A_{550}) of approximately 0.35.

Note: Detailed protocols for cultivation and preservation of *M. xanthus* are published elsewhere (Higgs and Merlie, 2008). The wild-type *M. xanthus* strain used in these studies is DZ2 (Campos and Zusman, 1975) which displays a slower developmental program than other commonly used wild-type strains.

(2) Pellet cells at 8000 × *g* for 10 min at room temperature (RT), remove supernatant and wash cells with 25 ml sterile demineralized (sd) H_2O. Resuspend cells in 1 ml sdH_2O, transfer to a 1.5-ml Eppendorf tube, pellet at 4000 × *g* for 5 min, and finally resuspend in 100 μl sdH_2O.

Cells can be used immediately or aliquoted in 50 μl volumes, flash frozen in liquid nitrogen, and stored at -80 °C.

Transformation of *M. xanthus* by electroporation

(3) Combine 1 μg plasmid (in less than 5 μl of low salt buffer) with 50 μl electrocompetent *M. xanthus*, transfer into a 0.1-mm electroporation cuvette, and electroporate at 0.65 kV, 400 Ω, 25 μF (GenPulser, BioRad). Immediately add 1 ml CYE-broth, transfer to a 2-ml Eppendorf tube and incubate the cell suspension for 4–5 h with shaking at 32 °C.

(4) Resuspend 100 μl, 300 μl, or 500 μl of the cell suspension in 3 ml molten CYE-top agar (CYE-broth, 0.74% (w/v) Difco^{TM} agar) tempered to ~50 °C and supplemented with 100 μg ml^{-1} kanamycin. Pour the top agar onto a CYE agar plate containing kanamycin at 100 μg ml^{-1}(CYE kan). Incubate plates at 32 °C for at least 5 days (in plastic boxes to avoid excessive drying of the agar plates).

Note: Cells are resuspended in top agar to isolate colonies arising from single cells and prevent cells from adjacent colonies from swarming into each other.

(5) Patch colonies onto a fresh CYE kan plate and incubate the plates at 32 °C for 2 days. Inoculate Kan^R colonies into 20 ml CYE-broth and grow to an A_{550} of ~0.35.

(6) Resuspend 200 μl of the cell culture into 3 ml CYE-top agar supplemented with 2.5% (w/v) galactose and pour the suspension onto a CYE agar plate containing 2.5% (w/v) galactose (CYE gal). Incubate the agar plates at 32 °C for at least 6 days.

(7) To confirm plasmid loss via homologous recombination, patch ~25 colonies onto CYE kan and CYE gal plates and incubate at 32 °C for 3 days. Putative positive colonies are kan^S gal^R. False positives (kan^R gal^R) may be as high as 50%.

(8) Determine whether plasmid loss generated the deletion or regenerated the wild type gene by PCR analysis. Prepare genomic DNA by colony lysis and PCR amplify using primers located in regions up- or downstream of the sequences cloned into the plasmid. At least three individual mutants should be tested in the initial phenotypic analysis.

Note: To PCR screen for the presence of point mutants, design two diagnostic forward primers containing either the wild-type or point mutated sequence at the extreme 3′ end, and a corresponding reverse primer ~200 bases downstream. Perform colony PCR using the mutant forward and reverse primer combination and, in a separate reaction, wild-type forward and reverse primer combination. Mutant strains should generate a

PCR product in the former, but not in the later, reactions. Confirm mutants by sequencing a PCR product spanning the mutated site using the mutant genome as template.

3. Phenotype Assays for Analysis of *M. xanthus* Development

3.1. Overview and general considerations

We are primarily interested in a set of TCS signaling proteins that function as negative regulators of the developmental program in *M. xanthus*. One shared feature amongst these TCS proteins is that they all lack an obvious effector domain and it is a considerable challenge to identify how they specifically modulate the developmental program. The following is a set of assays that we routinely use to assay exactly when and how the TCS mutants perturb the developmental program.

For a simple assay of gross developmental phenotype, cells can be spotted on nutrient limited agar plates (e.g., Fig. 14.1). For sporulation efficiency assays, or for assays where cells must be harvested, we prefer to induce development in submerged culture conditions (Kuner and Kaiser, 1982) (e.g., Fig. 14.3A). In this case, cells are allowed adhere to the bottom of tissue culture or Petri plates in rich media. To induce development, the media is replaced by starvation buffer. It should be noted though, that some mutants display significantly different developmental phenotypes under these two developmental conditions.

In general, developmental assays should be performed with cells that have been freshly streaked from freezer stocks onto rich media plates and not passed from plate to plate. Development is always induced from cells growing in vegetative broth cultures at 32 °C with shaking at 250 rpm to a density of approximately 4×10^8 cells ml^{-1} ($0.7A_{550}$). Wild-type and mutant strains should always be analyzed in parallel, inoculated from plate cultures of a similar age, and grown to a similar density in rich media before harvest for developmental assays.

3.2. Analysis of development on nutrient limited agar plates

Nutrient limited clone fruiting (CF) agar plates (10 m*M* MOPS (pH 7.6), 1 m*M* KH_2PO_4, 8 m*M* $MgSO_4$, 0.2% $(NH_4)_2SO_4$, 2% sodium citrate, 0.015% Casitone, 1.5% agar, plus 0.1% sodium pyruvate added after autoclaving) should be prepared the day before, and cured for 20 min at 32 °C just prior to use.

(1) Harvest 1 ml of vegetative cells and pellet at 7800 × *g* for 2 min at RT.

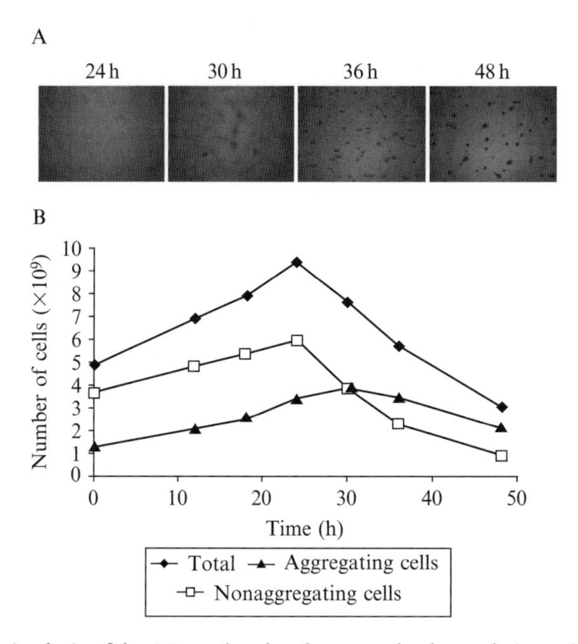

Figure 14.3 Analysis of the *M. xanthus* developmental subpopulations. (A) DZ2 wild-type *M. xanthus* cells developing under submerged culture for the indicated times. (B) Graphic display of the number of nonaggregating, aggregating, and total cells at the indicated time points during the developmental program. The total population increases for 24 h and then rapidly decreases likely due to programed cell death. Initially, nonaggregating cells comprise cells that are not yet stimulated to move into aggregation centers.

(2) Wash cells in 1 ml of MMC starvation buffer (10 m*M* MOPS (pH 7.6), 4 m*M* $MgSO_4$, 2 m*M* $CaCl_2$) and then resuspend to 4×10^9 cells ml^{-1} MMC. Cells must be fully resuspended, and no clumps of cells should be visible.
(3) Spot 10 μl of the cell suspension on CF agar plates. Cells will spread out during development; plate no more than five spots per plate. Allow spots to dry into media (~20 min), invert plates, and incubate in a sealed plastic container at 32 °C.
(4) Examine cell spots with a stereomicroscope and record pictures every 6–12 h for at least 3 days.

3.3. Analysis of development in submerged culture

This protocol is based on submerged cultures grown in 85 mm Petri plates, but it can be scaled up or down according to purpose of experiment. Prepare at least one plate for each time point that must be harvested.

(1) Harvest vegetatively growing cells from a broth culture and dilute to an A_{550} of 0.035 (2×10^7 cells ml^{-1}) in fresh rich medium. Prepare enough diluted culture for 16 ml cell suspension to each plate.
(2) Add 16 ml of diluted cells per Petri plate. Incubate at 32 °C in a sealed plastic container for 24 h. Cells will settle down onto the bottom surface of the well and form a semiadherent layer. The container should be completely flat to avoid growth of cells into uneven layers.
(3) Gently and completely remove the rich media by aspiration and add 16 ml of starvation buffer (MMC) to each well. It is important to avoid disruption of the layer of cells by adding the starvation buffer slowly to the side of the dish. Incubate at 32 °C in sealed plastic container.
(4) Record pictures at the desired time intervals and harvest cells by resuspending the layer of cells with a 20 ml glass pipette. Transfer the cell suspension to a 50-ml conical tube and pellet cells at $4620 \times g$ for 10 min. Cell pellets can be stored at −20 °C for subsequent assays.

3.4. Analysis of sporulation efficiency

Sporulation efficiencies can be analyzed at specific intervals during a developmental time course, or only at a developmental endpoint. Although the developmental endpoint is generally considered 72 h, the maximum sporulation efficiency is observed after 120 h. Sporulation efficiency should be recorded as the average of triplicate samples with the associated standard deviation. In addition to enumeration of heat and sonication resistant spores, the germination efficiency can also be assayed by plating spores on rich media and enumerating the resulting colonies.

(1) Induce development in submerged culture in 15.6 mm (24-well) tissue culture plates using 0.5 ml diluted cells per well. At the desired time point(s), harvest cells by resuspension with a pipet, transfer to a 1.5 ml Eppendorf tube, and pellet at $7800 \times g$ for 5 min. Cell pellets can be stored at −20 °C. All samples should be processed and counted on the same day.
(2) Resuspend cell pellets in 500 μl sterile sdH_2O, heat for 1 h at 50 °C, and sonicate two times 15 bursts (output 3, 30% duty Branson sonifier) with a microtip.
(3) Place 10 μl of cells onto a Helber Bacteria Counting Chamber (Hawksley, Lancing, UK). Count round and refractile spores. If necessary, dilute spores in water (1:2–1:10). Calculate total spores in the sample. In the Helber chamber, the smallest squares correspond to 5×10^{-5} μl.

(4) For germination assays, prepare three 100-fold dilutions of heat and sonication-treated spores in sterile water.
(5) Add 100 μl of diluted spores to 3 ml molten top agar (1.0%) prewarmed at 50 °C, vortex well and pour overnutrient rich media.
(6) Incubate at 32 °C for at least 5–8 days and count colonies.

3.5. Cell population analysis

During the developmental program, the total number of cells changes: first increasing as the cells finish dividing and then sharply decreasing likely due to PCD (Fig. 14.3B). The timing and proportion of cells undergoing PCD can be monitored throughout the developmental program by treating harvested cells with fluorescent dyes which are specifically permeable to live or dead cells (i.e., BacLight™ Bacterial Viability Kit, Invitrogen) and enumerating the differentially stained cell under a fluorescent microscope according to the manufacturer's instructions.

During development, the remaining cells also segregate into aggregating and nonaggregating subpopulations which correspond to cells destined to differentiate into spores within fruiting bodies and to cells that remain as peripheral rods, respectively (Fig. 14.3B). These two subpopulations can be separated based on a low-speed centrifugation (O'Connor and Zusman, 1991). Recently, we have begun to explore the role of TCS systems in influencing the proportions of cells in each subpopulation. The following is a protocol used to assay the number of cells in each population during a developmental time course.

(1) Induce development under submerged culture in 85 mm Petri plates. At the desired developmental time points, harvest the film of cells developing on the surface of the plate using a 20 ml glass pipette, and transfer the cell suspension to a 50 ml sterile conical tube.
(2) Centrifuge the cells at 50 × *g* for 5 min. Carefully remove the tube from the centrifuge to avoid disturbing the pellet (aggregating cells).
(3) Gently remove the supernatant (nonaggregating cells) using a 20-ml glass pipette and transfer to fresh 50 ml tube. Resuspend the pellet in an equivalent volume (16 ml) of starvation buffer (MMC).
(4) Disperse the resuspended aggregating cells by shaking the tube at 5 m s^{-1} for 45 s (FastPrep® 24 cell and tissue homogenizer, MP Biomedicals). The nonaggregating cell suspension does not need to be dispersed.
(5) For each population, place 10 μl of cells onto a counting chamber suitable for bacteria (Hawksley, cat no. Z30000, Lancing, UK), count the cells and calculate the total number of cells in each population.

4. Expression Analysis

4.1. Overview and general considerations

The *M. xanthus* development program is controlled by a temporally ordered cascade of gene and protein expression (Kroos *et al.*, 1986). In addition to the general developmental phenotype, observed phenotypic defects can be analyzed at the molecular level by examining the expression and/or accumulation of key developmental marker genes and proteins during the developmental program.

The molecular events driving the developmental program have been intensively investigated and a simplified model of the core molecular events is outlined here and in Fig. 14.4. Starvation, perceived via the stringent response (Singer and Kaiser, 1995), results in induction of A-signaling, a population sensing mechanism which prevents development until a threshold population is perceived (Kuspa *et al.*, 1992). Production and reception of the A-signal can be monitored by expression of the *spi* gene (Keseler and Kaiser, 1995). At sufficient population density, the expression of *mrpC*, encoding a CRP family transcriptional regulator is upregulated (Sun and Shi, 2001). MrpC plays a key role in both regulation of cell fate and control of the developmental program. First, MrpC acts as an antitoxin to inhibit MazF, a ribonuclease thought to be necessary for induction of PCD (Nariya and Inouye, 2008). Second, MrpC acts as a transcriptional regulator of itself (generating a positive feedback loop) (Sun and Shi, 2001). Third, MrpC acts as a transcriptional regulator of *fruA* (Ueki and Inouye, 2003), encoding a key developmental regulator (Ellehauge *et al.*, 1998) (below). Finally, in combination with FruA, MrpC is necessary for control of transcription of several genes expressed late during the developmental program near the onset of sporulation (Mittal and Kroos, 2009a,b).

MrpC-dependent expression of *fruA* is a key step in progression of the developmental program. FruA is an orphan DNA-binding RR of the TCS family which is proposed to be necessary to induce and coordinate aggregation and sporulation (Ellehauge *et al.*, 1998; Ogawa *et al.*, 1996). It is postulated that FruA becomes activated in response to C-signal, an extracellular signal produced in response to cell–cell contact (Kim and Kaiser, 1990a,b,c). Low levels of activated FruA are proposed to stimulate aggregation which in turn induces more cell–cell contact, more production of the C-signal, and therefore more activated FruA (Ellehauge *et al.*, 1998). Finally, it is proposed that higher levels of activated FruA induce transcription of the *dev* locus which is necessary for induction of sporulation (Viswanathan *et al.*, 2007). One of the products of the *dev* locus (DevT) is also necessary for stimulation of *fruA* transcription, providing a positive feedback loop on *fruA* expression (Boysen *et al.*, 2002). Thus, an ordered

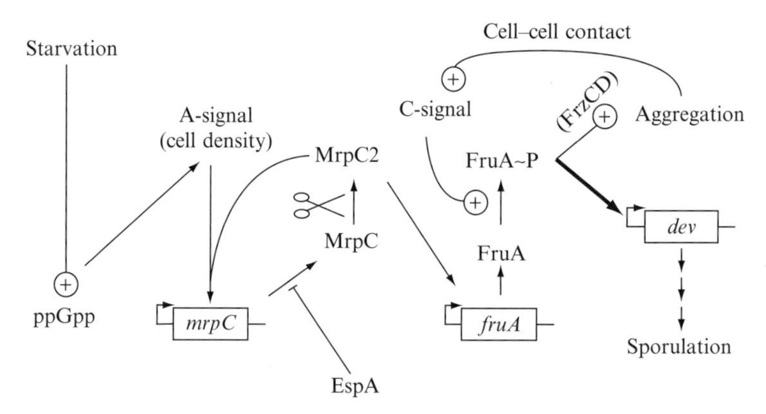

Figure 14.4 Schematic of core regulators of aggregation and sporulation during the *M. xanthus* developmental program. Developmental progression is controlled by a cascade of signaling events and a series of positive autoregulatory loops. See text for details. Except for EspA, all proteins, signals, and gene loci depicted are required to promote developmental progression. EspA is a negative regulatory TCS protein. Scissors: proteolytic cleavage event.

cascade of gene expression coupled to several positive feedback loops ensures that sporulation is coupled to aggregation. It is assumed that in nonaggregating cells destined to become peripheral rods, one or more of the positive feedback loops are not activated.

By comparing the expression patterns and of *spi*, *mrpC*, *fruA*, and *dev* and the protein accumulation patterns of MrpC and FruA between wild-type and the *espA* mutant, we demonstrated that in the *espA* mutant, MrpC accumulates earlier relative to wild type, but, during the same time frame, *mrpC* gene expression is not different from wild type (Higgs *et al.*, 2008). Consistently, in *espA* mutants, *spi* expression is not perturbed, but *fruA* expression and FruA accumulation are earlier than wild type. These analyses suggest EspA regulates the developmental program by decreasing translation of *mrpC* or stimulating degradation of MrpC (Higgs *et al.*, 2008). Thus, having gone from no clues as to the output of the EspA protein, this approach now directs our attention to targets for the EspA signaling system.

The following protocols describe the methods used analyze gene transcription patterns and protein accumulation patterns during the developmental program.

4.2. RT-PCR analysis of developmental markers

Gene expression patterns can be analyzed by real-time (RT) PCR analysis on cDNA generated from a developmental time course. For this protocol, cells are induced to develop under submerged culture in 85 mm Petri dishes. At each time point, cells are harvested, immediately pelleted at 4 °C and cell pellets are frozen in liquid nitrogen and stored at −80 °C. When all time points are harvested, RNA is isolated by the hot-phenol method and treated with DNase following protocols which have been described previously in detail (Mueller and Jakobsen, 2008). For each time point, cDNA is generated from 1 μg of RNA with Superscript III reverse transcriptase (RT) (Invitrogen) and random hexamer primers according to the manufacture's instructions. RT-PCR primers specific for each gene are designed using Primer3 (Rozen and Skaletsky, 2000) (http://frodo.wi.mit.edu/primer3/input.htm) with primer T_m set to optimal 64 °C, max 68 °C, GC clamp > 1, and a product size of 100–200 bp lying within the 50–200 bp after the start of the gene. PCR primers must first be tested for PCR efficiency using a dilution series of genomic DNA. RT-PCR is performed using SYBR GREEN PCR Master Mix kit according to manufacture's instructions on a 7300 RT-PCR system (Applied Biosystems). Each reaction is performed in triplicate. As controls, RT-PCR should be performed on mock cDNA reactions in which the RT was omitted, in samples in which cDNA was substituted with water, and on samples containing genomic DNA as template. The PCR program is 10 min at 95 °C, 15 s at 95 °C, followed by 40 cycles of: 1 min at 60 °C and 15 s at 95 °C. C_t values are assigned automatically and marker expression patterns are plotted as average C_t value versus hour of development.

4.3. Western blot analysis of developmental markers

Western blot analysis using antibodies specific to key developmental marker proteins is particularly important in cases where the markers are regulated by proteolysis, for example, MrpC cleavage into MrpC2 (Nariya and Inouye, 2005, 2006), or by modification that can be detected by Western Blot. Protein samples can be prepared by several methods (i.e., sonication, boiling lysis, trichloro acetic acid (TCA) precipitation). Often, these various methods must be tried to determine which method releases the highest amount of protein of good integrity. Except in the case of TCA precipitation, cell lysates must be prepared in the presence of protease inhibitors as *M. xanthus* is very rich in proteases. It should be noted that proteins are not released from spores by boiling lysis, or by sonication and for that reason, lysis by bead beating in the presence of zirconium beads is recommended. Finally, the patterns of protein accumulation over a developmental time course can be vastly different depending on whether samples are loaded using equal concentration of protein, equal numbers of cells, or equivalent proportions of cell culture. The later method is not recommended, because the cell number alters during the developmental program and many mutants will not follow the wild-type cell number patterns. We present here two methods which are the most robust for analysis of protein expression patterns.

4.3.1. Preparation of protein samples as equal protein concentration

(1) Induce cells to develop under submerged culture in 85 mm Petri dishes. Harvest cells at the required time points and centrifuge at 4620×*g* for 10 min. Cell pellets can be frozen at −20 °C until all samples are ready.
(2) Resuspend pellets in 0.5 ml starvation buffer (MMC) containing protease inhibitor and transfer to a bead beater tube containing 0.6 g of 0.1 mm zirconia/silica beads (BioSpec, Bartlesvillebead).
(3) Bead beat cells at 6.5 m s^{-1} for 45 s, and place on ice for 5 min. Repeat bead beating six times.
(4) Determine protein concentration of 5 μl of sample using Bradford reagent (Sigma) and BSA to generate a standard concentration curve.
(5) Prepare protein samples to 0.5 μg μl^{-1} by adding 2× protein sample buffer (0.125 *M* Tris–HCl (pH 6.8), 20% glycerol, 4% SDS, 10% 2 β-mercaptoethanol, 0.02% bromophenol blue). Boil protein samples at 99 °C for 5 min, store protein samples at −20 °C. Typically, 20 μg of protein lysate is used for Western blot analysis.

4.3.2. Preparation of protein samples as equal number of cells

(1) Harvest developing cells as above from two parallel cultures and immediately pellet one culture as per step 1 above, freeze pellets at −20 °C.
(2) From a parallel culture, disperse cells as described in step 4 of the cell counting protocol (above), and determine the total number of cells in each sample using a bacterial counting chamber.
(3) Resuspend each pellet to 4.3×10^6 cells μl^{-1} in 2× protein sample buffer containing protease inhibitor cocktail, and boil samples at 99 °C for 10 min. No clumps of cells or debris should be visible.
(4) Transfer ~200 μl of cell lysate to a bead beater tube containing 0.2 g of 0.1 mm zirconia/silica beads (BioSpec, Bartlesville) and bead beat the sample as described in step 3 above. Boil the samples again at 99 °C for 5 min and store the protein samples at −20 °C. This protocol is optimized to minimize the loss of material on the zirconium beads.

5. *In Vitro* Biochemical Analysis of TCS Proteins

In vitro biochemical analyses are essential to demonstrate that TCS homologous identified by sequence analysis are active His-Asp phosphorelay proteins, to confirm sites of phosphorylation, and to characterize phosphoryl flow in a TCS system. Versions of the proteins in which the putative phosphorylated residues are mutated are essential for interpretation of the results. We often first generate the mutated version of the gene *in vivo* as part of our genetic analyses, and then simply amplify the gene from the mutant genomic DNA for cloning into expression vectors.

For the biochemical assays, we always attempt to find conditions in which an affinity tagged recombinant protein can be overexpressed in a soluble form. We routinely clone TCS genes into several vectors [pET24 (C-term His_6-affinity tag), pET28 (C- and N-term His_6), pET32 (thioredoxin solubility tag), and pGEX4T-1 (N-term glutathione-*S*-transferase solubility tag)] (Novagen and GE Healthcare, respectively). These plasmids are tested in a small scale (20–200 ml) cultures in various *Escherichia coli* protein expression strains [BL21λDE3 (*lac* Pr/O T7 RNAP), Novagen; or GJ1158 (*proU* Pr T7 RNAP) (Bhandari and Gowrishankar, 1997)] in the presence or absence of accessory plasmids pLysS (T7 lysozyme, Novagen), pRP (tRNA for GC rich R and P codon usage, Novagen). Various induction conditions, including temperature (from 18 to 37 °C), time (2 h to overnight), and inducer concentration (e.g., 0.1–1 m*M* IPTG; 0.3 *M* NaCl) are tested. We have also had good success using the Studier autoinduction system (Studier, 2005). In each case, protein expression and solubility is tested by disrupting cells by sonication, and subjecting the lysate to a low speed 600–800 × *g* spin to separate soluble from inclusion body (IB) proteins. Fractions are harvested from each step, resuspended to equal cell concentration equivalents, and analyzed by SDS–PAGE and Coomassie stain. Only if these methods fail, do we resort to attempting to refold proteins which are only produced in IBs.

5.1. Protein overproduction and purification

Most of the His_6-affinity recombinant proteins are purified using an FPLC platform (ÄKTA™, GE Healthcare). Pellets from 2 l cell cultures are resuspended in HisLyse buffer (20 m*M* imidazol, 10 m*M* HEPES (pH 7.4), 500 m*M* NaCl), clarified at 100,000 × *g* and passed over a nickel affinity column (HisTrap™FF, GE Healthcare) using a flow rate of 0.8 ml min^{-1}. The column is washed excessively and the affinity tagged protein is eluted using a linear gradient of 20–500 m*M* imidazol in binding buffer. Elution fractions are examined by SDS–PAGE followed by Coomassie stain, and peak fractions of pure protein are pooled, placed into 7 kDa MWCO dialysis tubing, and dialyzed overnight at 4 °C against 2 l of our standard autophosphorylation TGND buffer (50 m*M* Tris (pH 8.0), 10% (v/v) glycerol, 150 m*M* NaCl, 1 m*M* DTT). Proteins are aliquoted and stored at −20 °C.

5.2. Refolding proteins from inclusion bodies

If all attempts to produce a soluble protein fail, we attempt to refold the protein from IBs. This method is the last resort because it is not necessarily possible to determine if the refolded protein will be active. However, the following protocol has been used successfully in our group to refold IBs to produce active protein.

(1) Generate 500 ml overexpression culture which has previously been shown to produce the target protein as IBs, and harvest the cells by centrifugation.
(2) Resuspend the cell pellet in 0.05 volumes lysis buffer (50 m*M* Tris (pH 8.0), 150 m*M* NaCl, 10 m*M* EDTA, 1 m*M* DTT). Disrupt the cells by French press two times at ~1300 bar.
(3) Harvest the IB by centrifugation at 800×*g* for 30 min, resuspend the IBs in an equivalent volume of lysis buffer, and French press as above.
(4) Harvest the IBs by centrifugation as above and determine the wet weight. Emulsify the IBs (1 ml for every 0.1 g IB) in preextraction buffer [0.6 *M* guanidinium–HCl (GuHCl) in lysis buffer] and stir for 30 min at RT. This step helps to remove contaminating proteins.
(5) Harvest the IBs by centrifugation as above and dissolve the IBs (2 ml for every 0.1 g IB) in extraction buffer (6 *M* GuHCl in lysis buffer). The samples may be stored at −20 °C at this step.
(6) Dilute 2 ml of the resolved IBs 1:5 with extraction buffer, transfer to 7 kDa MWCO dialysis tubing, and dialyze overnight against 2 l of refold buffer (2 *M* GuHCl in lysis buffer). Transfer the sample to 2 l of freshly prepared TGND buffer (above). During the final dialysis step, a portion of the protein may precipitate. Remove precipitated protein by centrifugation and transfer the refolded soluble protein to a fresh Eppendorf tube. Aliquote and store the protein at −20 °C.

5.3. Autophosphorylation of histidine kinases

To first biochemically characterize an HK homolog, we assay for autophosphorylation in the presence of ATP, confirm the residue which becomes phosphorylated, and determine the optimal conditions for autophosphorylation. These assays are preformed in the presence of [γ-^{32}P]-ATP; incorporation of the radiolabel onto the protein is detected by resolving the protein by SDS–PAGE followed by phosphorimager analysis. The site of phosphorylation is assumed to be an invariant histidine which is identified by sequence alignment with defined HKs. A mutant protein in which the invariant His is substituted should not label under conditions in which autophosphorylation of the wild-type HK is detected. The optimal time for autophosphorylation should be assessed and is generally in the range of

5 min to 2 h. The time point closest to signal saturation will be the incubation time of choice for phosphor transfer analysis.

In the case of hybrid HKs, it is possible that autophosphorylation leads phosphotransfer to the associated receiver domain followed by rapid hydrolysis to P_i (Freeman and Bassler, 1999; Timmen *et al.*, 2006). Alternatively, it is possible that the receiver domain prevents HK autophosphorylation (Chang and Winans, 1992). In either case, it may not be possible to detect a radiolabel on the HK unless the phosphoaccepting residue in the receiver domain is first mutated (Higgs *et al.*, 2008), or the receiver is deleted, respectively.

Autophosphorylation analyses are performed in parallel with wild-type and point mutated HKs as follows:

(1) Prepare the reaction as 10 μM of HK, 5 mM $MgCl_2$, 50 mM KCl, 0.025 μM [γ-^{32}P]-ATP (222 TBq mmol^{-1}), and 0.5 M ATP in a total volume of 10 μl (per reaction) of TGND (above). The ATP mix is added last to start the reaction.
(2) Remove 10 μl aliquots at the desired time points and quench the reactions by addition of 10 μl 2× protein sample buffer, resolve (unheated) samples by SDS–PAGE, cut the dye front off of the gel, and expose the gel to a phosphorimaging screen overnight.

Note: The majority of the radiolabel is at the dye front of the gel. Stop the electrophoresis before the dye front runs into the buffer to avoid generating large volumes of radioactive waste.

5.4. Autophosphorylation of response regulators

Some RRs are capable of autophosphorylation at the expense of small molecule phosphodonors, such as acetyl phosphate, phosphoramidate, and carbamoyl phosphate (McCleary and Stock, 1994). We routinely generate acetyl-[^{32}P] to determine whether receiver autophosphorylation can be detected. Lack of autophosphorylation does not mean the receiver is not active (many receivers do not autophosphorylate with acetyl phosphate), but in the cases where autophosphorylation can be detected, this method is an easy method to confirm a recombinant protein is active, confirm the phosphorylated residue, generate a radiolabeled substrate to assay HK phosphatase activity, and determine the half-life of the phosphorylated receiver (Jagadeesan *et al.*, 2009).

5.4.1. Production of acetyl-[^{32}P]

To generate acetyl-[^{32}P], incubate 0.16 μM [γ-^{32}P]-ATP (222 Tbq/mmol) and 1.5 U μl^{-1} acetate kinase in TKM-Buffer (25 mM Tris (pH 7.6), 60 mM K-acetate, 10 mM $MgCl_2$) in a total volume of 100 μl for 120 min at RT. Remove the acetate kinase by centrifugation at 8000×g

for 1 h at RT using an YM-10 Microcon column (Amicon). The flow-through contains acetyl-[^{32}P]; collect and store at 4 °C for a maximum of 1 week. It should be noted that the acetyl-[^{32}P] solution will contain residual [γ-^{32}P]-ATP.

5.4.2. Autophosphorylation of receiver domains

To assay for receiver autophosphorylation, dilute (per reaction) 10 μM RR in a total volume of 5 μl of TGND buffer and add 5 μl of the acetyl-[^{32}P] solution. Generate a parallel reaction containing the RR point mutated in the predicted site of phosphorylation. Incubate both reactions between 10 min and 2 h. Quench the reactions with 10 μl 2× protein sample buffer. Do not heat the samples. Resolve the reactions by SDS–PAGE, and cut the dye front away from the gel. In the case of small RRs, a 20% polyacrylamide gel should be used to separate the protein as much as possible from the dye front. The unincorporated acetyl-[^{32}P] produces an intense radioactive signal that may mask that from the RR. Expose the gel to a phosphorimager screen overnight.

5.5. Phosphotransfer analysis

Phosphotransfer reactions between TCS proteins are necessary to characterize the signal flow within a TCS system. Generally, these assays are performed using an autophosphorylated HK and a RR, but in the case of multistep phosphorelay or complex TCS systems, phosphotransfer from a radiolabeled RR to a histidine phosphotransferase (Burbulys *et al.*, 1991) or HK-like protein (Jagadeesan *et al.*, 2009) can be observed. Phosphotransfer between a HK and a RR can be assumed in based on one of the following observations: (1) detection of a radiolabeled signal on the RR after incubation in the presence of the autophosphorylated kinase. The signal should not be observed in similar reactions when either the HK or the RR is substituted with versions mutated in the phosphorylation site. (2) Little or no radioactive signal on the RR, and reduced or absent signal on the HK compared to a reaction in which the RR is omitted or the mutated RR is instead added (Skerker *et al.*, 2005). The latter case can be observed when the [γ-^{32}P]-ATP is depleted from the reaction either because the HK has additional phosphatase activity or because the kinase and intrinsic phosphatase activity of the RR are very rapid.

A simple phosphotransfer assay between a HK and a RR is performed by first generating radiolabeled HK as described above. Aliquot 10 μM RR for every reaction and add, in staged intervals, 10 μM phosphorylated HK to start the reactions. Incubation times are kept short (between 1 and 10 min) to reduce the possibility for nonspecific phosphotransfer (Skerker *et al.*, 2005). In one reaction, the RR should be replaced by buffer and quenched immediately to serve as a reference. Samples are processed and analyzed as above and the signal intensity on the HK in each reaction is compared to the HK signal in the reference sample.

5.6. Phosphatase analysis

Some HKs are bifunctional, displaying both kinase and phosphatase activities on a cognate RR. In most cases, the phosphatase activity leads to release of the phosphoryl group as P_i and does not require the conserved His in the HK (Atkinson and Ninfa, 1993; Hsing and Silhavy, 1997). However, the nature of the residue substituted for the conserved His can influence the phosphatase activity; for instance, substitution of EnvZ His with Asp destroys phosphatase activity, but substitution with Ala has little effect (Hsing and Silhavy, 1997). In addition, phosphatase activity may be stimulated in the presence of ATP, ADP, or nonhydrolyzable ATP analogs (Hsing and Silhavy, 1997; Jung and Altendorf, 1998; Keener and Kustu, 1988).

We routinely test for phosphatase activity if [^{32}P]-RR can be generated with [^{32}P]-acetate as described above. Phosphorylate 5 μM RR as described above and extensively wash the protein with TGND buffer supplemented with 5 mM $MgCl_2$ and 50 mM KCl using an YM-10 Microcon column (Amicon). This washing will dilute the RR to approximately 2–2.5 μM. Aliquot the [^{32}P]-RR into three reactions and add 5 μM HK plus 1 mM ADP, 5 μM HK, or buffer, respectively. Incubate all reactions for 20 min, quench, and analyze the samples as described previously.

ACKNOWLEDGMENT

Work in this laboratory was supported by the Max Planck Society.

REFERENCES

Atkinson, M. R., and Ninfa, A. J. (1993). Mutational analysis of the bacterial signal-transducing protein kinase/phosphatase nitrogen regulator II (NRII or NtrB). *J. Bacteriol.* **175,** 7016–7023.

Bhandari, P., and Gowrishankar, J. (1997). An *Escherichia coli* host strain useful for efficient overproduction of cloned gene products with NaCl as the inducer. *J. Bacteriol.* **179,** 4403–4406.

Boysen, A., Ellehauge, E., Julien, B., and Sogaard-Andersen, L. (2002). The DevT protein stimulates synthesis of FruA, a signal transduction protein required for fruiting body morphogenesis in *Myxococcus xanthus*. *J. Bacteriol.* **184,** 1540–1546.

Burbulys, D., Trach, K. A., and Hoch, J. A. (1991). Initiation of sporulation in *B. subtilis* is controlled by a multicomponent phosphorelay. *Cell* **64,** 545–552.

Campos, J. M., and Zusman, D. R. (1975). Regulation of development in *Myxococcus xanthus*: Effect of 3′:5′-cyclic AMP, ADP, and nutrition. *Proc. Natl. Acad. Sci. USA* **72,** 518–522.

Chang, C. H., and Winans, S. C. (1992). Functional roles assigned to the periplasmic, linker, and receiver domains of the *Agrobacterium tumefaciens* VirA protein. *J. Bacteriol.* **174,** 7033–7039.

Cho, K., and Zusman, D. R. (1999). Sporulation timing in *Myxococcus xanthus* is controlled by the *espAB* locus. *Mol. Microbiol.* **34,** 714–725.

Ellehauge, E., Norregaard-Madsen, M., and Sogaard-Andersen, L. (1998). The FruA signal transduction protein provides a checkpoint for the temporal co-ordination of intercellular signals in *Myxococcus xanthus* development. *Mol. Microbiol.* **30,** 807–817.

Freeman, J. A., and Bassler, B. L. (1999). Sequence and function of LuxU: A two-component phosphorelay protein that regulates quorum sensing in *Vibrio harveyi*. *J. Bacteriol.* **181,** 899–906.

Goldman, B., Bhat, S., and Shimkets, L. J. (2007). Genome evolution and the emergence of fruiting body development in *Myxococcus xanthus*. *PLoS One* **2,** e1329.

Hess, J. F., Oosawa, K., Kaplan, N., and Simon, M. I. (1988). Phosphorylation of three proteins in the signaling pathway of bacterial chemotaxis. *Cell* **53,** 79–87.

Higgs, P. I., and Merlie, J. P. Jr. (2008). *Myxococcus xanthus*: Cultivation, motility, and development. *In* "Myxobacteria: Multicellularity and Differentiation," (D. E. Whitworth, ed.), pp. 465–478. ASM Press, Washington, DC.

Higgs, P. I., Cho, K., Whitworth, D. E., Evans, L. S., and Zusman, D. R. (2005). Four unusual two-component signal transduction homologs, RedC to RedF, are necessary for timely development in *Myxococcus xanthus*. *J. Bacteriol.* **187,** 8191–8195.

Higgs, P. I., Jagadeesan, S., Mann, P., and Zusman, D. R. (2008). EspA, an orphan hybrid histidine protein kinase, regulates the timing of expression of key developmental proteins of *Myxococcus xanthus*. *J. Bacteriol.* **190,** 4416–4426.

Hoch, J. A., and Silhavy, T. J. (1995). Two-Component Signal Transduction. ASM Press, Washington, DC.

Hsing, W., and Silhavy, T. J. (1997). Function of conserved histidine-243 in phosphatase activity of EnvZ, the sensor for porin osmoregulation in *Escherichia coli*. *J. Bacteriol.* **179,** 3729–3735.

Jagadeesan, S., Mann, P., Schink, C. W., and Higgs, P. I. (2009). A novel "four-component" two-component signal transduction mechanism regulates developmental progression in *Myxococcus xanthus*. *J. Biol. Chem.* **284,** 21435–21445.

Jiang, M., Shao, W., Perego, M., and Hoch, J. A. (2000). Multiple histidine kinases regulate entry into stationary phase and sporulation in *Bacillus subtilis*. *Mol. Microbiol.* **38,** 535–542.

Julien, B., Kaiser, A. D., and Garza, A. (2000). Spatial control of cell differentiation in *Myxococcus xanthus*. *Proc. Natl. Acad. Sci. USA* **97,** 9098–9103.

Jung, K., and Altendorf, K. (1998). Truncation of amino acids 12-128 causes deregulation of the phosphatase activity of the sensor kinase KdpD of *Escherichia coli*. *J. Biol. Chem.* **273,** 17406–17410.

Keener, J., and Kustu, S. (1988). Protein kinase and phosphoprotein phosphatase activities of nitrogen regulatory proteins NTRB and NTRC of enteric bacteria: Roles of the conserved amino-terminal domain of NTRC. *Proc. Natl. Acad. Sci. USA* **85,** 4976–4980.

Keseler, I. M., and Kaiser, D. (1995). An early A-signal-dependent gene in *Myxococcus xanthus* has a sigma 54-like promoter. *J. Bacteriol.* **177,** 4638–4644.

Kim, S. K., and Kaiser, D. (1990a). C-factor: A cell-cell signaling protein required for fruiting body morphogenesis of *M. xanthus*. *Cell* **61,** 19–26.

Kim, S. K., and Kaiser, D. (1990b). Cell alignment required in differentiation of *Myxococcus xanthus*. *Science* **249,** 926–928.

Kim, S. K., and Kaiser, D. (1990c). Cell motility is required for the transmission of C-factor, an intercellular signal that coordinates fruiting body morphogenesis of *Myxococcus xanthus*. *Genes Dev.* **4,** 896–904.

Kroos, L., Kuspa, A., and Kaiser, D. (1986). A global analysis of developmentally regulated genes in *Myxococcus xanthus*. *Dev. Biol.* **117,** 252–266.

Kuner, J. M., and Kaiser, D. (1982). Fruiting body morphogenesis in submerged cultures of *Myxococcus xanthus*. *J. Bacteriol.* **151,** 458–461.

Kuspa, A., Plamann, L., and Kaiser, D. (1992). A-signalling and the cell density requirement for *Myxococcus xanthus* development. *J. Bacteriol.* **174,** 7360–7369.

Lee, B., Higgs, P. I., Zusman, D. R., and Cho, K. (2005). EspC is involved in controlling the timing of development in *Myxococcus xanthus*. *J. Bacteriol.* **187,** 5029–5031.

Magrini, V., Creighton, C., and Youderian, P. (1999). Site-specific recombination of temperate *Myxococcus xanthus* phage Mx8: Genetic elements required for integration. *J. Bacteriol.* **181,** 4050–4061.

Matsubara, M., Kitaoka, S. I., Takeda, S. I., and Mizuno, T. (2000). Tuning of the porin expression under anaerobic growth conditions by his-to-Asp cross-phosphorelay through both the EnvZ-osmosensor and ArcB-anaerosensor in *Escherichia coli*. *Genes Cells* **5,** 555–569.

McCleary, W. R., and Stock, J. B. (1994). Acetyl phosphate and the activation of two-component response regulators. *J. Biol. Chem.* **269,** 31567–31572.

Mittal, S., and Kroos, L. (2009a). A combination of unusual transcription factors binds cooperatively to control *Myxococcus xanthus* developmental gene expression. *Proc. Natl. Acad. Sci. USA* **106,** 1965–1970.

Mittal, S., and Kroos, L. (2009b). Combinatorial regulation by a novel arrangement of FruA and MrpC2 transcription factors during *Myxococcus xanthus* development. *J. Bacteriol.* **191,** 2753–2763.

Mueller, F.-D., and Jakobsen, J. S. (2008). *Myxococcus xanthus*: Expression analysis. *In* "Myxobacteria: Multicellularity and Differentiation," (D. E. Whitworth, ed.), ASM Press, Washington, DC.

Nariya, H., and Inouye, S. (2005). Identification of a protein Ser/Thr kinase cascade that regulates essential transcriptional activators in *Myxococcus xanthus* development. *Mol. Microbiol.* **58,** 367–379.

Nariya, H., and Inouye, S. (2006). A protein Ser/Thr kinase cascade negatively regulates the DNA-binding activity of MrpC, a smaller form of which may be necessary for the *Myxococcus xanthus* development. *Mol. Microbiol.* **60,** 1205–1217.

Nariya, H., and Inouye, M. (2008). MazF, an mRNA interferase, mediates programmed cell death during multicellular *Myxococcus* development. *Cell* **132,** 55–66.

O'Connor, K. A., and Zusman, D. R. (1991). Development in *Myxococcus xanthus* involves differentiation into two cell types, peripheral rods and spores. *J. Bacteriol.* **173,** 3318–3333.

Ogawa, M., Fujitani, S., Mao, X., Inouye, S., and Komano, T. (1996). FruA, a putative transcription factor essential for the development of *Myxococcus xanthus*. *Mol. Microbiol.* **22,** 757–767.

Parkinson, J. S. (1995). Genetic approaches for signaling pathways and proteins. *In* "Two-Component Signal Transduction," (J. A. Hoch and T. J. Silhavy, eds.), pp. 9–23. ASM Press, Washington, DC.

Rasmussen, A. A., and Sogaard-Andersen, L. (2003). TodK, a putative histidine protein kinase, regulates timing of fruiting body morphogenesis in *Myxococcus xanthus*. *J. Bacteriol.* **185,** 5452–5464.

Reichenbach, H. (1999). The ecology of the myxobacteria. *Environ. Microbiol.* **1,** 15–21.

Rosenberg, E., Keller, K. H., and Dworkin, M. (1977). Cell density-dependent growth of *Myxococcus xanthus* on casein. *J. Bacteriol.* **129,** 770–777.

Rozen, S., and Skaletsky, H. (2000). Primer3 on the WWW for general users and for biologist programmers. *In* "Bioinformatics Methods and Protocols: Methods in Molecular Biology," (S. Krawetz and S. Misener, eds.), pp. 365–386. Humana Press, Totowa, NJ.

Shi, X., Wegener-Feldbrugge, S., Huntley, S., Hamann, N., Hedderich, R., and Sogaard-Andersen, L. (2008). Bioinformatics and experimental analysis of proteins of two-component systems in *Myxococcus xanthus*. *J. Bacteriol.* **190,** 613–624.

Shimkets, L. J. (1999). Intercellular signaling during fruiting-body development of *Myxococcus xanthus*. *Annu. Rev. Microbiol.* **53,** 525–549.

Singer, M., and Kaiser, D. (1995). Ectopic production of guanosine penta- and tetraphosphate can initiate early developmental gene expression in *Myxococcus xanthus*. *Genes Dev.* **9,** 1633–1644.

Skerker, J. M., Prasol, M. S., Perchuk, B. S., Biondi, E. G., and Laub, M. T. (2005). Two-component signal transduction pathways regulating growth and cell cycle progression in a bacterium: A system-level analysis. *PLoS Biol.* **3,** e334.

Stock, A. M., Robinson, V. L., and Goudreau, P. N. (2000). Two-component signal transduction. *Annu. Rev. Biochem.* **69,** 183–215.

Studier, F. W. (2005). Protein production by auto-induction in high density shaking cultures. *Protein Expr. Purif.* **41,** 207–234.

Sun, H., and Shi, W. (2001). Genetic studies of *mrp*, a locus essential for cellular aggregation and sporulation of *Myxococcus xanthus*. *J. Bacteriol.* **183,** 4786–4795.

Timmen, M., Bassler, B. L., and Jung, K. (2006). AI-1 influences the kinase activity but not the phosphatase activity of LuxN of *Vibrio harveyi*. *J. Biol. Chem.* **281,** 24398–24404.

Ueki, T., and Inouye, S. (2003). Identification of an activator protein required for the induction of *fruA*, a gene essential for fruiting body development in *Myxococcus xanthus*. *Proc. Natl. Acad. Sci. USA* **100,** 8782–8787.

Ueki, T., Inouye, S., and Inouye, M. (1996). Positive-negative KG cassettes for construction of multi-gene deletions using a single drug marker. *Gene* **183,** 153–157.

Uhl, M. A., and Miller, J. F. (1996). Integration of multiple domains in a two-component sensor protein: The *Bordetella pertussis* BvgAS phosphorelay. *EMBO J.* **15,** 1028–1036.

Viswanathan, P., Ueki, T., Inouye, S., and Kroos, L. (2007). Combinatorial regulation of genes essential for *Myxococcus xanthus* development involves a response regulator and a LysR-type regulator. *Proc. Natl. Acad. Sci. USA* **104,** 7969–7974.

Whitworth, D. E., and Cock, P. J. (2008a). Two-component systems of the myxobacteria: Structure, diversity and evolutionary relationships. *Microbiology* **154,** 360–372.

Whitworth, D. E., and Cock, P. J. A. (2008b). Two-component signal transduction systems of the myxobacteria. *In* "Myxobacteria: Multicellularity and Differentiation," (D. E. Whitworth, ed.), pp. 169–189. ASM Press, Washington, DC.

Wireman, J. W., and Dworkin, M. (1977). Developmentally induced autolysis during fruiting body formation by *Myxococcus xanthus*. *J. Bacteriol.* **129,** 798–802.

Wu, S. S., and Kaiser, D. (1996). Markerless deletions of *pil* genes in *Myxococcus xanthus* generated by counterselection with the *Bacillus subtilis sacB* gene. *J. Bacteriol.* **178,** 5817–5821.

CHAPTER FIFTEEN

Two-Component Signaling to the Stress MAP Kinase Cascade in Fission Yeast

Susumu Morigasaki*,† *and* Kazuhiro Shiozaki*

Contents

Abstract

In the fission yeast *Schizosaccharomyces pombe*, the Mak2/3 sensor histidine kinases (HKs), the Mpr1 histidine-containing phosphotransfer (HPt) protein, and the Mcs4 response regulator (RR) constitute a multistep phosphorelay, which is connected to a stress-activated mitogen-activated protein kinase (MAPK) cascade. This hybrid signaling pathway senses H_2O_2 and transmits the stress signal by sequential phosphorylation of the component proteins, whose physical interactions play crucial roles to attain eventual activation of Spc1 MAPK. This chapter describes methodological details of the copurification assays in *S. pombe* cell lysate to detect the physical interactions between the Mpr1 HPt and Mcs4 RR proteins and between Mcs4 and the MAPK kinase kinases (MAPKKKs) of the Spc1 cascade. Unexpectedly, we found that the glycolytic enzyme glyceraldehyde-3-phosphate dehydrogenase (GAPDH) encoded by *tdh1*$^+$ is involved in the H_2O_2 signaling process, and its association

* Department of Microbiology, University of California, Davis, California, USA
† Graduate School of Biological Sciences, Nara Institute of Science and Technology, Ikoma, Nara, Japan

Methods in Enzymology, Volume 471
ISSN 0076-6879, DOI: 10.1016/S0076-6879(10)71015-6

with Mcs4 and MAPKKKs in cell lysate is also detectable by copurification assays. In response to H_2O_2, the catalytic cysteine residue of Tdh1 GAPDH is subjected to *S*-thiolation, of which detection protocol is described as well.

1. Introduction

Instead of a simple two-component system constituted of a histidine kinase (HK) and a response regulator (RR), its variant with added complexity, "multistep phosphorelay," is often utilized by fungi and plants (Grefen and Harter, 2004; Saito, 2001). In a multistep phosphorelay, the phosphotransfer from HK to RR is mediated by a third component, a "histidine-containing phosphotransfer (HPt)" protein. A eukaryotic phosphorelay was first demonstrated in the budding yeast *Saccharomyces cerevisiae* (Posas *et al.*, 1996). *S. cerevisiae* has only one HK, Sln1p, which serves as a transmembrane sensor for extracellular osmolarity (Maeda *et al.*, 1994). Unexpectedly, a homologous multistep phosphorelay in the fission yeast *Schizosaccharomyces pombe* was found to be a H_2O_2 sensing system composed of the Mak2/3 HKs, the Mpr1 HPt protein, and the Mcs4 RR (Buck *et al.*, 2001; Nguyen *et al.*, 2000) (Fig. 15.1A).

Both in *S. cerevisiae* and *S. pombe*, the multistep phosphorelay is linked to a mitogen-activated protein kinase (MAPK) cascade. Mcs4 RR in *S. pombe* binds and activates two homologous MAPK kinase kinases (MAPKKKs) Wis4 and Win1 (Samejima *et al.*, 1998; Shieh *et al.*, 1997; Shiozaki *et al.*, 1997), which phosphorylate Wis1 MAPK kinase (MAPKK; Warbrick and Fantes, 1991) (Fig. 15.1A). The phosphorylated, active Wis1 MAPKK in turn phosphorylates Spc1 MAPK. Spc1 (also known as Sty1; Millar *et al.*, 1995; Shiozaki and Russell, 1995) belongs to the stress-activated protein kinase (SAPK) family, whose members include Hog1p MAPK in *S. cerevisiae* (Brewster *et al.*, 1993) and mammalian p38 and JNK MAPKs (Kyriakis and Avruch, 2001). Once activated by phosphorylation, Spc1 MAPK enters the nucleus and induces a set of stress-resistance genes through the Atf1 transcription factor (Chen *et al.*, 2003; Gaits *et al.*, 1998; Nguyen *et al.*, 2002).

Although signaling in both phosphorelay and MAPK cascade is mediated by protein phosphorylation, their molecular mechanisms are quite different. A phosphorelay is initiated by HK autophosphorylation using ATP, and the phosphoryl group is then passed on to the cognate HPt and RR proteins by the His-Asp phosphotransfer reaction (Stock *et al.*, 2000). In contrast, each of the three protein kinases in a MAPK cascade catalytically phosphorylates their downstream substrates with ATP as phosphoryl donor, allowing signal amplification (Marshall, 1994; Nishida and

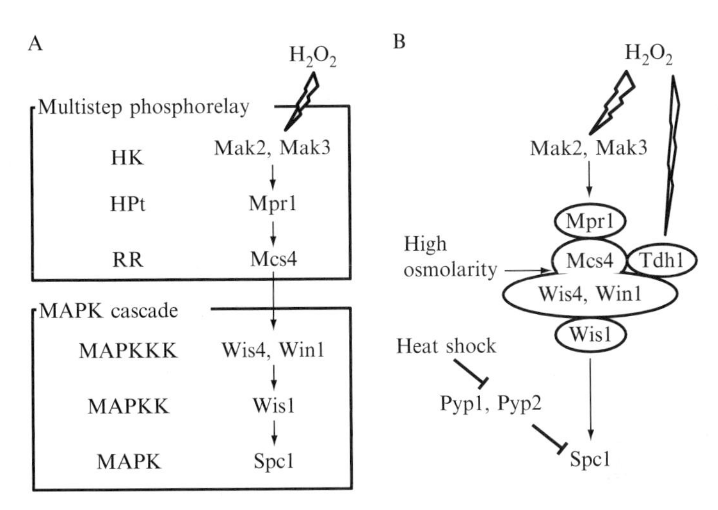

Figure 15.1 The *S. pombe* stress-signaling pathways that activate Spc1 MAPK. (A) A peroxide-sensing multistep phosphorelay is connected to the Spc1 MAPK cascade. (B) Mcs4 RR and two MAPKKKs, Wis4 and Win1, form a stable protein complex. The Mpr1 HPt protein, Wis1 MAPKK, and Tdh1 GAPDH also physically interact with the complex during stress response. The Mcs4-MAPKKK complex plays important roles in the activation of the Spc1 cascade upon H_2O_2 and osmostress, though Mak2/3 HKs and the Mpr1 HPt protein are not involved in osmostress sensing. Heat shock activates Spc1 through inactivation of tyrosine phosphatases, Pyp1 and Pyp2 (Nguyen and Shiozaki, 1999).

Gotoh, 1993). Nevertheless, signal transmission from one protein to another is dependent on their physical interaction in both phosphorelay and MAPK cascade mechanisms. Moreover, these two signaling modules in fission yeast are connected also by protein–protein interactions; Mcs4 RR and Wis4/Win1 MAPKKKs form a stable complex that interacts with the Mpr1 HPt protein as well as Wis1 MAPKK (Fig. 15.1B). Unexpectedly, we found that this MAPKKK complex also contains the glycolytic enzyme glyceraldehyde-3-phosphate dehydrogenase (GAPDH) encoded by $tdh1^+$ (Morigasaki *et al.*, 2008). In response to H_2O_2 stress, the catalytic cysteine residue of Tdh1 GAPDH is oxidized by *S*-thiolation, which promotes association between Tdh1 and Mcs4 RR. Importantly, Tdh1 is crucial to the physical interaction, and thus to the phosphotransfer, between the Mpr1 HPt and Mcs4 RR proteins. This chapter describes affinity-purification methods to detect the interaction network among Wis4/Win1 MAPKKKs, Mcs4 RR, Mpr1 HPt protein, and Tdh1 GAPDH, in addition to the assay for *S*-thiolation of Tdh1. General methods of *S. pombe* genetics and molecular biology have been reviewed previously (Alfa *et al.*, 1993; Moreno *et al.*, 1991).

2. Detection of Protein Interactions in the H_2O_2 Signaling Pathway

Inserting a DNA sequence encoding an epitope "tag" to the 5′ or 3′end of open reading frames (ORFs) on the *S. pombe* chromosomes is a popular technique to detect and purify endogenous proteins of interest without raising specific antibodies (Bähler *et al.*, 1998; Shiozaki and Russell, 1997). Furthermore, by constructing a strain in which two proteins are expressed with different tags, their physical association can be assayed by affinity-purification of one of the tagged proteins, followed by immunodetection of the other using antibodies against the epitope tags. However, there is no master condition optimal for detecting every protein–protein interaction, because individual proteins have diverse biochemical properties, and the strength of their association greatly differs depending on the types of interactions, such as hydrophobic and ionic interactions. Therefore, experimental conditions for interaction assays, especially buffer compositions, need to be optimized for each protein pair to be tested. Here we first describe a common procedure for affinity-purification and detection of tagged proteins from *S. pombe* cell lysate (Section 2.1). The subsequent sections present three different conditions established to detect association between Mcs4 RR and Wis4/Win1 MAPKKKs (Section 2.2), between the Mpr1 HPt and Mcs4 RR proteins (Section 2.3), and between Tdh1 GAPDH and Mcs4 or MAPKKKs (Section 2.4).

2.1. Affinity purification of tagged proteins from *S. pombe* cell lysate

S. pombe cells expressing tagged proteins are grown in liquid medium to an optical density (OD) at 600 nm of 0.2–0.4. Twenty OD cells ($\sim 2 \times 10^8$ cells) are harvested in a 1.5-ml microcentrifuge tube and immediately frozen in liquid nitrogen (Shiozaki and Russell, 1997). Cell pellets can be stored in a −80 °C freezer for several months. All the following steps to purify proteins are carried out at 4 °C or on ice. Before thawing cell pellet, lysis and washing buffers (see Sections 2.2 ~4 for their compositions) must be prepared and ligand-conjugated beads are equilibrated with lysis buffer as described in supplier's manuals/instructions.

Cell pellets are resuspended in 400 μl of lysis buffer, and 0.5-mm diameter glass beads are added to the meniscus of the cell suspension (~1 ml bed volume). Cells are disrupted with a vortex mixer at the full speed for 5 min or longer, depending on the mixer used; efficiency of cell disruption can be monitored by microscopy. To collect cell lysate, the bottom of the microcentrifuge tube is pierced with a 23-gauge needle, and the tube is placed on another 1.5-ml microcentrifuge tube. By brief centrifugation of the two-tube stack in a microcentrifuge, cell lysate is transferred to the lower tube. After removing the upper tube containing the glass beads, the lower tube with the cell lysate is centrifuged at 21,000×*g* for 10 min. The supernatant is transferred to a new 1.5-ml microcentrifuge tube. A 5-μl aliquot is taken and diluted 10-fold with H_2O for determination of the protein concentration using the Protein Assay Dye Reagent (Bio-Rad); lysate from 20-OD cells grown in rich medium typically contains ~4 mg protein after successful cell disruption by glass beads.

A 5-μl bed volume of appropriate affinity-purification resin is added to the cell extract; IgG-Sepharose beads (GE Healthcare) for Tandem affinity-purification (TAP)-tagged (Rigaut *et al.*, 1999) proteins and Ni^{2+}-nitrilotriacetic acid (NTA)-agarose beads (Qiagen) for hexahistidine-tagged proteins. The total volume is adjusted to 500 μl with lysis buffer. After the cell extract and the beads are gently mixed on a rocking table for 1 h, the beads are precipitated by brief centrifugation at 800×*g*. The supernatant is removed, and the beads are resuspended in 1 ml of washing buffer, followed by another brief centrifugation; this washing step is repeated four times.

Proteins bound to the beads are eluted by Laemmli's sample buffer for SDS–PAGE (Laemmli, 1970) under the conditions described for each interaction experiment in Sections 2.2–2.4. The samples are then briefly centrifuged at 800×*g* to precipitate the beads, and the supernatant is transferred to a fresh microcentrifuge tube to remove the beads completely. The eluate is stored at –20 °C or immediately analyzed by SDS–PAGE and immunoblotting to detect the purified proteins with antibodies against the epitope tags, such as anti-HA (12CA5; Roche), anti-*myc* (9E10; Covance), or anti-FLAG (M2; Sigma) mouse monoclonal antibodies.

2.2. Mcs4 RR–MAPKKK interactions

Association between Mcs4 RR and Wis4/Win1 MAPKKKs are detected using a *S. pombe* strain expressing Mcs4 with the 12-copy *myc* epitope tag, together with Wis4 or Win1 MAPKKKs tagged by three copies of the HA epitope followed by the TAP tag (HATAP; Morigasaki *et al.*, 2008). In addition, a strain expressing only Mcs4-12myc is used as a negative control.

MAPKKK tagged with HATAP is purified onto IgG-Sepharose beads following the procedure described in Section 2.1, and copurification of Mcs4-12myc is detected by anti-*myc* immunoblotting (Fig. 15.2A). The experiment uses Lysis Buffer A and Washing Buffer A, whose compositions are described below. Before adding NP-40, the rest of the buffers should be chilled on ice. Phosphatase inhibitors, β-glycerophosphate and *p*-nitrophenyl phosphate, are directly dissolved in the lysis buffer. Protease inhibitors are added just before use. *Lysis Buffer A*: 50 m*M* Tris–HCl, pH 8.0, 150 m*M* NaCl, 10% glycerol, 5 m*M* EDTA, 1 m*M* EGTA, 50 m*M* NaF, 0.1 m*M*

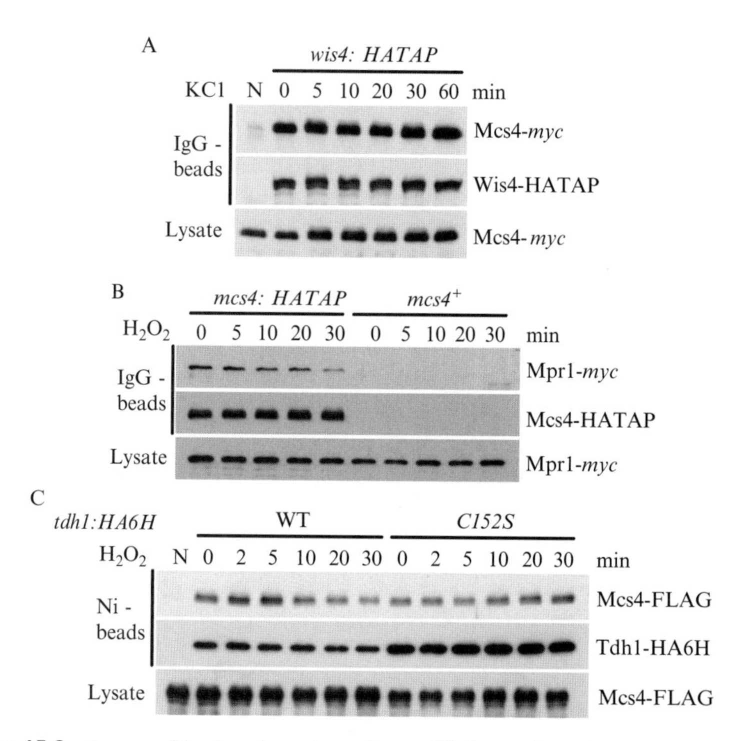

Figure 15.2 Immunoblotting detection of copurified proteins. HATAP- and HA6H-tagged proteins were detected with anti-HA (12CA5) monoclonal antibodies. FLAG- and *myc*-tagged proteins were detected with anti-FLAG (M2), anti-*myc* (9E10) monoclonal antibodies, respectively. (A) Copurification of Mcs4-myc with Wis4-HATAP affinity-purified onto IgG-Sepharose beads. N, a negative control with a strain expressing Mcs4-*myc* and untagged Wis4. High-osmolarity stress of 0.6 *M* KCl to the culture did not disturb the Mcs4–Wis4 interaction. (B) Copurification of Mpr1-*myc* with Mcs4-HATAP affinity-purified onto IgG-Sepharose beads. A strain expressing Mpr1-*myc* and untagged Mcs4 ("*mcs4*$^+$") was used as a negative control. (C) Copurification of Mcs4-FLAG with Tdh1-HA6H affinity-purified onto Ni^{2+}-NTA-agarose beads. N, a negative control with a strain expressing Mcs4-FLAG and untagged Tdh1. H_2O_2 transiently increased interaction of Mcs4 with wild-type Tdh1, but not with Tdh1 carrying a serine substitution of Cys-152 (C152S).

Na_3VO_4, 60 m*M* β-glycerophosphate, 15 m*M* *p*-nitrophenyl phosphate, 0.1% NP-40, 1 m*M* 2-mercaptoethanol, 1/200-volume of the Protease Inhibitor Cocktail for use with fungal and yeast extract (Sigma), 10 μg/ml leupeptin, 0.2 m*M* phenylmethylsulfonyl fluoride (PMSF). *Washing Buffer A*: 50 m*M* Tris–HCl, pH 8.0, 150 m*M* NaCl, 10% glycerol, 5 m*M* EDTA, 1 m*M* EGTA, 50 m*M* NaF, 0.1 m*M* Na_3VO_4, 0.1% NP-40, 1 m*M* 2-mercaptoethanol.

To minimize sample contamination with IgG derived from the IgG-Sepharose, proteins bound to the IgG-Sepharose can be eluted by incubation of the washed beads in Laemmli's sample buffer without 2-mercaptoethanol for 15 min at room temperature. After transferred to a new microcentrifuge tube, the eluate is incubated with 2-mercaptoethanol at 100 °C for 2 min.

2.3. Interaction between the Mpr1 HPt protein and Mcs4 RR

The buffer conditions used in Section 2.2 detect the specific association of Mcs4 RR with the MAPKKKs, but not with the Mpr1 HPt protein. It is plausible that the interaction of Mpr1 with Mcs4 RR is weak and transient, because Mpr1 needs to interface also with the receiver domain of Mak2/3 HKs to attain phosphotransfer from the HKs to Mcs4 RR (Santos and Shiozaki, 2001). This section describes the experimental condition developed to capture the Mpr1–Mcs4 association without overexpressing either of the proteins. The strain used expresses Mcs4 with the HATAP tag, together with Mpr1 tagged with 12 copies of the *myc* epitope; both are expressed from their endogenous chromosomal loci (Morigasaki *et al.*, 2008). In addition, a strain expressing only Mpr1-12myc is used as a negative control.

Mcs4 tagged with HATAP is purified onto IgG-Sepharose beads, following the procedure described in Section 2.1, and copurification of Mpr1-12myc is detected by anti-*myc* immunoblotting (Fig. 15.2B). The experiment uses Lysis Buffer B and Washing Buffer B, whose compositions are described below; potassium glutamate is used as a major salt to minimize the use of Na^+ and Cl^-, to which the Mpr1–Mcs4 interaction appears to be sensitive. Before adding NP-40, the rest of the buffers should be chilled on ice. β-glycerophosphate and *p*-nitrophenyl phosphate are directly dissolved in the lysis buffer. Protease inhibitors are added just before use. *Lysis Buffer B*: 50 m*M* HEPES–KOH, pH 7.8, 500 m*M* potassium glutamate, 10% glycerol, 0.1 m*M* Na_3VO_4, 15 m*M* *p*-nitrophenyl phosphate, 0.1% NP-40, 1/200-volume of the Protease Inhibitor Cocktail for use purification of histidine-tagged proteins (Sigma), 10 μg/ml leupeptin, 0.2 m*M* PMSF. *Washing Buffer B*: 50 m*M* HEPES–KOH, pH 7.8, 500 m*M* potassium glutamate, 10% glycerol, 0.1 m*M* Na_3VO_4, 15 m*M* *p*-nitrophenyl phosphate, 0.1% NP-40.

Proteins should be eluted from the IgG-Sepharose beads with >150 μl of Laemmli's sample buffer to avoid precipitation of dodecyl sulfate in the presence of potassium in Washing Buffer B. For complete solubilization, the samples mixed with Laemmli's sample buffer may be incubated at 30–40 °C for 10 min, followed by heating at 100 °C for 2 min. The samples should not be frozen, but can be kept at 4 °C for several days; samples stored at 4 °C are incubated again at 30–40 °C for 10 min before loading to SDS–PAGE gels.

2.4. Interaction of Tdh1 GAPDH with Mcs4 RR and MAPKKKs

In the course of our studies on Tdh1 GAPDH, nonspecific binding of Tdh1 to IgG-Sepharose was detected, which was further enhanced after stress treatment of *S. pombe* cells. Therefore, we constructed *S. pombe* strains in which the chromosomal *tdh1*$^{+}$ ORF was tagged with the sequence encoding two copies of the HA epitope followed by six histidine residues (HA6H; Shiozaki and Russell, 1997). With those strains, Tdh1-HA6H can be affinity-purified onto Ni^{2+}-NTA-agarose beads, followed by immunoblotting to detect the associated proteins, such as Mcs4 RR and Wis4/Win1 MAPKKKs tagged with either three copies of FLAG or 12 copies of *myc* epitopes (Fig. 15.2C). Strains expressing untagged Tdh1 are used as negative controls.

In these experiments, Tdh1-HA6H is purified using Ni^{2+}-NTA-agarose beads by the procedure described in Section 2.1, except that 1 mg or less protein should be incubated with 5 μl beads; using $>$ 1 mg protein results in a poor purification yield with unknown reasons. The experiment uses Lysis Buffer C and Washing Buffer C, whose compositions are described below; thiol reagents, such as 2-mercaptoethanol, and chelators, such as EDTA and EGTA, should be avoided to stabilize Ni^{2+}-NTA-agarose. Before adding NP-40, the rest of the buffers should be chilled on ice. β-glycerophosphate and *p*-nitrophenyl phosphate are directly dissolved in the lysis buffer. Protease inhibitors (EDTA-free) are added just before use. *Lysis Buffer C*: 50 m*M* Tris–HCl, pH 8.0, 10 m*M* imidazole, 10% glycerol, 50 m*M* NaF, 0.1 m*M* Na_3VO_4, 60 m*M* β-glycerophosphate, 15 m*M* *p*-nitrophenyl phosphate, 0.1% NP-40, 1/200-volume of the Protease Inhibitor Cocktail for purification of histidine-tagged proteins (Sigma), 10 μg/ml leupeptin, 0.2 m*M* PMSF. *Washing Buffer C*: 50 m*M* Tris–HCl, pH 8.0, 20 m*M* imidazole, 150 m*M* NaCl, 10% glycerol, 50 m*M* NaF, 0.1 m*M* Na_3VO_4, 0.1% NP-40.

Proteins are eluted from Ni-NTA-beads by adding 1/4-volume of 5$\times$ Laemmli's sample buffer with 2-mercaptoethanol, followed by heating at 100 °C for 2 min.

3. Detection of Cysteine *S*-Thiolation in Tdh1 GAPDH

Under oxidative environments, glutathione, and other low-molecular weight thiols form mixed disulfides with the sulfhydryl (–SH) group of cysteine residues in diverse proteins (Klatt and Lamas, 2000; Thomas *et al.*, 1995). Such modifications, known as *S*-thiolation, are believed to prevent irreversible oxidation of protein –SH groups to sulfinic/sulfonic acid but may also modulate protein function, like other posttranslational modifications. We found that Cys-152 of Tdh1 GAPDH in *S. pombe* is *S*-thiolated, which is important for the physical interaction and phosphotransfer between Mpr1 HPt and Mcs4 RR proteins during peroxide stress (Morigasaki *et al.*, 2008). The following protocol to detect *S*-thiolation of Tdh1 GAPDH was developed based on the previously reported methods to detect disulfide (S–S) bonds and cysteine sulfenate in proteins (Hirose *et al.*, 1988; Kim *et al.*, 2000).

All buffers are degassed before use and all steps are performed in the dark or under the weak light. *S. pombe* cells expressing Tdh1-HA6H are disrupted in Buffer G (6 *M* guanidine–HCl, 100 m*M* sodium phosphate, and 50 m*M* Tris–HCl, pH 8.0) containing 50 m*M* sodium iodoacetate (IAA) to block free –SH groups. Tdh1-HA6H is purified with Ni^{2+}-NTA-agarose by the procedure described in Section 2.1, except that Buffer G should be used as lysis and washing buffers. Disulfide (S–S) bond (or sulfenate, if any) in Tdh1-HA6H is then reduced with 10 m*M* DTT at 37 °C for 1 h. After 12-fold dilution with Buffer G, Tdh1-HA6H is recovered onto freshly added Ni^{2+}-NTA-agarose beads. DTT is removed by washing the beads with Buffer G, and reduced –SH groups are then labeled with 0.6 m*M* iodoacetyl-LC-biotin (Thermo Scientific) in Buffer G at room temperature for 1 h. The beads are washed three times with Buffer U (8 *M* urea, 100 m*M* sodium phosphate, 50 m*M* Tris–HCl, pH 8.0), and the Tdh1-HA6H protein is then eluted by adding the equal volume of 2× Laemmli's sample buffer at room temperature. Following SDS–PAGE, proteins are transferred onto a nitrocellulose membrane and probed by horseradish peroxidase-conjugated streptavidin (Thermo Scientific); oxidized Tdh1 is detected as positive signal by the chemiluminescence reaction (Fig. 15.3). A duplicate should be analyzed by anti-HA immunoblotting to detect the total Tdh1-HA6H protein recovered from the lysate. As a positive control, cells are disrupted in Buffer G without IAA, so that all free –SH groups are labeled by iodoacetyl-LC-biotin. For negative control, the DTT-reduced sample is

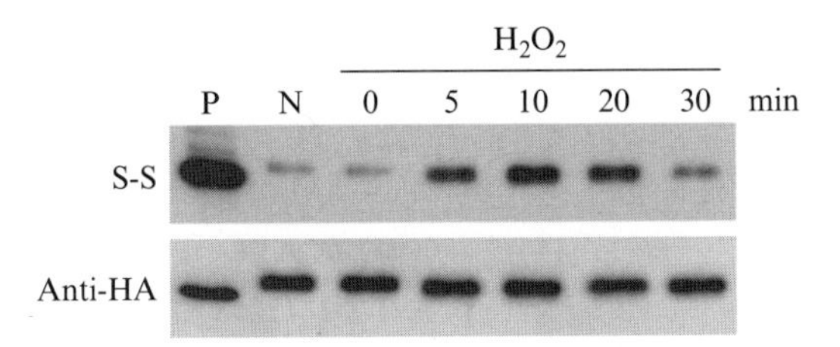

Figure 15.3 Detection of Tdh1 *S*-thiolation. *S*-thiolated cysteine was biotinylated, which was detected by horseradish peroxidase-conjugated streptavidin (upper panel). P, a positive control where all three cysteine residues in Tdh1 were biotinylated. N, a negative control in which all cysteine residues in Tdh1 were carboxymethylated using sodium iodoacetate. The Tdh1-HA6H protein was detected by anti-HA immunoblotting (lower panel).

treated with 50 m*M* IAA again to block newly formed SH groups, before the labeling step with iodoacetyl-LC-biotin.

ACKNOWLEDGMENTS

We thank members of the Shiozaki laboratory for support and discussion. We are indebted to the late Y. Kawasaki for advice in protein purification. This work was supported by a NIH grant (GM059788) to K. S., and Grant-in-Aid for Scientific Research (C) 21570198 awarded to S. M. from the Japan Society for the Promotion of Science. S. M. is a recipient of the International Research Fellowship from the Nara Institute of Science and Technology Global COE Program funded by the Ministry of Education, Culture, Sports, Science and Technology of Japan.

REFERENCES

Alfa, C., Fantes, P., Hyams, J., McLeod, M., and Warbrick, E. (1993). Experiments with Fission Yeast. A Laboratory Course Manual. Cold Spring Harbor Laboratory Press, Cold Spring Harbor, NY.

Bähler, J., Wu, J. Q., Longtine, M. S., Shah, N. G., McKenzie, A., Steever, A. B., Wach, A., Philippsen, P., and Pringle, J. R. (1998). Heterologous modules for efficient and versatile PCR-based gene targeting in *Schizosaccharomyces pombe*. *Yeast* **14,** 943–951.

Brewster, J. L., de Valoir, T., Dwyer, N. D., Winter, E., and Gustin, M. C. (1993). An osmosensing signal transduction pathway in yeast. *Science* **259,** 1760–1763.

Buck, V., Quinn, J., Soto Pino, T., Martin, H., Saldanha, J., Makino, K., Morgan, B. A., and Millar, J. B. A. (2001). Peroxide sensors for the fission yeast stress-activated mitogen-activated protein kinase pathway. *Mol. Biol. Cell* **12,** 407–419.

Chen, D., Toone, W. M., Mata, J., Lyne, R., Burns, G., Kivinen, K., Brazma, A., Jones, N., and Bähler, J. (2003). Global transcriptional responses of fission yeast to environmental stress. *Mol. Biol. Cell* **14,** 214–229.

Gaits, F., Degols, G., Shiozaki, K., and Russell, P. (1998). Phosphorylation and association with the transcription factor Atf1 regulate localization of Spc1/Sty1 stress-activated kinase in fission yeast. *Genes Dev.* **12,** 1464–1473.

Grefen, C., and Harter, K. (2004). Plant two-component systems: Principles, functions, complexity and cross talk. *Planta* **219,** 733–742.

Hirose, M., Takahashi, N., Oe, H., and Doi, E. (1988). Analyses of intramolecular disulfide bonds in proteins by polyacrylamide-gel electrophoresis following 2-step alkylation. *Anal. Biochem.* **168,** 193–201.

Kim, J. R., Yoon, H. W., Kwon, K. S., Lee, S. R., and Rhee, S. G. (2000). Identification of proteins containing cysteine residues that are sensitive to oxidation by hydrogen peroxide at neutral pH. *Anal. Biochem.* **283,** 214–221.

Klatt, P., and Lamas, S. (2000). Regulation of protein function by *S*-glutathiolation in response to oxidative and nitrosative stress. *Eur. J. Biochem.* **267,** 4928–4944.

Kyriakis, J. M., and Avruch, J. (2001). Mammalian mitogen-activated protein kinase signal transduction pathways activated by stress and inflammation. *Physiol. Rev.* **81,** 807–869.

Laemmli, U. K. (1970). Cleavage of structural proteins during the assembly of the head of bacteriophage T4. *Nature* **227,** 680–685.

Maeda, T., Wurgler-Murphy, S. M., and Saito, H. (1994). A two-component system that regulates an osmosensing MAP kinase cascade in yeast. *Nature* **369,** 242–245.

Marshall, C. J. (1994). MAP kinase kinase kinase, MAP kinase kinase and MAP kinase. *Curr. Opin. Genet. Dev.* **4,** 82–89.

Millar, J. B. A., Buck, V., and Wilkinson, M. G. (1995). Pyp1 and Pyp2 PTPases dephosphorylate an osmosensing MAP kinase controlling cell-size at division in fission yeast. *Genes Dev.* **9,** 2117–2130.

Moreno, S., Klar, A., and Nurse, P. (1991). Molecular genetic analysis of fission yeast *Schizosaccharomyces pombe. Methods Enzymol.* **194,** 795–823.

Morigasaki, S., Shimada, K., Ikner, A., Yanagida, M., and Shiozaki, K. (2008). Glycolytic enzyme GAPDH promotes peroxide stress signaling through multistep phosphorelay to a MAPK cascade. *Mol. Cell* **30,** 108–113.

Nguyen, A. N., and Shiozaki, K. (1999). Heat shock-induced activation of stress MAP kinase is regulated by threonine- and tyrosine-specific phosphatases. *Genes Dev.* **13,** 1653–1663.

Nguyen, A. N., Lee, A., Place, W., and Shiozaki, K. (2000). Multistep phosphorelay proteins transmit oxidative stress signals to the fission yeast stress-activated protein kinase. *Mol. Biol. Cell* **11,** 1169–1181.

Nguyen, A. N., Ikner, A. D., Shiozaki, M., Warren, S. M., and Shiozaki, K. (2002). Cytoplasmic localization of Wis1 MAPKK by nuclear export signal is important for nuclear targeting of Spc1/Sty1 MAPK in fission yeast. *Mol. Biol. Cell* **13,** 2651–2663.

Nishida, E., and Gotoh, Y. (1993). The MAP kinase cascade is essential for diverse signal transduction pathways. *Trends Biochem. Sci.* **18,** 128–131.

Posas, F., Wurgler-Murphy, S. M., Maeda, T., Witten, E. A., Thai, T., and Saito, H. (1996). Yeast HOG1 MAP kinase cascade is regulated by a multistep phosphorelay mechanism in the SLN1-YPD1-SSK1 "two-component" osmosensor. *Cell* **86,** 865–875.

Rigaut, G., Shevchenko, A., Rutz, B., Wilm, M., Mann, M., and Séraphin, B. (1999). A generic protein purification method for protein complex characterization and proteome exploration. *Nat. Biotechnol.* **17,** 1030–1032.

Saito, H. (2001). Histidine phosphorylation and two-component signaling in eukaryotic cells. *Chem. Rev.* **101,** 2497–2509.

Samejima, I., Mackie, S., Warbrick, E., Weisman, R., and Fantes, P. A. (1998). The fission yeast mitotic regulator *win1*$^{+}$ encodes an MAP kinase kinase kinase that phosphorylates and activates Wis1 MAP kinase kinase in response to high osmolarity. *Mol. Biol. Cell* **9,** 2325–2335.

Santos, J. L., and Shiozaki, K. (2001). Fungal histidine kinases. Science's STKE. http://stke.sciencemag.org/cgi/content/full/OC_sigtrans;2001/98/re1.

Shieh, J. C., Wilkinson, M. G., Buck, V., Morgan, B. A., Makino, K., and Millar, J. B. A. (1997). The Mcs4 response regulator coordinately controls the stress-activated Wak1-Wis1-Sty1 MAP kinase pathway and fission yeast cell cycle. *Genes Dev.* **11,** 1008–1022.

Shiozaki, K., and Russell, P. (1995). Cell-cycle control linked to extracellular environment by MAP kinase pathway in fission yeast. *Nature* **378,** 739–743.

Shiozaki, K., and Russell, P. (1997). Stress-activated protein kinase pathway in cell cycle control of fission yeast. *Methods Enzymol.* **283,** 506–520.

Shiozaki, K., Shiozaki, M., and Russell, P. (1997). Mcs4 mitotic catastrophe suppressor regulates the fission yeast cell cycle through the Wik1-Wis1-Spc1 kinase cascade. *Mol. Biol. Cell* **8,** 409–419.

Stock, A. M., Robinson, V. L., and Goudreau, P. N. (2000). Two-component signal transduction. *Annu. Rev. Biochem.* **69,** 183–215.

Thomas, J. A., Poland, B., and Honzatko, R. (1995). Protein sulfhydryls and their role in the antioxidant function of protein S-thiolation. *Arch. Biochem. Biophys.* **319,** 1–9.

Warbrick, E., and Fantes, P. A. (1991). The Wis1 protein-kinase is a dosage-dependent regulator of mitosis in *Schizosaccharomyces pombe. EMBO J.* **10,** 4291–4299.

CHAPTER SIXTEEN

Genetic and Biochemical Analysis of the SLN1 Pathway in *Saccharomyces cerevisiae*

Jan S. Fassler* *and* Ann H. West†

Contents

Abstract

The histidine kinase-based signal transduction pathway was first uncovered in bacteria and is a prominent form of regulation in prokaryotes. However, this type of signal transduction is not unique to prokaryotes; over the last decade two-component signal transduction pathways have been identified and characterized in diverse eukaryotes, from unicellular yeasts to multicellular land plants. A number of small but important differences have been noted in the architecture and function of eukaryotic pathways. Because of the powerful genetic approaches and facile molecular analysis associated with the yeast system, the SLN1 osmotic response pathway in *Saccharomyces cerevisiae* is particularly useful as a eukaryotic pathway model. This chapter provides an overview of genetic and biochemical methods that have been important in elucidating the stimulus-response events that underlie this pathway and in understanding the details of a eukaryotic His-Asp phosphorelay.

* Department of Biology, University of Iowa, Iowa City, Iowa, USA
† Department of Chemistry and Biochemistry, University of Oklahoma, Normal, Oklahoma, USA

Methods in Enzymology, Volume 471
ISSN 0076-6879, DOI: 10.1016/S0076-6879(10)71016-8

Abbreviations

βME	β-mercaptoethanol
BSA	bovine serum albumin
CBD	chitin binding domain
DTT	dithiothreitol
EDTA	ethylenediaminetetracetic acid
EMS	ethyl methane sulfonate
GST	glutathione-*S*-transferase
HK	histidine kinase
IPTG	isopropyl-β-D-thiogalactopyranoside
MOPS	3-morpholinopropanesulfonic acid
ONPG	*ortho*-nitrophenyl-β-galactoside
PMSF	phenylmethanesulfonyl fluoride
Rec	receiver domain of a response regulator
SC	synthetic complete
SC-aa	synthetic complete media lacking specific amino acids
SDS–PAGE	sodium dodecyl sulfate polyacrylamide gel electrophoresis
TCA	trichloroacetic acid
X-Gal	5-bromo-4-chloro-3-indolyl-β-D-galactopyranoside

1. Introduction

The SLN1 two-component signaling pathway of *Saccharomyces cerevisiae* is a branched multistep phosphorelay containing a single hybrid histidine kinase, a histidine-containing phosphotransfer protein, and two downstream response regulators (Fig. 16.1A). Like the well-studied bacterial two-component pathways, the activity of the SLN1 pathway responds to environmental conditions. Phosphorelay from the Sln1 histidine kinase to the cytoplasmic Ssk1 response regulator is important in dampening the activity of the HOG1 MAP kinase pathway under normal osmotic conditions (Maeda *et al.*, 1994; Posas and Saito, 1997; Reiser *et al.*, 2003) while phosphorelay from the Sln1 histidine kinase to the nuclear Skn7 response regulator appears to play an important role in the cellular response to cell wall perturbations (Shankarnarayan *et al.*, 2008).

Like most eukaryotic histidine kinases, Sln1 is a hybrid protein containing both kinase and receiver domains (Ota and Varshavsky, 1993) (Fig. 16.1B). In this configuration the first phosphotransfer step is between H576, the phoshorylatable histidine within the kinase domain of Sln1 to

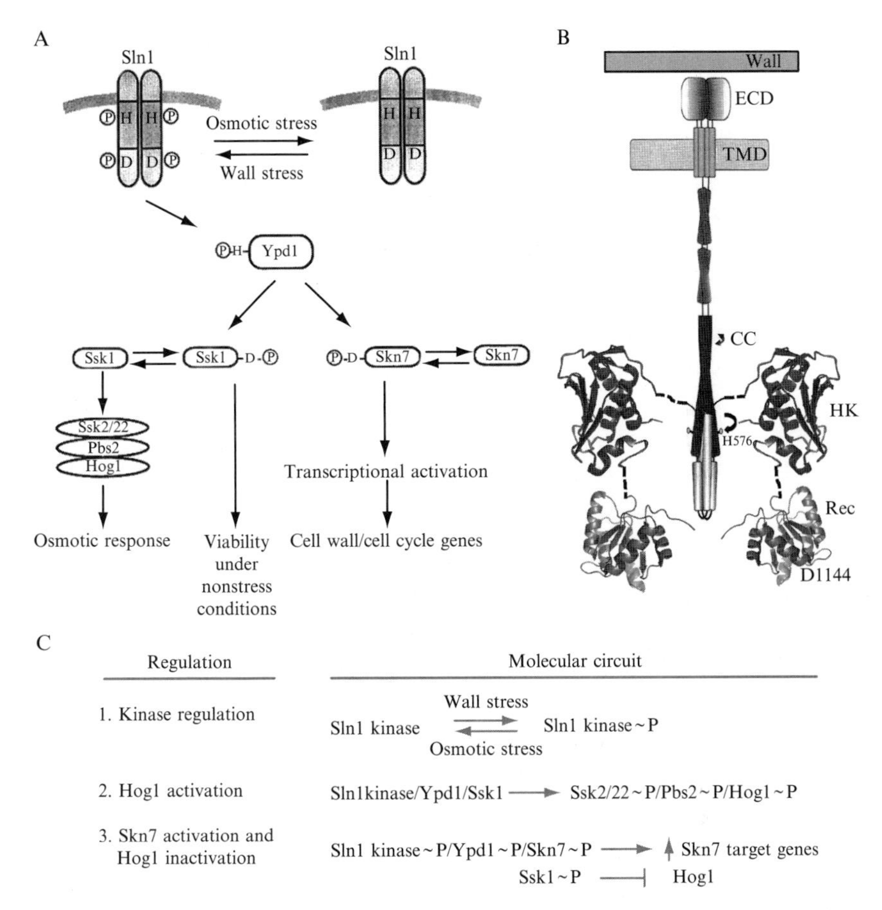

Figure 16.1 SLN1 pathway organization and regulation. (A) Schematic showing flow of phosphoryl groups through the SLN1 pathway. Plasma membrane associated Sln1 kinase is autophosphorylated on the H576 residue under normal growth conditions. Hyperactivation occurs under conditions causing weakening of the cell wall (wall stress). Hyperosmolarity and other conditions causing reduced turgor lead to a reduction in Sln1 kinase activity and accumulation of Sln1 in the dephosphorylated form. The phosphoryl group on H576 (H) is transferred to D1144 (D) in the receiver domain of Sln1 and then to H64 of the phosphotransferase, Ypd1, and finally to D554 and D427 in the receiver domains of the Ssk1 and Skn7 response regulators, respectively. When Ssk1 is unphosphorylated, it interacts with and stimulates the activity of the Ssk2 and Ssk22 MAPKKKs of the HOG1 pathway. Phosphorylation of Ssk1 renders it inactive with respect to the HOG1 pathway. Phosphorylation of Skn7 leads to activation of *SKN7*-dependent genes such as the mannosyltransferase, *OCH1*. (B) Schematic depicting the domain structure of Sln1. Two molecules of a Sln1 dimer are shown associated with the plasma membrane via two transmembrane domains (TMDs). An extensive extracellular domain (ECD) protrudes into the periplasmic space. This protein configuration is conducive to detection of changes in pressure against the membrane (turgor pressure) and changes in composition of the cell wall which may lead to altered rigidity.

D1144, the phosphoaccepting aspartate within the receiver domain of Sln1. The second phosphotransfer step is between Sln1 D1144 and histidine H64 in the phosphotransfer protein, Ypd1. The final phosphotransfer steps in the pathway occur between H64 of Ypd1 and the phosphoaccepting aspartate D554 in the receiver domain of the cytoplasmic response regulator, Ssk1 and between H64 of Ypd1 and aspartate D427 in the receiver domain of the nuclear response regulator, Skn7 (Fassler *et al.*, 1997; Li *et al.*, 1998; Posas *et al.*, 1996) (Fig. 16.1A).

The kinase activity of Sln1 is regulated in response to the environment. Under normal growth conditions, a modest level of kinase activity appears to be crucial for viability (Fig. 16.1A). Deletion of *SLN1* or mutation in any of the phosphorylatable residues in Sln1, Ypd1, or Ssk1 leads to inviability that is suppressible by inactivating mutations in components of the HOG1 MAP kinase pathway and by overexpression of HOG1 pathway phosphatases (Maeda *et al.*, 1994; Ota and Varshavsky, 1992; Wurgler-Murphy *et al.*, 1997). This type of evidence led to our current understanding of the SLN1–YPD1–SSK1 pathway as important in negatively regulating the activity of the HOG1 pathway in the absence of stress.

SLN1 pathway activity is diminished in the presence of elevated osmotic conditions (Ostrander and Gorman, 1999) leading to decreased turgor (Reiser *et al.*, 2003) (Fig. 16.1A and C). These conditions lead to the accumulation of SLN1 pathway components in the dephosphorylated form. Dephosphorylated Ssk1 interacts with and activates the kinase activity of the Ssk2 and Ssk22 MAPKKKs (Maeda *et al.*, 1995; Posas and Saito, 1998) of the HOG1 pathway thus setting in motion the signaling cascade that ultimately results in changes in expression of osmotic response genes including those involved in biosynthesis of the compatible osmolyte, glycerol (Albertyn *et al.*, 1994a,b).

The stimulus is transduced through several juxtamembrane domains including a coiled-coil domain (CC) which appears to be contiguous with the phosphoaccepting H576 residue. Rotation of the coiled-coil domain in response to a stimulus is thought to regulate access of H576 to the kinase domain (HK) and hence the level of kinase phosphorylation. The structure of the *Thermotoga maritima* TM0853 HK domain (PDB ID 2C2A) (Marina *et al.*, 2005) is depicted here. Transfer of the H576 phosphoryl group to the receiver domain (Rec) is a prerequisite for subsequent phosphotransfer to Ypd1 and the response regulators Ssk1 and Skn7 (not shown). The structure of the Sln1-Rec domain is shown here (PDB ID 2R25) (Zhao *et al.*, 2008) and the dashed lines represent linker regions of unknown structure. (C) Several types of regulation are embodied in the Sln1 pathway. (1) The Sln1 kinase itself is positively regulated by changes in the cell wall structure, and negatively regulated by hyperosmolarity and other situations leading to reduced turgor. (2) The HOG1 pathway is positively regulated by an inactive SLN1 pathway. Only the dephosphorylated form of the Ssk1 response regulator stimulates the Ssk2 and Ssk22 MAPKKKs of the HOG1 pathway. (3) Finally, the HOG1 pathway is inactivated and Skn7 is activated by the phosphorylated SLN1 pathway intermediates.

SLN1 pathway activity can also be increased (Fig. 16.1A and C). This was first shown by the isolation of mutations exhibiting elevated expression of *SKN7* target genes. These have been dubbed *sln1** activating mutants since they increase the activity of the Sln1 phosphorelay (Ault *et al.*, 2002; Fassler *et al.*, 1997; Tao *et al.*, 2002). Both the Skn7 and the Ssk1 response regulators appear to be more highly phosphorylated in these mutants. The elevated phosphorylation of Skn7 is responsible for the increase in expression of Skn7-dependent target genes (Li *et al.*, 1998, 2002) and the elevated phosphorylation of Ssk1 causes defects in the cellular response to hyperosmotic conditions due to changes in the kinetics of Hog1 phosphorylation (Fassler *et al.*, 1997). In principle, *sln1* activating mutations leading to increased phosphorylation of the Ssk1 and Skn7 response regulators could be attributable to a variety of mechanisms, including, for example, increased rate of Sln1 autophosphorylation or phosphotransfer activity, or diminished Sln1-directed phosphatase activity. In practice, *sln1* activating mutations map to both the Sln1 receiver domain and to the coiled-coil domain located between the second transmembrane domain and the kinase domain (Ault *et al.*, 2002; Fassler *et al.*, 1997; Tao *et al.*, 2002). The *sln1** P1148S activating allele is a Pro to Ser mutation of a helix capping proline that is conserved in the receiver domain of most response regulators and is equivalent to P61 of *Escherichia coli* CheY (Ault *et al.*, 2002; Fassler *et al.*, 1997). *In vitro* phosphotransfer analysis of the P1148S mutant revealed a shift in the phosphorelay equilibrium from Sln1 to Ypd1 but no change in the rate of hydrolysis of the aspartyl phosphate on the Sln1 receiver domain. Consistent with the increase in expression of *SLN1–SKN7*-dependent genes observed in the *sln1** P1148S mutant, *in vitro* phosphorelay assays revealed a twofold increase in accumulation of phosphorylated Skn7 with the mutant Sln1* receiver domain as the phosphodonor compared to the wild-type receiver (Fig. 16.5B) (Ault *et al.*, 2002).

The *sln1** T550I activating mutation located just upstream of H576 led to the identification of a functional coiled-coil (CC) domain in the linker region of Sln1 between the extracellular domain and the kinase domain (Fig. 16.1B). This region was recognizable as a coiled-coil using the LearnCoil algorithm (Singh *et al.*, 1998) that was developed to detect weak coiled-coils thought to be a general feature of histidine kinases occurring just upstream of the H box. The impact of the *sln1* T550I mutation and of the coiled-coil domain was studied using a functional kinase domain construct consisting of aa 537–950. Derivatives of that construct that lack aa 537–570 showed no evidence of autophosphorylation at 40 min, a period of time sufficient for autophosphorylation of a coiled-coil containing construct to plateau (Tao *et al.*, 2002). At 3 h, the mutant was phosphorylated to about 20% the extent of wild-type kinase (Ault, 2001). These observations are consistent with the inviable phenotype of the *sln1*ΔCC mutant (Tao *et al.*, 2002). The activation phenotype of the *sln1* T550I mutant correlates with the increased hydrophobicity afforded by the isoleucine residue at position 550. Additional mutations in the region that increased hydrophobicity at the "a" and "d" positions of the helix were likewise activating while mutations at other positions had no effect (Tao *et al.*, 2002). The absence of an effect of CC domain activating mutations on Sln1 homodimerization led to the hypothesis that the rotational flexibility of the coiled-coil in Sln1 is important for normal Sln1 activity (Tao *et al.*, 2002) (Fig. 16.1B). Consistent with this hypothesis, the coiled-coil domain of the strongly dimerizing mammalian C/EBP transcription factor rescued viability of the inviable *sln1ΔCC* mutant (Tao *et al.*, 2002), but like *sln1** mutants, this chimeric *SLN1* allele caused a salt-sensitive phenotype reflecting a defect in the ability of the kinase to be downregulated in response to osmotic stress (Tao *et al.*, 2002).

Although the isolation of activating mutations in *SLN1* suggest that Sln1 kinase activity can be stimulated, it has been difficult to define the environmental conditions that trigger an increase in Sln1 kinase activity. The study of *fps1* mutants led to the hypothesis that Sln1 kinase activity and/or signaling is stimulated by osmotic imbalance (Tao *et al.*, 1999). The *FPS1* gene encodes the major glycerol channel responsible for glycerol efflux (Tao *et al.*, 1999) and *fps1* mutants accumulate intracellular glycerol even in the absence of osmotic stress (Luyten *et al.*, 1995). Interestingly, although SLN1 pathway activity is elevated in the *fps1* mutant, transient osmotic imbalance caused, for example, by shifting an *FPS1*$^+$ strain from high to low osmotic environments does not activate the SLN1 pathway (Li, 2001; Shankarnarayan, 2007). Efforts to clarify the difference in physiology between transient and prolonged hypoosmotic stress led to the identification of a cell wall protein, encoded by the *CCW12* gene. Strains lacking the *CCW12* gene exhibit constitutive activation of the Sln1 kinase as do strains in which the Ccw12 protein is enzymatically removed from the wall (Shankarnarayan *et al.*, 2008). Interestingly, the *fps1* mutant is deficient in wall-associated Ccw12 protein suggesting that the activity of the Sln1 kinase is regulated by aspects of the environment that trigger specific wall perturbations (Shankarnarayan *et al.*, 2008).

Increased Sln1 kinase activity causes changes in gene expression via changes in the activity of the Skn7 response regulator. Skn7 is a transcription factor that binds to Skn7 response elements regardless of its phosphorylation state, but stimulates expression of *SLN1*-dependent genes only in response to phosphorylation of D427 in the Skn7 receiver domain. The nonphosphorylatable *skn7D427N* mutant fails to respond to elevated Sln1 signaling and the constitutive *skn7D427E* mutant increases expression of *SLN1*-dependent genes independent of *SLN1* (Li *et al.*, 1998, 2002).

The branched architecture of the pathway (Fig. 16.1A) in which a single histidine kinase signals to the Skn7 as well as the Ssk1 response regulator was suggested by genetic analysis and confirmed by biochemical analysis (Ketela *et al.*, 1998; Li *et al.*, 1998). Sln1 and Ypd1-dependent phosphorylation of the Skn7 response regulator was confirmed by reconstitution of the phosphorelay pathway using purified recombinant proteins consisting of single domains of each protein (Ault *et al.*, 2002; Li *et al.*, 1998). The radiolabeled phosphoryl group on the Sln1 kinase domain could be distributed in turn to the Sln1 receiver domain, the Ypd1 phosphorelay protein and to the receiver domain of Skn7. Furthermore, the distribution of the phosphoryl groups to each of these domains was dependent on the presence of domains participating in earlier steps in the pathway and on the phosphoaccepting residues in each domain (Li *et al.*, 1998). *In vitro* reconstitution of the Sln1 phosphorelay pathway allowed for biochemical characterization of *sln1* and *ypd1* mutants (Ault *et al.*, 2002; Janiak-Spens and West, 2000; Janiak-Spens *et al.*, 2000, 2005; Tao *et al.*, 2002) and for measurements of the phosphorylated lifetime of receiver domains (Janiak-Spens *et al.*, 1999, 2000). It also paved the way for structural analysis of pathway components (Xu and West, 1999; Xu *et al.*, 2003; Zhao *et al.*, 2008). In this chapter we describe genetic and biochemical methods for analysis of SLN1 pathway activity in response to relevant mutational and environmental factors.

2. Materials and Methods

2.1. *In vivo* methods to investigate Sln1 pathway activation

2.1.1. EMS mutagenesis and deletion collection screen for pathway activating mutants

The screen in which the *sln1** and *fps1* mutants were isolated involved random mutagenesis of the genome using the chemical mutagen, EMS, to identify mutations that elevate expression of a *SLN1*-dependent *lacZ* reporter gene (Tao *et al.*, 1999; Yu *et al.*, 1995). EMS mutagenesis is conducted according to the method of Lawrence (1991) with minor modifications. Grow a 20-ml culture of the starting strain[1] to saturation in dropout media to select for the continued presence of the plasmid carrying the *lacZ* reporter gene. Wash the cells twice with potassium phosphate buffer (0.0195 *M* KH_2PO_4, 0.0305 *M* K_2HPO_4, pH 7.0) and resuspend them at 1×10^8 cells/ml in the same buffer. Divide the cells into 5 ml aliquots and add 180 μl EMS (Sigma) to each. Incubate the treated cells at 30 °C for various times, stopping each reaction by addition of an equal volume of 10% sodium thiosulfate. Collect the cells by centrifugation, wash them twice with water, and resuspend the cells at a concentration of

[1] JF1567, the strain used in Tao *et al.* (1999), was transformed with a plasmid bearing a *SLN1* responsive *lacZ* reporter gene. JF1567 has recessive canavanine and cycloheximide resistance mutations (can^R and cyh^R) useful in the later screening steps of this protocol.

1×10^8 cells/ml. Perform serial dilutions and spread various dilutions on selective plates to give a projected yield of 300–400 cells/plate.

Select the EMS treatment time resulting in ~65% killing and spread these cells on 150–200 selective plates to generate between 50,000 and 100,000 colonies for screening.[2] Replica-plate colonies to selective media containing X-Gal.[3] Approximately 1% of colonies screened on X-Gal plates under these conditions exhibit a blue phenotype on X-Gal plates consistent with elevated expression of the *lacZ* reporter gene (Tao *et al.*, 1999). *KEX2* mutations account for more than 50% of the mutants in this type of screen (Tao *et al.*, 1999). Mutations in *KEX2* (encoding a serine protease) cause a nonspecific increase in the color of reporter bearing strains on X-Gal plates. Eliminate *kex2* mutants from further analysis by introducing a *KEX2* plasmid into the candidates via mass mating with a karyogamy deficient strain (*kar1*) of the opposite mating type (Georgieva and Rothstein, 2002). Using the recessive drug resistance alleles present in the starting strain, cytoductants retaining the desired nucleus that have acquired the (*LEU2* marked) *KEX2* plasmid can be selected.[4] Select only those candidates whose blue phenotype is not complemented by the *KEX2* plasmid. Tests of the remaining mutants to ensure that the phenotype is specific for *SLN1*-dependent reporter genes are described later in this section.

Because mutations in *KEX2* represent such a large fraction of the X-Gal positive mutants isolated by mutagenesis, it is useful to repeat the screen by examining individual mutants of the haploid deletion collection (Research Genetics). Conducting both types of screens is especially important when the deletion is lethal or when the deletion phenotype is not the same as the phenotype of point mutants. For example, deletion of the *SLN1* gene is lethal and would not be recovered in the deletion screen, but several different *sln1** mutations were identified in the EMS screen. The deletion collection is transformed with a *SLN1* responsive *lacZ* reporter plasmid such as P_{OCH1}-*lacZ* plasmid (pZL1320; Li *et al.*, 2002) using an adaptation of the microscale protocol of the Frozen-EZ yeast transformation kit (Zymo Research). Inoculate wells of a microtiter dish containing 0.25 ml of YPD medium plus 50 mg/l G418 using a sterile 96 pronged device. Incubate the microtiter dishes at 30 °C for 24–48 h without aeration. Remove the media using a multichannel pipettor being careful not to disturb the cells that have settled to the bottom of the wells. Wash cells in 0.1 ml EZ1 solution and

[2] A final plating density of between 300 and 400 colonies per plate is best for subsequent screening on X-Gal plates.

[3] X-Gal plates are prepared by addition of 1× M9 salts (Miller, 1972) and 50 mg/l X-Gal dissolved in *N,N*-dimethylformamide to standard synthetic complete (SC or SC-aa) media. X-Gal is added to media cooled to 65 °C immediately prior to pouring. The X-Gal concentration can be increased or decreased but should be titrated such that the starting strain is white or very pale blue.

[4] Selection is on SC media plates lacking the amino acid encoded by the *KEX2* plasmid and containing the appropriate drugs.

then pellet the cells using a microtiter plate microfuge at 1300 rpm for 5 min. Resuspend the pellet in 10 μl of EZ2 and 0.1 ml of EZ3 to which the plasmid DNA (5–6 μg/well) has been added. Incubate the dishes at 30 °C for 90 min and spread cells over one-fourth of a selective (SC-Ura) plate using a small sterile glass spreader. Incubate the plates several days at 30 °C until transformants became visible. An alternative approach in which pools of deletion strains are transformed can also be employed. In this approach, individual deletions are grown in the wells of a microtiter dish as before. However, prior to transformation, all the strains in a single microtiter dish are transferred using a sterile 96-prong device into a large Petri dish containing YPD. Use the Frozen-EZ yeast transformation kit (Zymo Research) to prepare competent cells and transform them with a *SLN1* responsive *lacZ* reporter plasmid. Spread the transformation mix on selective media.

Replica plate transformants from the deletion collection obtained using either method to X-Gal plates. Compare the color of transformants on X-Gal plates to the parental strain corresponding to the deletion collection in use. For example, transformants of strain BY4742 (Kelly *et al.*, 2001), the parent strain of the MATα haploid deletion collection, carrying the *SLN1* responsive *lacZ* reporter, are white on plates containing X-Gal (50 mg/l). Transformants that are blue on X-Gal plates should be chosen for further analysis. Conduct liquid β-galactosidase assays to confirm the phenotype and to establish the magnitude of the effect.

Quantitative β-galactosidase assays are performed using glass bead-lysed extracts. Grow a small culture in synthetic complete dropout media (e.g., SC-uracil to select for the presence of plasmids with a *URA3* marker) to saturation. Make a 1:50 dilution into 5 ml of fresh media and grow to a density of 1–2$\times 10^7$ cells/ml. Collect the cells by centrifugation and resuspend the pellets in 0.2 ml breaking buffer (0.1 *M* Tris, pH 8.0, 20% glycerol, 1 m*M* DTT; stored at −20 °C) plus 13 μl of 40 m*M* PMSF (PMSF is prepared in 95% ethanol and stored at −20 °C). Transfer the cell suspension to a 1.5-ml microtube and add 200 mg of glass beads. Store frozen at −80 °C for at least 15 min.[5] Thaw cells on ice, vortex briefly to mix and then alternate vortexing (15 s) with incubation on ice (15 s) between 8 and 15 times until cells are broken.[6] Pellet cell debris by microfugation on high at 4 °C for 15 min. Keep extracts on ice while measuring protein concentration using a Bradford protein assay (Biorad). The β-gal assay is conducted as described by Miller (1972, pp. 352–359) with minor modifications. Add 900–1000 ml of Z buffer (0.06 *M* $Na_2HPO_4 \cdot 7H_2O$), 0.04 *M* $NaH_2PO_4(H_2O)$, 0.01 *M* KCl, 0.001 *M* $MgSO_4(7H_2O)$, pH 7.0, with 0.05 *M*

[5] Incubation for 15 min at −80 °C facilitates breakage of the cells.

[6] The exact number of vortexing cycles will depend on the vortex and the strain. Cell breakage can be monitored by phase contrast microscopy. Ideally, cell breakage should exceed 80%.

βME added fresh for each use) to reaction tubes and add protein extract to generate 1 ml reaction mixtures. Incubate these tubes at 28 °C. Add 200 μl ONPG (4 mg/ml prepared in Z buffer, stored at 4 °C and prepared fresh weekly) to each tube and vortex briefly to begin the reaction, noting the time. When the reaction is just visibly yellow,[7] add 500 μl of 1 *M* Na_2CO_3 to each tube to stop the reaction. Note the time at which each reaction is stopped. Read the A_{420}. Calculate the activity using the formula: $A_{420}\times 378$/time (min)$\times$volume (ml)$\times$protein (mg/ml). Activities are reported as the average of 4–6 assay values using at least three independent transformants.

The gene of interest in deletion candidates chosen for their blue phenotype from among transformants of pooled deletion strains is identified by sequencing. Prepare genomic DNA and amplify the unique tag sequence built into each deletion cassette using a forward primer complementary to the Up Tag sequence (5′-CTACAGGTGCTCCAGAGA-3′) (http://chemogenomics.stanford.edu/supplements/04tag/protocols.html) and the Kan 250R reverse primer complementary to sequences within the KanMX4 cassette (5′-CTGCAGCGAGGAGCCGTAAT-3′). Examine the PCR product by agarose gel electrophoresis and purify it using the QIAquick Gel Extraction Kit (Qiagen). Obtain the sequence using the Kan 200 R primer (5′-CGTGCGGCCATCAAAATG-3′). The deletion tag sequence is located just downstream of the forward primer and upstream of the Kan MX4 sequence and is identified using the tag database (http://www-sequence.stanford.edu/group/yeast_deletion_project/downloads.html).

The specificity of the elevated reporter gene expression is examined by comparing the effect of the strain background on the activity of the *SLN1*-dependent pZL1320 (P_{OCH1}-*lacZ*) reporter versus the activity of control reporters such as pSL1156 in which *lacZ* is driven by the basal (− 154 to − 1) *OCH1* promoter which lacks *SLN1–SKN7* response elements and pZL1369 (Li *et al.*, 2002) which contains a single mutated Skn7 binding site.

2.1.2. Activation of the SLN1 pathway with zymolyase

In addition to pathway activation in *sln1**, *fps1* and *ccw12* genetic backgrounds, the SLN1 pathway can be activated with the cell wall digesting enzyme zymolyase. Zymolyase is a complex of enzymes isolated from *Arthrobacter luteus* with strong lytic activity for yeast cell walls (U.S. Biological 10^5 units/g). The complex consists of zymolyase A, β-1,3-glucan laminaripentaohydrolase and zymolyase B, an alkaline protease thought to be required for the action of zymolyase A by changing the structure of the

[7] The yellow color should be monitored closely and the reaction stopped promptly so that the A_{420} reading remains within the linear range of the spectrophotometer. Reaction times should not be less than 5 min, as these times will tend to be inaccurate. If necessary, repeat the reaction with less extract.

wall. Zymolyase is routinely used at high concentration to digest the walls from yeast cells. At high enzyme concentration cells must be resuspended in sorbitol to provide osmotic support for the fragile spheroplasts that result from removal of the cell wall. However, the very low zymolyase dose of 0.5–1 unit/ml that is sufficient for activation of the SLN1 pathway causes no loss in optical density or viability of cells. Grow cells to saturation in YPD[8] and dilute 1:50 in fresh media. Incubate cells at 30 °C with aeration until they reach early log phase ($\sim 10^7$ cells/ml). Adjust the pH of all cultures by adding Na_2HPO_4 to a final concentration of 20 m*M* (Reiser *et al.*, 2003). Add zymolyase to 0.5 or 1 unit/ml and incubate cultures at 30 °C for 30–60 min. Harvest cells by centrifugation and store pellets at −80 °C until needed.

2.1.3. SLN1 pathway activation assays

2.1.3.1. Northern analysis of SLN1-dependent genes The effect of zymolyase treatment on SLN1 pathway activity can be examined by Northern analysis. Grow small cultures to saturation and make a 1:50 dilution into 10–50 ml of fresh medium. Incubate at 30 °C with aeration until early log phase ($\sim 10^7$ cells/ml). Harvest cells by centrifugation and store pellets at −80 °C. Isolate total RNA using the acid phenol method (Ausubel *et al.*, 1989) with minor modifications. Resuspend the pellet in 0.4 ml TES (10 m*M* Tris, pH 7.5, 10 m*M* EDTA, 0.5% SDS) and add 0.4 ml acid phenol. Vortex 10 s. Incubate at 65 °C for 1 h, vortexing 5 s every 10 min. Move the tubes to ice and add 0.1 ml of TE (10 m*M* Tris, 1 m*M* EDTA), pH 7.5. Microfuge at high speed for 5 min and transfer the upper phase to a fresh microtube on ice. Add 0.4 ml hot acid phenol and vortex 20 s. Microfuge for 5 min and transfer the upper phase to new tubes on ice. Add 0.4 ml chloroform and vortex for 20 s. Microfuge again for 5 min and transfer the upper phase to new tubes. Add 1/10 volume of 3 *M* NaOAc and 1.0 ml of cold ethanol and allow the RNA to precipitate at −80 °C overnight. Microfuge for 5 min, aspirate, wash the pellet in 75% ethanol, repeat the spin, and aspirate the ethanol. Dry the pellet and resuspend the RNA in 100 μl RNase-free water by incubating at 65 °C for 5 min and vortexing. Store at −80 °C until needed. Thaw the samples on ice. Ensure complete resuspension of the samples by incubating samples at 65 °C for 2 min. Use 1–2 μl to measure the absorbance at 260 nm. An OD of 1 is equivalent to 40 μg/ml RNA. Ten milliliters of cells at 1×10^7 ml^{-1} yields ~250 μg of RNA. Check the quality of the RNA by running a sample on a standard (1× TAE) 1% agarose gel. Sharp ribosomal RNA bands of 1.8 and 2.5 kb and a more diffuse tRNA band at 0.2 kb should be visible. Samples

[8] Where plasmid retention is an issue, grow cultures in selective medium overnight and then subculture (1:50) into YPD for two generations prior to harvesting at log phase.

with reduced amounts of one rRNA species relative to the other, general smeariness or an accumulation of smaller molecular weight species should not be used further. Mix 10–20 μg of each total RNA sample with 1.3×RNA loading buffer (1× MOPS (0.2 *M* MOPS, 0.5 *M* sodium acetate, 0.01 *M* EDTA), 20% formaldehyde, 50% deionized formamide, 0.04 mg/ml ethidium bromide plus xylene cyanol, and bromphenol blue loading dyes), heat samples at 65 °C for 10 min, move to ice and load a 10% (v/v) formaldehyde 1% agarose gel prepared in 1× MOPS buffer. Run the gel at 75 V for several hours. Photograph the gel, wash it twice 5 min each in water and then set up a capillary blot (Ausubel *et al.*, 1989) to transfer the RNA to a nytran filter. UV crosslink the RNA to the blot by exposing the blot (RNA side up) to a germicidal lamp.[9] Prepare a ^{32}P labeled probe using random primers and the Prime-It random priming labeling kit (Stratagene) with appropriate PCR fragments as templates.[10] Store the probe at −20 °C. Prehybridize the blot in a small rotisserie bottle (Bellco Glass Inc.) using 4–6 ml of PerfectHyb (Sigma) hybridization buffer rotating for at least 1 h at 68 °C. Incubate the probe at 100 °C for 10 min, and then chill it on ice. Add the probe (~10 μl) to the rotisserie bottle and hybridize overnight at 68 °C. Wash the blot twice, 5 min each, in 30 ml of 2× SSC–0.1% SDS.[11] Wash the blot twice more in 0.5× SSC–0.1% SDS at 68 °C. Remove the blot from the rotisserie bottle, wrap it (RNA side down) in plastic film and place it in a phosphorimager cassette (or X-ray film) for 6–30 h. Visualize radiolabeled bands with a phosphorimager. Analyze bands using Quantity One software (Biorad). Strip the probe from the blot by incubating the blot in 0.2× SSC–1% SDS at 75 °C for 1 h followed by a rinse in 10 m*M* Tris (pH 7.5). Repeat prehybridization and hybridization procedures using a probe suitable for normalization.[12]

2.1.3.2. Osmotic activation of the HOG1 pathway SLN1 pathway activation leads to curtailment of HOG1 pathway signaling in response to salt (Fassler *et al.*, 1997). Hence, another assay for SLN1 pathway activation is to examine the effect of osmotic shock on Hog1 phosphorylation over time. To monitor Hog1 phosphorylation, grow cultures selectively overnight, dilute 1:50 into YPD and aerate at 30 °C to a cell density of 1×10^7 ml^{-1}.

[9] The appropriate time of exposure for optimal retention of RNA on the blot will depend on the age of the bulb and the distance from the bulb to the blot and should be determined empirically.

[10] Expression of the *SLN1* responsive *NCA3* gene is frequently used to monitor SLN1-SKN7 pathway expression because it is more abundant than *OCH1*. The *NCA3* probe is a 389 bp PCR product generated using NCA3 9F (5′-TTCCGCAGCTTTAATATTGTCTTCC-3′) and NCA3 398R (5′-TCACATGCATAAGAACAGTAGTAGCC-3′).

[11] SSC is diluted from a 20× stock (3 *M* NaCl, 0.3 *M* sodium citrate, pH 7).

[12] Normalization in zymolyase treatment experiments is performed using a probe corresponding to *CDC33*. The *CDC33* probe is a 636-bp PCR product generated using CDC33 3F (5′-GTCCGTTGAAGAAGTTAGCAAG-3′) and CDC33 639R (5′-CAAGGTGATTGATGGTTGAGGG-3′).

Use a 4*M* solution of NaCl prepared in YPD to adjust cultures to a final concentration of 0.4 *M* NaCl. Take 10 ml aliquots at 0, 2, 4, and 10 min after addition of NaCl. Harvest cells by filtration through a Millipore 0.45 μm HA filter (using full house vacuum). Quickly place filters into a microtube and rapidly freeze in a dry-ice ethanol bath. Store samples at −80 °C until preparation of the extract. Remove cells from filter into 1 ml of water and pellet the cells by microfugation. Remove the filter. To the pellet, add 100 μl 16% TCA and ~200 mg of glass beads. Vortex approximately 4–5 min. Add 400 μl of ice cold 5% TCA and vortex to mix. Transfer liquid to a fresh microtube and microfuge at high speed in the cold for 10 min. Aspirate the supernatant and resuspend the pellet in 50–100 μl loading buffer (150 m*M* Tris Base, 10 m*M* EDTA, 3% SDS, 5% βME, 15% glycerol, 0.01 mg/ml bromphenol blue). Boil samples for 5 min, microfuge samples at high speed for 5 min and load (~15 μl) onto two identical 10% SDS–PAGE gels (29:1 acrylamide:bisacrylamide, 10% running gel, 6% stacking gel). Transfer proteins to nitrocellulose membranes (0.45 μm; Schleicher and Schuell). Stain with Ponceau S staining solution (0.1% (w/v) Ponceau S in 5% (v/v) acetic acid), wash briefly in distilled water and scan or photograph for a record of relative loading of each lane and the quality of the protein extracts. Block the blot with TBST (10 m*M* Tris, pH 8.0, 150 m*M* NaCl, 0.05% Tween 20) and 4% BSA. Wash twice for 10 min in TBST and incubate for 2 h at 4 °C with RC20 recombinant antiphosphotyrosine antibody conjugated to horseradish peroxidase (Transduction Laboratories) diluted 1:2500 in TBST + 4% BSA. Wash twice with TBST. For Hog1 protein analysis, block the membrane in TBST with 5% skim milk for 1 h. Incubate with yC-20 anti-Hog1 antibody (Santa Cruz Biotechnology) diluted 1:2000 in TBST + 5% BSA overnight at 4 °C. Wash twice in TBST and then incubate with peroxidase conjugated bovine antigoat IgG (Santacruz Biotech sc2378) diluted 1:1000 in TBST. Wash the blot three times with 10 ml of TBST for 5 min each. Treat both blots with SuperSignal West Pico chemiluminescent substrate (Pierce) and expose to film to visualize the proteins.

2.2. *In vitro* Sln1 kinase and Sln1 pathway phosphorelay assays

2.2.1. Expression and purification of SLN1 pathway components from *E. coli*

2.2.1.1. GST-Sln1-HK and GST-Sln1-HK-Rec domains The Sln1 histidine kinase (HK) domain (aa 537–950) is expressed as a glutathione-*S*-transferase fusion protein in DH5α cells and is routinely used as the initial phosphodonor to downstream phosphorelay components. This protocol is a slightly modified version of the published procedure (Ault *et al.*, 2002; Li *et al.*, 1998). Grow 1 l of cells in LB medium containing 100 μg/ml ampicillin at

37 °C until late log phase (an optical density at 600 nm of ~0.9). Add IPTG to a final concentration of 0.2 m*M* and grow cells at a lower temperature (16 °C) for an additional 24 h Pellet cells by centrifuging at 4500×*g* for 10 min and resuspend in 4–5 ml/g (wet weight of cell pellet) in lysis buffer (50 m*M* Tris–HCl, pH 8.0, 150 m*M* NaCl, 1 m*M* EDTA, 1% Triton X-100, 0.1% *β*ME, 1 m*M* PMSF) supplemented with a 1× protease inhibitor cocktail[13] (0.1 *μ*g/ml chymostatin, 2 *μ*g/ml aprotinin, 1 *μ*g/ml pepstatin, 1.1 *μ*g/ml phosphoramidon, 7.2 *μ*g/ml E-64 protease inhibitor, 0.5 *μ*g/ml leupeptin, 2.5 *μ*g/ml antipain, 100 *μM* benzamidine, 100 *μM* sodium metabisulfite). If desired, the cell paste can be stored frozen at −20 °C. Lyse (thawed) cells by sonication[14] and clarify the cell lysate by centrifugation at 27,200×*g* for 30 min. Apply the supernatant to a 2-ml glutathione-Sepharaose 4B (GE Healthcare) column[15] that has been preequilibrated in lysis buffer. Wash sequentially with 10 column volumes of lysis buffer, then 5 column volumes of wash buffer (50 m*M* Tris–HCl, pH 8.0, 2 m*M* DTT, 1 m*M* EDTA), and finally with 5 column volumes of storage buffer[16] (50 m*M* Tris–HCl, pH 8.0, 100 m*M* KCl, 2 m*M* DTT, 1 m*M* EDTA, 10% glycerol). Leave about 2 ml of storage buffer above the resin bed and gently mix the slurry. Bead-bound Sln1-HK protein can be stored in aliquots at −20 °C and is fully active for subsequent autophosphorylation and phosphotransfer experiments. By retaining GST-Sln1-HK on the glutathione-Sepharose beads, one can conveniently remove this phosphodonor from phosphorecipient proteins by low-speed centrifugation.[17]

Another plasmid was constructed that expresses the Sln1 HK domain as well as the C-terminal receiver domain, Sln1-HK-Rec (aa 537–1220) (Ault, 2001; Janiak-Spens *et al.*, 2005). Purification is the same as described above for the Sln1-HK domain and is useful in preparing phosphorylated Ypd1 (see Section 2.2.2.3). (*Note*: the GST-SLN1-HK-Rec protein is susceptible to proteolytic degradation, thus it is important to include protease inhibitor cocktails for all purification and storage conditions.)

[13] Any commercial PIC cocktail appropriate for bacterial cells can be substituted here.

[14] Lysis can be achieved using a French press in place of sonication.

[15] Alternatively, batch purification method can work as well using a 50% slurry as described for full-length Skn7 (Section 2.2.1.1).

[16] GST-Sln1-HK can also be stored in reaction buffer (50 m*M* Tris–HCl, pH 8.0, 100 m*M* KCl, 10 m*M* $MgCl_2$, 2 m*M* DTT with the addition of 10–50% glycerol).

[17] For experiments in which the presence of GST might complicate analysis, use thrombin to cleave the protein from the GST moiety. Wash the bead bound protein four times in thrombin cleavage buffer (50 m*M* Tris, 150 m*M* NaCl, 10 m*M* $CaCl_2$), resuspend in the same buffer and add 5 units of biotinylated thrombin. Incubate at 25 °C for times from 1 to 6 h depending on the protein. Add 25 *μ*l of streptavidin beads during the final 15 min of the reaction. Recover the protein by transferring the reactions to microfuge filtration units and centrifuging briefly.

2.2.1.2. Full-length Ssk1 and Skn7 Full-length Ssk1 and Skn7 are expressed as GST fusions in pGEX vectors (Ault, 2001; Li *et al.*, 1998) in DH5α cells. Grow 1 l of culture to an OD_{600} between 0.6 and 1.0 in LB medium containing 100 μg/ml ampicillin. Add 0.2 m*M* IPTG.[18] Optimal induction of GST-Ssk1 protein occurs after 2 h at 37 °C and optimal induction of GST-Skn7 protein occurs after 20 h at 16 °C. The low-temperature incubation reduces the concentration of inclusion bodies and improves the yield of active protein.[19] Harvest cells by centrifugation and resuspend the pellet in 5 ml lysis buffer (50 m*M* Tris, pH 7.6, 150 m*M* NaCl, 1 m*M* EDTA, 10% glycerol, 1% Triton X-100, 0.1% βME, 1× protease inhibitor cocktail (Sigma)) and store frozen.[20] Thaw the cell pellets and add 1 m*M* PMSF. Lyse cells using a French press or by sonication.[21] Centrifuge for 10 min at 7000×*g* and to the supernatant add 3 ml of a 50% slurry of glutathione agarose beads. Incubate for 1 h at 4 °C. Wash the beads four times in lysis buffer and four times in storage buffer (50 m*M* Tris, pH 7.6, 50 m*M* KCl, 5 m*M* $MgCl_2$, 0.1% βME, 50% glycerol). Store bead-bound proteins at −20 °C. If desired, proteins can be eluted from the beads by incubation in elution buffer (100 m*M* glutathione, 50 m*M* Tris, pH 7.6, 50 m*M* KCl, 5 m*M* $MgCl_2$, 0.1% βME) with shaking at room temperature for 1 h.[22] Remove beads from the eluted protein by brief centrifugation in a microfiltration unit. Protein concentration can be determined based on analysis of band intensities on Coomassie stained SDS–polyacrylamide gels compared to stained gels with BSA concentration standards using the public domain NIH Image program (http://rsb.info.nih.gov/nih-image/).

2.2.1.3. Sln1, Ssk1, and Skn7 receiver domains The C-terminal response regulator domains from Sln1, Ssk1, and Skn7 can be expressed as cleavable tripartite fusion proteins consisting of an N-terminal receiver (Rec) domain, followed by the yeast VMA1 protein-splicing intein domain and a chitin binding domain (CBD) at the C-terminus. The CBD facilitates affinity purification, while the self-splicing activity of the intein domain can

[18] This relatively low concentration of IPTG is important in minimizing the number of truncation products in GST-Skn7 cultures. Since D427 is at the C-terminus of the protein and the GST tag at the N-terminus, truncation products are inactive.

[19] Induction of GST-Skn7 and GST-Ssk1 at 37 °C results in a low yield of full-length protein with many truncation products. GST-Skn7 that is induced at 37 °C is inactive in phosphotransfer assays. Performing the induction at 16 °C does not improve the yield of full-length protein but does improve the activity of the protein in the phosphotransfer assay (Ault *et al.*, 2002).

[20] Freezer storage can be at −20 or −80 °C.

[21] Add 1.25 mg lysozyme, 50 units of DNase (Boehringer Mannheim) and 10 μg of RNAse to lysis buffer prior to sonication. Set tubes in an ice water bath and do multiple rounds of sonication for 30 s each at 60 s intervals.

[22] A higher concentration of Skn7 is achieved by replacing batch-wise elution of GST-Skn7 with elution of the protein in a chromatography column using 50 m*M* glutathione. For a column consisting of three ml bead bound GST-Skn7, discard one ml of void volume and collect 0.5 ml aliquots for analysis by SDS–PAGE. The GST-Skn7 protein is concentrated in aliquots 3 and 4 (Ault *et al.*, 2002).

be induced at high DTT or βME concentrations to cleave the polypeptide at the junction between the receiver and intein domains. All protein purification steps described below should be carried out at 4 °C.

The Sln1 receiver domain (aa 1084–1220) is expressed from an inducible *tac* promoter (pCYB2 vector, IMPACT system, New England Biolabs) (Janiak-Spens *et al.*, 1999) in DH5α cells. Grow 1 l of cells in LB medium containing 100 μg/ml ampicillin at 37 °C until an optical density at 600 nm reaches 0.6. Add IPTG to a final concentration of 0.4 m*M* and grow cells an additional 8 h at room temperature. Harvest cells by centrifugation at 4500×*g* for 10 min and resuspend the cell pellet in wash buffer (0.1 *M* sodium phosphate, pH 7.0, 1 m*M* EDTA). Centrifuge again and resuspend the cells in lysis buffer (20 m*M* Tris–HCl, pH 7.5, 500 m*M* NaCl, 1 m*M* EDTA, 0.1% Triton X-100) at ~5 ml/g wet weight of the cell pellet. Lyse cells by sonication and clarify the cell lysate by centrifugation at 27,200×*g* for 30 min. Load the supernatant onto a 3-ml chitin bead column (New England Biolabs) equilibrated in lysis buffer at 4 °C using a peristaltic pump at a flow rate of about 0.5 ml/min. Wash with 20 column volumes of lysis buffer at a flow rate of about 1 ml/min. Then wash sequentially with 4–5 column volumes each of cleavage buffer (20 m*M* Tris–HCl, pH 7.5, 50 m*M* NaCl, 0.1 m*M* EDTA), cleavage buffer plus 5 m*M* ATP, 5 m*M* $MgCl_2$, then cleavage buffer plus 30 m*M* βME. Turn off column flow and incubate overnight at 4 °C to allow thiol-induced cleavage of the fusion protein. Elute with 3–4 column volumes of cleavage buffer and identify Sln1-Rec containing fractions by 15% SDS–polyacrylamide gel electrophoresis (SDS–PAGE). Pool fractions containing Sln1-Rec domain and apply to a Sephadex G50 (Sigma) gel filtration column (300 ml bed volume) equilibrated in 20 m*M* Tris–HCl, pH 7.5, 50 m*M* NaCl, 1 m*M* EDTA, 1 m*M* DTT. Protein concentration can be determined by absorbance at 280 nm ($\varepsilon = 4020\ M^{-1}\ cm^{-1}$). The purified protein can be stored frozen (−80 °C) in small aliquots containing 10% glycerol. Typical yields are 0.5–1.0 mg pure protein per liter cell culture.

Expression of the Ssk1 receiver domain (aa 495–712) from the pCYB2 vector in DH5α cells was low. Thus, we excised the Ssk1-Rec-intein-CBD fusion gene from the pCYB2 vector and subcloned the fragment into pET11a (Novagen) to give an inducible T7-based vector as described (Janiak-Spens *et al.*, 1999). The expression plasmid was transformed into BL21(DE3) cells. Grow 1 l of cells in LB medium containing 100 μg/ml ampicillin at 37 °C until an optical density at 600 nm of about 0.7. Cool the culture to 16 °C, then add IPTG to a final concentration of 1 m*M* and grow cells an additional 16 h at 16 °C. Harvest cells by centrifuging at 4500×*g* for 10 min and resuspend the cell pellet in wash buffer. Centrifuge again and resuspend the cells in lysis buffer containing 10% glycerol at ~6 ml/g wet weight of the cell pellet. A pinch of RNase (Sigma) and DNase (Sigma) is added, then lyse cells by French press operated at a cell pressure of

14,000 lb/in.^{2}. Clarify the cell lysate by centrifugation at 27,200×*g* for 30 min. Load the supernatant onto a 3-ml chitin bead column (New England Biolabs) equilibrated in lysis buffer containing 10% glycerol. Chitin affinity column chromatography is carried out as described above for the Sln1-Rec domain except that 10% glycerol is included in the cleavage buffer and Sephadex G75 resin (Sigma; 300 ml bed volume) equilibrated in 20 m*M* Tris–HCl, pH 8.0, 100 m*M* NaCl, 1 m*M* EDTA, 10% glycerol, 1 m*M* βME is used for the gel filtration chromatography step. Protein concentration can be determined by absorbance at 280 nm (ε = 25,440 M^{-1} cm^{-1}). The purified protein can be stored frozen (−20 °C) in small aliquots containing 10% glycerol. Typical yields are 1.0–1.5 mg pure protein per liter cell culture.

The vector expressing the Skn7 receiver domain (aa 361–622) was prepared as described above for the Ssk1-Rec domain (Janiak-Spens and West, 2000) and transformed into BL21(DE3) cells. Purification of the Skn7-Rec domain is similar to the Sln1-Rec domain as described above except an S-200 gel filtration column (300 ml bed volume; GE Healthcare) is used, equilibrated in 20 m*M* Tris–HCl, pH 8.0, 100 m*M* NaCl, 1 m*M* βME. Protein concentration can be determined by absorbance at 280 nm (ε = 6580 M^{-1} cm^{-1}). The purified protein can be stored frozen (−20 °C) in small aliquots containing 10% glycerol. Typical yields are 1–3 mg pure protein per liter cell culture.

2.2.1.4. Ypd1 phosphorelay protein The full-length Ypd1 protein (aa 1–167) can be constitutively expressed from a pUC12-derived vector in DH5α cells (Xu *et al.*, 1999). Grow a 1-l culture to saturation overnight at 37 °C in LB medium supplemented with 50 μg/ml ampicillin. Harvest cells by centrifuging at 4500×*g* for 10 min and resuspend the cell pellet in lysis buffer (0.1 *M* sodium phosphate, pH 7.0, 1 m*M* EDTA, 1.4 m*M* βME). Centrifuge again and resuspend the cells in lysis buffer at ~5 ml/g wet weight of the cell pellet. Lyse cells by sonication and clarify the cell lysate by centrifugation at 27,200×*g* for 30 min. Measure the volume of the supernatant. In a small glass beaker at 4 °C on a stir plate, while gently stirring, slowly add saturated ammonium sulfate to a final concentration of 55%. Collect the protein precipitate by centrifuging at 12,000×*g* for 10 min. Resuspend the protein pellet in 6–7 ml of dialysis buffer (20 m*M* BisTris, pH 6.5, 1.4 m*M* βME) and dialyze versus 2 l of the same buffer for 24 h with one change of buffer to remove the ammonium sulfate. Recover the protein solution from the dialysis bag and pellet any precipitated material. Filter the supernatant through a 0.2-μ*M* syringe filter and apply to a 5-ml HiTrap Q Sepharose Fast Flow (Pharmacia) anion-exchange column preequilibrated in 50 m*M* Bis–Tris, pH 6.5. Elute bound protein from the column with a linear salt gradient of 0–500 m*M* NaCl in 50 m*M* Bis–Tris, pH 6.5. Identify Ypd1-containing fractions by running aliquots

(12–15 μl) from fractions on 15% SDS–polyacrylamide gels. Pool fractions containing Ypd1, concentrate using Centricon 10 filter units (Amicon) to a volume $\leq$10 ml. Apply sample to a Sephadex G50 (Sigma) gel filtration column (350 ml bed volume) equilibrated in 50 m*M* sodium phosphate, pH 7.0, 1 m*M* EDTA, 1.4 m*M* βME. Identify fractions containing pure Ypd1 protein by SDS–PAGE and combine. Protein concentration can be determined by absorbance at 280 nm (ε = 15,280 M^{-1} cm^{-1}). Ypd1 can be stored frozen (-20 or -80 °C) in small aliquots containing 15% glycerol. Yields can be as high as 40 mg pure protein per liter cell culture.

Note: When we embarked on reconstituting the phosphorelay system *in vitro*, it was desirable to express and purify these domains without His- or other affinity tags that potentially could interfere with phosphotransfer assays. We have since discovered that most of these proteins can in fact be expressed as GST or His-tagged proteins with little or no loss in phosphotransfer activity. The Sln1-Rec domain and full-length Skn7 and YPD1 can also be expressed and purified as GST-fusion proteins using the procedure described in Section 2.2.1.1 (Ault *et al.*, 2002; Li *et al.*, 1998). Full-length Skn7, Ssk1-Rec, and Skn7-Rec domains can also be expressed and purified as His-tagged proteins (F. Janiak-Spens, D. Sparling, X. Zhao, and A. H. West, unpublished) (Dean, 2004; Mulford, 2009).

2.2.2. Preparation of phosphorylated proteins and phosphotransfer assays

The phosphotransfer reactions that comprise the full phosphorelay (Sln1-HK to Sln1-Rec to Ypd1 to Ssk1-Rec or Skn7-Rec) occur very rapidly, within the earliest 8 s time point that can be taken manually at the benchtop (Janiak-Spens and West, 2000) (Fig. 16.3B). Thus, a rapid quench flow instrument is required to carry out initial velocity studies and to measure rate constants for the individual steps (Janiak-Spens *et al.*, 2005). This approach is the subject of a separate chapter in this volume (Kaserer *et al.*, 2010). Here, we provide methods and strategies for obtaining phosphorylated proteins and for *in vitro* phosphotransfer assays, in which distribution of radiolabeled phosphoryl groups can be monitored over a time course of 5–60 min. These procedures have been particularly useful for characterizing mutant proteins (Ault, 2001; Ault *et al.*, 2002; Janiak-Spens and West, 2000; Janiak-Spens *et al.*, 2005; Li *et al.*, 1998; Tao *et al.*, 2002) and for examining half-lives of phosphorylated receiver domains (Janiak-Spens *et al.*, 1999, 2000).

2.2.2.1. Autophosphorylation of GST-tagged Sln1-HK bound to glutathione-coupled resin Incubate approximately 3 μM of bead-bound protein with 7 μM [γ-^{32}P]-ATP (30 Ci/mmol) in 50 m*M* Tris–HCl, pH 8.0, 100 m*M* KCl, 10 m*M* $MgCl_2$, 2 m*M* DTT, 20% glycerol for 30–60 min at room temperature in a total reaction volume of 100 μl. Gently pellet the resin-bound GST-Sln1-HK~P (100×g for 1 min), wash three times with 100 μl wash buffer (50 mM Tris–HCl, pH 8.0, 50 mM KCl, 10 mM $MgCl_2$, 1 mM DTT) to remove unincorporated radiolabel (Fig. 16.2A). Resuspend final pellet in 100 μl wash buffer. A typical autophosphorylation time course is shown in Fig. 16.2B. The phosphorylated kinase is stable for 8 h at room temperature. This preparation can be used immediately for phosphotransfer assays (e.g., see Fig. 16.2C) or stored frozen in aliquots at −20 °C for about a week. For some applications (e.g., mutant studies of the HK domain itself), it may be useful to elute the GST-Sln1-HK protein from the resin or remove the GST tag using thrombin (Fig. 16.2A).

To examine levels of autophosphorylation or to evaluate phosphotransfer from component to component in the pathway as described in subsequent sections, radiolabled reaction products are typically separated on 10–15% SDS–polyacrylamide gels.[23] The wet gel should be wrapped in plastic wrap and exposed to a phosphorimager screen for quantification of radiolabeled band intensity.

2.2.2.2. Phosphorylation of Sln1 receiver domain for half-life studies, mutant studies or phosphotransfer assays Phosphorylation of Sln1-Rec can be achieved by direct phosphotransfer from Sln1-HK~P (Fig. 16.3A). Incubate 7 μM GST-Sln1-HK bound to glutathione-Sepharose beads (as described in Section 2.2.2.1) with 7 μM [γ-^{32}P]-ATP (30 Ci/mmol), 60 μM unlabeled ATP,[24] and 30 μM Sln1-Rec in 50 mM Tris–HCl, pH 8.0, 100 mM KCl, 15 mM $MgCl_2$, 2 mM DTT, and 20% glycerol for 30 min at room temperature in a total volume of 100 μl. Centrifuge the mixture (1 min at 100×g) to pellet the resin-bound GST-Sln1-HK and recover P~Sln1-Rec in supernatant. This preparation should be used immediately for half-life studies (see Section 2.2.2.5) or, for example, *in vitro* phosphotransfer reactions involving Ypd1 (Fig. 16.3B).

2.2.2.3. Phosphorylation of Ypd1 for half-life studies, mutant studies or phosphotransfer assays Incubate GST-Sln1-HK-Rec (1 μM) bound to glutathione-Sepharose beads (as described in Section 2.2.1.1) with 7 μM [γ-^{32}P]-ATP (30 Ci/mmol), 100 μM unlabeled ATP, and 40 μM Ypd1 in 50 mM Tris–HCl, pH 8.0, 100 mM KCl, 15 mM $MgCl_2$, 2 mM DTT, and 20% glycerol for 90 min at room temperature in a total volume of 100 μl (Fig. 16.4). Centrifuge the mixture (1 min at 100×g) to pellet the resin-bound GST-Sln1-HK-Rec and recover P~Ypd1 in the supernatant. This preparation can then be used immediately for half-life studies or

[23] Do not boil samples prior to gel loading as this increases the rate of phosphate hydrolysis.
[24] Addition of unlabeled ATP is included to drive stoichiometric labeling of Sln1-Rec.

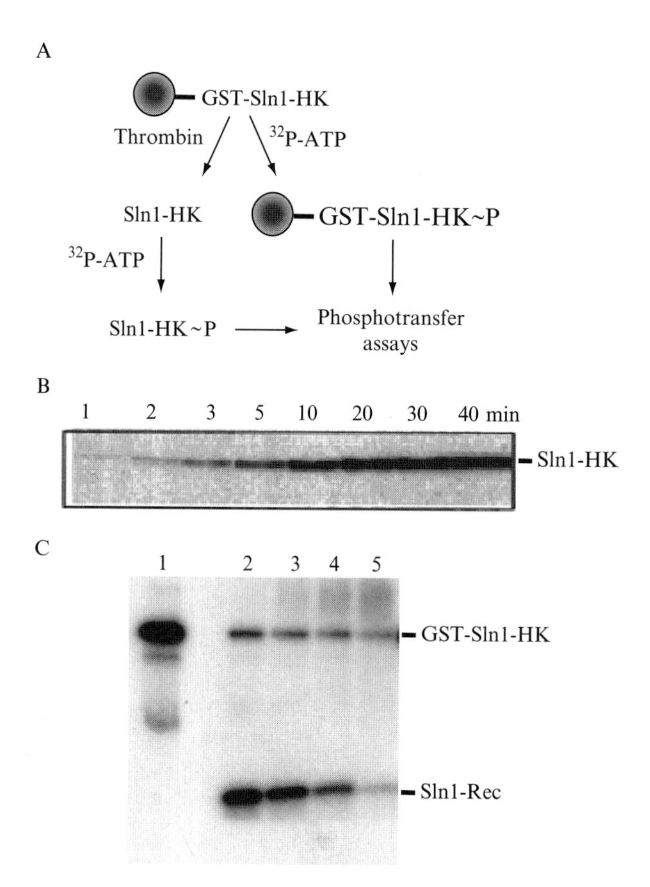

Figure 16.2 Autophosphorylation of Sln1-HK. (A) Schematic showing two alternate routes for working with resin-bound GST-Sln1-HK. The Sln1-HK domain can be cleaved from the GST tag and eluted from the resin using thrombin. The isolated Sln1-HK domain can then be autophosphorylated using radiolabeled ATP. This strategy is best for working with Sln1-HK mutants. A convenient alternative is to autophosphorylate GST-Sln1-HK while bound to resin. Bead-bound phospho-GST-Sln1-HK can then be used directly as a phosphodonor for *in vitro* phosphotransfer assays. (B) A representative autophosphorylation time-course for soluble Sln1-HK (reproduced from Tao *et al.*, 2002). Sln1-HK (2 μg) was incubated with [γ-^{32}P]ATP in reaction buffer and aliquots were removed at the times indicated and quenched in SDS loading buffer. Maximal levels of autophosphorylation are typically reached in 30–40 min. (C) An example of phosphotransfer from bead-bound GST-Sln1-HK to the Sln1-Rec domain (reproduced from Ault *et al.*, 2002). ^{32}P-labeled GST-Sln1-HK (2.5 μg) was incubated alone (lane 1) or with 1 μg Sln1-Rec for 5, 15, 30, or 60 min (lanes 2–5, respectively). Most of the radiolabel is transferred within 5 min and the gradual loss of radiolabel from Sln1-Rec is due to its intrinsic rate of dephosphorylation.

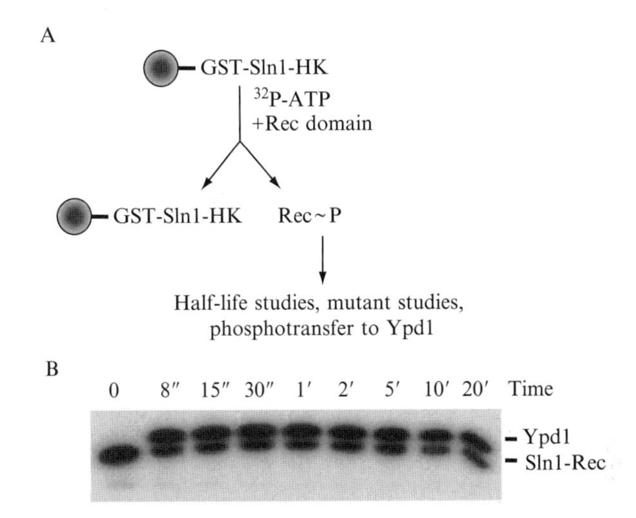

Figure 16.3 Phosphorylation of receiver domains. (A) Schematic showing strategy for coincubation of bead-bound GST-Sln1-HK and receiver domains in the presence of [γ-^{32}P]ATP for a period of time (~30 min) that allows for autophosphorylation of GST-Sln1-HK and rapid phosphotransfer to the receiver domain. The phosphorylated receiver domain can then be recovered in the supernatant by pelleting the bead-bound GST-Sln1-HK. The isolated P~Rec can then be used for various *in vitro* biochemical studies. (B) An example phosphotransfer experiment using Sln1-Rec~P as a phosphodonor to Ypd1 (reproduced from Janiak-Spens *et al.*, 2000). Isolated ^{32}P-labeled Sln1-Rec (1.6 μM) was incubated with 2 μM Ypd1 and aliquots were removed at the indicated time points. This time-course experiment showed that with respect to distribution of radiolabel for WT proteins, equilibrium is reached at the earliest time point taken (8 s). For mutant studies, it is useful to establish a time frame for optimal phosphotransfer.

phosphotransfer experiments involving receiver domains (see Ypd1 mutant study, e.g., in Fig. 16.4B).

2.2.2.4. Phosphorylation of Ssk1 and Skn7 receiver domains for half-life studies, mutant studies, or phosphotransfer assays The most efficient manner of phosphorylating the Ssk1 and Skn7 receiver domains is through the full phosphorelay (Sln1-HK to Sln1-R1 to Ypd1). However, one can phosphorylate Ssk1-Rec and Skn7-Rec directly using bead-bound Sln1-HK~P as a phosphodonor (as described for Sln1-Rec), but a longer incubation time is necessary (up to 90 min) (Fig. 16.3). This route reduces the number of phosphotransfer and intermediary purification steps. The phosphorylated receiver domain(s) should be used immediately due to its short half-life (Janiak-Spens *et al.*, 2000).

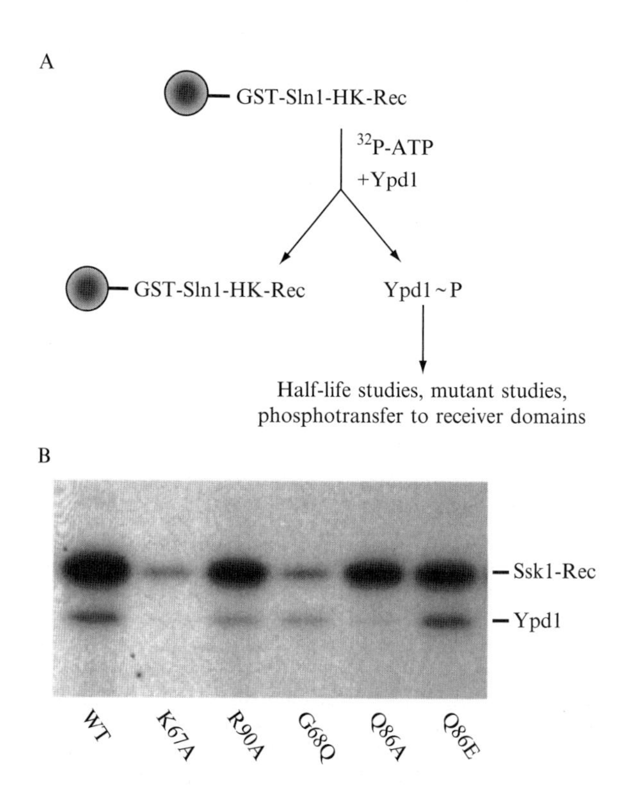

Figure 16.4 Phosphorylation of Ypd1. (A) Schematic depicting strategy for obtaining phosphoryated Ypd1 for subsequent *in vitro* biochemical studies. Bead-bound GST-Sln1-HK-Rec is coincubated with [γ-^{32}P]ATP and Ypd1 and the products separated by centrifugation. (B) An example phosphotransfer experiment in which isolated ^{32}P-labeled Ypd1 (WT) and a number of Ypd1 mutants were examined for their ability to transfer phosphoryl groups to the Ssk1-Rec domain (reproduced from Janiak-Spens *et al.*, 2000).

2.2.2.5. Measurement of the phosphorylated lifetime of SLN1 pathway components Response regulator proteins are activated and deactivated by phosphorylation, and the lifetime of the phosphorylated state often correlates with the duration of the output response. Thus, it is important to measure the half-life of the phosphorylated protein domains and to keep the half-life in mind when setting up *in vitro* time-dependent phosphotransfer experiments. One can obtain phosphorylated Sln1, Ssk1, or Skn7 receiver domains by incubation with ^{32}P-labeled GST-Sln1-HK as described in Sections 2.2.2.2 and 2.2.2.4 (Fig. 16.3A). The phosphodonor should be removed by centrifugation and the isolated phosphorylated receiver domain is then recovered in the supernatant. Incubate approximately 5 μM of P$\sim$receiver in 50 mM Tris–HCl, pH 8.0, 10 mM $MgCl_2$, 1 mM DTT in

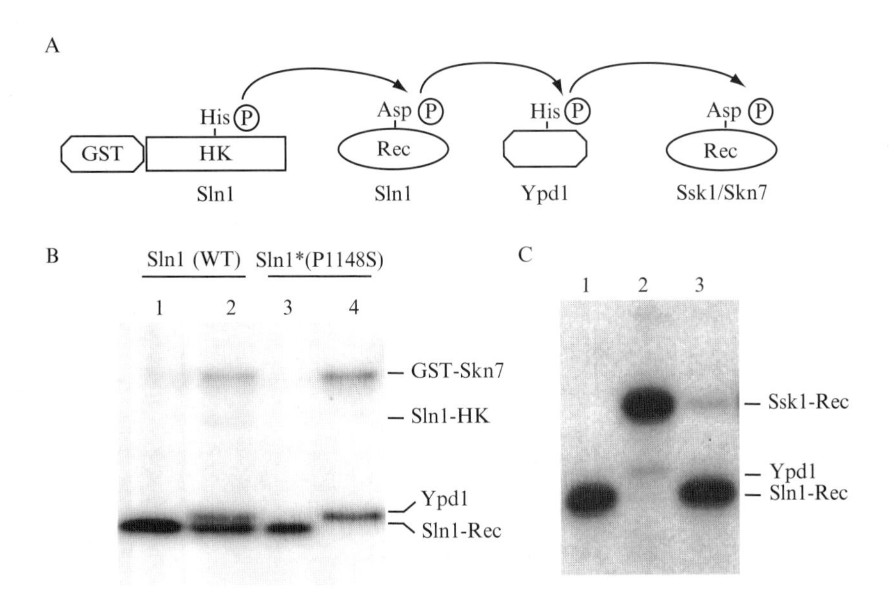

Figure 16.5 Phosphorelay experiments. (A) Schematic diagram representing the purified components of the SLN1 pathway. For phosphorelay experiments, one can start with ^{32}P-labeled GST-Sln1-HK or ^{32}P-labeled Sln1-Rec as the initial phosphodonor. Experiments are conducted with all phosphorelay components in the reaction mix in order to determine the total amount of radiolabel transmitted to the terminal phosphoreceiver (either Ssk1 or Skn7). (B) An example of a phosphorelay assay to examine phosphorylation of GST-Skn7 comparing WT Sln1-Rec to the Sln1* receiver mutant (P1148S). Equimolar concentrations of GST-Skn7, GST-Sln1-HK, Ypd1, and ^{32}P-labeled Sln1-Rec (WT) or Sln1*-Rec were incubated together for 15 min (lanes 2 and 4, respectively). Lanes 1 and 3 are the ^{32}P-labeled Sln1-Rec (WT) and Sln1*-Rec, respectively, incubated alone in reaction buffer. This *in vitro* experiment supported the *in vivo* genetic results that indicated that the P1148S mutation in the Sln1 receiver domain results in a hyperactive phenotype, that is, an increase in phosphorelay activity leading to a higher level of Skn7 phosphorylation. No reverse phosphotransfer from Sln1-Rec to GST-Sln1-HK was observed (lanes 2 and 4) (reproduced from Ault *et al.*, 2000). (C) An example phosphorelay experiment examining Ypd1-mediated phosphotransfer from Sln1-Rec to Ssk1-Rec. Equimolar concentrations of WT Ypd1 (lane 2) or a H64Q mutant Ypd1 (lane 3) was incubated with ^{32}P-labeled Sln1-Rec and Ssk1-Rec for 5 min. Lane 1 contains radiolabeled Sln1-Rec incubated alone. Using WT Ypd1, virtually all the radiolabel is transferred to Ssk1-Rec within 5 min (lane 2). In contrast, when the H64 residue within Ypd1 is mutated, little or no phosphotransfer is observed (reproduced from Janiak-Spens *et al.*, 2000).

a total volume of 50 μl at room temperature. Remove aliquots (4.5 μl) at various time points and mix with 4× stop buffer (0.25 *M* Tris–HCl, pH 6.8, 8% SDS, 60 m*M* EDTA, 40% glycerol, 0.008% bromophenol blue) to terminate the reaction. Freeze the aliquots at −20 °C until SDS–PAGE analysis. Determine the relative amounts of phosphorylated protein remaining at each time point by phosphorimager quantitation. Plot *ln*(% P~receiver remaining) versus time. The rate constant (k) for the dephosphorylation reaction is determined by least-squares fitting of the natural logarithm of the data to a linear relationship assuming first-order kinetics. The half-life of the phosphorylated receiver domain[25] is calculated accordingly ($t_{1/2} = \ln 2/k$). Phosphorylated Ypd1 can be obtained as described in Section 2.2.2.3 and the rate constant for dephosphorylation[26] is measured in the same manner as for receiver domains.

2.2.2.6. Phosphotransfer or phosphorelay experiments To monitor Sln1-Rec to Ypd1 phosphotransfer (e.g., see Fig. 16.3B), incubate P~Sln1-Rec with Ypd1 (approximately equimolar amounts) in a total volume of 15 μl in reaction buffer (50 mM Tris, pH 8.0, 30 mM KCl, 10 mM $MgCl_2$, 1 mM DTT) for 5 min at room temperature. Stop the reaction by adding 5 μl of 4× stop buffer (0.25 M Tris–HCl, pH 6.8, 8% SDS, 60 mM EDTA, 40% glycerol, 0.008% bromophenol blue). For time-course experiments, the reactions are prepared similarly except that the initial volume is 100 μl and aliquots (9 μl) can be removed at various time points, mixed with 4× stop buffer and keep on ice until gel analysis.

To assay phosphotransfer from P~Ypd1 to Ssk1-Rec, Skn7-Rec (or the reverse phosphotransfer to Sln1-Rec), follow the same procedure as above, except P~Ypd1 (1–2 μM) is typically incubated with slight molar excess of receiver domain (e.g., see Fig. 16.4B).

Phosphorelay experiments involving all protein components (Fig. 16.5A) in the reaction mix are typically carried out by incubating equimolar amounts (~2 μM) of either P~GST-Sln1-HK or P~Sln1-Rec as the initial phosphodonor, and Ypd1 and Ssk1-Rec (or Skn7-Rec) for 5 min in reaction buffer at room temperature. Representative phosphorelay experiments are shown in Fig. 16.5B and C.

[25] The half-lives of P~Sln1-Rec, P~Ssk1-Rec and P~Skn7-Rec are 13 ± 2 min, 13 ± 3 min, and 144 ± 6 min, respectively (Janiak-Spens *et al.*, 2000). However, in the presence of Ypd1, the lifetime of P~Ssk1-Rec is dramatically extended to 2280 ± 240 min presumably through complex formation that shields the phosphoaspartate from hydrolysis. This stabilization effect is specific to the Ssk1-Rec domain and is not observed for Sln1-Rec or Skn7-Rec.

[26] The half-life for P~Ypd1 is 4.1 ± 0.8 h (Janiak-Spens and West, 2000).

ACKNOWLEDGMENTS

We gratefully acknowledge funding from the NIH (GM59311 to A. H. W., GM068746 to Robert Deschenes and J. S. F., and GM56719 to J. S. F.), the Oklahoma Center for the Advancement of Science and Technology (A. H. W.), and the Center for Biocatalysis and Bioprocessing at the University of Iowa (J. S. F.) for the research described here.

REFERENCES

Albertyn, J., Hohmann, S., Thevelein, J. M., and Prior, B. A. (1994a). GPD1, which encodes glycerol-3-phosphate dehydrogenase, is essential for growth under osmotic stress in *Saccharomyces cerevisiae*, and its expression is regulated by the high-osmolarity glycerol response pathway. *Mol. Cell. Biol.* **14,** 4135–4144.

Albertyn, J., Hohmann, S., and Prior, B. A. (1994b). Characterization of the osmotic-stress response in *Saccharomyces cerevisiae*: Osmotic stress and glucose repression regulate glycerol-3-phosphate dehydrogenase independently. *Curr. Genet.* **25,** 12–18.

Ault, A. D. (2001). Analysis of the molecular mechanisms of signaling by Sln1, part of a two-component phosphorelay pathway in *Saccharomyces cerevisiae*. Ph.D. thesis, University of Iowa.

Ault, A. D., Fassler, J. S., and Deschenes, R. J. (2002). Altered phosphotransfer in an activated mutant of the *Saccharomyces cerevisiae* two-component osmosensor Sln1p. *Eukaryot. Cell* **1,** 174–180.

Ausubel, F. M., Brent, R., Kingston, R. E., Moore, D. E., Seidman, J. G., Smith, J. A., and Struhl, K. (1989). Current Protocols in Molecular Biology. Vol. 1. 2 vols, Wiley, New York.

Dean, S. (2004). Achieving Specificity in Yeast Stress Responses. Ph.D. thesis, University of Iowa.

Fassler, J. S., Gray, W. M., Malone, C. L., Tao, W., Lin, H., and Deschenes, R. J. (1997). Activated alleles of yeast SLN1 increase Mcm1-dependent reporter gene expression and diminish signaling through the Hog1 osmosensing pathway. *J. Biol. Chem.* **272,** 13365–13371.

Georgieva, B., and Rothstein, R. (2002). Kar-mediated plasmid transfer between yeast strains: Alternative to traditional transformation methods. *Methods Enzymol.* **350,** 278–289.

Janiak-Spens, F., and West, A. H. (2000). Functional roles of conserved amino acid residues surrounding the phosphorylatable histidine of the yeast phosphorelay protein YPD1. *Mol. Microbiol.* **37,** 136–144.

Janiak-Spens, F., Sparling, J. M., Gurfinkel, M., and West, A. H. (1999). Differential stabilities of phosphorylated response regulator domains reflect functional roles of the yeast osmoregulatory SLN1 and SSK1 proteins. *J. Bacteriol.* **181,** 411–417.

Janiak-Spens, F., Sparling, D. P., and West, A. H. (2000). Novel role for an HPt domain in stabilizing the phosphorylated state of a response regulator domain. *J. Bacteriol.* **182,** 6673–6678.

Janiak-Spens, F., Cook, P. F., and West, A. H. (2005). Kinetic analysis of YPD1-dependent phosphotransfer reactions in the yeast osmoregulatory phosphorelay system. *Biochemistry* **44,** 377–386.

Kaserer, A. O., Andi, B., Cook, P. F., and West, A. H. (2010). Kinetic studies of the yeast His-Asp phosphorelay signaling pathway. *Meth. Enzymol.* **471,** 59–75.

Kelly, D. E., Lamb, D. C., and Kelly, S. L. (2001). Genome-wide generation of yeast gene deletion strains. *Comp. Funct. Genomics* **2,** 236–242.

Ketela, T., Brown, J. L., Stewart, R. C., and Bussey, H. (1998). Yeast Skn7p activity is modulated by the Sln1p-Ypd1p osmosensor and contributes to regulation of the HOG pathway. *Mol. Gen. Genet.* **259,** 372–378.

Lawrence, C. W. (1991). Classical mutagenesis techniques. *Methods Enzymol.* **194,** 273–281.

Li, S. (2001). Molecular analysis of the Sln1 signaling pathway in yeast *Saccharomyces cerevisiae*. Ph.D. thesis, University of Iowa.

Li, S., Ault, A., Malone, C. L., Raitt, D., Dean, S., Johnston, L. H., Deschenes, R. J., and Fassler, J. S. (1998). The yeast histidine protein kinase, Sln1p, mediates phosphotransfer to two response regulators, Ssk1p and Skn7p. *EMBO J.* **17,** 6952–6962.

Li, S., Dean, S., Li, Z., Horecka, J., Deschenes, R. J., and Fassler, J. S. (2002). The eukaryotic two-component histidine kinase Sln1p regulates OCH1 via the transcription factor, Skn7p. *Mol. Biol. Cell* **13,** 412–424.

Luyten, K., Albertyn, J., Skibbe, W. F., Prior, B. A., Ramos, J., Thevelein, J. M., and Hohmann, S. (1995). Fps1, a yeast member of the MIP family of channel proteins, is a facilitator for glycerol uptake and efflux and is inactive under osmotic stress. *EMBO J.* **14,** 1360–1371.

Maeda, T., Wurgler-Murphy, S. M., and Saito, H. (1994). A two-component system that regulates an osmosensing MAP kinase cascade in yeast. *Nature* **369,** 242–245.

Maeda, T., Takekawa, M., and Saito, H. (1995). Activation of yeast PBS2 MAPKK by MAPKKKs or by binding of an SH3-containing osmosensor. *Science* **269,** 554–558.

Marina, A., Waldburger, C. D., and Hendrickson, W. A. (2005). Structure of the entire cytoplasmic portion of a sensor histidine-kinase protein. *EMBO J.* **24,** 4247–4259.

Miller, J. H. (1972). Experiments in Molecular Genetics. Cold Spring Harbor Laboratory, Cold Spring Harbor, NY.

Mulford, K. E. (2009). Characterization of the Oxidative Stress Regulation of Skn7 in *Saccharomyces cerevisiae*. M.S. thesis, University of Iowa.

Ostrander, D. B., and Gorman, J. A. (1999). The extracellular domain of the *Saccharomyces cerevisiae* Sln1p membrane osmolarity sensor is necessary for kinase activity. *J. Bacteriol.* **181,** 2527–2534.

Ota, I. M., and Varshavsky, A. (1992). A gene encoding a putative tyrosine phosphatase suppresses lethality of an N-end rule-dependent mutant. *Proc. Natl. Acad. Sci. USA* **89,** 2355–2359.

Ota, I. M., and Varshavsky, A. (1993). A yeast protein similar to bacterial two-component regulators. *Science* **262,** 566–569.

Posas, F., and Saito, H. (1997). Osmotic activation of the HOG MAPK pathway via Ste11p MAPKKK: Scaffold role of Pbs2p MAPKK. *Science* **276,** 1702–1705.

Posas, F., and Saito, H. (1998). Activation of the yeast SSK2 MAP kinase kinase kinase by the SSK1 two-component response regulator. *EMBO J.* **17,** 1385–1394.

Posas, F., Wurgler-Murphy, S. M., Maeda, T., Witten, E. A., Thai, T. C., and Saito, H. (1996). Yeast HOG1 MAP kinase cascade is regulated by a multistep phosphorelay mechanism in the SLN1-YPD1-SSK1 "two-component" osmosensor. *Cell* **86,** 865–875.

Reiser, V., Raitt, D. C., and Saito, H. (2003). Yeast osmosensor Sln1 and plant cytokinin receptor Cre1 respond to changes in turgor pressure. *J. Cell Biol.* **161,** 1035–1040.

Shankarnarayan, S. (2007). A Potential Role for the Cell Wall in the Regulation of the SLN1-SKN7 Pathway in the Yeast, *Saccharomyces cerevisiae*. Ph.D. thesis, University of Iowa.

Shankarnarayan, S., Malone, C. L., Deschenes, R. J., and Fassler, J. S. (2008). Modulation of yeast Sln1 kinase activity by the CCW12 cell wall protein. *J. Biol. Chem.* **283,** 1962–1973.

Singh, M., Berger, B., Kim, P. S., Berger, J. M., and Cochran, A. G. (1998). Computational learning reveals coiled coil-like motifs in histidine kinase linker domains. *Proc. Natl. Acad. Sci. USA* **95,** 2738–2743.

Tao, W., Deschenes, R. J., and Fassler, J. S. (1999). Intracellular glycerol levels modulate the activity of Sln1p, a *Saccharomyces cerevisiae* two-component regulator. *J. Biol. Chem.* **274,** 360–367.

Tao, W., Malone, C. L., Ault, A. D., Deschenes, R. J., and Fassler, J. S. (2002). A cytoplasmic coiled-coil domain is required for histidine kinase activity of the yeast osmosensor, SLN1. *Mol. Microbiol.* **43,** 459–473.

Wurgler-Murphy, S. M., Maeda, T., Witten, E. A., and Saito, H. (1997). Regulation of the *Saccharomyces cerevisiae* HOG1 mitogen-activated protein kinase by the PTP2 and PTP3 protein tyrosine phosphatases. *Mol. Cell. Biol.* **17,** 1289–1297.

Xu, Q., and West, A. H. (1999). Conservation of structure and function among histidine-containing phosphotransfer (HPt) domains as revealed by the crystal structure of YPD1. *J. Mol. Biol.* **292,** 1039–1050.

Xu, Q., Nguyen, V., and West, A. H. (1999). Purification, crystallization and preliminary X-ray diffraction analysis of the yeast phosphorelay protein YPD1. *Acta Crystallogr. D Biol. Crystallogr.* **55,** 291–293.

Xu, Q., Porter, S. W., and West, A. H. (2003). The yeast YPD1/SLN1 complex: Insights into molecular recognition in two-component signaling systems. *Structure (Camb.)* **11,** 1569–1581.

Yu, G., Deschenes, R. J., and Fassler, J. S. (1995). The essential transcription factor, Mcm1, is a downstream target of Sln1, a yeast "two-component" regulator. *J. Biol. Chem.* **270,** 8739–8743.

Zhao, X., Copeland, D. M., Soares, A. S., and West, A. H. (2008). Crystal structure of a complex between the phosphorelay protein YPD1 and the response regulator domain of SLN1 bound to a phosphoryl analog. *J. Mol. Biol.* **375,** 1141–1151.

CHAPTER SEVENTEEN

Analysis of Mitogen-Activated Protein Kinase Phosphorylation in Response to Stimulation of Histidine Kinase Signaling Pathways in *Neurospora*

Carol A. Jones *and* Katherine A. Borkovich

Contents

Abstract

In eukaryotes, two-component regulatory systems have been demonstrated to regulate phosphorylation of mitogen-activated protein kinases (MAPKs). Here, we describe a method implementing preparation of a protein extract under denaturing conditions, followed by Western analysis using MAPK antibodies that can be used to observe the effects of components of two-component signaling pathways or other proteins on the phosphorylation status of MAPKs.

Department of Plant Pathology and Microbiology, University of California, Riverside, California, USA

Methods in Enzymology, Volume 471
ISSN 0076-6879, DOI: 10.1016/S0076-6879(10)71017-X

The protein extraction method presented may also be used to concentrate cellular proteins for additional applications, such as metabolic labeling or analysis of other posttranslational modifications.

1. Introduction

Histidine kinases and aspartate-containing response regulator proteins are elements of two-component signaling pathways. Two-component phosphorelay systems have been identified in archaea, bacteria, protists, fungi, and plants, but to-date have not been found in any animal, including humans (Bahn, 2008; Borkovich *et al.*, 2004; Park *et al.*, 2009; Wolanin *et al.*, 2002). In the simplest two-component systems, the phosphorelay begins with autophosphorylation of a conserved histidine residue in the histidine kinase in response to some environmental signal (Posas *et al.*, 1996). This same phosphate is then transferred onto an aspartate within the response regulator domain of a different protein. The phosphorylation status of the response regulator protein determines its ability to regulate gene expression or to activate or inhibit other downstream effector proteins.

A more complex version of the pathway described above is the so-called multicomponent pathway (Borkovich *et al.*, 2004; Catlett *et al.*, 2003; Chang and Stewart, 1998; Wolanin *et al.*, 2002). Such systems include hybrid histidine kinases (proteins containing both histidine kinase and response regulator domains) and histidine phosphotransfer proteins (HPTs). Multicomponent pathways begin with autophosphorylation of a histidine residue in the hybrid histidine kinase. However, instead of being transferred to a separate protein, the phosphate is passed in an intramolecular reaction to the aspartate residue on the response regulator domain in the hybrid kinase. This phosphate is subsequently transferred to a HPT protein and then to a terminal response regulator. In eukaryotic filamentous fungi such as *Neurospora crassa*, all histidine kinases are of the hybrid type. *Neurospora* possesses 11 hybrid histidine kinases, one HPT and two response regulator proteins (Borkovich *et al.*, 2004; Galagan *et al.*, 2003).

In *Neurospora* and other eukaryotes, multicomponent signaling pathways have been found to activate mitogen-activated protein (MAP) kinase cascades (Catlett *et al.*, 2003; Chang and Stewart, 1998; West and Stock, 2001). These cascades consist of three serine/threonine protein kinases that act sequentially. The first step involves the phosphorylation of a MAP kinase kinase kinase (MAPKKK), which leads to the phosphorylation of a MAP kinase kinase (MAPKK) that triggers the phosphorylation of the downstream MAP kinase (MAPK). The phosphorylation status of the MAPK determines its ability to regulate other proteins. Interestingly, when an upstream MAPKKK or MAPKK phosphorylates the next MAPKK or MAPK in the pathway, it is not the same phosphate (as is the case in histidine kinase signaling pathways), but a new ATP that is hydrolyzed for each phosphorylation event (Choi *et al.*, 2008).

Neurospora possesses nine MAPKKK/MAPKK/MAPK proteins (three of each type) that are thought to make up three MAPK modules. These three modules correspond to (1) the osmosensing (OS-2 MAPK cascade; see below), (2) the cell integrity (MAK-1 MAPK cascade), and (3) the pheromone response/filamentation (MAK-2 MAPK cascade) pathways that have been well studied in yeast (Borkovich *et al.*, 2004; Posas *et al.*, 1998).

In *Neurospora*, the osmotic sensitive (*os*) mutants correspond to a hybrid histidine kinase (*os-1/nik-1*) and components of a downstream MAPK cascade (*os-4* MAPKKK, *os-5* MAPKK and *os-2* p38 class MAPK; Emerson and Emerson, 1958; Livingston, 1969; Selitrennikoff *et al.*, 1981; Zhang *et al.*, 2002). In addition, our laboratory has demonstrated that mutants lacking the response regulator *rrg-1* have many phenotypes in common with *os* mutants. *os* mutants are characterized by fragile conidia that tend to lyse and these strains are often female-sterile (Jones *et al.*, 2007). These mutants also share the phenotypes of osmotic sensitivity and resistance to certain fungicides, traits that are believed to result from the inability to properly stimulate glycerol production or other counter solutes (Ellis *et al.*, 1991; Fujimura *et al.*, 2000; Kanetis *et al.*, 2008; Pillonel and Meyer, 1997; Zhang *et al.*, 2002).

Fludioxonil is a phenylpyrrole fungicide that is effective against *Neurospora*, and various genetic screens have produced several mutants (including the *os* mutants discussed above) that are resistant to this and other fungicides (Fujimura *et al.*, 2000; Grindle and Temple, 1982; Ochiani *et al.*, 2001; Selitrennikoff *et al.*, 1981). Fludioxonil has been approved for use against a broad spectrum of fungal plant pathogens to prevent fruit spoilage (Förster, *et al.*, 2007; Kanetis *et al.*, 2008). Fludioxonil appears to target the same pathway as the hyperosmotic stress response pathway, leading to the activation of the OS-2 MAPK cascade (Zhang *et al.*, 2002).

In addition to roles in osmotic stress and fungicide resistance, we have recently demonstrated that RRG-1 and the downstream OS-2 MAPK are involved in regulation of the circadian rhythm that controls asexual spore formation (conidiation) in *Neurospora*. Although the clock is not required to mount a rapid response during hyperosmotic conditions, Δ*rrg-1* mutants display a period shortening defect and a delay in the conidiation rhythm when grown on race tubes (Vitalini *et al.*, 2007).

This chapter presents a method to assay MAPK phosphorylation after exposure of *Neurospora* cells to hyperosmotic stress or fungicide. The method involves freezing cells in liquid nitrogen, followed by bead lysis in an ethanol solution. Proteins are subjected to SDS–PAGE and gels electroblotted onto a membrane. MAPK antibodies are used to visualize phosphorylated and total MAPK protein. This method has been used successfully to demonstrate regulation of phosphorylation of the OS-2 MAPK by two-component signaling proteins in *Neurospora* (Jones *et al.*, 2007; Vitalini *et al.*, 2007).

2. Growth of Cultures and Exposure to Hyperosmotic Conditions or Fungicide

All strains were cultured on Vogel's minimal medium (VM; Davis and de Serres, 1970; Vogel, 1964). Shaking liquid cultures were exposed to hyperosmotic stress induced by NaCl or the fungicide fludioxonil during the last 10 min of growth. Cells were collected by filtration and cell pads frozen in liquid nitrogen.

2.1. Reagents

- *Stock solutions*
  50 μg/ml biotin in 50% ethanol (biotin solution)
  10 mg/ml fludioxonil (dissolved in 100% DMSO)
  4 M NaCl (dissolved in water).
- *50× concentrated VM medium salts (per liter)*
  126.8 g sodium citrate • $2H_2O$
  250 g KH_2PO_4
  100 g NH_4NO_3
  10 g $MgSO_4 \cdot 7H_2O$
  5 g $CaCl_2 \cdot 2H_2O$ (predissolve first in water and then added slowly)
  5 ml biotin solution
  5 ml trace elements solution (see below)
  After bringing solution to final volume, transfer to bottle. Add 5 ml of chloroform (acts as fungicide; sinks to bottom of bottle) to the solution. Make sure not to pipet any of the chloroform when withdrawing 50× VM solution for use.
- *Trace elements solution (for 100 ml)*
  5 g citric acid • H_2O
  5 g $ZnSO_4 \cdot 7H_2O$
  1 g $Fe(NH_4)_2(SO_4) \cdot 6H_2O$
  0.25 g $CuSO_4 \cdot 5H_2O$
  0.05 g $MnSO_4 \cdot H_2O$
  0.05 g H_3BO_3

0.05 g $NaMoO_4 \cdot 2H_2O$
Filter sterilize and store in the dark at 4 °C (Vogel, 1964).

- *VM medium (per liter)*
  20 ml 50× concentrated VM salts
  15 g sucrose
  10 g agar (for propagation of conidia in flasks)
  Autoclave to sterilize.

2.2. Growth and collection of cultures

1. Inoculate a 250-ml Erlenmeyer flask containing 50 ml VM agar medium with each strain to be analyzed. Incubate flasks at 30 °C in the dark for 3 days, followed by 2–5 days at 25 °C in the light.
2. Suspend vegetative tissue from the VM agar cultures using 50 ml of sterile ice-cold water by swirling. The material is filtered into a sterile 250 ml Erlenmeyer flask prepared with a Handi wipe towel® (Clorox Corp., Oakland, CA) affixed to the rim with a rubberband (entire assembly autoclaved prior to use) to separate the hyphal filaments (retained by filter) from the conidia (flow through to flask).
3. Transfer conidial suspension to a 50-ml conical tube and centrifuge at 2500 rpm for 3–5 min at room temperature. Resuspend the pellet in 20–40 ml of ice-cold sterile water and repeat centrifugation. Resuspend the final pellet in 200–400 μl of sterile water.
4. Dilute the suspension (1:100) and count using a hemacytometer to quantify the number of conidia. Adjust the concentration to 1×10^8 conidia/ml. Store conidia at 4 °C in the dark for up to 1 day before use.
5. Sixteen-hour shaking liquid cultures were generated by inoculating 20 ml of liquid VM to a final concentration of 1×10^6 conidia/ml. Cultures were incubated in the dark at 30 °C with shaking at 200 rpm for 16 h.
6. During the last 10 min of growth, treat cultures with the same volume of either NaCl (final concentration 0.1 or 0.8 *M*) or water (control for hyperosmotic stress) or fludioxonil dissolved in dimethyl sulfoxide (DMSO; final concentration 100 μg/ml) or DMSO (control for fungicide exposure).
7. Collect cultures by filtration. This can be accomplished using a funnel, vacuum flask and filter paper or by filtering and then pressing out the liquid using a Handi wipe towel® (Clorox Corp.). The tissue should be quick-frozen in liquid nitrogen. The tissue may be stored at −80 °C until protein is extracted; however, lengthy storage of cell pads prior to protein extraction is not recommended and best results are obtained using freshly collected tissue.

3. Mitogen-Activated Protein Kinase Assay

A procedure for detection of MAPK phosphorylation in protein extracts is described below. This process is being described with reference to the MAPK OS-2, but it has also been used for detection of all three MAPKs in *N. crassa* and should be translatable to other species. The assay involves protein extraction and western blot analysis using commercially available antibodies. The antibodies have been developed against yeast Hog1p MAPK and mammalian p38 MAPK proteins, but cross-react with their highly conserved counterparts in *Neurospora* and other fungi, including the OS-2 homologs PdOS-2 from *Penicillium digitatum* (Kanetis *et al.*, 2008), SakA from *Aspergillus nidulans* (Kawasaki *et al.*, 2002), OSC-1 from *Colletotrichum lagenarium* (Kojima *et al.*, 2004), and BmHog1 from *Colletotrichum heterostrophus* (Yoshimi *et al.*, 2005). One antibody reacts with all forms of OS-2, whether phosphorylated or not, while the other only recognizes the phosphorylated version of the MAPK.

3.1. Reagents

- For preparation of cell extracts:
  1% sodium dodecyl sulfate (SDS) in water
  3 M phenylmethylsulfonyl fluoride (PMSF) in 95% ethanol. Prepare from a stock solution of 200 m*M* PMSF (that was dissolved in isopropanol).
- For SDS–PAGE electrophoresis:

2× Buffer A (per liter):

181.5 g Tris base
4 g SDS
Adjust the pH to 8.8 (before adding SDS). Filter the solution and store at 4 °C.

1:3 Buffer A:

Dilute 2× Buffer A solution with water as indicated (1:3).

2× Buffer B (per liter):

60.4 g Tris base
4 g SDS
Adjust pH to 6.8 (before adding SDS). Filter solution and store at 4 °C.

Premixed acrylamide/bisacrylamide 37.5:1, 40% solution (#1490, EMD Chemicals, Madison, WI).

10% ammonium persulfate (APS; diluted in water). Make fresh before use.
Tetramethylethylenediamine (TEMED; #BP150, Fisher BioReagents, Pittsburgh, PA).
10% SDS (w/v in water)

SDS sample buffer (per 10 ml):

2 ml glycerol
2 ml 10% SDS
1.25 ml 2× Buffer B
6 mg bromphenol blue
4.25 ml water
Store 10× SDS sample buffer at room temperature. Add 1/20 volume of β-mercaptoethanol to an aliquot of this solution immediately before use.

10% SDS–PAGE resolving gel (for 12.5 ml):

3.1 ml water
3.13 ml acrylamide/bisacrylamide 37.5:1, 40% solution
6.25 ml 2× Buffer A
60 μl 10% APS
7 μl TEMED

SDS–PAGE stacking gel (for 5 ml):

2 ml water
500 μl acrylamide/bisacrylamide 37.5:1, 40% solution
2.5 ml 2× Buffer B
60 μl 10% APS
7 μl TEMED

10× running buffer (per liter):

30 g Tris base
143.75 g glycine
5 g SDS
Dilute 1:10 with water to make 1× running buffer.

- For Western blot analysis:

Transfer buffer (per liter):

3 g Tris base
14.1 g glycine
200 ml methanol
Bring to final volume with water. This solution can be reused three to four times. Store in an airtight container in the dark.

10× Tris-buffered saline (TBS) (per liter):

24.2 g Tris base
80 g NaCl
Adjust pH to 7.6 with HCl, then bring to final volume. Autoclave to sterilize for long-term storage. Store solution at room temperature. Dilute 1:10 with water to make 1× TBS solution for washes of the membranes to be probed with the phospho-p38 antibody.

TBSM:

Prepare by dissolving 5 g dry milk in 100 ml TBS. Use immediately to block membrane/dilute secondary antibody (for phospho-p38 antibody probed membranes) or store overnight at 4 °C.

TBST:

Add 50 μl of Tween-20 for every 100 ml of TBS. Use this solution to prepare TBSTB.

TBSTB:

Prepare by dissolving 5 g bovine serum albumin (BSA; Fraction V) in 100 ml TBST. Use immediately to dilute the phospho-p38 primary antibody or store overnight at 4 °C.

5× phosphate-buffered saline (PBS) (per liter):

41.2 g Na_2HPO_4
10.2 g NaH_2PO_4
20 g NaCl
Adjust pH to 7.4 with NaOH then bring to final volume. Autoclave to sterilize for long-term storage. Store solution at room temperature. Dilute 1:5 with water to make 1× PBS solution for washes of the membranes to be probed with the Hog1p antibody.

PBSM:

Prepare by dissolving 5 g dry milk in 100 ml PBS. Use immediately to block the membrane/dilute secondary antibody (for Hog1p antibody probed membranes) or store overnight at 4 °C.

PBST:

Add 50 μl of Tween-20 for every 100 ml of PBS. Use this solution to prepare PBSTB.

PBSTB:

Prepare by dissolving 5 g BSA (Fraction V) in 100 ml PBST. Use immediately to dilute the Hog1p primary antibody or store overnight at 4 °C.

Phospho-p38 rabbit primary antibody (#9211; Cell Signaling Technology, Inc., Beverly, MA), diluted 1:600 in TBSTB just before use.

Hog1p rabbit primary antibody (#sc-9079; Santa Cruz Biotechnology, Inc., Santa Cruz, CA), diluted 1:600 in PBSTB just before use.

Peroxidase-conjugated goat antirabbit IgG secondary antibody (Sigma Chemical Company, St. Louis, MO), diluted 1:6000 in TBSM or PBSM (depending on primary antibody diluent).

3.2. Preparation of protein extracts

1. Grind (~1 min) the frozen *Neurospora* tissue into a fine powder in liquid nitrogen using a mortar and pestle. Add additional liquid nitrogen to the mortar as needed.
2. Transfer 20–80 mg of ground tissue into a plastic 2-ml screw-top centrifuge tube containing 1 ml of chilled 3 m*M* PMSF in 95% ethanol and 0.2 g of 0.5 mm glass beads. Vortex briefly to suspend the tissue in solution. Store samples on ice briefly until several have been processed.
3. Vortex samples three times for 60 s with 60 s rests on ice in between.
4. Chill extracts at −20 °C for at least 16 h (it is fine to leave the samples for several days in the freezer). This step causes the protein to precipitate out of solution.
5. Centrifuge the samples at 14,000 rpm for 10 min at 4 °C in a microcentrifuge.
6. Remove as much of the supernatant as possible without disturbing the pellet. Evaporate the remaining ethanol by spinning the tubes for 30 min in a vacuum concentrator (Eppendorf, Westbury, NY). This step precipitates the proteins on the beads. At this point, the samples may be reconstituted (see next step) or stored at −20 °C up to several days before continuing.
7. Reconstitute samples in 200–250 μl of 1% SDS and heat at 85 °C for 5 min (vortex every 1–2 min). Centrifuge tubes at 14,000 rpm for 3–5 min at room temperature in a microcentrifuge.
8. Remove and save the protein supernatant in a new microcentrifuge tube.
9. Reconstitution (steps 7 and 8) may be repeated once and the supernatants can be combined.
10. Subject the combined supernatants to a final heating, centrifugation, and supernatant recovery as described above to remove residual cellular debris.
11. To ensure protein stability, add SDS sample buffer at a ratio of one volume sample buffer to two volumes protein as soon as possible.

We typically add sample buffer to the majority of the extract, only leaving enough extract to successfully perform the protein assay.

12. Determine the concentration of protein in extracts using the BCA micro protein assay (Thermo Scientific, Pierce, Rockford, IL) according to the manufacturer's instructions. The absorbance of assay tubes is measured at 562 nm. The standard curve consists of 0, 2, 4, 6, 8, and 10 μg BSA. Make all protein dilutions in 1% SDS.

3.3. SDS–PAGE electrophoresis and transfer of protein to membrane

1. Pour the 10% resolving gel between two glass plates, leaving approximately the top 3 cm empty. Using a Pasteur pipet, overlay the resolving gel with bubbles generated by shaking the 1:3 Buffer A solution. Follow by overlaying the 1:3 Buffer A solution itself. This helps keep the gel moist and yields a sharp interface at the top edge of the gel. Let the resolving gel polymerize for at least 45 min.
2. Rinse away any residue (such as unpolymerized acrylamide) with water. Remove excess water from above resolving gel by blotting with a piece of filter paper.
3. Pour the stacking gel and insert a comb. Allow the stacking gel to polymerize 5–15 min.
4. Rinse wells that were generated in the stacking gel by the comb with water several times to remove any unpolymerized acrylamide and then rinse the wells with 1× running buffer. Assemble gel in electrophoresis apparatus.
5. Protein samples (prepared as described above) should be heated at 85–100 °C and then centrifuged at room temperature for 5 min at 14,000 rpm before loading onto a 10% SDS–PAGE gel. If the gel is not run the same day, store the samples at −20 °C or lower until use. Reheat and centrifuge the samples before use. Do not use extracted protein that has not been stored in sample buffer, as the sample buffer inhibits degradation of the OS-2 MAPK protein.
6. Duplicate gels can be run at 15–70 mA (gels usually take 3–16 h to run depending on the voltage; never exceed 250 V), as previously described (Krystofova and Borkovich, 2005) in 1× running buffer.
7. The protein is transferred from the SDS–PAGE gel to a nitrocellulose membrane (0.45 μm, GE Water and Process Technologies, Trevose, PA) in a Trans-blot Cell Western blot apparatus (Bio-Rad, Hercules, CA) at 215 mA for 2–3 h in transfer buffer.

3.4. Western blot analysis using MAPK antibodies

1. After transfer of the protein onto the nitrocellulose membranes is complete allow the duplicate membranes to dry. One membrane will be used to determine whether the OS-2 protein is equally expressed in all samples, while the other will be used to measure the level of phosphorylated OS-2.
2. The membranes can be washed and probed immediately after drying, or stored for up to several weeks. Wash the membranes three times for 5 min each in TBS (for the phospho-p38 antibody probed membranes) or PBS (for the Hog1p antibody probed membranes) with shaking at room temperature.
3. Block the membranes in TBSM (phospho-p38 antibody-probed membranes) or PBSM (Hog1p antibody-probed membranes) for 1 h with shaking at room temperature. Blocking inhibits nonspecific binding of antibody to the membrane later in the procedure.
4. Wash the membranes again 3× 5 min each in TBS or PBS wash buffer, as appropriate, with shaking at room temperature.
5. Primary antibodies should be diluted as indicated above, in TBSTB or PBSTB. These antibodies are commercially available and were generated against mammalian (p38) or *S. cerevisiae* (Hog1p) MAPKs, but recognize forms of the 42 kDa OS-2 protein. The anti-phospho-p38 antibody can be used to detect phospho-OS-2. Anti-Hog1p reacts with both unphosphorylated and phosphorylated forms of OS-2, and serves as a control for the total amount of OS-2 protein in the cell. Incubate membranes overnight at 4 °C in the primary antibodies with gentle shaking.
6. Rinse away the excess antibody with three 5 min washes (with shaking at room temperature) in TBS or PBS, depending on primary antibody diluent.
7. Incubate the membranes for 1 h with the peroxidase-conjugated goat antirabbit IgG secondary antibody diluted in TBSM or PBSM (as appropriate) with shaking at room temperature.
8. Wash the membranes three times in wash buffer as described above (step 4) to remove excess secondary antibody.
9. Place the membranes on a clean transparency and treated with chemiluminescent chemicals. Chemiluminescent detection may be performed using the SuperSignal West Pico kit (Thermo Scientific, Pierce), mixing equal volumes of the Peroxide and Luminol Enhancer solutions (as previously described; Krystofova and Borkovich, 2005). Make sure the chemicals cover the membranes evenly. Rotate the transparency by hand for one minute to allow the mixture to flow evenly over the membrane and then place the transparency in the dark for 4 additional minutes. Briefly allow the excess chemiluminescent regents to drain off the membrane and then cover with a clear transparency before imaging.

3.5. Results

The results for a typical OS-2 phosphorylation assay are shown in Fig. 17.1. Levels of OS-2 protein are similar under all conditions in all strains, except in Δ*os-2* mutants, which serve as a negative control for both OS-2 protein and phospho-OS-2. Phosphorylation of OS-2 is stimulated in cultures treated with 0.8 *M* NaCl or 100 μg/ml fludioxonil (fungicide) in wild-type cells. Δ*rrg-1* (response regulator) mutants are unable to induce OS-2-phosphate levels in response to NaCl or fludioxonil above the basal level observed in wild type (Jones *et al.*, 2007). Δ*os-1* (hybrid histidine kinase) mutants exhibit a hybrid response, with an increase in response to NaCl, but not to fungicide. Together, these results show that a divergence exists between the hyperosmotic stress and fungicide response pathways. The hyperosmotic stress response pathway is likely more complex than the fungicide response pathway, possibly involving an additional osmotic sensitive factor. Of note, the ability of *rrg-1*^{D921N} mutants (mutated at the predicted site of phosphorylation) to induce OS-2 phosphorylation in a manner similar to wild type suggests that the unphosphorylated form of RRG-1 is the one that activates the OS-2 MAPK cascade (Jones and Borkovich, unpublished data; Jones *et al.*, 2007).

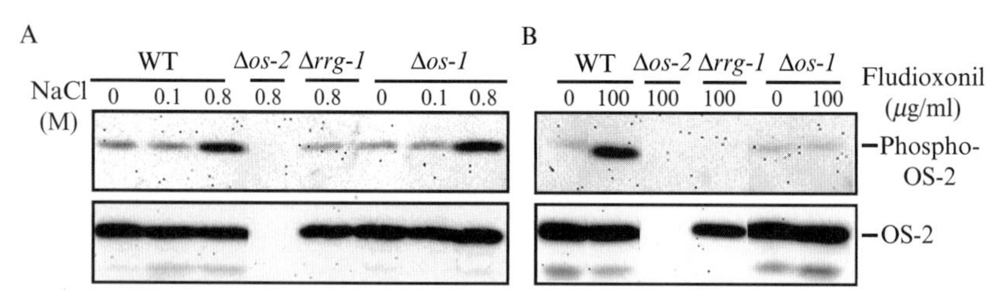

Figure 17.1 OS-2 phosphate levels of the Δ*os-1* and Δ*rrg-1* gene replacement mutants. (A) Analysis of OS-2 protein levels and phosphorylation in response to hyperosmotic conditions. Wild-type (74A), Δ*os-1*, Δ*rrg-1*, and Δ*os-2* strains were grown for 16 h in liquid submerged VM cultures. The cultures were brought to 0, 0.1, or 0.8 *M* NaCl and incubated with shaking for the last 10 min of growth. The cells were collected, whole cell extracts were prepared and then subjected to Western analysis. The top panel is a Western blot using anti-p38 that recognizes only phosphorylated OS-2, while the bottom panel is a Western blot using anti-Hog1p that reacts with all forms of the OS-2 protein. (B) Analysis of OS-2 protein levels and phosphorylation during fungicide exposure. Sixteen-hour liquid submerged VM cultures of the strains indicated in (A) were treated with 0 or 100 μg/ml fludioxonil for the last 10 min of incubation. Whole cell extracts were prepared and subjected to Western analysis with the antibodies described in (A).

4. Adapting the MAPK Assay

The MAPK detection method described above for the p38-class *Neurospora* MAPK OS-2 has also been used to detect the OS-2 homolog PdOS-2 from *P. digitatum* (Kanetis *et al.*, 2008). A similar MAPK assay has been implemented to observe the level of phosphorylated MAK-1 and MAK-2, the other two MAPKs found in *Neurospora* (Pandey *et al.*, 2004; Park *et al.*, 2008). MAK-1 (50 kDa) and MAK-2 (39 kDa) are p44/42/Erk class MAPKs. These two proteins are recognized by commercially available antibodies that were generated against mammalian (p44/42) MAPK sequences. Both phosphorylated and nonphosphorylated MAK-1 and MAK-2 react with one antibody (anti-p44/42 #9102; 1:200 dilution in TBSTB; Cell Signaling Technology, Inc.), while another antibody recognizes only the phosphorylated forms of the two proteins (antiphospho-p44/42 #9101; 1:200 dilution in TBSTB; Cell Signaling Technology, Inc.). An advantage of this assay for MAK-1 and MAK-2 is that both proteins can be detected on the same Western blot, due to their large difference in molecular mass.

5. Discussion

The method discussed in this paper may theoretically be modified to observe the phosphorylation status of any MAPK or the level of any cellular protein for which an antibody is available. We have used this technique to detect epitope-tagged proteins from lysed cells of *Neurospora* strains that have been transformed with a tagged gene construct (Jones and Borkovich, unpublished results). Furthermore, the protein extraction method discussed above can be used to generate extremely concentrated protein solutions. The concentration of the extract can be controlled by adjusting the volume of 1% SDS used to reconstitute the proteins.

The cell lysis method has been used to extract proteins from many types of *Neurospora* tissues, including various plate and shaking liquid cultures. We commonly use tissues harvested from overnight shaking liquid VM cultures. A 20-ml culture generates an adequate quantity of tissue from strains that grow similarly to wild type. However, larger cultures were sometimes needed to obtain enough tissue from slow-growing mutants.

Some Western blot procedures will tolerate stripping of a membrane after probing with the first antibody and then reprobing with a second antibody. Such a procedure would reduce the number of blots needed. However, in our experience, probing the same blot with the phospho and protein MAPK antibodies, as opposed to use of duplicate blots, is not recommended, due to issues with incomplete removal of the first antibody and/or removal of protein from the blot during the stripping procedure. In particular, we have found that the Hog1p antibody is difficult to remove completely.

The primary antibody solution can be reused two to three times to probe a membrane. Although this practice can be used to conserve relatively expensive primary antibody during preliminary studies, freshly prepared antibody solutions are preferred for published experiments, as their actual titer is known. If reusing a primary antibody solution, it is important that it be diluted in PBSTB or TBSTB (not a milk solution) and stored at −20 °C between uses. Thaw the antibody solution at room temperature before reuse.

The results in Fig. 17.1 show that levels of the OS-2 MAPK protein do not fluctuate greatly over time or under different experimental conditions. However, the phosphorylation status of OS-2 is altered in response to fungicide treatment or a hyperosmotic environment (see Fig. 17.1; Jones *et al.*, 2007) and regularly oscillates throughout the day (Vitalini *et al.*, 2007) in wild-type strains. Future work will focus on analysis of additional upstream regulators, including additional hybrid histidine kinases, that regulate OS-2 phosphorylation in *Neurospora*.

ACKNOWLEDGMENTS

We thank Dr. Gyungsoon Park for helpful discussions and assistance with fine-tuning the MAK-1 and MAK-2 detection protocol.

REFERENCES

Bahn, Y. S. (2008). Master and commander in fungal pathogens: The two-component system and the HOG signaling pathway. *Eukaryot. Cell* **7,** 2017–2036.

Borkovich, K. A., Alex, L. A., Yarden, O., Freitag, M., Turner, G. E., Read, N. D., Seiler, S., Bell-Pedersen, D., Paietta, J., Plesofsky, N., Plamann, M., Goodrich-Tanrikulu, M., *et al.* (2004). Lessons from the genome sequence of *Neurospora crassa*: Tracing the path from genomic blueprint to multicellular organism. *Microbiol. Mol. Biol. Rev.* **68,** 1–108.

Catlett, N. L., Yoder, O. C., and Turgeon, B. G. (2003). Whole-genome analysis of two-component signal transduction genes in fungal pathogens. *Eukaryot. Cell* **2,** 1151–1161.

Chang, C., and Stewart, R. C. (1998). The two-component system. Regulation of diverse signaling pathways in prokaryotes and eukaryotes. *Plant Physiol.* **117,** 723–731.

Choi, H. S., Kim, J. R., Lee, S. W., and Cho, K. H. (2008). Why have serine/tyrosine kinases been evolutionarily selected in eukaryotic signaling cascades? *Comput. Biol. Chem.* **32,** 218–221.

Davis, R. H., and de Serres, F. J. (1970). Genetic and microbiological research techniques for *Neurospora crassa*. *Methods Enzymol.* **17A,** 79–143.

Ellis, S. W., Grindle, M., and Lewis, D. H. (1991). Effect of osmotic stress on yield and polyol content of dicarboximide-sensitive and -resistant strains of *Neurospora crassa*. *Mycol. Res.* **95,** 457–464.

Emerson, S., and Emerson, M. R. (1958). Production, reproduction, and reversion of protoplast-like structures in the osmotic strain of *Neurospora crassa*. *Proc. Natl. Acad. Sci. USA* **44,** 668–671.

Förster, H., Driever, G. F., Thompson, D. C., and Adaskaveg, J. E. (2007). Postharvest decay management for stone fruit crops in California using the "reduced-risk" fungicides fludioxonil and fenhexamid. *Plant Dis.* **91,** 209–215.

Fujimura, M., Ochiai, N., Ichiishi, A., Usami, R., Horikoshi, K., and Yamaguchi, I. (2000). Sensitivity to phenylpyrrole fungicides and abnormal glycerol accumulation in *os* and *cut* mutant strains of *Neurospora crassa*. *J. Pestic. Sci.* **25,** 31–36.

Galagan, J. E., Calvo, S. E., Borkovich, K. A., Selker, E. U., Read, N. D., Jaffe, D., FitzHugh, W., Ma, L. J., Smirnov, S., Purcell, S., Rehman, B., Elkins, T., *et al.* (2003). The genome sequence of the filamentous fungus *Neurospora crassa*. *Nature* **422,** 859–868.

Grindle, M., and Temple, W. (1982). Fungicide-resistance of *os* mutants of *Neurospora crassa*. *Neurospora Newslett.* **29,** 16–17.

Jones, C. A., Greer-Phillips, S. E., and Borkovich, K. A. (2007). The response regulator RRG-1 functions upstream of a mitogen-activated protein kinase pathway impacting asexual development, female fertility, osmotic stress, and fungicide resistance in *Neurospora crassa*. *Mol. Biol. Cell* **18,** 2123–2136.

Kanetis, L., Forster, H., Jones, C. A., Borkovich, K. A., and Adaskaveg, J. E. (2008). Characterization of genetic and biochemical mechanisms of fludioxonil and pyrimethanil resistance if field isolates of *Penicillium digitatum*. *Phytopathology* **98,** 205–214.

Kawasaki, L., Sanchez, O., Shiozaki, K., and Aguirre, J. (2002). SakA MAP kinase is involved in stress signal transduction, sexual development and spore viability in *Aspergillus nidulans*. *Mol. Microbiol.* **45,** 1153–1163.

Kojima, K., Takano, Y., Yoshimi, A., Tanaka, C., Kikuchi, T., and Okuno, T. (2004). Fungicide activity through activation of a fungal signalling pathway. *Mol. Microbiol.* **53,** 1785–1796.

Krystofova, S., and Borkovich, K. A. (2005). The heterotrimeric G-protein subunits GNG-1 and GNB-1 form a Gbetagamma dimer required for normal female fertility, asexual development, and Galpha protein levels in *Neurospora crassa*. *Eukaryot. Cell* **4,** 365–378.

Livingston, L. R. (1969). Locus-specific changes in cell wall composition characteristic of osmotic mutants of *Neurospora crassa*. *J. Bacteriol.* **99,** 85–90.

Ochiai, N., Fujimura, M., Motoyama, T., Ichiishi, A., Usami, R., Horikoshi, K., and Yamaguchi, I. (2001). Characterization of mutations in the two-component histidine kinase gene that confer fludioxonil resistance and osmotic sensitivity in the *os-1* mutants of *Neurospora crassa*. *Pest Manag. Sci.* **57,** 437–442.

Pandey, A., Roca, M. G., Read, N. D., and Glass, N. L. (2004). Role of a mitogen-activated protein kinase pathway during conidial germination and hyphal fusion in *Neurospora crassa*. *Eukaryot. Cell* **3,** 348–358.

Park, G., Pan, S., and Borkovich, K. A. (2008). Mitogen-activated protein kinase cascade required for regulation of developmental and secondary metabolism in *Neurospora crassa*. *Eukaryot. Cell* **7,** 2113–2122.

Park, G., Jones, C., and Borkovich, K. (2009). Signal transduction pathways. *In* "Cellular and Molecular Biology of Filamentous Fungi," (K. A. Borkovich and D. J. Ebbole, eds.), (in press).

Pillonel, C., and Meyer, T. (1997). Effect of phenylpyrroles on glycerol accumulation and protein kinase activity of *Neurospora crassa*. *Pest Manag. Sci.* **49,** 229–236.

Posas, F., Wurgler-Murphy, S. M., Maeda, T., Witten, E. A., Thai, T. C., and Saito, H. (1996). Yeast HOG1 MAP kinase cascade is regulated by a multistep phosphorelay mechanism in the SLN1-YPD1-SSK1 "two-component" osmosensor. *Cell* **86,** 865–875.

Posas, F., Takekawa, M., and Saito, H. (1998). Signal transduction by MAP kinase cascades in budding yeast. *Curr. Opin. Microbiol.* **1,** 175–182.

Selitrennikoff, C., Lilley, B., and Zucker, R. (1981). Formation and regeneration of protoplasts derived from a temperature-sensitive osmotic strain of *Neurospora crassa*. *Exp. Mycol.* **5,** 155–161.

Vitalini, M. W., de Paula, R. M., Goldsmith, C. S., Jones, C. A., Borkovich, K. A., and Bell-Pedersen, D. (2007). Circadian rhythmicity mediated by temporal regulation of the activity of p38 MAPK. *Proc. Natl. Acad. Sci. USA* **104,** 18223–18228.

Vogel, H. J. (1964). Distribution of lysine pathways among fungi: Evolutionary implications. *Am. Nat.* **98,** 435–446.

West, A. H., and Stock, A. M. (2001). Histidine kinases and response regulator proteins in two-component signaling systems. *Trends Biochem. Sci.* **26,** 369–376.

Wolanin, P. M., Thomason, P. A., and Stock, J. B. (2002). Histidine protein kinases: Key signal transducers outside the animal kingdom. *Genome Biol.* **3,** REVIEWS3013.

Yoshimi, A., Kojima, K., Takano, Y., and Tanaka, C. (2005). Group III histidine kinase is a positive regulator of Hog1-type mitogen-activated protein kinase in filamentous fungi. *Eukaryot. Cell* **4,** 1820–1828.

Zhang, Y., Lamm, R., Pillonel, C., Lam, S., and Xu, J.-R. (2002). Osmoregulation and fungicide resistance: The *Neurospora crassa os-2* gene encodes a *HOG1* mitogen-activated protein kinase homologue. *Appl. Environ. Mircobiol.* **68,** 532–538.

CHAPTER EIGHTEEN

Biochemical Characterization of Plant Hormone Cytokinin-Receptor Histidine Kinases Using Microorganisms

Takeshi Mizuno *and* Takafumi Yamashino

Contents

Abstract

Results of recent studies on the model higher plant *Arabidopsis thaliana* have led us to learn about the generality and versatility of two-component systems (TCS) in eukaryotes. In the plant, TCS are crucially involved in certain signal transduction mechanisms underlying the regulation of plant development in response to a subset of plant hormones, namely, cytokinin and ethylene.

Laboratory of Molecular Microbiology, School of Agriculture, Nagoya University, Chikusa-ku, Nagoya, Japan

Methods in Enzymology, Volume 471
ISSN 0076-6879, DOI: 10.1016/S0076-6879(10)71018-1

Results of extensive plant genomics revealed that these hormone-responsive TCS are evolutionarily conserved in many other plants, including mosses, grasses, crops, and trees. In particular, the conserved cytokinin-responsive TCS is typical in the sense that the signaling pathway consists of cytokinin-receptor histidine kinases (HK), histidine-containing phosphotransfer (HPt) factors, and downstream phosphoaccepting response regulators (RR), which together act as His-to-Asp multistep phosphorelay components, and which together modulate the downstream network of cytokinin-responsive gene regulation. The ethylene-responsive TCS is atypical in that ethylene-receptor HKs appear to directly interact with the downstream mitogen-activated protein kinase (MAPK) cascade. The ethylene-responsive HKs have already been introduced in the previous edition of *Methods in Enzymology* [Schaller, G. E., and Binder, B. M. (2007). Biochemical characterization of plant ethylene receptors following transgenic expression in yeast. *Methods Enzymol.* 422, 270–287]. Hence, here we focus on the cytokinin-receptor HKs, which are capable of functioning in microorganisms, such as *Escherichia coli* and *Saccharomyces cerevisiae*. Some versatile protocols useful for analyzing plant TCS factors by employing these microorganisms will be introduced.

1. Introduction

The so-called two-component system (TCS) is a paradigm of signal transduction in the model plant *Arabidopsis thaliana* (Mizuno, 2004, 2005). As has been well established in prokaryotes, central to TCS is a histidine protein kinase (HK), which serves as a sensor (or receptor) for a certain environmental stimulus (e.g., EnvZ, ArcB, RcsC, see Fig. 18.1, left-hand side) (Mizuno, 1998; Stock *et al.*, 2000). *A. thaliana* has 11 nuclear-encoded HKs (Hwang *et al.*, 2002), among which five HKs act redundantly as ethylene-receptors (Bleecker, 1999; O'Malley *et al.*, 2005; Schaller and Kieber, 2002). Ethylene is a gaseous plant hormone, which plays crucial roles in fruit ripening and flower senescence (Abeles *et al.*, 1992). However, ethylene-receptor HKs (ETR1, ETR2, ERS1, ERS2, EIN4) are atypical in that they directly modulate the function of a downstream component CTR1, which does not belong to the TCS family (Benavente and Alonso, 2006; Chang and Shockey, 1999; Woeste and Kieber, 1998). In fact, CTR1 is homologous to a mitogen-activated protein kinase kinase kinase (MAPKKK). Therefore, the ethylene-responsive signal transduction system most likely does not rely on the typical His-to-Asp phosphorelay. Also, it is the subject of debate whether or not the histidine kinase activity of the ethylene-receptors is essential for the signal transduction (Bleecker, 1999; Chang and Stadler, 2001; Schaller and Bindez, 2007; Wang *et al.*, 2003). In addition to those ethylene-receptors HKs, *A. thaliana* possesses three other HKs (AHK2, AHK3, AHK4/CRE1) (Higuchi *et al.*, 2004;

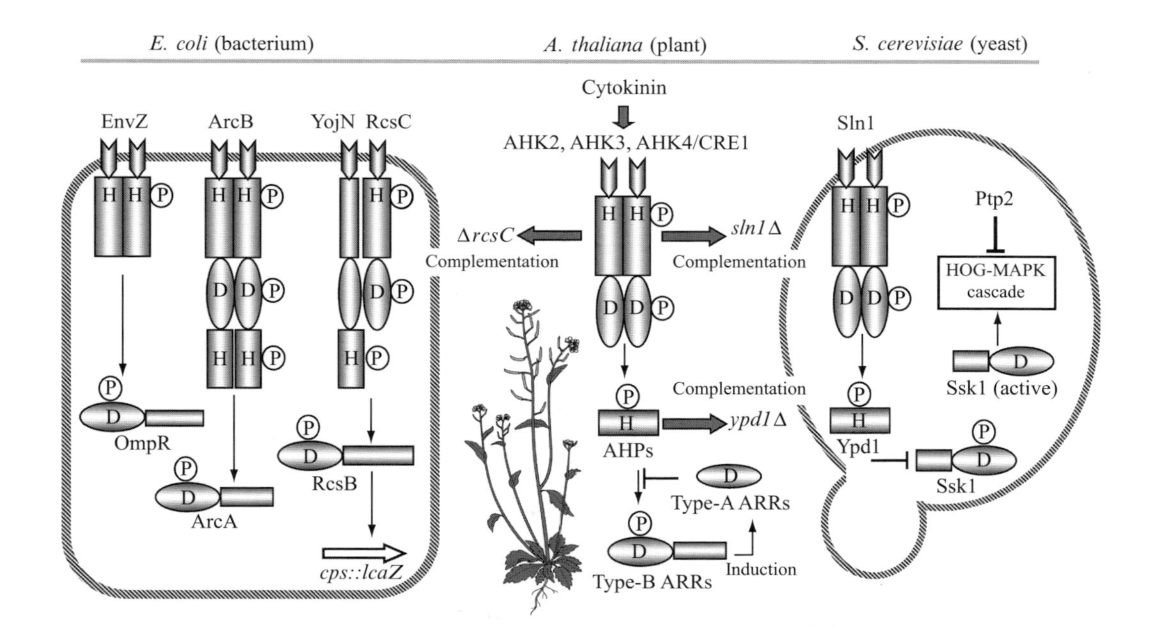

Figure 18.1 A schematic representation of typical TCS components, which were found and characterized in model organisms, including *E. coli* (bacterium), *A. thaliana* (higher plant), and *S. cerevisiae* (yeast). They are highly homologous to each other in their primary amino acid sequences, as schematically illustrated. Other details are given in the text.

Nishimura *et al.*, 2004; Ueguchi *et al.*, 2001a), which redundantly serve as receptors for cytokinins (Suzuki *et al.*, 2001; Yamada *et al.*, 2001), a class of plant hormones implicated in nearly all aspects of plant growth and development, including cell division, differentiation, and shoot initiation (Mok and Mok, 2001). As illustrated in Fig. 18.1 (central), this cytokinin-responsive signaling pathway is typical in that the cytokinin-receptor HKs phosphorylate the downstream components, namely, a set of histidine-containing phosphotransfer (HPt) factors (AHPs) (Miyata *et al.*, 1998; Suzuki *et al.*, 1998). Subsequently, the phosphorylated AHPs serve as phosphodonor for a set of response regulators (ARRs) (Imamura *et al.*, 1998, 1999; Kiba *et al.*, 1999). The cytokinin-mediated AHK–AHP–ARR phosphorelay signaling network has been characterized, and a number of comprehensive review articles are currently available (Aoyama and Oka, 2003; Heyl and Schmülling, 2003; Kakimoto, 2003; Mizuno, 2004, 2005; Müller and Sheen, 2007; Sheen, 2002; To and Kieber, 2008). A given AHK–AHP–ARR phosphorelay signaling pathway resembles the well-known *Escherichia coli* ArcB–ArcA (Ishige *et al.*, 1994; Matsubara *et al.*, 2000; Tsuzuki *et al.*, 1995) and RcsC–YojN–RcsB multistep phosphorelay pathways (Hagiwara *et al.*, 2003; Majdalani and Gottesman, 2007; Takeda *et al.*, 2001), both of which are also reminiscent of the *Saccharomyces cerevisiae* Sln1–Ypd1–Ssk1 osmoresponsive phosphorelay pathway (Hohmann, 2002; Saito and Tatebayashi, 2004; Wurgler-Murphy and Saito, 1997). In particular, note that AHKs have structural designs very similar to RcsC (*E. coli*) and Sln1 (*S. cerevisiae*) in that they are hybrid HKs containing a histidine kinase domain and a receiver domain in a single molecule (Fig. 18.1). Note also that these multistep phosphorelay pathways commonly involve HPt intermediates (Ishige *et al.*, 1994; Posas *et al.*, 1996); therefore, they are considered to be sophisticated versions of the canonical TCS, like the prototype osmoregulatory EnvZ–OmpR system (Fig. 18.1, left-hand side) (Mizuno and Mizushima, 1990).

As mentioned above, cytokinins are a class of plant growth hormones implicated in nearly all aspects of plant development (Mok and Mok, 2001). Cytokinins include a variety of derivatives of N^6-substituted aminopurines and their riboside, which are represented by *trans*-zeatin (*t*-zeatin) and isopentenyl-adenine (*i*P) (Kakimoto, 2003; Kamada-Nobusada and Sakakibara, 2009). Here, this cytokinin-responsive AHK–AHP–ARR phosphorelay network is summarized briefly (Fig. 18.1, central). This phosphorelay pathway is initiated by the redundant cytokinin-receptor HKs (AHK2, AHK3, AHK4/CRE1) (Inoue *et al.*, 2001; Ueguchi *et al.*, 2001b). These HKs in the plasma membrane sense the hormonal signal by directly binding external cytokinin (Miwa *et al.*, 2007; Suzuki *et al.*, 2001; Yamada *et al.*, 2001). This stimulus induces the autophosphorylation activity of HKs at the canonical phosphoaccepting histidine (His, H) residue. Subsequently, AHKs phosphorylate a set of downstream HPt intermediates

(AHPs) at their invariant phosphoaccepting histidine (H) residue (Hutchison *et al.*, 2006; Tanaka *et al.*, 2004). Phospho-AHPs move into the nucleus and donate the phosphoryl group to response regulators (ARRs) containing a canonical phosphoaccepting aspartate (Asp, D) residue in the receiver domain (Imamura *et al.*, 1998, 2003). The phosphorylated form of type-B ARRs serve as active transcriptional regulators through binding to specific target genes (Hosoda *et al.*, 2002; Imamura *et al.*, 2003; Mason *et al.*, 2005; Sakai *et al.*, 2001), resulting in rapid modulation of expression profiles of many cytokinin-controlled genes, which include the ones encoding a set of type-A ARRs (Kiba *et al.*, 2005; Rashotte *et al.*, 2003). The cytokinin-inducible type-A ARRs are believed to act as negative feedback regulators that attenuate exaggerated cytokinin responses by competitively titrating phospho-AHPs (Kiba *et al.*, 2003; To *et al.*, 2004). After all, it should be emphasized that the cytokinin-receptor-less (or cytokinin-impotent) mutant plants are viable, but very dwarf and sterile, implying that the AHK-mediated phosphorelay circuitry plays pivotal and essential roles in plant development and survival throughout the life cycle (Argyros *et al.*, 2008; Higuchi *et al.*, 2004; Ishida *et al.*, 2008; Nishimura *et al.*, 2004).

In addition to the ethylene or cytokinin-receptors HKs, *A. thaliana* has three more HKs (AtHK1, CKI1, AHK5), each of which has a structural design similar to that of a cytokinin-receptor hybrid HK. These HKs have also been characterized to some extent, although their respective primary stimuli are only partially defined. AtHK1 has been proposed as an osmotic sensor, which appears to be involved in cold and/or drought stress responses (Tran *et al.*, 2007; Urao *et al.*, 1999; Wohlbach *et al.*, 2008). CKI1 has originally been implicated in a cytokinin-mediated signal transduction pathway, as judged by its ability to enhance (or induce) the *in vitro* shoot formation from explants (Kakimoto, 1996). It was then suggested that this HK is crucial for a process of female gametophyte development (Hejatko *et al.*, 2003; Pischke *et al.*, 2002). AHK5 (also known as CKI2) is involved in regulation of root elongation and/or stomatal cell opening in response to abscisic acid (another plant hormone, generally referred as to ABA) in a manner dependent on the ethylene-receptor ETR1 (Desikan *et al.*, 2008; Iwama *et al.*, 2007). This is indicative of the existence of a higher order of hormone-responsive TCS signaling network, into which certain multi-signals are integrated through a set of HKs.

Notably, recent results involving extensive plant genomics revealed that most of these *Arabidopsis* HKs are conserved in many other plant species, including moss, grass, crop, and tree, if not all (e.g., for rice, see Pareek *et al.*, 2006; Schaller *et al.*, 2007). These varieties of plant species are apparently divergent from each other in morphological, physiological, ecological, and agronomical properties. Therefore, it will be of interest to experimentally clarify the biological functions of HKs and RRs in different plant species.

For instance, legumes (e.g., *Glycine max*, soybean) have a unique ability to engage in beneficial symbiosis with nitrogen-fixing bacteria, which allows the host plant to utilize atmospheric nitrogen. In fact, it was recently demonstrated for the legume *Lotus japonicus* that the cytokinin-receptor HK (an AHK ortholog) is directly involved in root nodule organogenesis (Ishida *et al.*, 2009; Murray *et al.*, 2007; Tirichine *et al.*, 2007). Based on this general background we will introduce some versatile and sophisticated protocols useful for analyzing plant HKs and RRs in microorganisms.

2. Characterization of Plant TCS Components in *E. coli*

2.1. Overview and logic behind

First of all, the following four facts (a–d) should be emphasized. (a) *Arabidopsis* cytokinin-receptor AHKs are capable of functioning in the model bacterium *E. coli* cells in a manner dependent on the externally applied cytokinin in growth medium (Suzuki *et al.*, 2001; Yamada *et al.*, 2001). (b) The intact *E. coli* cells expressing a given plant AHK can be directly used for the cytokinin-binding assay (Romanov *et al.*, 2005). (c) Furthermore, the *E. coli* cytoplasmic membranes containing a large amount of *E. coli* ArcB HK serve as an enzyme source, which can phosphorylate *Arabidopsis* phosphotransfer intermediate AHPs *in vitro* (Suzuki *et al.*, 1998). (d) Finally, the radioactive phospho-AHPs, thus prepared, are capable of transferring their phosphoryl groups to response regulator ARRs *in vitro* (Imamura *et al.*, 2003). By adopting the *E. coli* system, one may characterize given TCS components (HK, HPt, and RR) from any plant species.

As schematically shown in Fig. 18.2, the *E. coli* RcsC–YojN–RcsB phosphorelay system is unique in that it consists of three components: namely, a membrane-located hybrid HK (RcsC), an HPt factor (YojN), and a transcriptional factor RR (RcsB) (Takeda *et al.*, 2001). This phosphorelay signaling system responds to certain stresses on the cell envelope, especially, at a growth temperature below 25 °C (Hagiwara *et al.*, 2003). As a result, the expression of the capsular polysaccharide synthesis (*cps*) operon is markedly induced by the activated phospho-RcsB transcription factor, giving rise to mucous-like *E. coli* colonies on solid medium. This RcsC–YojN–RcsB system can be monitored with an appropriate *E. coli* strain containing a chromosome-integrated *cps-promoter::lacZ* gene encoding β-galactosidase (β-Gal) on solid medium containing 5-bromo-4-chloro-3-indolyl-β-D-galactoside (X-Gal), or through quantitative measurement of the enzyme activity with intact *E. coli* cells grown in liquid medium. These measurements are absolutely dependent on the primary sensor HK RcsC; therefore, *E. coli* mutants lacking the *rcsC* gene (designated Δ*rcsC*) exhibits

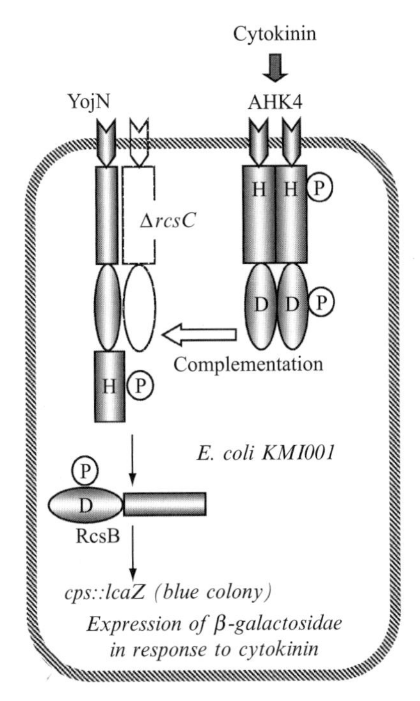

Figure 18.2 A schematic explanation as to how one can easily assess the activity of a cytokinin-receptor plant HK in the special mutant *E. coli* strain. Note that it is also possible to adopt this strategy to characterize any plant HK other than the cytokinin-receptors. Note also that the transformed *E. coli* cells can be directly used for the cytokinin-binding assay. Other details are given in the text.

no β-Gal activity even in the presence of stimuli. Based on the fact that the structural design of *Arabidopsis* AHK4/CRE1 is very similar to that of RcsC, Suzuki *et al.* (2001) demonstrated that the Δ*rcsC* mutation was fully complemented when the AHK4/CRE1 cDNA was cloned into an *E. coli* expression plasmid vector, and introduced into the Δ*rcsC* cells. An important fact was that the complementation was dependent on the externally applied cytokinin in the growth medium. This indicates that the plant hormone receptor HK is capable of functioning in the *E. coli* cells in a manner dependent on the plant hormone cytokinin. Based on this, this versatile method has been established to examine the activity of any plant HKs by adopting the simple *E. coli* system. Indeed, this method was applicable also for AHK2 and AHK3 (Miwa *et al.*, 2007; Romanov *et al.*, 2006; Spichal *et al.*, 2004), and for homologous cytokinin-receptor HKs from other plant species (e.g., *L. japonicus* and *Zea mays*) (Tirichine *et al.*, 2007; Yonekura-Sakakibara *et al.*, 2004). Furthermore, other *Arabidopsis* CKI1 and AHK5 HKs have the ability to complement the Δ*rcsC* mutation, although in a cytokinin-independent manner (Iwama *et al.*, 2007; Yamada *et al.*, 2001).

These results suggest that AHK4/CRE1 is properly inserted in the *E. coli* cytoplasmic membranes with the N-terminal domain extruding toward the medium, so that its extracellular domain directly binds to cytokinin present in the medium, and that its intracellular catalytic HK domain interacts with the noncognate YojN protein (Miwa *et al.*, 2007). Taking advantage of this assumption, Romanov *et al.* (2005) further developed a rapid method to quantitatively analyze the interaction of AHK4/CRE1 with various kinds of cytokinin derivatives with the use of AHK4/CRE1 expressing intact *E. coli* cells.

The *E. coli* system provides another useful way to characterize the *in vitro* phosphotransfer interaction between *Arabidopsis* AHPs and ARRs. Indeed, Suzuki *et al.* (1998) developed an *in vitro* method to prepare the radioactive phospho-AHPs with the use of isolated *E. coli* membranes enriched with the ArcB hybrid HK (see Fig. 18.1). The isolated *E. coli* membranes have the ability to phosphorylate the purified AHP proteins *in vitro*. The resultant radioactive phospho-AHP, thus prepared, can be used as a phosphodonor for the purified ARR proteins to examine the phosphotransfer reaction between them (Imamura *et al.*, 2003). It is reasonable to assume that this method can also be used with HPt and RR components from any other plant species.

2.2. Materials for the *E. coli* system

(1) To assay the activity of a given plant HK, a special *E. coli* mutant strain is required (see Fig. 18.2), namely, strain KMI001 [Δ*rcsC ara thi* Δ(*pro-lac*) Δ(*wzc-wca*)]. A derivative of KMI001 (named KMI002), which lacks the intrinsic *cps* operon, can also be used. These *E. coli* strains carry the artificial *cps-promoter::lacZ* fusion gene on the chromosome (Hagiwara *et al.*, 2003; Takeda *et al.*, 2001). The reporter gene encodes β-Gal under the control of the *cps*-promoter, which is regulated through the RcsC–YojN–RcsB phosphorelay signal transduction in response to an yet unidentified environmental signal. Note that KMI001 and KMI002 lack the *rcsC* gene, which encodes a sensor HK essential for activation of the *cps*-promoter. Hence, the mutant cells cannot induce the *cps::lacZ* gene. However, if a plant HK gene on an appropriate *E. coli* expression plasmid vector was introduced into this strain, it would be capable of complementing the defect, so that the transformed *E. coli* cells would express the *cps::lacZ* gene, the enzyme product of which could be easily assayed by conventional laboratory methods (Sambrook and Russell, 2001, see vol. 3, 17.97).

(2) To express a given plant HK gene, any versatile *E. coli* expression vector could be used (Sambrook and Russell, 2001, see vol. 3, A3.2).

However, to clone a plant HK cDNA, a low-copy-number plasmid with a relatively weak promoter is preferable (e.g., commercially available plasmid vectors, pSTV28/29, with the *lac* promoter), because the expression of plant membrane protein in *E. coli* is often deleterious (for pSTV28/29, see the catalog at http://www.takara-bio.com/research.htm). The *E. coli* expression vectors pSTV28/29 have replication origins from pACYC184, so that they are compatible with another common plasmid like pBR322 in *E. coli*. Therefore, it is also possible to characterize functional interactions between a given HK and its cognate downstream component (e.g., HPt factor) by coexpressing both the plant HK and HPt genes in *E. coli*, as has been demonstrated (Suzuki *et al.*, 2001).

(3) To prepare *E. coli* cytoplasmic membranes containing a large amount of the ArcB HK, the *E. coli* strain DAC903 (*Δlac169 araD rpsL relA thiA flbB arcB::Cm*^{R})-carrying plasmid pIA001 containing the *arcB* gene is useful (Matsushika and Mizuno, 1998; Tsuzuki *et al.*, 1995). The isolated membranes from this strain exhibit the ability to efficiently phosphorylate a given plant HPt factor (e.g., AHPs) *in vitro* (Suzuki *et al.*, 1998). The purified radioactive phospho-HPt protein can be used as a phosphodonor to a given purified RR (e.g., ARRs) through an *in vitro* phosphotransfer reaction (Imamura *et al.*, 2003).

(4) To grow *E. coli* cells, conventional M9 synthetic medium or Luria-Bertani (LB) complex medium can be used. Details of these *E. coli* media are in the conventional laboratory manual (Sambrook and Russell, 2001, see vol. 3, A2.2).

3. Protocol for Histidine Kinase Assays in *E. coli*

3.1. Semi-quantitative protocol

(1) As a typical example, see Miwa *et al.* (2007, Figs. 1 and 2 therein).

(2) *E. coli* KMI002 cells carrying pSTV28-AHK3 harboring the full length AHK3 cDNA were grown overnight in LB-medium containing 40 m*M* glucose and 50 m*M* phosphate buffer (pH 7.2) at 37 °C.

(3) Fully grown *E. coli* cells are diluted with fresh glucose-LB-medium to approximately 10^6 cells/ml.

(4) Diluted *E. coli* cells are directly spotted onto glucose-LB-agar-plates containing X-Gal (40 μg/ml), the selection marker of plasmid (chloramphenicol, 30 μg/ml), and appropriate concentrations of *t*-zeatin (final concentrations of 0.1–100 μ*M*).

(5) Plates are incubated for 48 h at 25 °C.

(6) Inspection reveals whether or not the spotted cells show blue color in a manner dependent on cytokinin in the medium, and pictures are taken.
(7) The cytokinin-dependent intensity of blue color reflects the histidine kinase activity of a given plant HK.

3.2. Quantitative protocol

(1) As a typical example, see Yamada *et al.* (2001, Fig. 1 therein).
(2) *E. coli* KMI001 carrying pINΔEH-AHK4 containing the full length AHK4 cDNA is cultivated overnight in glucose-LB-medium at 37 °C.
(3) A fully grown *E. coli* culture (60 μl) is inoculated into fresh glucose-LB-medium containing 40 m*M* glucose and 50 m*M* phosphate buffer (pH 7.2) (1.5 ml). Varied concentrations of cytokinins are added to the medium (e.g., *t*-zeatin, *i*P) (final concentrations of 0.01–10 μ*M*), in addition to the selection marker of plasmid (ampicillin, 50 μg/ml).
(4) *E. coli* cells are grown overnight by vigorous shaking at 25 °C.
(5) An aliquot (100–200 μl) of cells are collected by centrifugation, and then suspended in 500 μl of Z-buffer in a test tube. Z-buffer contains 4.8 g NaH_2PO_4, 21.48 g $Na_2HPO_4 \cdot 12H_2O$, 0.75 g KCl, 0.246 g $MgSO_4 \cdot 7H_2O$ in 1 l of H_2O.
(6) Optical density ($OD_{600\ nm}$) of the suspension is measured and recorded.
(7) A drop of toluene (10 μl) is added to the tube and the tube is vigorously mixed.
(8) After being incubated for 1 h at 37 °C, β-Gal activity is measured at 30 °C, as follows. ONPG (*o*-nitrophenyl-β-galactopyranoside) is freshly prepared with 100 m*M* sodium phosphate buffer (4 mg/ml). To start the reaction, the ONPG solution (100 μl) is added to the tube. When the color of reaction mixture turns yellow, 1 *M* Na_2CO_3 (1 ml) is added to stop the reaction.
(9) Record the reaction times as precisely as possible.
(10) The optical densities of the reaction mixtures measured at both $OD_{420\ nm}$ and $OD_{550\ nm}$.
(11) β-Gal activity is calculated as relative units by using the following formula: $1000\times\ [(OD_{420\ nm} - OD_{550\ nm}\times 1.75)/t(\text{min})\ \text{vol.(ml)}\ OD_{600\ nm}]$.
(12) For more details of the β-Gal assay, see the conventional laboratory manual (Sambrook and Russell, 2001, vol. 3, 17.97).
(13) The activity reflects the histidine kinase activity of a given plant HK, which functions in a manner dependent on the concentrations of cytokinin added in the medium.

4. Protocol of Cytokinin-Binding Assay with Intact *E. coli* Cells

(1) As a typical example, see Romanov *et al.* (2005, Table 1 therein). The following protocol was adopted from the reference with slight modifications.
(2) *E. coli* KMI001 carrying pINΔEH-AHK4 is cultivated to logarithmic phase in M9-medium with 0.1% casamino acids, supplemented with carbenicillin (25 μg/ml) at 37 °C.
(3) An aliquot of the cell suspension is mixed with and without highly tritium-labeled *t*-zeatin (*t*-[2-^{3}H] zeatin, 1–2 pmol, 20,000–40,000 cpm).
(4) After being incubated on ice for 30 min, the *E. coli* cells are collected by centrifugation at 13,000 rpm for 2 min in the cold.
(5) The supernatants are carefully removed by a vacuum aspiration and the remaining cells are suspended in ethanol (99%, 0.24 ml). Make sure that the supernatants do not contain protein.
(6) The cell pellet is extracted with ethanol by shaking for 1 h, followed by a brief centrifugation.
(7) The radioactivity of 0.2 ml of ethanol extract is measured with scintillation counting (Amersham Bioscience, UK).
(8) The normalized radioactivity (cpm) reflects the cytokinin-binding activity of AHK4/CRE1.
(9) Note that a competitive-binding assay with a variety of cytokinin derivatives is also informative.

5. Preparation of Radioactive Phospho-HPt Factor, and *In Vitro* Assay of Phosphotransfer to RR

(1) *E. coli* cytoplasmic membranes containing a large amount of ArcB HK protein are isolated, as follows (Tokishita *et al.*, 1990). *E. coli* strain DAC903 carrying plasmid pIA001 was grown in 2 l of LB-medium (Tsuzuki *et al.*, 1995). The cells are harvested during the logarithmic growth phase are treated on ice in a phosphate buffer (pH 7.0) with lysozyme (0.25 mg/ml) to prepare spheroplasts. After the addition of DNase I (50 μg/ml of reaction mixture), the spheroplasts are passed through an Aminco French pressure cell (FA#073) at 250 kg/cm^2, and then homogenized to shear the released DNA. The cell lysate is centrifuged at 11,400×*g* for 15 min to remove intact cells. It is then

further centrifuged at 127,000×*g* for 2 h, and the collected membranes are suspended in a small volume of buffer, 50 m*M* Tris–acetate (pH 7.8), 1 m*M* DTT, 2% glycerol, 250 m*M* sucrose. The membranes are then layered over a 4.4-ml linear gradient, 30–44% (w/w), of sucrose dissolved in the same buffer, and centrifuged at 60,000×*g* for 15 h at 4 °C. The sucrose density gradient is divided into five fractions, from the bottom to top. Fractions containing the cytoplasmic membranes with pale brown color are collected and diluted with a large volume of a buffer (TGS-buffer) containing 50 m*M* Tris–acetate (pH 7.8), 1 m*M* DTT, 10% glycerol, 100 m*M* sucrose. The cytoplasmic membranes are collected by centrifugation of 127,000×*g* for 2 h, and then suspended in a small volume of TGS-buffer. Then, F1-ATPase should be stripped from the cytoplasmic membranes, as follows. The isolated cytoplasmic membranes (2 mg of protein) are treated with 1 ml of TGS-buffer containing 4 *M* urea for 30 min on ice, and then recovered by centrifugation at 50,000 rpm for 30 min. The pellet is washed once with 1 ml of TGS-buffer and finally with 1 ml of TEDG-buffer containing 50 m*M* Tris–HCl (pH 8.0), 0.35 m*M* EDTA, 2 m*M* DTT, 10% glycerol. The urea-treated cytoplasmic membranes are suspended in 0.5 ml of the same buffer and stored at −80 °C.

(2) Plant HPt and RR proteins are purified, according to conventional procedures, for instance, by employing appropriate plasmid vectors for protein production in *E. coli*. As typical instances, see Suzuki *et al.* (1998) and Imamura *et al.* (2003), in which AHP1 and ARR11-receiver domain proteins were purified and subjected to *in vitro* phosphotransfer assay, as follows.

(3) Urea-treated membranes (10 mg of protein) are incubated with the purified AHP1 polypeptide (0.4 μg) at 37 °C in the presence of 0.05 m*M* [γ-^{32}P]ATP (10,000 cpm/pmol), 200 m*M* KCl and 5 m*M* $MgCl_2$ in TEDG-buffer. The reaction mixture is immediately applied to a Sephadex G-75 column previously equilibrated with TEDG-buffer. Fractions containing the labeled AHP1 protein, which are essentially free from used membranes and ATP, are collected. Existence of the radioactive phospho-AHP1 is confirmed by a conventional sodium dodecylsulfate polyacrylamide gel electrophoresis (SDS–PAGE), followed by autoradiography. Then, one can demonstrate that a given HPt factor has the ability to receive a radioactive phosphoryl group from ArcB HK.

(4) The purified radioactive phospho-AHP1 polypeptides are incubated with purified ARR11 polypeptides for short periods in TEDG-buffer (30 μl) containing 50 m*M* KCl and 7 m*M* $MgCl_2$ on ice. After incubation, the samples are immediately subjected to SDS–PAGE, followed by autoradiography. Then, one can demonstrate that a given HPt factor has the ability to transfer a phosphoryl group to a given RR component.

6. Characterization of Plant TCS Components in *S. cerevisiae*

6.1. Overview and logic behind

The model yeast *S. cerevisiae* has a single Sln1(HK)–Ypd1 (Hpt)–Ssk1 (RR) phosphorelay system, which is mainly involved in the adaptation to changes in medium osmolarity (Fig. 18.1, right-hand side) (Hohmann, 2002; Saito and Tatebayashi, 2004; Wurgler-Murphy and Saito, 1997). This TCS pathway acts as a gas-and-brake system for the downstream eukaryotic mitogen-activated protein kinase (MAPK) cascade, involving HOG1-MAPK. At moderate medium osmolarity, the osmosensor HK Sln1 signals the downstream HPt Ypd1, the phosphorylated form of which acts as a brake against the further downstream RR Ssk1. The phospho-Ssk1 is inactive, and the nonphosphorylated form of Ssk1 activates the downstream HOG-MAPK cascade. At high medium osmolarity, the HK activity of Sln1 is downregulated, thereby accumulating the nonphosphorylated active form of Ssk1, which stimulates the production of the intracellular osmotic solute glycerol by activating the stress-responsive HOG-MAPK cascade. However, the exaggerated MAPK cascade is attenuated by the inducible Ptp2 phosphatase specific to MAP kinase(s), through a feedback inhibition. Certain yeast mutants lacking either the SLN1 or YPD1 genes are lethal, presumably due to hyper-activation of the HOG-MAPK cascade even in medium of modest osmolarity (Fig. 18.3). However, when these mutants (*snl1*Δ or *ypd1*Δ) were transformed with an appropriate plasmid carrying the PTP2 gene fused to the GAL1-promoter, which is induced by galactose but not glucose, the transformed mutant cells were conditional lethal (see Fig. 18.3A and B). In other words, the cells cannot grow on the glucose-containing medium, while they can form colonies on the galactose-containing solid medium, because the GAL1-induced Ptp1 phosphatase attenuated the exaggerated HOG-MAPK cascade. Many plant researchers took advantage of these sophisticated yeast HK mutants, and tested whether certain plant HK genes can complement the conditional lethal phenotype of *sln1*Δ. Indeed, the attempts were successful in most of the cases (e.g., AHK4/CRE1, CKI1, AtHK1) (Fig. 18.3C) (Inoue *et al.*, 2001; Kakimoto, 1996; Murray *et al.*, 2007; Reiser *et al.*, 2003; Ueguchi *et al.*, 2001b). For instance, the AHK4/CRE1 gene cloned onto an appropriate yeast expression plasmid vector was introduced into *sln1*Δ. The resulting transformed cells were viable even on the glucose-containing medium, but only when cytokinin was supplied in the medium. Similarly, a plant AHP gene could also rescue the conditional lethal phenotype of *ypd1*Δ, although in a cytokinin-independent manner (Suzuki *et al.*, 1998).

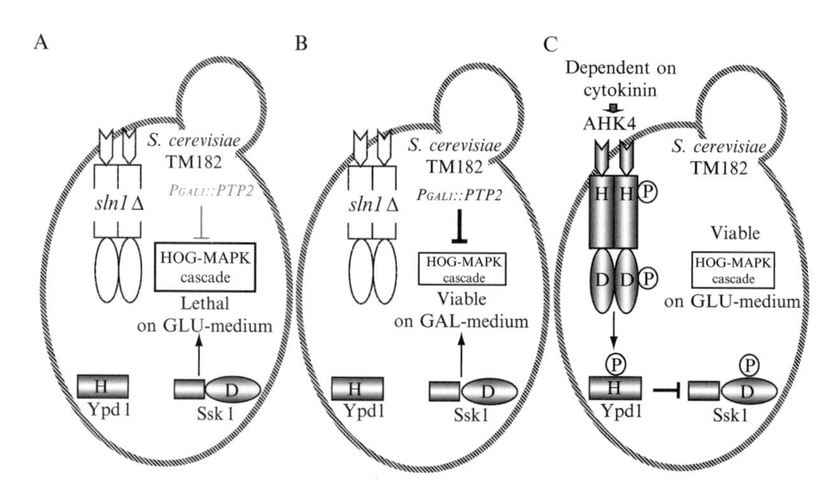

Figure 18.3 A schematic explanation as to how one can easily assess the activity of a cytokinin-receptor plant HK in the special mutant *S. cerevisiae* strain. Note that it is also possible to adopt this strategy to characterize any plant HK other than the cytokinin-receptors, although they would function in a manner independent of cytokinin. Other details are given in the text.

Schizosaccharomyces pombe is also an eukaryotic microorganism used to characterize plant HKs by expressing the functional form of plant proteins in the yeast cells. This fission yeast also has the TCS-MAPK hybrid cascade, which is analogous to that of the budding yeast (Aoyama *et al.*, 2000; Santos and Shiozaki, 2001). The fission yeast has three HKs, which respond mainly to oxidative stresses, and which are distinct from the budding yeast Sln1. Otherwise, the downstream signaling components, including HPt and RR factors, are highly homologous between the fission and budding yeasts. Furthermore, Yamada *et al.* (2001) succeeded in preparing the *S. pombe* membranes containing a large amount of functional AHK4/CRE1 proteins, which served for the *in vitro* cytokinin-binding assay, by which it was proved for the first time that AHK4/CRE1 directly binds specifically a set of cytokinin derivatives with a very high affinity. The details of this *S. pombe* system are not included in this chapter, due to limited space. However, a brief outline will be reported below, for those who are interested in the protocol.

6.2. Materials for the yeast system

(1) To assay the activity of a given plant HK, a special *S. cerevisiae* mutant strain is required. This is strain TM182 [*MATa uea3 leu2 his3 sln1::hisG*] carrying plasmid pSSP25 (Maeda *et al.*, 1994). Plasmid pSSP25 (URA3 marker) carries the GAL1p::PTP2 gene, with the PTP2 gene under the control of the GAL1-promoter. This strain lacking the SLN1 HK gene

(*sln1*Δ) is lethal in a standard glucose-containing medium, and the lethality is suppressed by the induction of PTP2 phosphatase encoded by the GAL1p::PTP2 gene, so that the cells can grow only in medium containing the inducer galactose, as explained above. However, if a given plant HK gene complemented the defect of *sln1*Δ, the transformed *S. cerevisiae* cells could grow even on a glucose-based medium. Note that unknown suppression mutations occur frequently in this mutant, so that a single colony with the appropriate nature should be selected each time on galactose-containing medium.

(2) To assay the activity of a given plant HPt, another special *S. cerevisiae* mutant strain is required, namely, CUY1 (*MATa ura3 leu2 his3 ypd1::LEU2*) carrying pGP22 (HIS3 marker) encompassing the GAL1p::PTP2 gene (Posas *et al.*, 1996). Other details are the same as those given in the above paragraph.

(3) To express either a given plant HK or HPt gene in yeast, any versatile *E. coli*-cloning and *S. cerevisiae*-expression binary vector could be used (Sambrook and Russell, 2001, see vol. 3, A3.4). To clone and express a plant HK cDNA; however, a low-copy-number stable plasmid with a relatively weak promoter is preferable (e.g., commercially available plasmid vector pAUR123, with the ADH1 promoter, and with the CEN4/ARS1 replication origin) (Takara, Kyoto, Japan, see the catalog at http://www.takara-bio.com/research.htm).

(4) To grow *S. cerevisiae* cells, media, and growth conditions are used according to the conventional laboratory manuals of yeast genetics (Amberg *et al.*, 2005; Guthrie and Fink, 1994). For instance, complete synthetic (CS) medium can be used (Sambrook and Russell, 2001, see vol. 3, A2.9). Cytokinins (*t*-zeatin, *i*P, and any others to be tested) are dissolved in dimethylsulfoxide (DMSO) at 5000 times of the final concentrations used in yeast media.

7. Protocol of Histidine Kinase Assay in *S. cerevisiae*

(1) As a typical example, see Inoue *et al.* (2001, Fig. 4 therein).

(2) *S. cerevisiae* TM182 carrying a plasmid harboring the cytokinin-receptor *AHK4/CRE1* gene (e.g., p415CYC-CRE1, LEU marker) and pSSP25 (URA3 marker) is streaked on solid CS-galactose-medium (without uracil and leucine, of cause), and the agar-plates are incubated at 30 °C for 48 h.

(3) A colony is picked and suspended in 1 ml of liquid CS-glucose-medium.

(4) The cell density is measured with a hemocytometer, and varied densities of cells with the same medium are prepared (10^4–10^6 cells/ml).
(5) An aliquot (10 μl) of the cell suspensions, serially diluted, is spotted on solid CS-glucose-medium, with and without cytokinin (10 μM). The same is done on solid CS-galactose-medium, as a reference.
(6) The plates are incubated at 30 °C for 48 h.
(7) PTP2-dependent growth in the absence of cytokinin is confirmed by comparing the growth differences on the +Gal and +Glu plates. The cytokinin (or CRE1)-dependent viability is then assessed on +Glu plates.

8. Protocol of HPt Factor Assay in *S. cerevisiae*

(1) As a typical example, see Suzuki *et al.* (1998, Fig. 5 therein).
(2) *S. cerevisiae* CUY1 carrying a plasmid harboring the *AHP1* gene (e.g., pCUY15-AHP1) is streaked on solid CS-galactose-medium (without uracil and histidine, in this case), and the agar-plates are incubated at 30 °C for 48 h.
(3) Other details are the same as those described above (steps 3–7).
(4) Compare the growth differences on the +Gal and +Glu plates. Then, assess the AHP1-dependent viability of *S. cerevisiae* cells. In this case, note that the complementation could be observed in a cytokinin-independent manner.

9. Protocol of Cytokinin-Binding Assay by Using *S. pombe* Membranes Enriched in AHK4/CRE1

(1) As a typical example, see Yamada *et al.* (2001, Fig. 2 therein), in which the *S. pombe* membranes enriched in AHK4/CRE1 were prepared and used for cytokinin-binding assay.
(2) The *S. pombe* strain used is JY741 [h^- *leu1-32 ade6-M216 ura4-D18*], carrying pREP-AHK4 (Suzuki *et al.*, 2001). The entire AHK4/CRE1 cDNA was cloned downstream of the $nmt1^+$ promoter on pREP1 (an *S. pombe* versatile expression plasmid vector). *S. pombe* cells are grown at 30 °C in EMM-medium (a standard synthetic complete medium for *S. pombe*).
(3) *S. pombe* cells are harvested, and then fractionated into soluble and insoluble membrane fractions by means of appropriate conventional

procedures. The isolated membranes can be used for an *in vitro* cytokinin-binding assay.

(4) For the binding assay, a highly radioactive cytokinin derivative is needed (some might be coavailable). Otherwise, ^{3}H-labeled isopentenyl-adenine (^{3}H-*i*P) can be synthesized from common ^{3}H-AMP. First, ^{3}H-labeled isopentenyl-adenosine (^{3}H-*i*PA) is synthesized from ^{3}H-AMP and dimethyl-allyl-diphosphate (DAP) by incubation with an isopentenyl transferase. The product is treated with an alkaline phosphatase to remove the phosphoryl group at the position of C^5 of the adenosine moiety. The resulting highly active ^{3}H-*i*PA is concentrated (e.g., 5×10^4 dpm/pmol). To obtain ^{3}H-*i*P, the ^{3}H-*i*PA product is treated with 0.1 *M* HCl at 100 °C, and stored at −25 °C.

(5) The AHK4-containing *S. pombe* membranes (50 μg) are incubated with varied concentrations of ^{3}H-*i*P in buffer (100 μl), comprised of 10 m*M* MES (pH 6.0), 1 m*M* $MgCl_2$, 0.05 $MnCl_2$, 0.1 m*M* $CaCl_2$, 10% glycerol, 0.1 mg bovine serum albumin. When necessary, an excess amount (>200-fold) of nonradioactive (cold) *i*P was also added.

(6) The mixture is incubated for 30 min at 25 °C. The bound and free ^{3}H-*i*P are separated by a filter-equipped minicentrifuge tube (Ultrafree 0.1 μm filter, Millipore) following centrifugation. The radioactivity on the filter is quantified by a scintillation counter. Specific binding is determined by subtracting the binding in the presence of cold-*i*P from the total binding.

ACKNOWLEDGMENTS

We thank Dr. T. Kakimoto (Osaka University, Japan) for valuable suggestions as to the yeast protocols. This study was supported by Grants-in-Aid for Scientific Research on Priority Areas, and by a Grant-in-Aid for the GCOE Programs (Systems Biology) from the Ministry of Education, Culture, Sports, Science, and Technology of Japan.

REFERENCES

Abeles, F. B., Morgan, P. W., and Saltveit, M. E. Jr. (1992). *Ethylene in Plant Biology* Academic Press, San Diego.

Amberg, D. C., Burke, D., and Strathern, J. N. (2005). Methods in Yeast Genetics: A Cold Spring Harbor Laboratory Course Manual. Cold Spring Harbor Laboratory Press, Cold Spring Harbor, NY.

Aoyama, T., and Oka, A. (2003). Cytokinin signal transduction in plant cells. *J. Plant Res.* **116,** 221–231.

Aoyama, K., Aiba, H., and Mizuno, T. (2000). Spy1, a histidine-containing phosphotransfer (HPt) signaling protein regulates fission yeast cell cycle through the Mcs4 response regulator. *J. Bacteriol.* **182,** 4868–4874.

Argyros, R. D., Mathews, D. E., Chiang, Y. H., Palmer, C. M., Thibault, D. M., Etheridge, N., Argyros, D. A., Mason, M. G., Kieber, J. J., and Schaller, G. E. (2008). Type B response regulators of *Arabidopsis* play key roles in cytokinin signaling and plant development. *Plant Cell* **20,** 2102–2116.

Benavente, L. M., and Alonso, J. M. (2006). Molecular mechanisms of ethylene signaling in *Arabidopsis. Mol. Biosyst.* **2,** 165–173.

Bleecker, A. B. (1999). Ethylene perception and signaling: An evolutionary perspective. *Trends Plant Sci.* **4,** 269–274.

Chang, C., and Shockey, J. A. (1999). The ethylene-response pathway: Signal perception to gene regulation. *Curr. Opin. Plant Biol.* **2,** 352–358.

Chang, C., and Stadler, R. (2001). Ethylene hormone receptor action in *Arabidopsis. Bioessays* **23,** 619–627.

Desikan, R., Horák, J., Chaban, C., Mira-Rodado, V., Witthöft, J., Elgass, K., Grefen, C., Cheung, M. K., Meixner, A. J., Hooley, R., Neill, S. J., Hancock, J. T., *et al.* (2008). The histidine kinase AHK5 integrates endogenous and environmental signals in *Arabidopsis* guard cells. *PLoS One* **18,** e2491.

Guthrie, C., and Fink, G. R. (1994). Guide to yeast genetics and molecular and cell biology, pp. 933. Academic Press, San Diego.

Hagiwara, D., Sugiura, M., Oshima, T., Mori, H., Yamashino, T., and Mizuno, T. (2003). Genome-wide analyses revealing a signaling network of the RcsC-YojN-RcsB phosphorelay system in *Escherichia coli. J. Bacteriol.* **185,** 5735–5746.

Hejatko, J., Pernisova, M., Eneva, T., Palme, K., and Brzobohaty, B. (2003). The putative sensor histidine kinase CKI1 is involved in female gametophyte development in *Arabidopsis. Mol. Genet. Genomics* **269,** 443–453.

Heyl, A., and Schmülling, T. (2003). Cytokinin signal perception and transduction. *Curr. Opin. Plant Biol.* **6,** 480–488.

Higuchi, M., Pischke, M. S., Mahonen, A. P., Miyawaki, K., Hashimoto, Y., Seki, M., Kobayashi, M., Shinozaki, K., Kato, T., Tabata, S., Helariutta, Y., Sussman, M. R., *et al.* (2004). In planta functions of the *Arabidopsis* cytokinin receptor family. *Proc. Natl. Acad. Sci. USA* **101,** 8821–8826.

Hohmann, S. (2002). Osmotic stress signaling and osmoadaptation in yeasts. *Microbiol. Mol. Biol. Rev.* **66,** 300–372.

Hosoda, K., Imamura, A., Katoh, E., Hatta, T., Tachiki, M., Yamada, H., Mizuno, T., and Yamazaki, T. (2002). Molecular structure of the GARP family of plant Myb-related DNA-binding motifs of the *Arabidopsis* response regulators. *Plant Cell* **14,** 2015–2029.

Hutchison, C. E., Li, J., Argueso, C., Gonzalez, M., Lee, E., Lewis, M. W., Maxwell, B. B., Perdue, T. D., Schaller, G. E., Alonso, J. M., Ecker, J. R., and Kieber, J. J. (2006). The *Arabidopsis* histidine phosphotransfer proteins are redundant positive regulators of cytokinin signaling. *Plant Cell* **18,** 3073–3087.

Hwang, I., Chen, H.-I., and Sheen, J. (2002). Two-component signal transduction pathways in *Arabidopsis. Plant Physiol.* **129,** 500–515.

Imamura, A., Hanaki, N., Umeda, H., Nakamura, A., Suzuki, T., Ueguchi, C., and Mizuno, T. (1998). Response regulators implicated in His-Asp phosphotransfer signaling in *Arabidopsis. Proc. Natl. Acad. Sci. USA* **95,** 2691–2696.

Imamura, A., Hanaki, N., Nakamura, A., Suzuki, T., Taniguchi, M., Kiba, T., Ueguchi, C., Sugiyama, T., and Mizuno, T. (1999). Compilation and characterization of *Arabidopsis thaliana* response regulators implicated in His-Asp phosphorelay signal transduction. *Plant Cell Physiol.* **40,** 733–742.

Imamura, A., Kiba, T., Tajima, Y., Yamashino, T., and Mizuno, T. (2003). In vivo and in vitro characterization of the ARR11 response regulator implicated in the His-to-Asp phosphorelay signal transduction in *Arabidopsis thaliana. Plant Cell Physiol.* **44,** 122–131.

Inoue, T., Higuchi, M., Hashimoto, Y., Seki, M., Kobayashi, M., Kato, T., Tabata, S., Shinozaki, K., and Kakimoto, T. (2001). Identification of CRE1 as a cytokinin receptor from *Arabidopsis. Nature* **409,** 1060–1063.

Ishida, K., Yamashino, T., Yokoyama, A., and Mizuno, T. (2008). Three hype-B response regulators, ARR1, ARR10, and ARR12 play essential but redundant roles in cytokinin signal transduction through the life cycle of *Arabidopsis thaliana. Plant Cell Physiol.* **49,** 47–57.

Ishida, K., Yamashino, T., and Mizuno, T. (2009). A Genome-wide compilation of the two-component systems in *Lotus japonicus. DNA Res.* **16,** 237–247.

Ishige, K., Nagosawa, S., Tokishita, S., and Mizuno, T. (1994). A novel device of bacterial signal transducers. *EMBO J.* **13,** 5195–5202.

Iwama, A., Yamashino, T., Tanaka, Y., Sakakibara, H., Kakimoto, T., Sato, S., Kato, T., Tabata, S., Nagatani, A., and Mizuno, T. (2007). AHK5 His-kinase regulates root elongation through an ETR1-dependent integrated abscisic acid and ethylene signaling pathway in *Arabidopsis thaliana. Plant Cell Physiol.* **48,** 375–380.

Kakimoto, T. (1996). CKI1, a histidine kinase homolog implicated in cytokinin signal transduction. *Science* **274,** 982–985.

Kakimoto, T. (2003). Perception and signal transduction of cytokinins. *Annu. Rev. Plant Biol.* **54,** 605–627.

Kamada-Nobusada, T., and Sakakibara, H. (2009). Molecular basis for cytokinin biosynthesis. *Phytochemistry* **70,** 444–449.

Kiba, T., Taniguchi, M., Imamura, A., Ueguchi, C., Mizuno, T., and Sugiyama, T. (1999). Differential expression of genes for response regulators in response to cytokinins and nitrate in *Arabidopsis thaliana. Plant Cell Physiol.* **40,** 767–771.

Kiba, T., Yamada, H., Sato, S., Kato, T., Tabata, S., Yamashino, T., and Mizuno, T. (2003). The type-A response regulator, ARR15, acts as a negative regulator in the cytokinin-mediated signal transduction in *Arabidopsis thaliana. Plant Cell Physiol.* **44,** 868–874.

Kiba, T., Naitou, T., Koizumi, N., Yamashino, T., Sakakibara, H., and Mizuno, T. (2005). Combinatorial microarray analysis revealing *Arabidopsis* genes implicated in cytokinin responses through the His-to-Asp phosphorelay circuitry. *Plant Cell Physiol.* **46,** 339–355.

Maeda, T., Wurgler-Murphy, S. M., and Saito, H. (1994). A two-component system that regulates an osmosensing MAP kinase cascade in yeast. *Nature* **369,** 242–245.

Majdalani, N., and Gottesman, S. (2007). Genetic dissection of signaling through the Rcs phosphorelay. *Methods Enzymol.* **423,** 349–362.

Mason, M. G., Mathews, D. E., Argyros, D. A., Maxwell, B. B., Kieber, J. J., Alonso, J. M., Ecker, J. R., and Schaller, G. E. (2005). Multiple type-B response regulators mediate cytokinin signal transduction in Arabidopsis. *Plant Cell* **17,** 3007–3018.

Matsubara, M., Kitaoka, S., Takeda, S., and Mizuno, T. (2000). Tuning of the porin expression under anaerobic growth conditions by His-to-Asp cross-phosphorelay through both the EnvZ-osmosensor and ArcB-anaerosensor in *Escherichia coli. Genes Cells* **5,** 555–569.

Matsushika, A., and Mizuno, T. (1998). A dual-signaling mechanism mediated by the ArcB hybrid sensor kinase containing the histidine-containing phosphotransfer domain in *Escherichia coli. J. Bacteriol.* **180,** 3973–3977.

Miwa, K., Ishikawa, K., Terada, K., Yamada, H., Suzuki, T., Yamashino, T., and Mizuno, T. (2007). Identification of amino acid substitutions that render the *Arabidopsis* cytokinin receptor histidine kinase AHK4 constitutively active. *Plant Cell Physiol.* **48,** 1809–1814.

Miyata, S., Urao, T., Yamaguchi-Shinozaki, K., and Shinozaki, K. (1998). Characterization of genes for two-component phosphorelay mediators with a single HPt domain in *Arabidopsis thaliana. FEBS Lett.* **437,** 11–14.

Mizuno, T. (1998). His-Asp phosphotransfer signal transduction. *J. Biochem.* **123,** 555–563.

Mizuno, T. (2004). Plant response regulators implicated in signal transduction and circadian rhythm. *Curr. Opin. Plant Biol.* **7,** 499–505.

Mizuno, T. (2005). Two-component phosphorelay signal transduction systems in plants: From hormone response to circadian rhythms. *Biosci. Biotechnol. Biochem.* **69,** 2263–2276.

Mizuno, T., and Mizushima, S. (1990). Signal transduction and gene regulation through the phosphorylation of two regulatory components: The molecular basis for the osmotic regulation of porin genes. *Mol. Microbiol.* **4,** 1077–1082.

Mok, D. W. S., and Mok, M. C. (2001). Cytokinin metabolism and action. *Annu. Rev. Plant Physiol. Plant Mol. Biol.* **52,** 89–118.

Müller, B., and Sheen, J. (2007). Cytokinin signaling pathway. *Sci. STKE* **9,** 407.

Murray, J. D., Karas, B. J., Sato, S., Tabata, S., Amyot, L., and Szczyglowski, K. (2007). A cytokinin perception mutant colonized by *Rhizobium* in the absence of nodule organogenesis. *Science* **315,** 101–104.

Nishimura, C., Ohashi, Y., Sato, S., Kato, T., Tabata, S., and Ueguchi, C. (2004). Histidine kinase homologs that act as cytokinin receptors possess overlapping functions in the regulation of shoot and root growth in *Arabidopsis*. *Plant Cell* **16,** 1365–1377.

O'Malley, R. C., Rodriguez, F. I., Esch, J. J., Binder, B. M., O'Donnell, P., Klee, H. J., and Bleecker, A. B. (2005). Ethylene-binding activity, gene expression levels, and receptor system output for ethylene receptor family members from *Arabidopsis* and tomato. *Plant J.* **41,** 651–659.

Pareek, A., Singh, A., Kumar, M., Kushwaha, H. R., Lynn, A. M., and Singla-Pareek, S. L. (2006). Whole-genome analysis of Oryza sativa reveals similar architecture of two-component signaling machinery with Arabidopsis. *Plant Physiol.* **142,** 380–397.

Pischke, M. S., Jones, L. G., Otsuga, D., Fernandez, D. E., Drews, G. N., and Sussman, M. R. (2002). An Arabidopsis histidine kinase is essential for megagametogenesis. *Proc. Natl. Acad. Sci. USA* **99,** 15800–15805.

Posas, F., Wurgler-Murphy, S. M., Maeda, T., Witten, E. A., Thai, T. C., and Saito, H. (1996). Yeast HOG1 MAP kinase cascade is regulated by a multistep phosphorelay mechanism in the SLN1-YPD1-SSK1 "two-component" osmosensor. *Cell* **86,** 865–875.

Rashotte, A. M., Carson, S. D. B., To, J. P. C., and Kieber, J. J. (2003). Expression profiling of cytokinin action in Arabidopsis. *Plant Physiol.* **132,** 1998–2011.

Reiser, V., Raitt, D. C., and Saito, H. (2003). Yeast osmosensor Sln1 and plant cytokinin receptor Cre1 respond to changes in turgor pressure. *J. Cell Biol.* **161,** 1035–1040.

Romanov, G. A., Spíchal, L., Lomin, S. N., Strnad, M., and Schmülling, T. (2005). A live cell hormone-binding assay on transgenic bacteria expressing a eukaryotic receptor protein. *Anal. Biochem.* **347,** 129–134.

Romanov, G. A., Lomin, S. N., and Schmülling, T. (2006). Biochemical characteristics and ligand-binding properties of *Arabidopsis* cytokinin receptor AHK3 compared to CRE1/AHK4 as revealed by a direct binding assay. *J. Exp. Bot.* **57,** 4051–4058.

Saito, H., and Tatebayashi, K. (2004). Regulation of the osmoregulatory HOG MAPK cascade in yeast. *Biochemistry* **136,** 267–272.

Sakai, H., Honma, T., Aoyama, T., Sato, S., Kato, T., Tabata, S., and Oka, A. (2001). ARR1, a transcription factor for genes immediately responsive to cytokinins. *Science* **294,** 1519–1521.

Sambrook, J., and Russell, D. W. (2001). Molecular Cloning: A Laboratory Manual. Cold Spring Harbor Laboratory, Cold Spring Harbor, NY.

Santos, J. L., and Shiozaki, K. (2001). Fungal histidine kinases. *Sci. STKE* **4,** RE1.

Schaller, G. E., and Binder, B. M. (2007). Biochemical characterization of plant ethylene receptors following transgenic expression in yeast. *Methods Enzymol.* **422,** 270–287.

Schaller, G. E., and Kieber, J. J. (2002). Ethylene. *In* "The Arabidopsis Book," (C. Somerville and E. Meyerowitz, eds.), American Society of Plant Biologists, Rockville, MD.

Schaller, G. E., Doi, K., Hwang, I., Kieber, J. J., Khurana, J. P., Kurata, N., Mizuno, T., Pareek, A., Shiu, S., Wu, P., and Yip, W. K. (2007). Nomenclature of two-component signaling elements of rice. *Plant Physiol.* **143,** 555–557.

Sheen, J. (2002). Phosphorelay and transcription control in cytokinin signal transduction. *Science* **296,** 1650–1652.

Spichal, L., Yu, N., Riefler, M., Mizuno, T., Romanov, G., Strnad, M., and Schmulling, T. (2004). Two cytokinin receptors of *Arabidopsis thaliana*, Cre1/AHK4 and AHK4, differ in their ligand specificity in bacterial assay. *Plant Cell Physiol.* **45,** 1299–1305.

Stock, A. M., Robinson, V. L., and Goudreau, P. N. (2000). Two-component signal transduction. *Annu. Rev. Biochem.* **69,** 183–221.

Suzuki, T., Imamura, A., Ueguchi, C., and Mizuno, T. (1998). Histidine-containing phosphotransfer (HPt) signal transducers implicated in His-to-Asp phosphorelay in *Arabidopsis*. *Plant Cell Physiol.* **39,** 1258–1268.

Suzuki, T., Miwa, K., Ishikawa, K., Yamada, H., and Mizuno, T. (2001). The *Arabidopsis* sensor His-kinase, AHK4, can respond to cytokinin. *Plant Cell Physiol.* **42,** 107–113.

Takeda, S., Fujisawa, Y., Matsubara, M., and Mizuno, T. (2001). A novel feature of the multistep phosphorelay in *Escherichia coli*: A revised model of the RcsC-YojN-RcsB signaling pathway implicated in capsular synthesis and swarming behavior. *Mol. Microbiol.* **40,** 440–450.

Tanaka, Y., Suzuki, T., Yamashino, T., and Mizuno, T. (2004). Comparative studies of the AHP histidine-containing phosphotransmitters implication in His-to-Asp phosphorelay in *Arabidopsis thaliana*. *Biosci. Biotechnol. Biochem.* **68,** 462–465.

Tirichine, L., Sandal, N., Madsen, L. H., Radutoiu, S., Albrektsen, A. S., Sato, S., Asamizu, E., Tabata, S., and Stougaard, J. (2007). A gain-of-function mutation in a cytokinin receptor triggers spontaneous root nodule organogenesis. *Science* **315,** 104–107.

To, J. P., and Kieber, J. J. (2008). Cytokinin signaling: Two-components and more. *Trends Plant Sci.* **13,** 85–92.

To, J. P., Haberer, G., Ferreira, F. J., Deruere, J., Mason, M. G., Schaller, G. E., Alonso, J. M., Ecker, J. R., and Kieber, J. J. (2004). Type-A Arabidopsis response regulators are partially redundant negative regulators of cytokinin signaling. *Plant Cell* **16,** 658–671.

Tokishita, S., Yamada, H., Aiba, H., and Mizuno, T. (1990). Transmembrane signal transduction and osmoregulation in *Escherichia coli*: II. The osmotic sensor, EnvZ, located in the isolated cytoplasmic membrane displays its phosphorylation and dephosphorylation abilities as to the activator protein, OmpR. *J. Biochem.* **108,** 488–493.

Tran, L. S., Urao, T., Qin, F., Maruyama, K., Kakimoto, T., Shinozaki, K., and Yamaguchi-Shinozaki, K. (2007). Functional analysis of AHK1/ATHK1 and cytokinin receptor histidine kinases in response to abscisic acid, drought, and salt stress in Arabidopsis. *Proc. Natl. Acad. Sci. USA* **104,** 20623–20628.

Tsuzuki, T., Ishige, K., and Mizuno, T. (1995). Phosphotransfer circuitry of the putative multi-signal transducer, ArcB, of *Escherichia coli*. *Mol. Microbiol.* **18,** 953–962.

Ueguchi, C., Koizumi, H., Suzuki, T., and Mizuno, T. (2001a). Novel family of sensor histidine kinase genes in *Arabidopsis thaliana*. *Plant Cell Physiol.* **42,** 231–235.

Ueguchi, C., Sato, S., Kato, T., and Tabata, S. (2001b). The AHK4 gene involved in the cytokinin-signaling pathway as a direct receptor molecule in *Arabidopsis thaliana*. *Plant Cell Physiol.* **42,** 751–755.

Urao, T., Yakubov, B., Satoh, R., Yamaguchi-Shinozaki, K., Seki, M., Hirayama, T., and Shinozaki, K. (1999). A transmembrane hybrid-type histidine kinase in Arabidopsis functions as an osmosensor. *Plant Cell* **11,** 1743–1754.

Wang, W., Hall, A. E., O'Malley, R., and Bleecker, A. B. (2003). Canonical histidine kinase activity of the transmitter domain of the ETR1 ethylene receptor from Arabidopsis is not required for signal transmission. *Proc. Natl. Acad. Sci. USA* **100,** 352–357.

Woeste, K., and Kieber, J. J. (1998). The molecular basis of ethylene signalling in Arabidopsis. *Philos. Trans. R. Soc. Lond. B Biol. Sci.* **353,** 1431–1438.

Wohlbach, D. J., Quirino, B. F., and Sussman, M. R. (2008). Analysis of the *Arabidopsis* histidine kinase ATHK1 reveals a connection between vegetative osmotic stress sensing and seed maturation. *Plant Cell* **20,** 1101–1117.

Wurgler-Murphy, S. M., and Saito, H. (1997). Two-component signal transducers and MAPK cascade. *Trends Biochem. Sci.* **22,** 172–176.

Yamada, H., Suzuki, T., Terada, K., Takei, K., Ishikawa, K., Miwa, K., Yamashino, T., and Mizuno, T. (2001). The *Arabidopsis* AHK4 histidine kinase is a cytokinin-binding receptor that transduces cytokinin signals across the membrane. *Plant Cell Physiol.* **42,** 1017–1023.

Yonekura-Sakakibara, K., Kojima, M., Yamaya, T., and Sakakibara, H. (2004). Molecular characterization of cytokinin-responsive histidine kinases in maize. Differential ligand preferences and response to cis-zeatin. *Plant Physiol.* **134,** 1654–1661.

CHAPTER NINETEEN

Characterization of Pseudo-Response Regulators in Plants

Woe-Yeon Kim,[*,‡,1] Patrice A. Salomé,[†,1] Sumire Fujiwara,[*] David E. Somers,[*] *and* C. Robertson McClung[†]

Contents

Abstract

A small family of clock-regulated pseudo-response regulators (PRRs) plays a number of critical roles in the function of the plant circadian clock. The

[*] Department of Plant Cellular and Molecular Biology, Ohio State University, Columbus, Ohio, USA
[†] Department of Biological Sciences, Dartmouth College, Hanover, New Hampshire, USA
[‡] Division of Applied Life Science (BK21 Program) and Environmental Biotechnology National Core Research Center, Gyeongsang National University, Jinju, Korea

[1] These authors have contributed equally

Methods in Enzymology, Volume 471
ISSN 0076-6879, DOI: 10.1016/S0076-6879(10)71019-3

regulation of the PRRs is complex and entails both transcriptional and post-translational regulation. PRR proteins engage in a number of important protein–protein interactions, some of which are modulated by modifications including phosphorylation. PRR stability is also tightly controlled. This chapter provides methods for studying both the *PRR* genes and their encoded proteins.

1. The *Arabidopsis* Circadian Clock

Circadian rhythms, the subset of rhythms with a period of approximately one solar day, are widespread in nature. These rhythms are driven by endogenous, self-sustaining clocks. All eukaryotic circadian oscillators studied to date are based on multiple interlocked negative feedback loops (Bell-Pedersen *et al.*, 2005) and the Arabidopsis clock is no exception (Harmer, 2009; McClung, 2008).

Five pseudo-response regulators (PRRs) play important roles in clock function. *TIMING OF CAB EXPRESSION1* (*TOC1*, also known as *PSEUDO-RESPONSE REGULATOR1*, *PRR1*) was the first clock gene identified in plants on the basis of a mutant circadian phenotype, altered period length (Millar *et al.*, 1995). *TOC1/PRR1* is the founding member of a small family of five *PRR* genes whose transcript abundance oscillates with circadian period; *PRR9*, *PRR7*, *PRR5*, *PRR3*, and finally *PRR1/TOC1* expression peaks in succession at ~2-h intervals, with *PRR9* peaking early in the morning and *TOC1* peaking in the early evening (Matsushika *et al.*, 2000; Strayer *et al.*, 2000). Like *TOC1*, the other four *PRR* genes lack the conserved aspartic acid found in the receiver domain of classical response regulators and so are unlikely to function via a conventional phosphorelay (Mizuno and Nakamichi, 2005). The modified receiver domain is termed a pseudo-receiver (PsR) domain. Reverse genetic analysis established that each *PRR* gene is important for clock function because loss of *PRR* function alters period length (McClung, 2006; Mizuno and Nakamichi, 2005). In addition, several *PRRs* (*PRR5*, *PRR7*, and *PRR9*) are positive regulators of flowering time (gain of function leads to early flowering) and are positive regulators of light sensitivity as measured by hypocotyl elongation (Matsushika *et al.*, 2007; Mizuno and Nakamichi, 2005; Nakamichi *et al.*, 2005). However, there is no good model explaining the biochemical function of the PRR proteins (Harmer, 2009; McClung, 2008; Mizuno and Nakamichi, 2005).

The Arabidopsis clock has multiple loops, with PRRs involved in each (Fig. 19.1). In the central loop, TOC1 is a positive regulator of two partially redundant single Myb-domain transcription factors, CIRCADIAN CLOCK ASSOCIATED1 (CCA1) and LATE ELONGATED HYPOCOTYL (LHY) (Alabadí *et al.*, 2001). Although TOC1 lacks defined

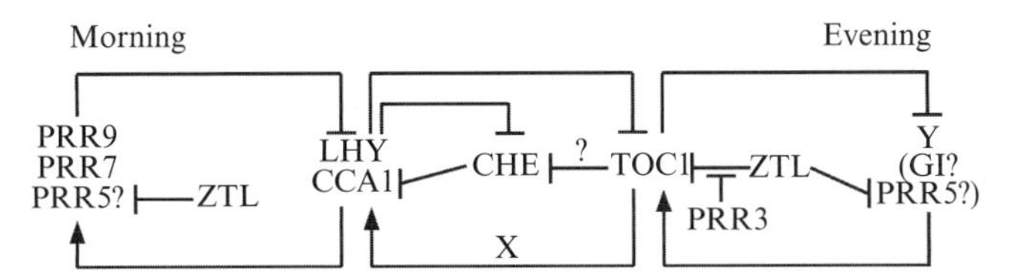

Figure 19.1 Model of the *Arabidopsis* circadian clock. PRR proteins play prominent roles in the *Arabidopsis* circadian clock, which consists of multiple interlocked negative feedback loops. This model is oversimplified to emphasize the basic architecture and the prominence of the PRRs; not all known components are shown and more components remain to be identified. This model includes some speculations, indicated with question marks, as positions of and interactions among some components remain inconclusively determined.

DNA-binding domains, it is recruited to the *CCA1* promoter, possibly through interaction with another DNA-binding protein(s). CCA1 HIKING EXPEDITION (CHE), a TCP transcription factor, binds to a canonical TCP binding site in the *CCA1* promoter, and negatively regulates *TOC1* (Pruneda-Paz *et al.*, 2009). TOC1 and CHE interact, but the significance of this interaction remains incompletely known (Pruneda-Paz *et al.*, 2009). CCA1 and LHY form the negative arm of this feedback loop, binding to the *TOC1* promoter to inhibit expression (Alabadí *et al.*, 2001, 2002; Mizoguchi *et al.*, 2002).

In the second interlocked loop, also called the "evening" loop because of the time of day at which the expression of loop members is maximal, TOC1 represses a component, Y, that may include GIGANTEA (GI), which activates TOC1 (Locke *et al.*, 2005, 2006). In the third interlocked loop, also called the "morning" loop, CCA1 and LHY are positive regulators of two TOC1 relatives, *PRR7* and *PRR9* (Harmer and Kay, 2005; Mizuno and Nakamichi, 2005). The *prr7prr9* double mutant exhibits dramatic period lengthening and is conditionally arrhythmic (Farré *et al.*, 2005; Nakamichi *et al.*, 2005; Salomé and McClung, 2005). PRR5 is also implicated in this loop; the *prr5prr7prr9* triple mutant is completely arrhythmic and PRR5/7/9 are considered to be negative regulators of *CCA1/LHY* because *CCA1* is constitutively transcribed in that triple mutant (Nakamichi *et al.*, 2005). Thus, each of these four PRRs (TOC1, PRR5, PRR7, and PRR9) regulate *CCA1* and *LHY* expression, with three (PRR5, PRR7, and PRR9) negative regulators and TOC1 a positive regulator.

Transcriptional regulation is important in clock function, but it is clear that posttranscriptional regulation is an essential constituent of the clock mechanism (Gallego and Virshup, 2007). In particular, the temporally regulated proteasomal degradation of specific clock proteins is necessary for progression through the oscillation. The stability of a number of plant clock proteins, including GI (David *et al.*, 2006), LHY (Song and Carré,

2005), ZEITLUPE (ZTL) (Kim *et al.*, 2003), and members of the TOC1/PRR family (Farré and Kay, 2007; Fujiwara *et al.*, 2008; Ito *et al.*, 2007; Kiba *et al.*, 2007; Más *et al.*, 2003; Para *et al.*, 2007) is clock regulated. Most is known about TOC1, which peaks in abundance at dusk. An E3 ubiquitin ligase SCF complex including the F-box protein ZTL is crucial for clock-regulated proteasomal degradation of TOC1 (Han *et al.*, 2004; Más *et al.*, 2003). ZTL also targets PRR5 for proteasomal degradation through direct interaction with the PsR domain of PRR5 (Kiba *et al.*, 2007). The mechanism(s) regulating the stability of PRR7 and PRR9 remain less fully described, although proteasome activity is implicated (Farré and Kay, 2007; Ito *et al.*, 2007). Unlike the other PRRs, which regulate *CCA1* and *LHY* expression, PRR3 regulates the stability of TOC1 via protein–protein interaction (Fujiwara *et al.*, 2008; Para *et al.*, 2007). This mechanism is notably similar to that by which ZTL protein abundance oscillations, despite noncycling mRNA abundance, are a consequence of periodic stabilization through interaction with GI protein, which cycles in abundance as a consequence of rhythmic transcription and mRNA accumulation (Kim *et al.*, 2007).

Critical to the investigation of PRR function is the ability to monitor PRR protein abundance and localization. We have taken advantage of PRR–GFP protein fusions, using the GFP moiety both as a fluorescent tag and as a protein tag for immunological detection.

2. Detection of PRR Proteins

2.1. Immunodetection of GFP-tagged PRR proteins from Arabidopsis extracts

To examine the circadian dynamics of the PRR proteins *in planta* and at endogenous levels we used stably transformed Arabidopsis plants expressing PRR:GFP fusions driven by the endogenous PRR promoter (Fujiwara *et al.*, 2008). Genomic fragments including the full promoter (starting at −1431, −5116, −1194, and −1541 bp for *PRR3*, *PRR5*, *PRR7*, and *PRR9*, respectively), 5′UTR and coding sequences up to the last codon before the STOP codon, were amplified by PCR with ExTaq (Takara Bio USA, Madison, WI) and subcloned first into pGEM-T Easy (Promega, Madison, WI), transferred into the Gateway Entry vector pENTR-1A, and finally placed upstream of the GFP variant mGFP6 in the binary vector pMDC206 (Curtis and Grossniklaus, 2003) by LR recombination (Invitrogen, Carlsbad, CA). pPRR:PRR:GFP constructs were transformed into Arabidopsis Columbia plants by floral dip (Bechtold *et al.*, 1993) via *Agrobacterium tumefaciens* strain GV3101. Analysis was performed in select stably transformed lines.

For TOC1–YFP, PRR3–GFP, PRR5–GFP, PRR7–GFP, and PRR9–GFP detection, Arabidopsis seedlings were grown under light–dark cycles (12 h light: 12 h darkness) for 8–10 days and harvested rapidly into prelabeled aluminum foil packets and placed immediately into liquid nitrogen. Seeds were sown directly onto filter paper, overlaying MS media (GIBCO BRL, Cleveland, OH) +3% sucrose solidified with 0.8% agar, to facilitate easy harvest. For transient expressions in *Nicotiana benthamiana* using Agrobacterium-mediated coinfiltration, all constructs were 35S cauliflower mosaic virus (CaMV) promoter-driven (Kim *et al.*, 2007; Voinnet *et al.*, 2003).

The following protocols also describe the techniques to detect phosphorylated forms of the PRR proteins (Fujiwara *et al.*, 2008). The critical feature is the adjustment of the acrylamide:bisacrylamide ratio from the "standard" 37.5:1 to a ratio of 149:1. This ratio may be used for any percentage gel, and we find that an 8% acrylamide gel works best for the PRR proteins.

2.2. Protein extract preparation for PRR family members

(1) Resuspend frozen ground tissues in extraction buffer (50 m*M* Tris–HCl, pH 7.5, 150 m*M* NaCl, 0.5% Nonidet P-40, 1 m*M* EDTA, 3 m*M* DTT, 1 m*M* phenylmethylsulfonyl fluoride, 5 μg/ml leupeptin, 1 μg/ml aprotinin, 1 μg/ml pepstatin, 5 μg/ml antipain, 5 μg/ml chymostatin, 2 m*M* Na_3VO_4, 2 m*M* NaF, 2 m*M* glycerol phosphate, 50 μM MG132, 50 μM MG115, 50 μM ALLN; all reagents were purchased from Sigma) by vortexing for 10 s (v/v, 1:1). Keep the tubes on ice.

(2) Centrifuge the extracts at 16,000×*g* for 5 min at 4 °C.

(3) Keep the tubes on ice and quickly transfer the supernatant to prechilled new tubes.

(4) Add 6× sample buffer (SDS Reducing buffer; 125 m*M* Tris–HCl, pH 6.8, 50% glycerol, 4% (w/v) SDS, 0.02% (w/v) Bromophenol blue, and freshly added 10% (v/v) β-mercaptoethanol) to the supernatant, warm the sample to room temperature for 2 min and centrifuge at 1000×*g* for 1 min at 4 °C before loading. Heating to higher than room temperature may cause severe degradation of PRR proteins in the total lysate. Do not boil samples.

2.3. PRR protein detection: immunoblots

(1) We use 8% SDS–PAGE (acrylamide:bisacrylamide, 149:1) gels to detect phosphorylated PRR proteins (GFP or YFP tagged) migrating in the range of approximately 80–106 kDa. It is best to run the gel with low voltage at 4 °C. Running at a higher voltage (e.g., 120 V)

may cause dumbbell-like band patterns making it difficult to detect closely spaced bands, especially with larger proteins. We use 60 V for 30 min until samples go into the lower gel, and then change to 90 V to get the best result.

(2) Equilibrate the gel and soak the polyvinylidene difluoride (PVDF) membrane [(Bio-rad) briefly pretreated with 100% methanol)], filter paper, and fiber pads in transfer buffer for 10 min. Always wear gloves when handling membranes to prevent contamination.
(3) Arrange gel sandwich for blotting. It is important to remove any air bubbles between the gel and the membrane to ensure complete and even transfer. Use a glass tube or roller to gently roll air bubbles out.
(4) Transfer proteins at 100 V for 1 h at 4 °C.
(5) To prepare 5% blocking solution, dissolve 5 g of nonfat dried milk in 100 ml of 1× TBST (10 m*M* Tris–HCl, pH 8.0, 150 m*M* NaCl, 0.05% Tween-20) for 30 min.
(6) Soak the membrane in 5% blocking solution and incubate at room temperature for 1 h with agitation.
(7) Incubate with primary anti-GFP antibody (Abcam, ab6556, 1:2000) in the blocking solution for 1–2 h at room temperature or overnight at 4 °C.
(8) Wash the membrane three times with 1× TBST for 10 min each.
(9) Incubate with anti-rabbit secondary IgG, HRP-Linked (GE Healthcare, NA934, 1:3000) in the blocking solution for 1 h at room temperature.
(10) Wash the membrane three times for 10 min each with 1× TBST.
(11) Detect signals with Pierce West Pico solution (Pierce, 34080) using standard X-ray film (MIDSCI, St Louis, MO) or a chemiluminescence detector.

2.4. Immunoprecipitation/coimmunoprecipitation

(1) To prepare antibody-bound resin, mix 30 μl of protein A-agarose 50% slurry (Invitrogen, 15918-014), anti-GFP antibody (mouse monoclonal, Molecular Probes, A11120,1:250), and 40 μl of 1× immunoprecipitation (IP) buffer (50 m*M* Tris–HCl, pH 7.5, 150 m*M* NaCl, 0.5% Nonidet P-40, 1 m*M* EDTA, 3 m*M* DTT).
(2) Incubate at least 1 h at 4 °C with gentle agitation.
(3) Resuspend 1 ml of ground tissues in 1 ml IP buffer containing 1 m*M* phenylmethylsulfonyl fluoride, 5 μg/ml leupeptin, 1 μg/ml aprotinin, 1 μg/ml pepstatin, 5 μg/ml antipain, 5 μg/ml chymostatin, 2 m*M* Na_3VO_4, 2 m*M* NaF, 2 m*M* Glycerol phosphate, 50 μ*M* MG132, 50 μ*M* MG115, and 50 μ*M* ALLN by vortexing for 10 s (v/v, 1:1) and spin at 16,000×*g* for 5 min at 4 °C.

(4) Incubate 1–2 mg of soluble protein extracts (ca. 950 μl) from #3 and the resin from #2 with gentle agitation for 1 h at 4 °C.
(5) Centrifuge at 1000×*g* at 4 °C for 1 min to collect immune complexes.
(6) Wash the protein A-agarose beads three times with 1 ml of ice-cold IP buffer by gently inverting the tube 10 times.
(7) Wash the pellet once more with 1× PBS (0.2 *M* phosphate, pH 7.4; 1.5 *M* NaCl).
(8) Suspend pellet in 30 μl of 2× SDS–PAGE sample buffer. Heat samples at 90 °C for 60–90 s prior to SDS–PAGE and immunoblotting. Many of the PRR proteins are quite labile and it is important to heat briefly and below 93 °C to minimize degradation. However, PRR proteins isolated by immunoprecipitation are less labile than in crude protein extracts, so higher temperatures at this stage are both necessary (to dissociate the immuncomplexes) and not as damaging.
(9) To prepare a positive control (pre-IP input), mix 90 μl of the protein extract from #3 and mix with 30 μl of 6× sample buffer, warm the sample at room temperature for 2 min and centrifuge at 1000×*g* for 1 min at 4 °C before loading. As noted earlier, heating higher than room temperature may cause severe degradation of PRRs in the total lysate.

2.5. Determination of phosphorylation states of PRR family members

(1) Prepare protein extracts either in 1× λ-phosphatase buffer (New England Biolabs; supplied by company) supplemented with 2.5 m*M* $MnCl_2$, 0.5% Triton X-100, and 0.4% Nonidet P-40 for λ-phosphatase treatment or in New England Biolabs buffer 3 supplemented with 0.5% Triton X-100 and 0.4% Nonidet P-40 for calf intestinal alkaline phosphatase (CIP) treatment. Modified IP buffer without EDTA supplemented with 2.5 m*M* $MnCl_2$ for λ-phosphatase or 10 m*M* $MgCl_2$ for CIP can be used. In all cases supplement with protease inhibitors (2.5 μg/ml Antipain, 2.5 μg/ml Chymostatin, 1 μg/ml Pepstatin, 5 μg/ml Leupeptin, 5 μg/ml Aprotinin, 1 m*M* phenylmethylsulfonyl fluoride) and proteasome inhibitors (50 μM MG132, 50 μM MG115, and 50 μM ALLN) to prohibit nonspecific degradation during extraction. Depending on the protein, λ-phosphatase may work better than CIP, or *vice versa*.
(2) Incubate 50-μl aliquots of the resulting protein extracts with 400 U of λ-protein phosphatase (New England Biolabs) with or without phosphatase inhibitors (NaF/Na_3VO_4) at 30 °C for 15 min or with 10–20 U of CIP (New England Biolabs) with or without phosphatase inhibitors (NaF/Na_3VO_4) at 37 °C for 15 min.

(3) Stop the reaction by adding phosphatase inhibitors cocktail (2 m*M* Na_3VO_4, 2 m*M* NaF, 2 m*M* glycerol phosphate).
(4) Add 20 μl of 6× loading buffer to the reaction, heat at room temperature for 2 min and centrifuge at 16,000×*g* for 1 min at room temperature.
(5) Load samples into 8% SDS–PAGE (acrylamide:bisacrylamide, 149:1) and analyze by immunoblotting as described above. As noted earlier, running the gel at low voltage and at 4 °C will help to improve the resolution of closely migrating forms of the protein. Maximum resolution is usually observed when the protein of interest has migrated to the center of the gel.

3. Localization of PRR Proteins

3.1. Localization of GFP-tagged PRR proteins in leek cells

One of the first steps to elucidate protein function is to determine in which subcellular compartment the protein resides and acts. CCA1 and LHY, two Myb-domain transcription factors, localize to the nucleus (Carré and Kim, 2002; Yakir *et al.*, 2009), where they bind to the promoters of their target genes. The genetic interactions of the PRR family members with known nuclear-localized clock components such as CCA1 and LHY suggest a role in the nucleus if these interactions are direct. Consistent with this hypothesis, protein localization programs such as PSORT (http://psort.ims.u-tokyo.ac.jp/form.html) predict nuclear localization for each of the PRRs. TOC1/PRR1 has been known for some time to be nuclear-localized. Overexpression of a TOC1:GFP fusion demonstrated a clear nuclear location of TOC1 in tobacco BY-2 cells (Strayer *et al.*, 2000). Similarly, a TOC1:GUS fusion protein showed nuclear localization in transgenic Arabidopsis (Para *et al.*, 2007). Most recently, TOC1 was localized to DNA at the *CCA1* (but not *LHY*) promoter by chromatin IP (Pruneda-Paz *et al.*, 2009).

Despite the evident importance of PRR localization, experimental evidence has only recently been forthcoming for other family members. We set out to localize GFP fusions of the four remaining circadian PRR family members, first in leek and then in Arabidopsis (Fujiwara *et al.*, 2008). All constructs examined in leek used the constitutive CaMV 35S promoter to drive expression of PRR fused in frame to a N-terminal GFP. *PRR* cDNAs were cloned into the entry vector pENTR-2B (Invitrogen) and recombined with the Gateway^{TM} destination vector pK7WGF2 (Karimi *et al.*, 2002). Each construct was introduced into leek cell by biolistics (Sanford *et al.*, 1993).

We found that PRR3, PRR5, PRR7, and PRR9 proteins localized to the nucleus of leek cells (Fig. 19.2), and we detected differences in the subcellular localization of some PRR family members. For instance,

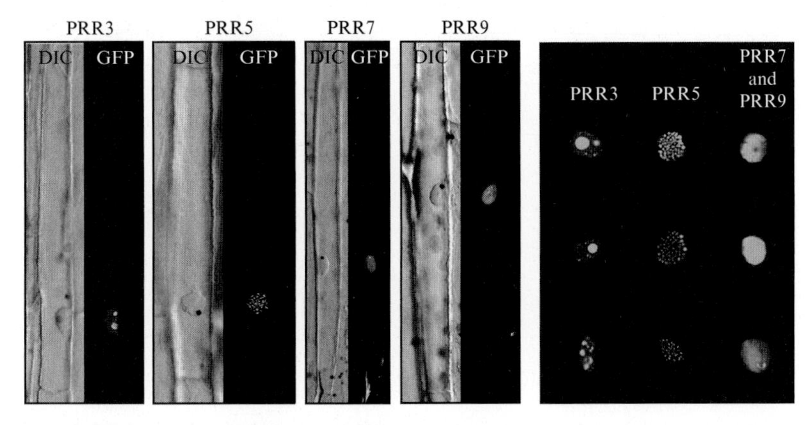

Figure 19.2 Nuclear localization of PRR family members in transient assays in leek cells. Leek epidermal cells were bombarded with gold particles coated with DNA for binary constructs expressing a N-terminal GFP fusion to the PRR, placed under the control of the 35S promoter (*p35S:GFP:PRR*). Left panel: examples of individual cells expressing the PRR:GFP fusions are shown on the left; the position of the nucleus can be clearly seen by differential interference contrast microscopy (DIC). Right panel: individual nuclei from three cells expressing *p35S:GFP:PRR3*, *p35S:GFP:PRR5*, or *p35S:GFP:PRR9*. The localization of TOC1 is very similar to PRR5 (Strayer *et al.*, 2000); as shown in the left panels, PRR7 and PRR9 show the same evenly distributed nuclear localization pattern.

PRR3 is found most strongly in the nucleolus of leek cells, while PRR7 and PRR9 appear evenly distributed in the nucleus. PRR5 offered an interesting case in that, although clearly nuclear-localized, the GFP fusion tends to form aggregates, oftentimes called subnuclear foci. The number and size of these foci are in direct proportion with the GFP signal, suggesting that accumulation of PRR5 to subnuclear foci might be an artifact from high expression.

3.1.1. Protocol: transient expression of 35S:GFP:PRR clones in leek cells

(1) Prepare 100 mm diameter petri plates containing MS medium supplemented with 1–2% sucrose.
(2) Buy leeks at your local food store. Wider leeks are preferred as the epidermis will not curl as much once on the plate.
(3) Cut a 3 cm section from the inner, white parts of the leek, discarding green leaves.
(4) Gently peel off the single-cell layer on the inner side of the leek white section, and deposit on the surface of the medium, inner (adaxial) side against the medium to prevent curling.

(5) Mix 2 μg of DNA per *p35S:GFP:PRR* fusion with 10 μl gold particles (1.5–3 μm; ~60 mg/l), 50 μl 2.5 *M* $CaCl_2$, and 20 μl 0.1 *M* spermidine and vortex for 1 min.
(6) Quickly spin gold particles down (5–10 s, at most, in a microfuge operated below 4000 rpm).
(7) Wash gold particles once in 70% ethanol, then twice in 100% ethanol.
(8) Resuspend gold particles in 10–20 μl of 100% ethanol, and apply to the macrocarrier disk.
(9) Bombard leek peels at 1000 psi under 27 in. of Hg vacuum. If using a PDS-1000/He instrument (Bio-Rad, Hercules, CA), gap distance should be at 3 mm.
(10) Incubate bombarded tissue in the dark overnight, but no longer than 16–20 h, as strong overexpression of the proteins might lead to artifactual localization.
(11) Mount tissue in water for visualization on a confocal microscope (GFP excitation: 488 nm; GFP emission 498–561 nm). Look for gold particles under back-lighting to find bombarded zones, then switch to GFP settings for GFP detection.

3.2. Immunolocalization of GFP-tagged PRR proteins in Arabidopsis plants

To confirm the nuclear localization of the PRR proteins under endogenous conditions, we turned to stably transformed Arabidopsis plants expressing PRR:GFP fusions driven by the endogenous *PRR* promoter (Fujiwara *et al.*, 2008). Because PRR proteins are inabundant, the GFP signal was generally weak and, in our hands, disappeared rather quickly due to GFP photobleaching. We therefore decided to use the GFP moiety as an epitope for immunolocalization, rather than as a fluorescent molecule, taking advantage of the signal amplification provided by the primary and secondary antibodies. A number of protocols exist for immunolocalization of proteins in Arabidopsis seedlings and ours, detailed below, is adapted from two protocols described earlier (Guo *et al.*, 2001; Zachgo *et al.*, 2000). The most critical part of the procedure is to partially digest away the cell wall and revert the cross-linking so that primary and secondary antibodies are allowed access to the epitopes. Proteinase K digestion proved effective for our own purposes. Although others have employed a heat treatment of fixed seedlings rather than proteinase digestion (Vitha *et al.*, 2001), we did not compare the effectiveness of the two treatments. It is likely that another protein (other than GFP) will behave differently when subjected to paraformaldehyde fixation.

(1) Grow seedlings under light–dark cycles (12 h light: 12 h darkness) for 8–10 days.

(2) Fix seedlings for 3 h in 4% paraformaldehyde in 1× PBS at 4 °C. Eppendorf tubes (1.5 ml) work well for young seedlings; alternatively, 12- or 24-well plates can be used.
(3) Wash three times in 1× PBS, 0.1% Triton X-100 for 15 min, and three times in water for 15 min.
(4) Digest cell walls in 1% cellulase, 1% macerozyme in 1× PBS for 45 min with shaking. Wash three times in 1× PBS, 0.01% Triton X-100.
(5) Incubate seedlings in blocking buffer (5% BSA in 1× PBS) for 2 h at room temperature, followed by incubation with the primary antibody (rabbit anti-GFP, Abcam, Cambridge, UK) diluted at 1:250 in 2% BSA, 1× PBS with gentle shaking overnight at 4 °C.
(6) Wash seedlings four times, 15 min each, in 1× PBS, 0.01% Triton X-100.
(7) Incubate with secondary antibody for 2 h at room temperature (in this case: goat antirabbit, Alexa Fluor-488 conjugated, Molecular Probes, Eugene, OR), diluted to 1:400 in 2% BSA, 1× PBS.
(8) Wash four times 15 min in 1× PBS, 0.01% Triton X-100.
(9) Mount seedlings in Mount Quick (Electron Microscopy Sciences, Washington, PA) and detect signal on confocal microscope using same GFP settings as for transient expression in leek. Note that, when dry, Mount Quick will become autofluorescent under the conditions used for GFP localization. For long-term storage and retrieval of slides, a good choice of mounting medium is ProLong antifade reagent.

4. Exploring the Circadian Phenotypes of *PRR* Mutants

4.1. Cotyledon movement analysis

High-throughput forward genetic mutant screens based on rhythmic luciferase expression are very powerful tools for the isolation of novel genes involved in circadian rhythms (Southern and Millar, 2005; Welsh *et al.*, 2005). Such a screen in Arabidopsis made use of the *LHCB1*1* (*CAB2*) promoter, and led to the identification of, a number of mutants, including *timing of cab expression1* (*toc1*) and *zeitlupe* (*ztl*) (Millar *et al.*, 1995; Somers *et al.*, 2000).

The ability to perform reverse genetics in which one might wish, for example, to assess sets of targeted T-DNA insertion for alleles that confer circadian defects would be greatly facilitated by an assay that did not require the introduction of a transgene. One obvious class of mutants warranting investigation includes loss of function alleles of homologues to known clock genes like *CCA1*, *LHY*, *TOC1*, and *ZTL*. Another class of mutants includes those identified from screens not based on circadian phenotypes, such as flowering time, hypocotyl elongation, or hormone signaling. For many years the only circadian rhythm studied was leaf movement in plants (McClung, 2006). Such studies emphasized pulvinar movements in many legume species, but species that lack pulvini, like Arabidopsis, often express a circadian rhythm in the movements of cotyledons (Millar *et al.*, 1995).

In Arabidopsis, cotyledon movement presumably is driven by daily rhythmic changes in cell elongation in the petiole. An inexpensive surveillance camera is sufficient to record cotyledon movement, coupled to a computer system to control image capture. Seeds are surface-sterilized and sown on MS medium supplemented with sucrose. After 3–4 days of stratification at 4 °C in the dark, plates are released under the desired entraining conditions: light–dark cycles of any photoperiod or hot–cold cycles for thermoentrainment. The number of days in entraining cycles before recording movement will depend on the genotype and growth conditions (thermocycles slow down growth), but averages 4–6 days. Seedlings reach the optimal stage for transfer when cotyledons are well expanded, and primary leaves are just starting to emerge at the apex. We then cut out a cube of solid medium around the seedling, and gently transfer the seedling in agar to the wells of clear 24 square-well plates. We tape Whatmann paper to the back of the plate to increase contrast between the seedling and the plate. The lid of the plate is held into place with 1-in. wide surgical tape to allow gas exchange. Recording can be started immediately with your favorite program, such as Kujamorph (http://millar.bio.ed.ac.uk/JnlPage.htm) or NKTRACE (Onai *et al.*, 2004). We routinely assay seedlings for 1 week. We analyze circadian data with fast Fourier transform-nonlinear least-squares (FFT-NLLS) analysis (Plautz *et al.*, 1997; Straume *et al.*, 1991) and with BRASS (http://millar.bio.ed.ac.uk/PEBrown/BRASS/BrassPage.htm).

A number of variations can be applied during seedling growth. For mutants with long hypocotyls, it might be necessary to first release them into continuous light for 3–4 days before starting an entraining regime (Hicks *et al.*, 1996). The first day in continuous light will inhibit hypocotyl elongation and allow seedlings to fit much more easily in the wells of the plates. Seedlings can then be entrained for 2 days before recording movement. Thermocycles can also be applied while cotyledon movement is being recorded; differences in the two temperatures should be kept between 0 and 6 °C to limit the amount of condensation on the lid of the plate around transitions from one temperature to the other. Temperatures that are too low (below 12 °C) will stall cotyledon movement, although we know that the underlying circadian clock is still rhythmic (Gould *et al.*, 2006). Finally, the temperature during the assay need not be 22 °C, but can be set lower or higher. It is important to adjust the timing of transfer of the plates to the new temperature. Ideally, a step up in temperature should be made to coincide with subjective dawn, while a step down in temperature should follow subjective dusk.

Cotyledon movement is of course not limited to Arabidopsis. For example, we are now measuring cotyledon movement from *Brassica rapa* seedlings grown directly on soil (P. Liu, Q. Xiu, X. Xu, and CRM, unpublished); cotyledon movement is likely to be applied successfully to many dicotyledonous species.

4.2. Generating versatile LUCIFERASE (LUC) fusion constructs

Although recording movement of cotyledons is fast and does not require the introduction of a transgene into the genotype to be tested, it also suffers from being several levels removed from the central oscillator controlling the rhythm. To more directly measure clock gene expression, the firefly luciferase has become the gold standard for circadian biologists (see Welsh *et al.*, 2005) in Neurospora, mammalian systems and plants (Gooch *et al.*, 2008; Millar *et al.*, 1992a, 1995; Morgan *et al.*, 2003; Stanewsky, 2007; Welsh *et al.*, 2004; Wilsbacher *et al.*, 2002). Circadian research in green algae *Synechococcus elongatus* PCC 7942 has opted for a bacterial version of the gene, encoded by the *luxAB* operon from *Vibrio harveyi* (Kondo *et al.*, 1993; Liu *et al.*, 1995); the necessary genes for the synthesis of the substrate have been introduced into the Synechococcus genome to circumvent lethality issues with exogenous substrates. How easily transgenic seedlings can be selected determines whether primary transgenic plants can be assayed. The original firefly LUC construct, in the pPZP series backbone (Hajdukiewicz *et al.*, 1994), conferred resistance to gentamycin and required several weeks growth on selective media to allow unambiguous determination of resistance. To make selection of transgenic plants easier and more unambiguous, we replaced the gentamycin resistance cassette with either neomycin phosphotransferase II (derived from pBI101, Jefferson *et al.*, 1987) or phosphinotricine phosphotransferase (from 35SpBARn, LeClere and Bartel, 2001) for kanamycin and BASTA selection, respectively. Then, to facilitate the cloning of promoter fragments, we introduced a Gateway^{TM} recombination cassette (comprised of the chloramphenicol resistance gene and the ccdB gene for counter-selection of intact destination vector molecules, derived from the Gateway^{TM} destination vector pK7WGF2, Karimi *et al.*, 2002) upstream of luciferase. The resulting clones, PS517 (for kanamycin selection) and PS643 (for BASTA selection) were recombined with entry clones bearing the promoters of the clock genes *CCA1*, *LHY*, *TOC1*, *PRR7*, and *PRR9*. Maps for the PS517 and PS643 vectors are shown in Fig. 19.3.

Primary transformants can now be selected unambiguously within 7–10 days. Seeds are first sterilized 1–2 h in 95% ethanol, air-dried, and then plated out on MS medium *not* supplemented with sucrose, and containing either 50 μg/ml kanamycin and 500 μg/ml carbenicillin (for PS517-derived constructs) or 10 μg/ml glufosinate ammonium and 500 μg/ml carbenicillin (for PS643 derivatives).

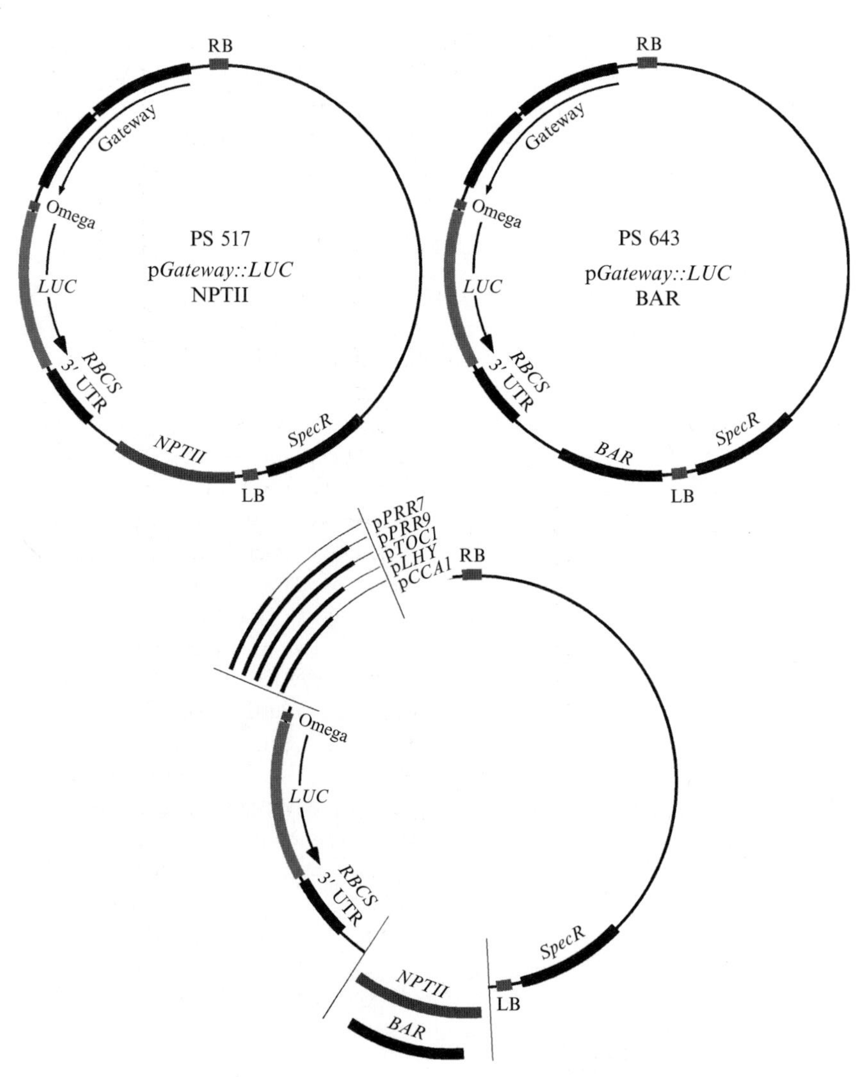

Figure 19.3 A set of versatile vectors to characterize the expression pattern of the *PRR* family members. PS517 was constructed from pPZΩLUC+ by replacing the plant selection marker (gentamycin) with kanamycin resistance. The same gentamycin selection marker was replaced with BASTA resistance (from 35SpBARn) to create PS643. The Gateway™ recombination cassette was PCR-amplified from pK7WG2D and cloned between the *Bam*HI and *Hin*dIII sites upstream of luciferase. Promoters from the clock genes *CCA1*, *LHY*, and *TOC1*, as well as the *TOC1*-related genes *PRR7* and *PRR9*, were subcloned into the entry vector pENTR11 or pENTR-2B, and placed upstream of luciferase by LR recombination with either PS517 or PS643.

4.3. Transient expression in *Nicotiana benthamiana*

The generation and selection of transgenic plants is time-consuming, and one does not wish to discover after months of wait that the circadian reporter expressed by one's transgenic seedlings does not actually exhibit rhythmic expression. Likewise, in the context of a promoter resection study, a quick look at the behavior of the constructs made thus far can accelerate the identification of sequences of interest (for rhythmicity or high amplitude expression) and direct further cloning while awaiting phenotypic confirmation in stable transgenic lines. Arabidopsis cell suspensions can in theory be transfected, but handling and maintaining the cultures can turn into a time-consuming effort. In addition, many plant cell lines are not intrinsically rhythmic, greatly limiting their use for quick assays.

A convenient option is the use of *N. benthamiana. Nicotiana benthamiana* plants grow quickly, and a few pots can be sown every week until needed without requiring much growth space. An Agrobacterium suspension bearing the construct of interest is infiltrated into leaves from the abaxial (bottom) side of the leaves. The bacterial suspension is made competent to transfer the T-DNA fragment of the construct by induction with acetosyringone the day before infiltration. Most transgenes expressed in this fashion tend to be silenced within 2–3 days as the plant generates small RNAs targeting the highly expressed transgenes, but coinfiltration with a viral suppressor gene like p19 or HC-Pro will sequester siRNAs away from the luciferase mRNA (Voinnet *et al.*, 2003). Quite surprisingly, luciferase activity can be readily detected for at least 1–2 weeks when leaf cuttings of the infiltrated regions are transferred to 96-well plates and provided with luciferin. With this method, we have tested the *CCA1*, *LHY*, and *TOC1* promoters (in the PS643 backbone, Fig. 19.4), as well as the *CCA1*, *LUX*, *GI*, *CCR2* (*GRP7*), and *CAT3* promoters in other vector backbones with great success. Coinfiltration with p19 or HC-Pro is not required to detect clear rhythms; it does however tend to help decrease variation between cuttings from different leaves. The fact that clock-regulated Arabidopsis promoters oscillate in Nicotiana argues that the regulatory modules found in these promoters are conserved among species, consistent with genomic and transcriptomic analyses (Michael *et al.*, 2008; Zdepski *et al.*, 2008).

4.4. Transient rhythmic assay in *Nicotiana benthamiana*

(1) Start a small overnight 10 ml culture of the agrobacterial strains carrying the luciferase constructs and the viral suppressor (p19 or HC-Pro). We currently use ASE1 with good success.

(2) In the morning, use the overnight culture to inoculate a 30-ml culture, diluting the agrobacteria 1:200. In the evening, induce cell culture with acetosyringone, applied to a final concentration of 150 μM. Allow formation of the necessary pilus overnight.

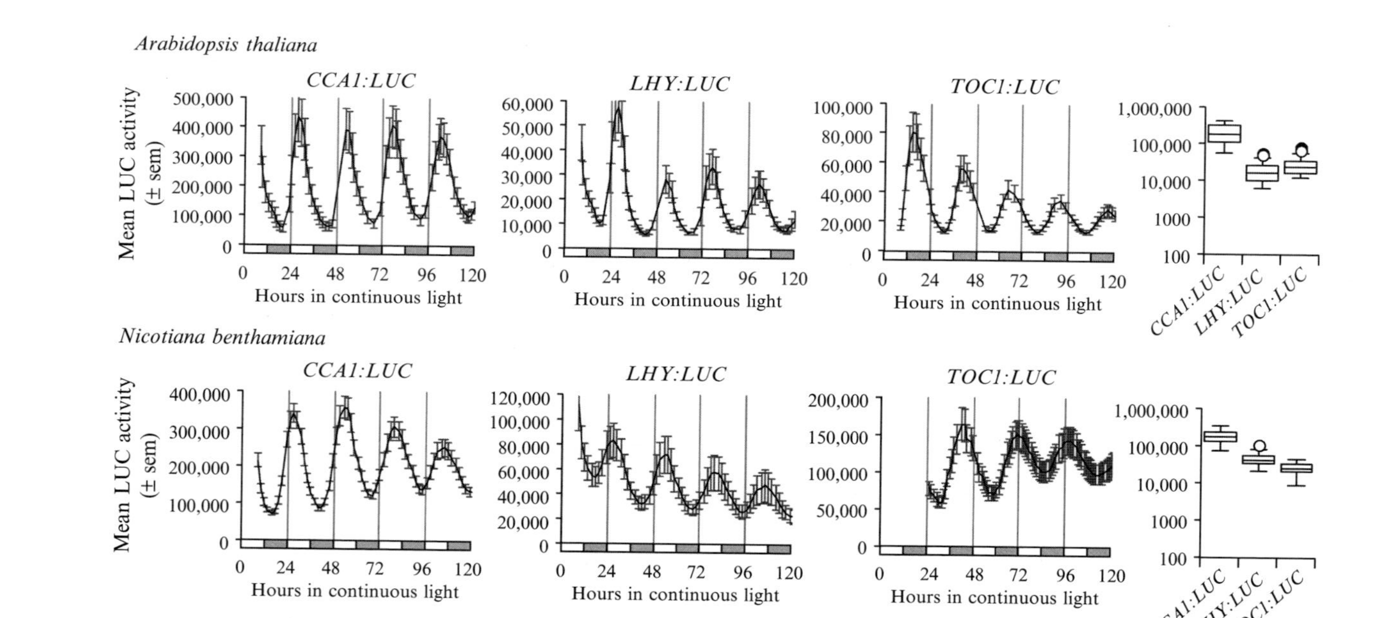

Figure 19.4 A transient expression assay in *Nicotiana benthamiana* for circadian experiments. Binary constructs bearing the luciferase fusions to the *CCA1*, *LHY*, or *TOC1* promoters were described previously (Salomé and McClung, 2005). They all share the same backbone (PS643), and confer BASTA resistance in plants. A typical circadian profile of the three constructs in stably transformed Arabidopsis seedlings is shown in the upper panels. The lower panels show representative mean traces for the same binary constructs, introduced into *Nicotiana benthamiana* leaves via Agrobacterium-mediated infiltration (Voinnet *et al.*, 2003).

(3) Collect induced cells by centrifugation (15 min, 2500 rpm, at room temperature). Wash pellet carefully with resuspension solution (10 m*M* $MgCl_2$, 10 m*M* MES, pH 5.7, 150 μM acetosyringone; filter sterilize), to remove all traces of antibiotics. Resuspend cells in 30 ml of resuspension solution, and add fresh acetosyringone at final concentration of 150 μM.
(4) Allow cells to incubate for 2–3 h at room temperature, without shaking.
(5) Mix cells bearing the luciferase construct with the viral suppressor, 3:1 (volume/volume), and proceed to infiltrate suspension from the abaxial side of leaves. Gently scrape the epidermis of the leaf (without cutting through the leaf) to ease entry of the suspension, and push the suspension into the leaves with a 5-ml syringe without needle. A good infiltration will generate a ring of macerated tissue 1–3 cm in diameter.
(6) Transfer infiltrated plants back to the plant room for 2–3 days. On the day of LUC assay, detach infiltrated leaves, and cut out small squares or circles of leaf tissue, and transfer to the well of a 96-well plate, already containing MS medium (with or without added sucrose) and luciferin.
(7) Keep plate under one more entraining cycle to allow burn off of early expression luciferase, and start recording luciferase activity as for a normal Topcount™ assay.

4.5. Luciferase assays on a Topcount™

Although many labs rely on imaging to measure luciferase activity (Southern and Millar, 2005; Welsh *et al.*, 2005), another powerful and high-throughput circadian assay is based on a 96-well plate format (Southern and Millar, 2005). We have been using a Topcount™ Microplate Scintillation and Luminescence Counter (Perkin Elmer) with six detectors. With a reading time for each well set to 10 s, one full 96-well plate can be read in about 2 min, and a stack of 11 plates in constant light will run through the machine in 90 min.

96-well plates are fed to the Topcount luminometer from "stackers," long towers made of aluminum that sit outside on the sampling chamber. Two models are available, for 20 or 40 plates. The current version of stackers has solid walls of aluminum surrounding the plates inside, for increased sturdiness; this however causes a problem when one wishes to assay plants in the light. To permit light to reach the plates, we have cut holes on all sides without compromising the physical integrity of the stacker columns. For each sample plate, we position three clear 96-well plates above (from Costar, catalog number 3795), to allow light to reach the seedlings. A "stop plate" must be placed after the last sample plate: this plate should have two barcode stickers on the right side, to be recognized by the luminometer as the last of a stack of plates. Together with this protocol, we have put together a step-by-step guide with screenshots for setting up the assay on the Topcount computer; it is available on request.

Just as for cotyledon movement assays, seeds are surface-sterilized and plated on MS medium, which can be supplemented with 1–2% sucrose. After stratification for 3–4 days at 4 °C, plates are released into entraining conditions (light dark cycles or temperature cycles) for 7–10 days, or until primary leaves are just starting to show between cotyledons. In contrast to cotyledon movement, where we transfer both seedling and the medium onto which the seedling had germinated, only seedlings are transferred to the wells of the plate (opaque, white or black, to prevent light contamination between wells; white plates are purchased from Perkin Elmer, Optiplate-96 catalog number 6005299); it is therefore advantageous to make medium with only 0.6–0.8% agar, so as to facilitate the removal of seedlings with their roots intact, although robust rhythms will still be obtained when the root is snapped off. Each well of the 96-well plates contain 200 μl of the same medium used for seedling growth during entrainment, as well as 30 μl of a 2.5-m*M* D-luciferin solution (potassium salt, purchased from Biosynth or PJK). Plates are then sealed with adhesive sealant (Perkin Elmer, catalog number 6005185); 2–3 holes should be made above each plant with a needle to allow gas exchange. Luciferase protein is quite stable and accumulates prior to introduction of the substrate luciferin, which destabilizes luciferase activity. This allows luciferase activity to accurately represent *de novo* translation; because the transcript is unstable, luciferase activity accurately tracks *de novo* transcription (Millar *et al.*, 1992b; Welsh *et al.*, 2005). Therefore, introduction of luciferin results in a transient pulse of anomalously highlight production which should be allowed to dissipate prior to measurement of *de novo* activity: we therefore routinely entrain seedlings within 96-well plates for one more entraining cycle before release into constant conditions. For some experiments where the assay temperature is distinct from the temperature during entrainment, 1 or 2 more days of entrainment can be used to acclimate seedlings to the new conditions. For instance, for a temperature compensation experiment at 30 °C, one does not wish to shift seedlings directly from 22 to 30 °C. Rather, it is a good idea to expose seedlings within 96-well plates to the new temperature, while still enjoying light–dark cycles. The transition to the higher temperature should be timed with dawn; after 1–2 days under this new regime, plates can be moved to the Topcount for luciferase activity recordings. Conversely, for temperature compensation experiments run at 12–16 °C, the transition to the new temperature should coincide with dusk. These precautions ensure that one is looking at the effects of the new temperature on the pace of the clock, and not the combined effects of the new temperature and the temperature shift associated with the single transfer, which in and of itself would be similar to a temperature pulse for a phase response curve.

It is often necessary to rescue a seedling from a well of a 96-well plate at the end of a run, and we have found that seedlings can be readily transferred to soil. Often the cotyledons have stuck to the adhesive seal; one can scrape the leaves off the adhesive seal with a razor blade while gently pulling the seal off the wells. Some epidermal tissue will remain on the seal, but seedlings will survive. Another possibility is to cut out the seal above the well, and transfer the whole seedling, still attached to the seal. After a few days of growth on soil, seedlings will have grown enough to move their "plastic hat" away from the center of the rosette. We have had almost 100% success with either method, rendering rescue of critical seedlings a very easy process.

5. Concluding Remarks

The availability of increasingly sophisticated reagents and methods is facilitating mechanistic studies of the circadian clock. One can be optimistic that the next few years will yield much deeper understanding of the roles played by the PRR proteins in the circadian clock mechanism.

Acknowledgments

This work was supported by grants from the National Science Foundation to D.E.S. (MCB-0544137 and IOS-0748749) and to C.R.M (MCB-0343887) and from MEST/NRF to W.-Y. K. (EB-NCRC 20090091494 and World Class University Program R32-10148, South Korea).

References

Alabadí, D., Oyama, T., Yanovsky, M. J., Harmon, F. G., Más, P., and Kay, S. A. (2001). Reciprocal regulation between *TOC1* and *LHY/CCA1* within the *Arabidopsis* circadian clock. *Science* **293,** 880–883.

Alabadí, D., Yanovsky, M. J., Más, P., Harmer, S. L., and Kay, S. A. (2002). Critical role for CCA1 and LHY in maintaining circadian rhythmicity in *Arabidopsis*. *Curr. Biol.* **12,** 757–761.

Bechtold, N., Ellis, J., and Pelletier, G. (1993). *In planta Agrobacterium* mediated gene transfer by infiltration of adult *Arabidopsis thaliana* plants. *C. R. Acad. Sci. Paris Life Sci.* **316,** 1194–1199.

Bell-Pedersen, D., Cassone, V. M., Earnest, D. J., Golden, S. S., Hardin, P. E., Thomas, T. L., and Zoran, M. J. (2005). Circadian rhythms from multiple oscillators: Lessons from diverse organisms. *Nat. Rev. Genet.* **6,** 544–556.

Carré, I. A., and Kim, J.-Y. (2002). MYB transcription factors in the *Arabidopsis* circadian clock. *J. Exp. Bot.* **53,** 1551–1557.

Curtis, M. D., and Grossniklaus, U. (2003). A Gateway cloning vector set for high-throughput functional analysis of genes in planta. *Plant Physiol.* **133,** 462–469.

David, K. M., Armbruster, U., Tama, N., and Putterill, J. (2006). *Arabidopsis* GIGANTEA protein is post-transcriptionally regulated by light and dark. *FEBS Lett.* **580,** 1193–1197.

Farré, E. M., and Kay, S. A. (2007). PRR7 protein levels are regulated by light and the circadian clock in Arabidopsis. *Plant J.* **52,** 548–560.

Farré, E. M., Harmer, S. L., Harmon, F. G., Yanovsky, M. J., and Kay, S. A. (2005). Overlapping and distinct roles of PRR7 and PRR9 in the *Arabidopsis* circadian clock. *Curr. Biol.* **15,** 47–54.

Fujiwara, S., Wang, L., Han, L., Suh, S. S., Salomé, P. A., McClung, C. R., and Somers, D. E. (2008). Post-translational regulation of the circadian clock through selective proteolysis and phosphorylation of pseudo-response regulator proteins. *J. Biol. Chem.* **283,** 23073–23083.

Gallego, M., and Virshup, D. M. (2007). Post-translational modifications regulate the ticking of the circadian clock. *Nat. Rev. Mol. Cell Biol.* **8,** 139–148.

Gooch, V. D., Mehra, A., Larrondo, L. F., Fox, J., Touroutoutoudis, M., Loros, J. J., and Dunlap, J. C. (2008). Fully codon-optimized luciferase uncovers novel temperature characteristics of the *Neurospora* clock. *Eukaryot. Cell* **7,** 28–37.

Gould, P. D., Locke, J. C. W., Larue, C., Southern, M. M., Davis, S. J., Hanano, S., Moyle, R., Milich, R., Putterill, J., Millar, A. J., and Hall, A. (2006). The molecular basis of temperature compensation in the *Arabidopsis* circadian clock. *Plant Cell* **18,** 1177–1187.

Guo, F. Q., Wang, R., Chen, M., and Crawford, N. M. (2001). The Arabidopsis dual-affinity nitrate transporter gene AtNRT1.1 (CHL1) is activated and functions in nascent organ development during vegetative and reproductive growth. *Plant Cell* **13,** 1761–1777.

Hajdukiewicz, P., Svab, Z., and Maliga, P. (1994). The small, versatile *pPZP* family of *Agrobacterium* binary vectors for plant transformation. *Plant Mol. Biol.* **25,** 989–994.

Han, L., Mason, M., Risseeuw, E. P., Crosby, W. L., and Somers, D. E. (2004). Formation of an SCF^{ZTL} complex is required for proper regulation of circadian timing. *Plant J.* **40,** 291–301.

Harmer, S. L. (2009). The circadian system in higher plants. *Annu. Rev. Plant Biol.* **60,** 357–377.

Harmer, S. L., and Kay, S. A. (2005). Positive and negative factors confer phase-specific circadian regulation of transcription in Arabidopsis. *Plant Cell* **17,** 1926–1940.

Hicks, K. A., Millar, A. J., Carré, I. A., Somers, D. E., Straume, M., Meeks-Wagner, D. R., and Kay, S. A. (1996). Conditional circadian dysfunction of the *Arabidopsis early-flowering 3* mutant. *Science* **274,** 790–792.

Ito, S., Nakamichi, N., Kiba, T., Yamashino, T., and Mizuno, T. (2007). Rhythmic and light-inducible appearance of clock-associated pseudo-response regulator protein PRR9 through programmed degradation in the dark in *Arabidopsis thaliana. Plant Cell Physiol.* **48,** 1644–1651.

Jefferson, R. A., Kavanagh, T. A., and Bevan, M. W. (1987). GUS fusions: β-glucuronidase as a sensitive and versatile gene fusion marker in higher plants. *EMBO J.* **6,** 3901–3907.

Karimi, M., Inzé, D., and Depicker, A. (2002). GATEWAY vectors for Agrobacterium-mediated plant transformation. *Trends Plant Sci.* **7,** 193–195.

Kiba, T., Henriques, R., Sakakibara, H., and Chua, N.-H. (2007). Targeted degradation of PSEUDO-RESPONSE REGULATOR5 by a SCFZTL complex regulates clock function and photomorphogenesis in *Arabidopsis thaliana. Plant Cell* **19,** 2516–2530.

Kim, W.-Y., Geng, R., and Somers, D. E. (2003). Circadian phase-specific degradation of the F-box protein ZTL is mediated by the proteasome. *Proc. Natl. Acad. Sci. USA* **100,** 4933–4938.

Kim, W.-Y., Fujiware, S., Suh, S.-S., Kim, J., Kim, Y., Han, L., David, K., Putterill, J., Nam, H. G., and Somers, D. E. (2007). ZEITLUPE is a circadian photoreceptor stabilized by GIGANTEA in blue light. *Nature* **449,** 356–360.

Kondo, T., Strayer, C. A., Kulkarni, R. D., Taylor, W., Ishiura, M., Golden, S. S., and Johnson, C. H. (1993). Circadian rhythms in prokaryotes: Luciferase as a reporter of circadian gene expression in cyanobacteria. *Proc. Natl. Acad. Sci. USA* **90,** 5672–5676.

LeClere, S., and Bartel, B. (2001). A library of Arabidopsis 35S-cDNA lines for identifying novel mutants. *Plant Mol. Biol.* **46,** 695–703.

Liu, Y., Golden, S. S., Kondo, T., Ishiura, M., and Johnson, C. H. (1995). Bacterial luciferase as a reporter of circadian gene expression in cyanobacteria. *J. Bacteriol.* **177,** 2080–2086.

Locke, J. C. W., Southern, M. M., Kozma-Bognar, L., Hibberd, V., Brown, P. E., Turner, M. S., and Millar, A. J. (2005). Extension of a genetic network model by iterative experimentation and mathematical analysis. *Mol. Sys. Biol.* **1,** 0013,10.1038/msb4100018.

Locke, J. C. W., Kozma-Bognár, L., Gould, P. D., Fehér, B., Kevei, É., Nagy, F., Turner, M. S., Hall, A., and Millar, A. J. (2006). Experimental validation of a predicted feedback loop in the multi-oscillator clock of *Arabidopsis thaliana. Mol. Syst. Biol.* **2,** 59,10.1038/msb4100102.

Más, P., Kim, W.-Y., Somers, D. E., and Kay, S. A. (2003). Targeted degradation of TOC1 by ZTL modulates circadian function in *Arabidopsis thaliana. Nature* **426,** 567–570.

Matsushika, A., Makino, S., Kojima, M., and Mizuno, T. (2000). Circadian waves of expression of the APRR1/TOC1 family of pseudo-response regulators in *Arabidopsis thaliana*: insight into the plant circadian clock. *Plant Cell Physiol.* **41,** 1002–1012.

Matsushika, A., Murakami, M., Ito, S., Nakamichi, N., Yamashino, T., and Mizuno, T. (2007). Characterization of circadian-associated pseudo-response regulators: I. Comparative studies on a series of transgenic lines misexpressing five distinctive PRR genes in *Arabidopsis thaliana. Biosci. Biotechnol. Biochem.* **71,** 527–534.

McClung, C. R. (2006). Plant circadian rhythms. *Plant Cell* **18,** 792–803.

McClung, C. R. (2008). Comes a time. *Curr. Opin. Plant Biol.* **11,** 5213–5520.

Michael, T. P., Mockler, T. C., Breton, G., McEntee, C., Byer, A., Trout, J. D., Hazen, S. P., Shen, R., Priest, H. D., Sullivan, C. M., Givan, S. A., Yanovsky, M., *et al.* (2008). Network discovery pipeline elucidates conserved time-of-day-specific cis-regulatory modules. *PLoS Genet.* **4,** e14,10.1371/journal.pgen.0040014.

Millar, A. J., Short, S. R., Chua, N.-H., and Kay, S. A. (1992a). A novel circadian phenotype based on firefly luciferase expression in transgenic plants. *Plant Cell* **4,** 1075–1087.

Millar, A. J., Short, S. R., Hiratsuka, K., Chua, N.-H., and Kay, S. A. (1992b). Firefly luciferase as a reporter of regulated gene expression in higher plants. *Plant Mol. Biol. Rep.* **10,** 324–337.

Millar, A. J., Carré, I. A., Strayer, C. A., Chua, N.-H., and Kay, S. A. (1995). Circadian clock mutants in *Arabidopsis* identified by luciferase imaging. *Science* **267,** 1161–1163.

Mizoguchi, T., Wheatley, K., Hanzawa, Y., Wright, L., Mizoguchi, M., Song, H.-R., Carré, I. A., and Coupland, G. (2002). *LHY* and *CCA1* are partially redundant genes required to maintain circadian rhythms in Arabidopsis. *Dev. Cell* **2,** 629–641.

Mizuno, T., and Nakamichi, N. (2005). Pseudo-response regulators (PRRs) or true oscillator components (TOCs). *Plant Cell Physiol.* **46,** 677–685.

Morgan, L. W., Greene, A. V., and Bell-Pedersen, D. (2003). Circadian and light-induced expression of luciferase in *Neurospora crassa. Fungal Genet. Biol.* **38,** 327–332.

Nakamichi, N., Kita, M., Ito, S., Sato, E., Yamashino, T., and Mizuno, T. (2005). PSEUDO-RESPONSE REGULATORS, PRR9, PRR7 and PRR5, together play essential roles close to the circadian clock of *Arabidopsis thaliana. Plant Cell Physiol.* **46,** 686–698.

Onai, K., Okamoto, K., Nishimoto, H., Morioka, C., Hirano, M., Kami-ikc, N., and Ishiura, M. (2004). Large-scale screening of *Arabidopsis* circadian clock mutants by a high-throughput real-time bioluminescence monitoring system. *Plant J.* **40,** 1–11.

Para, A., Farré, E. M., Imaizumi, T., Pruneda-Paz, J. L., Harmon, F. G., and Kay, S. A. (2007). PRR3 is a vascular regulator of TOC1 stability in the *Arabidopsis* circadian clock. *Plant Cell* **19,** 3462–3473.

Plautz, J. D., Straume, M., Stanewsky, R., Jamison, C. F., Brandes, C., Dowse, H. B., Hall, J. C., and Kay, S. A. (1997). Quantitative analysis of *Drosophila period* gene transcription in living animals. *J. Biol. Rhythms* **12,** 204–217.

Pruneda-Paz, J. L., Breton, G., Para, A., and Kay, S. A. (2009). A functional genomics approach reveals CHE as a novel component of the *Arabidopsis* circadian clock. *Science* **323,** 1481–1485.

Salomé, P. A., and McClung, C. R. (2005). *PRR7* and *PRR9* are partially redundant genes essential for the temperature responsiveness of the *Arabidopsis* circadian clock. *Plant Cell* **17,** 791–803.

Sanford, J. C., Smith, F. D., and Russell, J. A. (1993). Optimizing the biolistic process for different biological applications. *Methods Enzymol.* **217,** 483–509.

Somers, D. E., Schultz, T. F., Milnamow, M., and Kay, S. A. (2000). ZEITLUPE encodes a novel clock-associated PAS protein from *Arabidopsis. Cell* **101,** 319–329.

Song, H.-R., and Carré, I. A. (2005). DET1 regulates the proteasomal degradation of LHY, a component of the *Arabidopsis* circadian clock. *Plant Mol. Biol.* **57,** 761–771.

Southern, M. M., and Millar, A. J. (2005). Circadian genetics in the model higher plant, *Arabidopsis thaliana. Methods Enzymol.* **393,** 23–35.

Stanewsky, R. (2007). Analysis of rhythmic gene expression in adult Drosophila using the firefly luciferase reporter gene. *Methods Mol. Biol.* **362,** 131–142.

Straume, M., Frasier-Cadoret, S. G., and Johnson, M. L. (1991). Least squares analysis of fluorescence data. *In* "Topics in Fluorescence Spectroscopy—Vol. 2, Principles," (J. R. Lakowicz, ed.), pp. 177–240. Plenum Press, New York.

Strayer, C., Oyama, T., Schultz, T. F., Raman, R., Somers, D. E., Más, P., Panda, S., Kreps, J. A., and Kay, S. A. (2000). Cloning of the *Arabidopsis* clock gene *TOC1*, an autoregulatory response regulator homolog. *Science* **289,** 768–771.

Vitha, S., McAndrew, R. S., and Osteryoung, K. W. (2001). FtsZ ring formation at the chloroplast division site in plants. *J. Cell Biol.* **153,** 111–120.

Voinnet, O., Rivas, S., Mestre, P., and Baulcombe, D. (2003). An enhanced transient expression system in plants based on suppression of gene silencing by the p19 protein of tomato bushy stunt virus. *Plant J.* **33,** 949–956.

Welsh, D. K., Yoo, S.-H., Liu, A. C., Takahashi, J. S., and Kay, S. A. (2004). Bioluminescence imaging of individual fibroblasts reveals persistent, independently phased circadian rhythms of clock gene expression. *Curr. Biol.* **14,** 2289–2295.

Welsh, D. K., Imaizumi, T., and Kay, S. A. (2005). Real-time reporting of circadian-regulated gene expression by luciferase imaging in plants and mammalian cells. *Methods Enzymol.* **393,** 269–288.

Wilsbacher, L. D., Yamazaki, S., Herzog, E. D., Song, E. J., Radcliffe, L. A., Abe, M., Block, G., Spitznagel, E., Menaker, M., and Takahashi, J. S. (2002). Photic and circadian expression of luciferase in mPeriod1-luc transgenic mice *in vivo. Proc. Natl. Acad. Sci. USA* **99,** 489–494.

Yakir, E., Hilman, D., Kron, I., Hassidim, M., Melamed-Book, N., and Green, R. M. (2009). Posttranslational regulation of CIRCADIAN CLOCK ASSOCIATED1 in the circadian oscillator of Arabidopsis. *Plant Physiol.* **150,** 844–857.

Zachgo, S., Perbal, M.-C., Saedler, H., and Schwarz-Sommer, S. (2000). *In situ* analysis of RNA and protein expression in whole mounts facilitates detection of floral gene expression dynamics. *Plant J.* **23,** 697–702.

Zdepski, A., Wang, W., Priest, H. D., Ali, F., Alam, M., Mockler, T. C., and Michael, T. P. (2008). Conserved daily transcriptional programs in *Carica papaya. Trop. Plant Biol.* **1,** 236–245.

CHAPTER TWENTY

Reversible Histidine Phosphorylation in Mammalian Cells: A Teeter-Totter Formed by Nucleoside Diphosphate Kinase and Protein Histidine Phosphatase 1

Thomas Wieland,* Hans-Jörg Hippe,† Katrin Ludwig,‡ Xiao-Bo Zhou,§ Michael Korth,§ *and* Susanne Klumpp‡,✠

Contents

Abstract

Regulation of protein phosphorylation by kinases and phosphatases is involved in many signaling pathways in mammalian cells. In contrast to prokaryotes and lower eukaryotes a role for the reversible phosphorylation of histidine residues

* Institut für Experimentelle und Klinische Pharmakologie und Toxikologie, Medizinische Fakultät Mannheim, Universität Heidelberg, Mannheim, Germany
† Medizinische Klinik III, Universitätsklinikum, Universität Heidelberg, Heidelberg, Germany
‡ Institut für Pharmazeutische und Medizinische Chemie, Westfälische Wilhelms-Universität, Münster, Germany
§ Institut für Pharmakologie für Pharmazeuten, Universitätsklinikum Hamburg-Eppendorf, Hamburg, Germany
✠ Deceased June 17, 2009

Methods in Enzymology, Volume 471
ISSN 0076-6879, DOI: 10.1016/S0076-6879(10)71020-X

is just emerging. The β subunit of heterotrimeric G proteins, the metabolic enzyme adenosine 5′-triphosphate-citrate lyase (ACL), and the Ca^{2+}-activated K^+ channel $K_{Ca}3.1$ have been identified as targets for nucleoside diphosphate kinase (NDPK) acting as protein histidine kinase and the so far only identified mammalian protein histidine phosphatase (PHPT-1). Herein, we describe the analysis of the phosphorylation and dephosphorylation of histidine residues by NDPK and PHPT-1. In addition, experimental protocols for studying the consequences of heterotrimeric G protein activation via NDPK/G$\beta\gamma$ mediated phosphorelay, the regulation of ACL activity and of $K_{Ca}3.1$ conductivity by histidine phosphorylation will be presented.

1. Introduction

The importance of reversible phosphorylation of proteins as a widespread, posttranslational regulatory mechanism of cellular functions is nowadays well recognized, and thus the two regulatory components of such systems, kinases and phosphatases, are interesting targets for drug development. In mammalian systems, however, kinases and phosphatases are believed to mainly target protein serine, threonine, and tyrosine residues, and such regulatory systems have been extensively studied. Nevertheless, the phosphorylation of histidine residues (P-His) was already observed in bovine liver in the early 1960's (Boyer *et al.*, 1962), but only sporadic reports during the following decades have addressed this issue in vertebrate systems. In contrast, in prokaryotes (Hess *et al.*, 1988) as well as in lower eukaryotes, like yeast, fungi, and plants (Kennelly and Potts, 1996; Swanson *et al.*, 1994) two or multicomponent signaling systems were discovered in which protein histidine kinases and histidine phosphatases are important mediators of cellular responses such as bacterial chemotaxis. Homology screens did not uncover similar systems and thus a role of P-His, besides its existence as phosphorylated intermediate in enzymatic reactions, was for quite sometime believed to be of minor importance for vertebrates.

Therefore, when Wagner and Vu (1995) reported the phosphorylation of the histidine residue 760 of the enzyme ATP-citrate lyase (ACL; EC 2.3.3.8) by another metabolic enzyme commonly known as nucleoside diphosphate kinase (NDPK; EC 2.7.4.6), the data were perceived with great skepticism. The correct biochemical name of NDPK is NTP/NDP transphosphorylase. It catalyzes the transfer of terminal phosphate groups from 5′-triphosphate- to 5′-diphosphate-nucleotides. In man 10 genes have been identified to be a part of the NDPK family. Nine genes are named NME 1–9. The 10th gene encoding a truncated NDPK is named RP2. The enzymatic activity has been unequivocally demonstrated for the NME1–4 gene products NDPK A–D which are also often called Nm23-H1–H4

(H stands for Human) with respect to the function of NDPKs as tumor suppressors (Boissan *et al.*, 2009; Steeg *et al.*, 2008). ACL has meanwhile been confirmed as a substrate of NDPK A acting as protein histidine kinase, and at least two additional proteins, the G protein β subunit (Cuello *et al.*, 2003) and the intermediate conductance potassium channel $K_{Ca}3.1$ (Srivastava *et al.*, 2006) are complexed with and phosphorylated specifically by NDPK B at residues His-266 and His-358, respectively. Therefore, at least the NDPK isoforms A and B have been proven to act as protein histidine kinases and the similarity to systems in lower organisms was further increased by the identification of the mammalian protein histidine phosphatase (PHPT-1, also named PHP) by two independent groups (Ek *et al.*, 2002; Klumpp *et al.*, 2002). The phosphorylated forms of all three afore mentioned substrates of the NDPK protein histidine kinase activity are targets of this phosphatase *in vitro* and in living cells (Klumpp *et al.*, 2003; Mäurer *et al.*, 2005; Srivastava *et al.*, 2008) and thus novel roles for reversible histidine phosphorylation in vertebrates are emerging (for review see Klumpp and Krieglstein, 2009).

In this chapter, we will therefore describe methods to analyze the phosphorylation and dephosphorylation of histidine residues by NDPK and PHPT-1 *in vitro* and provide experimental protocols for studying the consequences of heterotrimeric G protein activation via NDPK/G$\beta\gamma$ mediated phosphorelay (Hippe *et al.*, 2007), the regulation of ACL activity (Klumpp *et al.*, 2009; Krieglstein *et al.*, 2008; and of $K_{Ca}3.1$ conductivity (Srivastava *et al.*, 2008) by histidine phosphorylation and dephosphorylation.

2. Analysis of Phosphorylation and Dephosphorylation of Histidine Residues *In Vitro*

2.1. Expression and purification of NDPK and PHPT-1

NDPK A and B (Srivastava *et al.*, 2006; Wagner *et al.*, 1997) as well as PHPT-1 (Bäumer *et al.*, 2007; Srivastava *et al.*, 2008) can be expressed and purified from *Escherichia coli* and the purified enzymes can be used to study their enzymatic activity as well as the consequences of their interaction with their common target proteins *in vitro*. In the following, we will therefore describe the purification of fusion proteins from *E. coli* lysates and standardized activity assays.

2.1.1. Expression and purification of recombinant human NDPK, standardized activity assays

The coding sequence of human NDPK A or B was subcloned into the *Eco*RI site of the pGEX2T vector (GE Healthcare) and expression is performed in the *E. coli* BL21DE3 strain. A single colony of transformed bacteria is picked from an agar plate and cultured overnight at 37 °C in a shaking incubator in 10 ml Y2T medium containing 100 μg/ml ampicillin. Two milliliters of this culture is added to 200 ml of fresh Y2T/ampicillin medium and incubated further until the OD_{600} reaches 0.5. For induction of protein expression, isopropyl-β-thiogalactoside (IPTG, final concentration 1 m*M*) is added and the incubation is continued for 1 h at room temperature. The bacteria are harvested by centrifugation (10 min at 10,000×*g* at 4 °C) and the pellet is resuspended in 3 ml ice-cold phosphate buffered saline (PBS, 137 m*M* NaCl, 2.7 m*M* KCl, 8 m*M* Na_2HPO_4, 1.8 m*M* KH_2PO_4, pH 7.4) containing 1 m*M* dithiothreitol and 0.1% Triton X-100. The bacteria are lyzed on ice by ultrasonic homogenization using five pulses for 30 s each, with 2 min intermittent cooling phases. The lysate is centrifuged for 15 min at 25,000×*g* to pellet cell debris and particles. The clear supernatant is added to 500 μl of a glutathione sepharose slurry (GE Healthcare) in PBS and incubated on a shaker for min at 4 °C and then transferred to disposable column. The beads are washed with 20 ml PBS and then eluted with 10 ml PBS containing 10 m*M* glutathione. The eluate is then concentrated by using Vivaspin 20 centrifugal concentrators (Satorius). Typically, 1 ml of recombinant GST-fusion protein at concentration of 500–600 μg/ml is obtained.

2.1.2. Assay for NDPK activity

For determination of the enzymatic NDPK activity, 3–10 ng protein of recombinant protein is used in reaction buffer (100 μl) consisting of 50 m*M* triethanolamine, pH 7.5, 0.2 m*M* cAMP, 0.2 m*M* ATP, 6 m*M* $MgCl_2$, 2 m*M* EDTA, 2 m*M* DTT, 0.2 m*M* GDP, and 1 μCi of [8-^{3}H]GDP. The samples are incubated for 10 min at 37 °C in a water bath and the reaction is terminated by adding 10 μl of a 10% (w/v) SDS solution. The guanine nucleotides are separated by thin layer chromatography. First, for each sample 2 μl of a GMP–GDP–GTP solution (each 3 m*M*) is spotted as a carrier near the bottom of a PEI cellulose plate, then 10 μl of the sample is spotted in four steps (2.5 μl each) onto the carrier. The plate is developed in a reservoir with a freshly prepared soluble phase consisting of equal parts of 1 *M* LiCl and 2 *M* formic acid. The separated guanine nucleotides are visualized with UV light at 254 nm, the areas containing GTP, GDP, and GMP are cut off and 4 ml scintillation liquid is added. To elute the nucleotides the samples are shaken vigorously for 2 h. The radioactivity is determined in a liquid scintillation spectrometer. As described in earlier *Methods in Enzymology* chapter (Lutz *et al.*, 2004), this assay can also be used to quantify NDPK activity in cell lysates or homogenates, membranes or cytosol. For such applications protein amounts of 0.3–1 μg is good point to start with.

2.1.3. Expression and purification of human recombinant PHPT-1

The coding sequence of human PHPT-1 was subcloned into the expression vector pET16b (Novagen) providing an N-terminal 10× His-tag extension (Bäumer *et al.*, 2007). For expression of PHPT-1, transformed *E. coli* BL21 (DE3) is inoculated in 50 ml of LB medium supplemented with 50 μg/ml ampicillin overnight at 37 °C in a shaking incubator. 5 ml of this culture are added to 1 l of LB/ampicillin medium and grown further at 37 °C until OD_{600} reaches 0.6. For induction of protein expression, IPTG (final concentration 10 μ*M*) is added and the cells are further incubated for 5 h at 37 °C in the shaking incubator before being harvested by centrifugation (15 min, 3000×*g*, 4 °C). The pellet is resuspended in 10 ml of a buffer 1 consisting of 20 m*M* Tris–HCl, pH 7.9, 5 m*M* imidazole, and 250 m*M* NaCl. Cells are lyzed on ice by sonification using 10 × 20 s pulses with 10 s cooling phases in between. The lysate is centrifuged (30 min, 20,000×*g*, 4 °C) and the supernatant is used for isolation of PHPT-1. 1.5 ml Ni-NTA agarose (Qiagen) slurry is poured into a Bio-Rad column (diameter 1.5 cm, length 15 cm). The resin is allowed to settle and equilibrated with 10 column volumes of buffer 1. The supernatant is loaded onto the column a with a flow rate of approximately 1 ml/min. The column is extensively washed with about 40 column volumes buffer 2 (20 m*M* Tris–HCl, pH 7.9, 50 m*M* imidazole, 500 m*M* NaCl). PHPT-1 is eluted with buffer 3 (20 m*M* Tris–HCl, pH 7.9, 500 m*M* imidazole) in 1 ml fractions. Those containing the highest amount of protein are pooled, then dialyzed against 2 l of buffer 4 (25 m*M* Tris–HCl, pH 7.9, 2 m*M* β-mercaptoethanol) for 1 h and for a second time against 5 l overnight at 4 °C. The eluate is concentrated using Amicon Ultra centrifugation devices (Millipore) up to a concentration of 5–10 mg/ml.

2.1.4. Measurement of PHPT-1 activity

Recombinant bacterial CheA, which autophosphorylates on His48, is expressed His-tagged protein in *E. coli* and purified exactly as described by Swanson *et al.* (1993). The phosphorylation of 8.5 μg CheA is performed a reaction mixture (50 μl) consisting of 50 m*M* HEPES, pH 8.0, 1 m*M* $MgCl_2$, 1 μ*M* ATP containing 100 μCi [γ-^{32}P]ATP (3000 Ci/mmol, Perkin-Elmer), for 3 h at 37 °C. After removal of unincorporated [γ-^{32}P] ATP by Centrisep spin columns (Applied Biosystems) [^{32}P-His48]CheA is stable for several days at 4 °C and can be used as standard protein to determine P-His-phosphatase activity of PHPT-1 and other phosphatases.

Dephosphorylation is carried out in a volume of 40 μl for 30 min at 37 °C with 0.2 μg purified PHPT-1, 200 pmol [^{32}P-His48]CheA, in a reaction mix containing 25 m*M* triethanolamine–HCl, pH 7.5, 10 m*M* $MgCl_2$, and 0.1% β-mercaptoethanol. The incubation is terminated by adding 10 μl of 0.5 *M* EDTA immediately followed by addition of 150 μl methanol/acetone (1:1) to precipitate the proteins. Note, that protein precipitation by strong acids such as trichloroacetic acid has to be avoided due to the acid lability of P-His. Samples are centrifuged (5 min, 14,000×*g*, room temperature). The supernatant is quantified for $^{32}P_i$ content by liquid scintillation counting after addition of a suitable scintillation cocktail.

Alternatively, PHPT-1 activity can be determined by autoradiography which does not require precipitation of not-hydrolyzed [^{32}P-His48]CheA and thus is performed with a smaller amount of the substrate (0.4 μg) in a reduced volume of 15 μl. In this case, the reaction is terminated by addition of 5 μl of fourfold concentrated sample buffer (Laemmli, 1970) As the phosphoamidate bond of P-His is heat sensitive, boiling of the samples should be avoided. Proteins are separated by discontinuous SDS–PAGE on gels containing 10% (w/v) acrylamide. The dye front containing free nucleotides is cut off from the gel. The gel is rinsed for 5 min in a 10% (v/v) glycerol/water mixture, dried, and autoradiographed. As phosphoamidate bonds are sensitive to acid, staining of the proteins by methods using an acidic fixation, for example, Coomassie Brilliant Blue staining, should be avoided. The use of prestained molecular weight markers can be recommended.

2.2. Analysis of P-His-phosphorylation and dephosphorylation in NDPK and PHPT-1 substrate proteins

2.2.1. Phosphorylation and dephosphorylation of Gβ

Apparently, phosphorylation of Gβ on His-266 by NDPK B is more complex than the phosphorylation of the other known protein substrates. As mentioned before (Cuello *et al.*, 2003), a reconstitution of Gβ phosphorylation cannot be achieved by an interaction of purified recombinant proteins as a so far unknown scaffold protein is involved. Therefore, Gβ phosphorylation is usually detected in cell membranes or in a NDPK B/G$\beta\gamma$ complex enriched, for example, from retinal rod outer segments, the preparation of which has been described in detail in a previous volume of *Methods in Enzymology* (Lutz et al., 2004). [γ-^{32}P]GTP (6000 Ci/mmol, Perkin-Elmer) is the best substrate for the phosphorylation of Gβ. It offers the advantage that it is a poor substrate for many protein kinases using ATP and therefore offers a simple method to visualize phospho-NDPK and phospho-Gβ with low background phosphorylation (Fig. 20.1). For that reason the assay described herein can also be used to visualize the phosphorylation and dephosphorylation of $K_{Ca}3.1$ by NDPK and PHPT-1, respectively (Edward Skolnik, New York, personal communication). The assay is usually performed in Eppendorf tubes in a total volume of 10–20 μl in a reaction mixture containing 50 m*M* triethanolamine–HCl, pH 7.4, 150 m*M* NaCl, 2 m*M* $MgCl_2$, 1 m*M* EDTA, and 1 m*M* dithiothreitol,

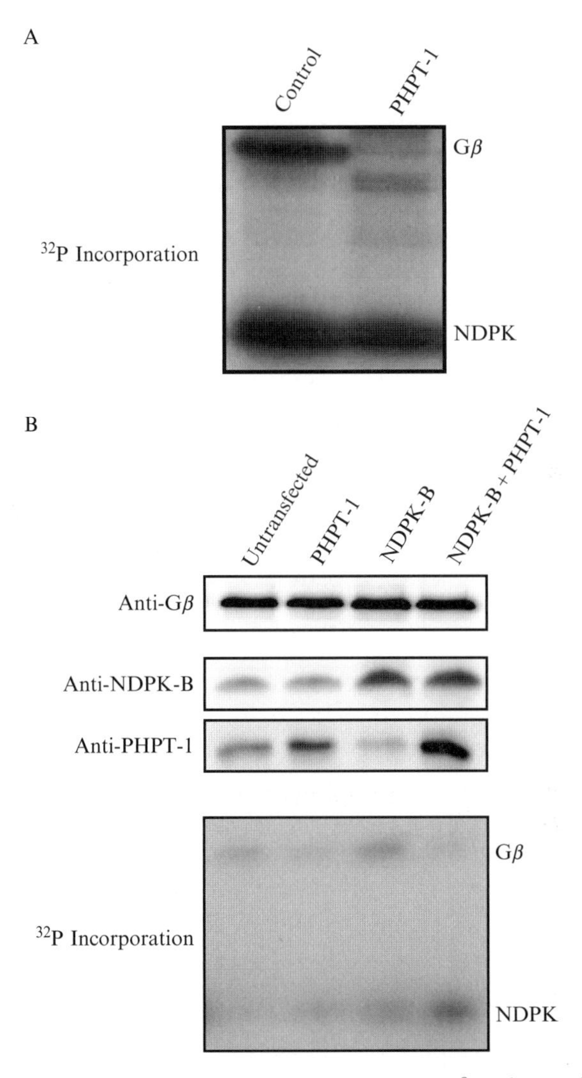

Figure 20.1 Dephosphorylation of His-P on the Gβ subunit by PHPT-1. (A) A complex formed by $G\beta_1\gamma_1$-dimers of transducin and NDPK B was purified as described (Lutz *et al.*, 2004) and 10 μg of protein were phosphorylated with [γ-^{32}P]GTP for 10 min at 37 °C. The phosphorylated proteins were subsequently treated with solvent (Control) or 2.5 μg PHPT-1 for 30 min. Proteins were separated on 15% SDS–PAGE and autoradiographed. (B) Five micrograms of membrane of H10 cells (Lutz *et al.*, 2004) obtained from control cells (Untransfected), cells sevenfold overexpressing PHPT-1 (PHPT-1), threefold overexpressing human NDPK B (NDPK B) or cells threefold overexpressing NDPK B, and sevenfold PHP (NDPK B + PHPT-1) were subjected to SDS–PAGE and the content of Gβ, NDPK B, and PHPT-1 was visualized by immunoblot analysis (upper panels) using antibodies against Gβ (T-20, Santa Cruz), NDPK B (MC-381, Kamya), or PHPT-1 (Klumpp *et al.*, 2002). The same membranes (5 μg) were phosphorylated with [γ-^{32}P]GTP for 5 min. An autoradiogram after SDS–PAGE is shown in the lower panel. Phosphorylated Gβ and phosphorylated NDPK are indicated.

10 n*M* [γ-^{32}P]GTP (Perkin-Elmer). Usually 5 μl of the fourfold concentrated reaction mixture is given into a tube on ice. Before incubation the tubes are warmed for 5 min at 30 °C and the protein, for example, 10 μg NDPK B/G$\beta\gamma$ enriched fraction from transducin or 5 μg of a membrane suspension in 10 m*M* triethanolamine–HCl, pH 7.4, is added and incubated for the desired periods of time (see Fig. 20.1). Note that especially in membranes the phosphorylation of Gβ is transient due to the transfer of high energetic P-His onto GDP and the consumption of the formed GTP by GTPases and thus incubation times in range 1–5 min can be recommended (Fig. 20.1B). The optimal conditions for obtaining good phosphorylation results have to be adjusted in each system by varying incubation times and protein amount (see Lutz *et al.*, 2004). If a subsequent analysis of the dephosphorylation by PHPT-1 is desired (Fig. 20.1A) the reaction should be terminated by the addition EDTA (5 m*M* final concentration). As both the kinase activity of the NDPK and the retransfer of p-His onto GDP require the presence of divalent cations, this preserves the amount of incorporated phosphate in absence of the PHPT-1, which do not require the presence of divalent cations. Addition of 2.5 μg of recombinant PHPT-1 per sample is sufficient to obtain complete dephosphorylation of the PHPT-1 substrate (Fig. 20.1A) after an incubation time of 30 min at 37 °C.

The reaction is terminated by the addition of 10 μl of threefold concentrated sample buffer (Laemmli, 1970), followed by a 1 h incubation at room temperature. Analysis is performed by SDS–PAGE as described above.

2.2.2. Phosphorylation and dephosphorylation of ATP-citrate lyase

ACL is an enzyme involved in acetylcholine and cholesterol synthesis, fatty acid and energy metabolism, and was the first vertebrate protein substrate of NDPK and PHPT-1 to be identified (Klumpp *et al.*, 2003; Wagner and Vu, 1995). The protein is a homotetramer formed of 110 kDa subunits, and autophosphorylates at His-760 in the catalytic domain using ATP as substrate to form an enzymatic intermediate (Cottam and Srere, 1969; Elshourbagy *et al.*, 1990). Noiman and Shaul (1995) reported that phosphorylation by ATP in the presence of an excess of EDTA (e.g., 5 m*M* final concentration) results in an *exclusive* labeling of P-His-containing proteins as kinases acting on serine, threonine, or tyrosine residues essentially require the presence of divalent cations such as Mg^{2+}. The method used for searching PHPT-1 substrates took advantage of this selective labeling: Phosphorylation of soluble rabbit liver extract with [γ-^{32}P]ATP in the presence of EDTA resulted only in three labeled proteins, one of them being dephosphorylated after addition of PHPT-1. This protein was identified as ACL. Note, that in contrast to soluble extracts of tissues or cell culture lysates, autophosphorylation of purified ACL can only be observed in the presence of Mg^{2+}. Autophosphorylation of ACL is carried out in Eppendorf tubes in a total volume of 10 μl in a reaction mixture consisting of 25 m*M* Tris–HCl, pH 7.5, and 1 μ*M* ATP including 2 μCi [γ-^{32}P]ATP at 37 °C for 15 min. The use of 0.1 μg purified ACL in the presence of 5 m*M* $MgCl_2$, 100 μg soluble liver extract or 15–25 μg of extract of cultured cells, both in the presence of 5 m*M* EDTA, can be recommended. To study dephosphorylation of [^{32}P-His]ACL 0,6 μg recombinant PHPT-1 is added in a volume of 5 μl containing 25 m*M* Tris–HCl, pH 7.5, and 5 m*M* EDTA. Samples are incubated at 37 °C for 30 min and the reaction is terminated by addition of 5 μl fourfold concentrated sample buffer without heating. Analysis is performed by SDS–PAGE (12.5%, w/v, acrylamide) as described above and labeled ACL is detected by autoradiography.

3. Functional Analysis of NDPK/PHPT-1 Regulated Systems in Living Cells

3.1. Regulation of basal cardiac cAMP synthesis by the NDPK B/G$\beta\gamma$ phosphorelay

It has been demonstrated that the NDPK B/G$\beta\gamma$ mediated phosphorelay contributes to basal, G protein coupled receptor (GPCR)-independent cAMP formation in cardiac myocytes and thereby regulates contractility (Hippe *et al.*, 2003, 2007). We therefore describe here two experimental settings which allow the quantification of the contribution of this phosphorelay to cAMP formation. The first is the overexpression of a mutant of the G protein β_1-subunit, in which the intermediately phosphorylated His-266 is substituted by leucine (Gβ_1H266L), and thus is no substrate for the NDPK B mediated phosphorylation. The second is the siRNA-induced knockdown of NDPK B. The loss in cAMP formation in cultured cardiomyocytes in both settings reflects the reduced Gα_s activation via the NDPK B/G$\beta\gamma$ phosphorelay. Note however, that the siRNA-mediated depletion might effect other cellular functions of NDPK B as well, whereas the overexpression of Gβ_1H266Lγ_2-dimers selectively disturbs the NDPK B-induced phosphorelay without interfering with the GPCR-induced G protein activation (Hippe *et al.*, 2007).

3.1.1. Generation and amplification of recombinant adenoviruses

To allow for formation of recombinant G$\beta_1\gamma_2$-dimers in the target cells, biscistronic adenoviral shuttle vectors can be generated. These vectors encode wild-type (WT) Gβ_1 (Gβ_1WTγ_2) or H266L-mutated Gβ_1 (Gβ_1H266Lγ_2) in conjunction with Gγ_2. First, the cDNA encoding WT Gβ_1 or the H266L mutant (Hippe *et al.*, 2007) is subcloned into pShuttle-IRES-hrGFP-1 vector (Stratagene). Second, Gβ_1 in conjunction with the internal ribosome entry site (IRES) is isolated from pShuttle by *Bst*XI/*Nhe*I digest and then subcloned into the pAdTrack-CMV shuttle vector containing a GFP reporter driven by an additional CMV promoter. Finally, Gγ_2 was inserted into this vector construct downstream the IRES, allowing the expression of two proteins from the same promoter. A shuttle vector encoding PHPT-1 is obtained by subcloning the PHPT-1 cDNA into the *Not*I/*Hin*dIII sites of pAdTrack-CMV vector. All constructs have to be confirmed by restriction digest analysis and DNA sequencing. As control vector the use of pAdTrack-CMV vector, which will produce an adenovirus encoding for EGFP (AdGFP) only, can be recommended. Homologous recombination between the shuttle vectors and the virus genome (pAdEasy) is performed in *E. coli* strain BJ5183 (Luo *et al.*, 2007). Detailed protocols for this procedure as well as the amplification and purification are published (Luo *et al.*, 2007) and can be obtained at www.coloncancer.org/adeasy.htm. Thus, they are described here only briefly: Linearized recombinant viral DNA is transfected into HEK293 cells at 50–70% confluency in a culture dish (ø 60 mm) using Lipofectamine 2000 Reagent (Invitrogen) or Polyfect (Qiagen). After 7–14 days the cells are harvested and cell extracts are prepared by three cycles of freezing (in liquid nitrogen) and thawing (in a water bath of 37 °C) to achieve total disruption of infected cells. The cell debris is removed by centrifugation (730×*g*, 7 min, at room temperature) and the supernatant is dispensed on 70–80% confluent culture dishes (ø 15 cm) with HEK293 cells. After 2 days cells are collected again using the procedure described before and the supernatant is distributed to five 70–80% confluent dishes with HEK293 cells. In the next step, the supernatant of these cells obtained again after 48 h is distributed to 25 culture dishes. In the final step, the cells are harvested by centrifugation (730×*g*, 7 min, room temperature) just before the general cell lyzes is about to start (usually 36–48 h after infection). The resulting pellet is resuspended in 2 ml of the supernatant and the cell suspension is freezed and thawed (see above) three times and is incubated with 35 m*M* $MgCl_2$, 6 μg/ml *DNaseI* and 100 μg/ml *RNaseA* for 30 min at 37 °C, followed by a second centrifugation (1500×*g*, 30 min, 4 °C). The supernatant is used for purification of the viral particles by a CsCl gradient centrifugation (100,000×*g*, 20 h, 4 °C). The virus suspension is dialyzed twice against PBS (3 h, 20 h), and then against virus storage buffer (140 m*M* NaCl, 1 m*M* KCl, 10 m*M* Tris–HCl, pH 7.4, 2 m*M* $MgCl_2$) for another 4 h. The virus suspension is diluted 1:1 with glycerol and stored at −20 °C. Viral stocks (about 10^9 biological active viral particles per μl) can be used for at least 6 months. Note that the amount of viruses (MOI, *m*ultiplicity *o*f *i*nfection, ratio of the number of biological active viruses to number of cells) required to obtain sufficient infection rates has to be optimized for each target cell type.

3.1.2. Isolation and culture of neonatal rat cardiomyocytes

Neonatal rat cardiac myocytes (NRCM) were isolated from 1- to 3-day-old Sprague–Dawley neonates by serial collagenase/pancreatin digestion. Rat hearts from 30 to 50 pups were minced and subjected to a serial collagenase digestion to release single cells. The digestion step in the digestion buffer (116 m*M* NaCl, 20 m*M* HEPES, 0.8 m*M* Na_2HPO, 5.6 m*M* glucose, 5.4 m*M* KCl, 0.8 m*M* $MgSO_4$, pH 7.35, 0.6 mg/ml of pancreatin, and 0.53 mg/ml collagenase type II) is repeated five times. The cell suspensions from each digestion are combined and centrifuged at 730×*g* for 10 min. The pellet of cells is resuspended in complete culture medium (Dulbecco's minimal essential medium (DMEM) supplemented with 10% fetal calf serum, 100 μg/ml streptomycin, and 100 U/ml penicillin). The cells are filtered through 45 μm mesh and preplated for 1.5–2 h to allow the differential attachment of nonmyocyte (fibroblasts, endothelial, and vascular smooth muscle) cells. The nonattached cells, that is, the cardiac myocyte fraction, are pelleted and replated in 12 well-tissue culture dishes at a density at 250,000 cells per well in complete culture medium with 0.1 m*M* bromodeoxyuridine (BrdU) to prevent overgrowth by the remaining fibroblasts. NRCM are used for experiments 4 days after isolation, when they spontaneously and synchronously contract at approximately 300 beats/min.

For immunoblot analysis, cardiac myocytes were lyzed in lysis buffer (10 m*M* Tris–HCl, pH 7.45, 0.1 m*M* EDTA, 1 m*M* PMSF, 1% Triton X-100, 0.5% sodium deoxycholate, and protease inhibitors (protease inhibitor cocktail, Sigma)).

3.1.3. Adenoviral overexpression of WT or phosphorylation-deficient $G\beta_1\gamma_2$-dimers in cardiomyocytes

To analyze the impact of the phosphorylation-deficient $G\beta_1$-mutant ($G\beta_1$H266L) on the activation of $G\alpha_s$ in intact cardiac myocytes cAMP production is measured following overexpression of $G\beta_1WT\gamma_2$ or $G\beta_1H266L\gamma_2$ together with $G\alpha_s$ by adenoviral gene transfer. Spontaneously beating NRCM are infected 4 days after isolation in 12 well-tissue culture dishes at a density of 250,000 cells per well cultured in complete culture medium. NRCM were washed once with serum-free DMEM. Viral stocks are diluted in 300 μl serum-free DMEM per well to obtain the desired MOI. NRCM are incubated with the virus-containing medium for 20 min at 37 °C. Then 700 μl DMEM containing 4% FCS is added per well, and the cells are analyzed after additional 48 h in culture. Figure 20.2A shows the overexpression of WT and mutant $G\beta_1$ with or without overexpression of $G\alpha_s$ at MOI of 10 bav per cell and the suppression of basal, receptor-independent cAMP content induced by $G\beta_1H266L\gamma_2$ compared to $G\beta_1WT\gamma_2$.

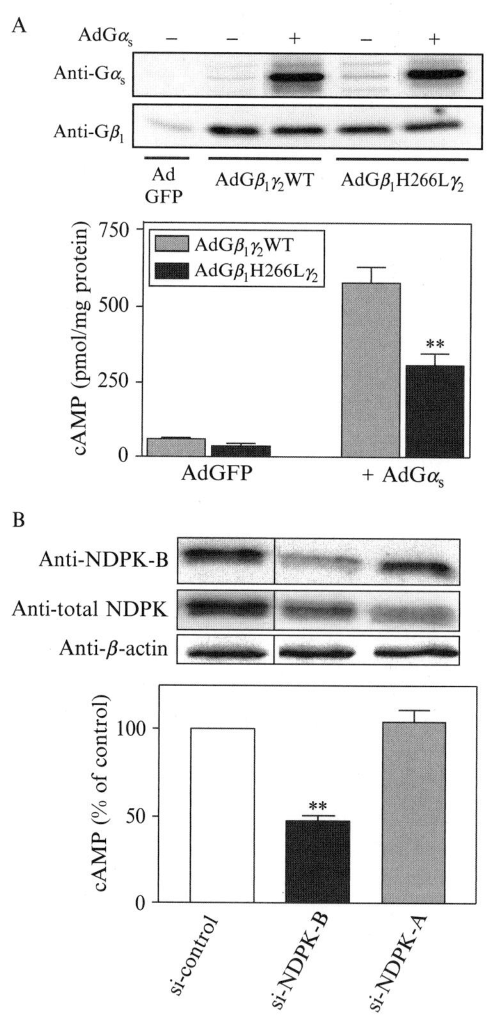

Figure 20.2 Suppression of basal cAMP production in cardiac myocytes by expression of $G\beta_1H266L\gamma_2$ or depletion of NDPK B. (A) NRCM were coinfected with $AdG\alpha_s$ and AdGFP or $AdG\beta_1\gamma_2$, either WT or H266L-mutant at MOI 10. Representative immunoblots of the expression of $G\beta_1$ and $G\alpha_s$ 48 h after infection are shown. (B) NRCM were sequentially, double transfected with scrambled siRNA (si-Control) or specific siRNA (si-) against NDPK A or B as indicated. Representative immunoblots of total NDPK and specific NDPK B expression in NRCM 72 h after the second transfection are shown. Expression levels of β-actin served as loading controls. Basal cAMP content in NRCM was measured in the presence of 1 m*M* IBMX and 1 μM propranolol. Data are means ± S.E.M., $n > 4$, **, $P < 0.01$. Both, inhibition of the phosphorelay by expressing $G\beta_1H266L\gamma_2$ and depletion of NDPK B reduced the basal cAMP content by about 50%.

3.1.4. siRNA-induced depletion of NDPK B in NRCM

Knockdown of NDPK A or B in NRCM is performed by a sequential, double transfection with siRNA duplexes. siRNAs are directed against the following target sequences: NDPK A: GGA TTC CGC CTG GTT GGT T (Fan *et al.*, 2003), NDPK B#1: GGG GTT CCG CCT GGT GGC C, NDPK B: AACTGATTGACT ATA AGT CTT (Kapetanovich *et al.*, 2005). Control siRNA (si-Control) was a scrambled siRNA (Ambion). Spontaneously beating NRCM were transfected 2–3 days after isolation in 12 well-plates at a density of 250,000 cells/well cultured in complete culture medium without antibiotics using Lipofectamine 2000 reagent (Invitrogen). For transfection, 100 pmol siRNA is diluted in 100 μl Opti-MEM I (Invitrogen) and combined with 2 μl Lipofectamine 2000 (Invitrogen) prediluted in 100 μl Opti-MEM I (Invitrogen). After gently mixing and incubation for 20 min, the siRNA-Lipofectamine complexes (200 μl) are added to each well containing myocytes in 1 ml culture medium without antibiotics and incubated at 37 °C and 5% CO_2. To achieve maximally effective knockdown, a second transfection is performed after 24 h precisely as described for the first transfection. The following day, the medium is changed to fresh culture medium including antibiotics, and the cells are analyzed 72 h after the second transfection. Figure 20.2B shows the specifically reduced expression levels of the targeted NDPK isoform. Depletion of NDPK B, but not of NDPK A, results in significantly diminished cAMP levels which correlate to the reduction in NDPK B expression levels.

3.1.5. Determination of basal cAMP synthesis in NRCM

To quantify $G\alpha_s$ activation by the NDPK B/$G\beta\gamma$ phosphorelay cAMP accumulation in intact NRCM is determined a competitive enzyme immunoassay. Myocytes are cultured in 12 well-tissue culture dishes containing complete culture medium as described above. Twelve hours before starting the cAMP accumulation assay (i.e., 36 h after adenoviral infection or 72 h after the second siRNA transfection) the medium is switched to serum free medium. The accumulation of cAMP is determined in 1 ml per well of an assay buffer containing serum-free DMEM, 20 m*M* HEPES, pH 7.4, 1 m*M* isobutylmethylxanthine (IBMX) and 100 μM propranolol, an inverse agonist of β-adrenoceptors, which inhibits constitutive receptor activity and blocks the binding of eventually remaining catecholamines. Myocytes are allowed to equilibrate in the assay buffer for 30 min at 37 °C. Thereafter, the assay medium is removed, cells are lyzed by addition of 400 μl 0.1 *M* HCl per well, scraped off and transferred into an Eppendorf tube. After centrifugation at 20,000×*g* for 15 min at 4 °C, 100 μl of the supernatant is used for the competitive enzyme immunoassay for cAMP (R&D System, Wiesbaden, Germany) precisely following to the manufacturer's instructions. The acid-insoluble cell pellets are neutralized with 0.1 *M* NaOH and the protein content is determined according to Bradford (1976). Production of cAMP is normalized to the protein content.

3.2. Regulation of ACL activity by PHPT-1

It has been hypothesized that an increased activity of PHPT-1 accelerates the dephosphorylation of ACL at His-760 in the active site and, subsequently, reduces its cellular activity. As ACL is crucial for fat and cellular energy metabolism, a loss of ACL activity, for example, by siRNA-mediated knockdown is associated with reduced viability of, for example, neuronal cells due to enhanced apoptosis (Krieglstein *et al.*, 2008). Similarly, the reduction of ACL activity by PHPT-1 overexpression in primary cultures of hippocampal or cortical neurons from embryonic rats, in the human neuroblastoma cell line (SH-SY5Y) as well as in the cholinergic murine neuroblastoma cell line SN56.B5 (SN56) causes a reduction of ACL activity and impairs cell viability (Klumpp *et al.*, 2009; Krieglstein *et al.*, 2008). We therefore describe herein the culture of SN56 cells, the adenoviral overexpression of PHPT-1 and assays to detect the phosphorylation status of ACL (see above) as well as ACL activity in SN56 cell extracts.

3.2.1. Culture of SN56 neuroblastoma cells and adenoviral infection

The SN56 cells are cultured in Dulbecco's modified Eagle Medium (DMEM) supplemented with 10% (v/v) FCS and 50 μg/ml gentamycin in a humidified atmosphere containing 5% CO_2 at 37 °C. The medium is changed every 2–3 days and the cells are splitted when confluent. Therefore, the medium is removed and the cells are washed once with prewarmed sterile PBS before 2 ml of trypsin–EDTA (0.05% trypsin, 0.02% EDTA) is added for 5 min at 37 °C. To stop trypsinization, 4 ml of culture medium is added to the flask. Cells are collected and centrifuged (130×*g*, 7 min, room temperature). The pellet is resuspended in 5 ml culture medium and 1 ml is transferred to a new culture flask with 10 ml culture medium. Passages 25–36 are used for experiments. For adenoviral infection, 600,000 SN56 cells were plated on culture dishes (ø 60 mm). After 1 day in culture the medium is replaced by Opti-MEM containing active viral particles (AdGFP, AdPHPT-1) at an MOI of 1.

3.2.2. Preparation of cell extracts and analysis of PHPT-1 expression and ACL activity

Cell extracts are prepared after washing cells with ice-cold PBS first. Thereafter, the cells are resuspended in homogenization buffer containing, 130 m*M* Tris–HCl, pH 6.8, 10% glycerol 1 m*M* PMSF, 7 μg/ml trypsin inhibitor (Sigma), 1 μM calpain inhibitor (Sigma), and are briefly sonificated. The homogenate is separated by centrifugation (13,000×*g*, 10 min, 4 °C), and the supernatant containing soluble ACL as well as PHPT-1 is used for further analysis. For immunoblot detection of PHPT-1 expression, 20–40 μg of extract protein are separated by a SDS–PAGE (15%, w/v, arylamide) and later transferred electrophoretically to a nitrocellulose membrane. A purified polyclonal antibody (Klumpp *et al.*, 2002) is used at 1:400 dilution in 10 m*M* Tris–HCl, pH 7.6, 150 m*M* NaCl, 0.1% Tween, 0.1% bovine serum albumin overnight at 4 °C. The detection of the primary antibody by a secondary horseradish–peroxidase-conjugated antirabbit antibody and the use of an enhanced chemiluminescence reagent can be recommended. The phosphorylation of ACL (15 μg extract protein/10 μl assay volume) is detected as described above (*Phosphorylation and Dephosphorylation of ATP-citrate lyase*).

ACL activity is monitored using a method introduced by Linn and Srere (1979) in which a malate dehydrogenase (EC 1.1.1.37) coupled reaction is used to assay ACL activity via the oxidation of NADH and the thereby caused decrease in absorbance at 340 nm. The reaction mixture A containing 100 m*M* Tris–HCl, pH 7.4, 1.5 m*M* ATP, and 5–15 μl cell extract is incubated for 10 min at 30 °C in an Eppendorf tube. Reaction mixture B containing 16 m*M* potassium citrate, 1 m*M* dithiothreitol, 10 m*M* $MgCl_2$, 2.5 U malate dehydrogenase, 0.2 m*M* coenzyme A, and 0.14 m*M* NADH is warmed in parallel also for 10 min at 30 °C. Mixture B is given into a quartz cuvette, H_2O and the preincubated mixture A is added to give final volume of 150 μl and mixed. The absorbance at 340 nm is read every 10 s in a suitable spectrophotometer for a period of 10 min. The decrease in absorbance is plotted against the incubation time. The slope in the linear range of the resulting curve is directly correlated to ACL activity.

As shown in Fig. 20.3, the adenoviral overexpression of PHPT-1 in SN56 cells largely reduces the phosphorylation level of ACL which correlates to a loss of about 40% of ACL activity in the cell extract.

3.3. Regulation of $K_{Ca}3.1$ activity by NDPK B and PHPT-1

$K_{Ca}3.1$ channels belong to a subfamily of Ca^{2+}-activated K^+ channels with an intermediate conductance (syn. SK4, $K_{Ca}4$, IK_1, IK_{Ca} KCNN4, "Gardos channels"). They are expressed predominantly in peripheral cells, including those of the hematopoietic system, colon, lung, placenta, pancreas, salivary glands, and proliferating vascular smooth muscle (Chou *et al.*, 2008; Wulff *et al.*, 2007). Activation of the channels mediates K^+ efflux setting up a negative membrane potential, which is required to establish a favorable electrochemical gradient for Ca^{2+} influx. $K_{Ca}3.1$ channels are important for diverse physiological responses in a variety of cell types, including osmotic and volume regulation in red blood cells, mitogen-dependent

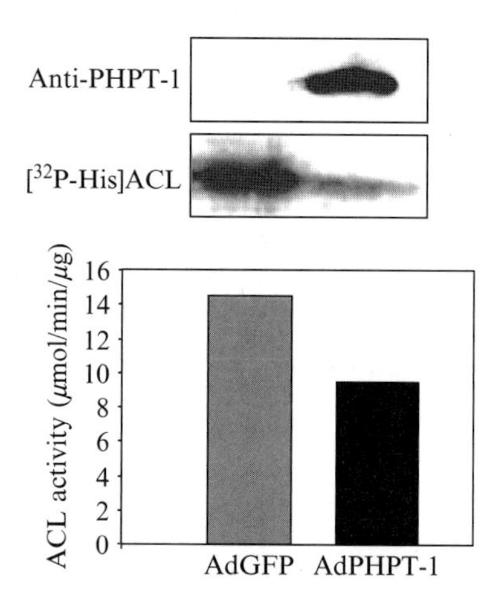

Figure 20.3 Overexpression of PHPT-1 in SN56 cells reduced ACL phosphorylation and activity. SN56 cells were infected with AdGFP or AdPHPT-1 and cell extracts were prepared 72 h later. PHPT-1 was visualized by immunoblot analysis of 30 μg of cell extracts (upper panel). Phosphorylation of ACL was visualized by incubation of 15 μg cell extracts with [γ-^{32}P]ATP, 10% SDS–PAGE and autoradiography (middle panel). ACL activity was quantified by the malate dehydrogenase-dependent NADH oxidation (lower panel). The overexpression of PHPT-1 resulted in decreased phosphorylation levels of ACL which correlated with its reduced activity.

activation of T-lymphocytes, Cl^- secretion of exocrine epithelial cells, and control of proliferation of T- and B-lymphocytes, vascular smooth muscle cells, and some cancer cell lines (Chou *et al.*, 2008; Wulff *et al.*, 2007). Intermediate conductance $K_{Ca}3.1$ channels can be distinguished pharmacologically from calcium-activated K^+ channels with small (SK channels) or big (BK channels) conductance by their sensitivity to blockade by the clotrimazole-derivative 1-[(2-chlorophenyl) diphenylmethyl]-1*H*-pyrazole (TRAM-34) and by their insensitivity to apamin and iberiotoxin. $K_{Ca}3.1$ channels associate with calmodulin and the channels open only after Ca^{2+} has bound to this regulatory protein. The intracellular Ca^{2+} concentration is the most evident regulator of $K_{Ca}3.1$ channels, but other intracellular factors like protein kinases can also regulate the activity of $K_{Ca}3.1$ channels. Protein kinase A (Pellegrino and Pellegrini, 1998) and protein kinase C (Wulf and Schwab, 2002) have been shown to activate $K_{Ca}3.1$ channels. As mentioned in Section 1, it was recently reported that specifically NDPK B is complexed with the channel and is able to activate the channel by phosphorylation of His-358 located in the C-terminal tail (Srivastava *et al.*, 2006). Moreover, PHPT-1 is apparently also part of the complex and can reverse the channel activation by P-His-dephosphorylation (Srivastava *et al.*, 2008).

To demonstrate this regulation of $K_{Ca}3.1$ channels by NDPK B and PHPT-1, we used the heterologous expression system of HEK293 cells, which are devoid of endogenous $K_{Ca}3.1$, and therefore do not display a TRAM-34-sensitive K^+ current before transfection with an eukaryotic expression vector in which the cDNA encoding $K_{Ca}3.1$ (pCMV-Sport/$K_{Ca}3.1$) was subcloned. Electrophysiological measurements are carried out in $K_{Ca}3.1$ expressing HEK293 cells either in the whole-cell configuration or in inside-out patches (Figs. 20.4 and 20.5). In the whole-cell configuration the regulation of the KCa3.1 by NDPK B and PHPT-1 is studied after cotransfection with the eukaryotic expression vectors pAdTrack/NDPK B and pAdTrack/PHPT-1 (Fig. 20.4). In the cell-free inside-out patches the purified recombinant proteins are applied directly to the intracellular side of the cell membrane (Fig. 20.5).

3.3.1. Culture and transfection of HEK293 cells

HEK293 cells are cultured in MEM-EARLE medium (Biochrom, Berlin, Germany) containing 10% fetal calf serum, 2 m*M* L-glutamine, 100 U/ml penicillin and 100 μg/ml streptomycin at 37 °C, and 6% CO_2. The cells are passaged every 3–4 days, up to the 50th passage. Briefly, when the cells reach 70–80% confluency, the medium is removed and the cells are washed once with 10 ml (for T75 flask) prewarmed sterile PBS. Thereafter, 2 ml of a trypsin–EDTA (0.05% trypsin with 0.02% EDTA) solution is added for a few minutes until cells detach. To stop trypsinization, 10 ml of culture medium is added to the flask. The cells are gently but thoroughly resuspended by pipetting up and down. Two milliliters of the cell suspension is transferred to a new culture flask with 20 ml culture medium, and kept in a humidified incubator at 37 °C and 6% CO_2. One day prior to transfection, 10^5 cells are plated in a 35-mm tissue culture dish. The cells are transiently transfected with a mixture of DNA plasmids by calcium phosphate precipitation. One microgram of each plasmid cDNA is diluted in water in a total volume of 90 μl. Ten microliters of 2.5 *M* $CaCl_2$ solution is added to the DNA and mixed by pipetting up and down twice. Then 100 μl 2× BES-buffered solution (BBS: 50 m*M* *N,N*-bis[2-hydroxyethyl]-2-aminoethanesulfonic acid, 1.5 m*M* Na_2HPO_4, 280 m*M* NaCl) is added and vortexed for 10 s. The sample is then incubated for 17 min at room temperature to allow for transfection complex formation. Thereafter, the sample (200 μl) is transferred to the culture dish (2 ml medium) in which the cells are growing at 40–60% confluence. Immediately before, the cells are washed once with culture medium. The cells are then cultured at 35 °C and 3% CO_2. For transfection of the cells, the following four combinations of DNA plasmids are used: (1) pCMV-Sport/$K_{Ca}3.1$ + pcDNA3/EGFP, (2) pCMV-Sport/$K_{Ca}3.1$ + pcDNA3/EGFP + pAdTrack/NDPK B, (3) pCMV-Sport/$K_{Ca}3.1$ + pcDNA3/EGFP + pAdTrack/NDPK B(H118N), and (4), pCMV-Sport/$K_{Ca}3.1$ + pcDNA3/EGFP + pAdTrack/NDPK B + pAdTrack/PHPT-1. 18 h after the transfection, the cells are washed twice with PBS, then 2 ml culture medium is added and the cells are cultivated for another 48–72 h

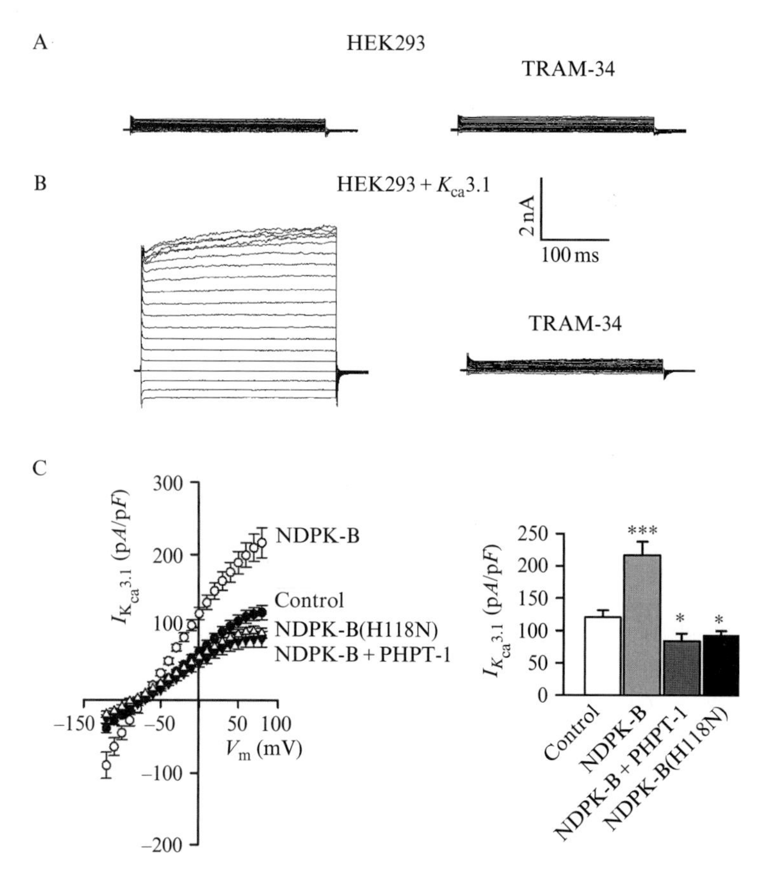

Figure 20.4 NDPK B stimulates and PHPT-1 inhibits $K_{Ca}3.1$ channel currents in HEK293 cells. Whole-cell currents were elicited from a holding potential of −70 mV by depolarizing the cells every 5 s for 300 ms from −120 to +80 mV in 10 mV increments. (A) Original current traces before and after application of 100 n*M* TRAM-34, recorded in nontransfected HEK293 cells. Note, that HEK293 cells do not exhibit endogenous TRAM-34-sensitive currents. (B) Original current traces recorded in HEK293 cells transiently transfected with $K_{Ca}3.1$ channels and inhibition of the currents by 100 n*M* TRAM-34. (C) HEK293 cells were transiently transfected with $K_{Ca}3.1$ channels alone (control) or cotransfected with NDPK B, NDPK B (H118N), or NDPK B + PHPT-1 encoding plasmids. The *I*/*V* curves (left) were obtained by plotting current densities (*pA*/*pF*) against the corresponding voltages (from −120 to +80 mV). Means ± S.E.M. of 39 cells (control), 31 cells (NDPK B), 42 cells (PHPT-1), and 45 cells (NDPK B (H118N)) are shown. Note that all currents represent TRAM-34-sensitive currents ($IK_{Ca}3.1$). Bars (right) represent mean current densities ± S.E.M. at a membrane potential of +80 mV (data from the *I*/*V* curve). The pipette solution contained 1 μM free Ca^{2+}. *$P < 0.05$, ***$P < 0.001$ versus control.

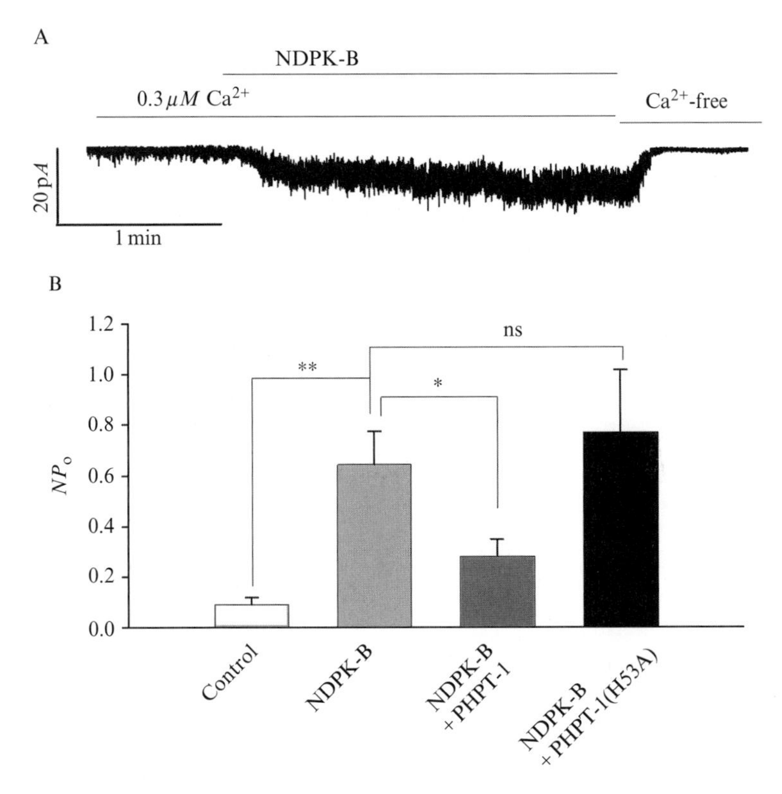

Figure 20.5 NDPK B enhances and PHPT-1 reverses the $K_{Ca}3.1$ channel activity in cell-free membrane patches. Channel open probabilities (NP_o) were recorded in inside-out patches excised from HEK293 cells transiently transfected with $K_{Ca}3.1$ channels in symmetrical K^+ (145 m*M*) solution. The membrane potential is constantly kept at −40 mV. $[Ca^{2+}]_i > 0.3$ μ*M*. Recombinant proteins of NDPK B, PHPT-1, or PHPT-1 (H53A) were applied directly to the cytosolic surface of a single patch. (A) Original single-channel recording showing the activation of $K_{Ca}3.1$ channel activity by 10 μg ml^{-1} NDPK B and the calcium sensitivity of the channels in an inside-out patch. (B) the patches were superfused with 10 μg/ml NDPK B followed by 10 μg/ml either active PHPT-1 or the inactive PHPT-1(H53A) mutant. Bars represent mean NP_o values ± S.E.M. of 10 patches (control and NDPK B), and of five patches treated with NDPK B + PHPT-1 and NDPK B + PHPT-1(H53A), respectively. $^*P < 0.05$, $^{**}P < 0.01$ versus control; ns, not significant.

at 37 °C and 6% CO_2. On the day of the experiment, the cells are first washed several times with physiological saline solution (PSS; 10 m*M* HEPES, 127 m*M* NaCl, 5.9 m*M* KCl, 2.4 m*M* $CaCl_2$, 1.2 m*M* $MgCl_2$, 11 m*M* glucose, pH 7.4), and then transferred in the culture dish to the stage of an inverted microscope (Zeiss Axiovert, 200) for electrophysiological measurements. The transfection efficiency varies between 40% and 70% as judged by the expression of EGFP in transfected cells.

3.3.2. Recording techniques

Standard patch-clamp recording techniques are used to measure currents in the inside-out, or the whole-cell patch-clamp configuration (Hamill *et al.*, 1981). Patch electrodes are pulled from borosilicate glass capillaries (MTW 150F; world Precision Instruments, Inc., Sarasota, FL) using a DMZ-Universal Puller (Zeitz-Instrumente Vertriebs GmbH, Martinsried, Germany) and filled with prefiltered solutions of different composition (see below). Currents are recorded at room temperature with an EPC-7 amplifier (HEKA Elektronik, Lambrecht, Germany) connected via a 16-bit A/D interface to a pentium IBM clone computer. The signals are low-pass filtered (1 kHz) before 5 kHz digitization. Data acquisition and analysis are performed with an ISO-3 multitasking patch-clamp program (MFK M. Friedrich, Niedernhausen, Germany). Pipette resistance range from 2 to 3 MΩ in whole-cell, and 8–9 MΩ in the excised-patch experiments. Electrode offset potentials are always zero-adjusted before a Giga-seals is formed. To avoid dirt on the pipette tip, a slight positive pressure is applied to the pipette interior before the pipette is moved into the bath solution, but pressure is released shortly before the pipette reaches the cell surface. After the pipette tip is gently pressed against the cell membrane, Giga-seals are obtained within several seconds when a negative pressure is applied to the pipette interior. Sometimes Giga-seals develop spontaneously without suction. To reduce the difficulty of obtaining high resistance Giga-seals, only freshly pulled and clean pipettes are used. To reduce electrical noise to a minimum, the experiment is carried out only when the seal resistance is higher than 10 GΩ.

3.3.3. Recording of whole-cell currents

For recording whole-cell currents, the membrane under the pipette tip is disrupted by negative pressure and as a consequence, transient capacitance currents occur at the beginning, and at the end of small test-pulses. After establishment of the whole-cell configuration, the membrane capacitance and series resistance are compensated (60–80%). Whole-cell currents are elicited by applying 300-ms step depolarizations to potentials ranging from −120 to +80 mV in 10-mV increments from a holding potential of −70 mV. Currents are sampled by ISO-3 and saved in the computer for later data analysis. Current–voltage (*I*/*V*) relationships are obtained by plotting current density (pA/pF) measured at the end of a depolarizing pulse against the corresponding voltage. For whole-cell recordings, the bath is superfused with PSS. The pipette solution contains 10 m*M* HEPES, 100 m*M* K^+-gluconate, 24 m*M* KCl, 1.2 m*M* $MgCl_2$, 5 m*M* EGTA, 11 m*M* glucose, and 3 m*M* ATP, pH 7.4. The free Ca^{2+} concentration is adjusted to 1 μM by adding 4.75 m*M* $CaCl_2$ (Fabiato, 1988; Mermi *et al.*, 1991).

To dissect K^+ currents conducted by $K_{Ca}3.1$ channels from total outward currents, transfected HEK293 are superfused with 100 n*M* TRAM-34, a specific $K_{Ca}3.1$ channel blocker. $K_{Ca}3.1$ channel currents are obtained by subtracting the residual currents after inhibition by TRAM-34 from whole-cell outward currents (Fig. 20.4B and *I*/*V* curves in C). Overexpression of NDPK B enhances $K_{Ca}3.1$ channel activity whereas expression of the inactive form (NDPK B(H118N)) or of PHPT-1 leads to a significant suppression of the current (Fig. 20.4C). In accordance with previously published data in CHO and human CD4 T-lymphocytes (Srivastava *et al.*, 2006, 2008), these data demonstrate that $K_{Ca}3.1$ channel activity is controlled by NDPK B and PHPT-1 at endogenous expression levels of these proteins.

3.3.4. Single-channel recording

To obtain single-channel recordings from inside-out patches of HEK293 cells expressing $K_{Ca}3.1$ channels, a Giga-seal is formed first. Then the pipette tip is slowly withdrawn from the cell surface until the bridge between pipette tip and cell surface ruptures, leaving a small vesicle protruding from the pipette tip. The pipette tip is then passed briefly (1–2 s) through the air–water interface, which disrupts the protruding membrane of the vesicle and forms spontaneously an inside-out membrane patch. After establishment of the inside-out configuration, the membrane potential is constantly clamped at −40 mV and single-channel currents generated by spontaneous openings of $K_{Ca}3.1$ channels are recorded. To verify that single-channel currents are due to the opening of $K_{Ca}3.1$ channels, the bath solution (see below) is changed to a Ca^{2+}-free solution which immediately stops channel activity (Fig. 20.5A). Ca^{2+}-activated K^+ channels other than $K_{Ca}3.1$ are apparently functionally not expressed in HEK293 cell plasma membranes. The average channel open probability (NP_o) in inside-out patches is determined by the following equation:

$$NP_o = \frac{(\sum_{j=1}^{N} t_j j)}{T}$$

where P_o is the open probability, T is the duration of the measurement, t_j is the time spent with $j > 1, 2, \ldots, N$ channels open, and N is the maximum number of channels seen. To obtain mean NP_o values, membrane patches are equilibrated for at least 3–5 min before the recombinant proteins are applied for another 3–5 min. A multibarreled perfusion pipette placed by a hydraulic micromanipulator 200 μm away from the patch is used to switch between superfusion solutions. The flow rate of 3 μl/min generated by a syringe pump (TSE 200, Bad Homburg, Germany) resulted in a flow velocity of 1 cm/s through an 80-μm orifice. For inside-out recordings, the extracellular (pipette) solution contains 5 m*M* HEPES, 145 m*M*

KCl, 1.2 m*M* $MgCl_2$, 1 m*M* $CaCl_2$, pH 7.4, and the bath solution (cytosolic surface of the patch) contains 5 m*M* HEPES, 139 m*M* KCl, 1.2 m*M* $MgCl_2$, 5 m*M* EGTA, 11 m*M* glucose, 3 m*M* ATP, pH 7.3. The free Ca^{2+} concentration is adjusted to 0.3 μM by adding 3.8 m*M* $CaCl_2$ (Fabiato, 1988; Mermi *et al.*, 1991). Superfusion of the cytosolic side of the patch with recombinant NDPK B (10 μg/ml) enhances NP_o several-fold, whereas the subsequent application of NDPK B together with PHPT-1 (10 μg/ml) strongly attenuates the response. Application of NDPK B plus the inactive mutant phosphatase PHPT-1(H53A) (10 μg/ml) enhances NP_o comparable to NDPK B alone (Fig. 20.5B).

REFERENCES

Bäumer, N., Mäurer, A., Krieglstein, J., and Klumpp, S. (2007). Expression of protein histidine phosphatase in *Escherichia coli*, purification, and determination of enzyme activity. *Methods Mol. Biol.* **365,** 247–260.

Boissan, M., Dabernat, S., Peuchant, E., Schlattner, U., Lascu, I., and Lacombe, M. L. (2009). The mammalian Nm23/NDPK family: From metastasis control to cilia movement. *Mol. Cell. Biochem.* **329,** 51–62.

Boyer, P. D., Deluca, M., Ebner, K. E., Hultquist, D. E., and Peter, J. B. (1962). Identification of phospohistidine in digests from a probable intermediate of oxidative phosphorylation. *J. Biol. Chem.* **237,** PC3306–PC3308.

Bradford, M. M. (1976). A rapid and sensitive method for the quantitation of microgram quantities of protein utilizing the principle of protein-dye binding. *Anal. Biochem.* **72,** 248–254.

Chou, C. C., Lunn, C. A., and Murgolo, N. J. (2008). K_{Ca}3.1: Target and marker for cancer, autoimmune disorder and vascular inflammation? *Expert Rev. Mol. Diagn.* **8,** 179–187.

Cottam, G. L., and Srere, P. A. (1969). Nature of the phosphorylated residue in citrate cleavage enzyme. *Biochem. Biophys. Res. Commun.* **35,** 895–900.

Cuello, F., Schulze, R. A., Heemeyer, F., Meyer, H. E., Lutz, S., Jakobs, K. H., Niroomand, F., and Wieland, T. (2003). Activation of heterotrimeric G proteins by a high energy phosphate transfer via nucleoside diphosphate kinase (NDPK) B and Gβ subunits. Complex formation of NDPK B with G$\beta\gamma$ dimers and phosphorylation of His-266 in Gβ. *J. Biol. Chem.* **278,** 7220–7226.

Ek, P., Pettersson, G., Ek, B., Gong, F., Li, J. P., and Zetterqvist, O. (2002). Identification and characterization of a mammalian 14-kDa phosphohistidine phosphatase. *Eur. J. Biochem.* **269,** 5016–5023.

Elshourbagy, N. A., Near, J. C., Kmetz, P. J., Sathe, G. M., Southan, C., Strickler, J. E., Gross, M., Young, J. F., Wells, T. N., and Groot, P. H. (1990). Rat ATP citrate-lyase. Molecular cloning and sequence analysis of a full-length cDNA and mRNA abundance as a function of diet, organ, and age. *J. Biol. Chem.* **265,** 1430–1435.

Fabiato, A. (1988). Computer programs for calculating total from specified free or free from specified total ionic concentrations in aqueous solutions containing multiple metals and ligands. *Methods Enzymol.* **157,** 378–417.

Fan, Z., Beresford, P. J., Oh, D. Y., Zhang, D., and Lieberman, J. (2003). Tumor suppressor NM23-H1 is a granzyme A-activated DNase during CTL-mediated apoptosis, and the nucleosome assembly protein SET is its inhibitor. *Cell.* **112,** 659–672.

Hamill, O. P., Marty, A., Neher, E., Sakmann, B., and Sigworth, F. J. (1981). Improved patch-clamp techniques for high-resolution current recording from cells and cell-free membrane patches. *Pflugers Arch.* **391,** 85–100.

Hess, J. F., Bourret, R. B., and Simon, M. I. (1988). Histidine phosphorylation and phosphoryl group transfer in bacterial chemotaxis. *Nature* **336,** 139–143.

Hippe, H. J., Lutz, S., Knorr, K., Vogt, A., Jakobs, K. H., Wieland, T., and Niroomand, F. (2003). Activation of heterotrimeric G proteins by a high energy phosphate transfer via nucleoside diphosphate kinase (NDPK) B and Gβ subunits. Specific activation of $G_s\alpha$ by a NDPK B-G$\beta\gamma$ complex in H10 cells. *J. Biol. Chem.* **278,** 7227–7233.

Hippe, H. J., Luedde, M., Lutz, S., Köhler, H., Eschenhagen, T., Frey, N., Katus, H. A., Wieland, T., and Niroomand, F. (2007). Regulation of cardiac cAMP synthesis and contractility by nucleoside diphosphate kinase B/G protein $\beta\gamma$ dimer complexes. *Circ. Res.* **100,** 1191–1199.

Kapetanovich, L., Baughman, C., and Lee, T. H. (2005). Nm23H2 facilitates coat protein complex II assembly and endoplasmic reticulum export in mammalian cells. *Mol. Biol. Cell* **16,** 835–848.

Kennelly, P. J., and Potts, M. (1996). Fancy meeting you here! A fresh look at "prokaryotic" protein phosphorylation. *J. Bacteriol.* **178,** 4759–4764.

Klumpp, S., and Krieglstein, J. (2009). Reversible phosphorylation of histidine residues in proteins from vertebrates. *Sci. Signal.* **2,** pe13.

Klumpp, S., Hermesmeier, J., Selke, D., Baumeister, R., Kellner, R., and Krieglstein, J. (2002). Protein histidine phosphatase: A novel enzyme with potency for neuronal signaling. *J. Cereb. Blood Flow Metab.* **22,** 1420–1424.

Klumpp, S., Bechmann, G., Mäurer, A., Selke, D., and Krieglstein, J. (2003). ATP-citrate lyase as a substrate of protein histidine phosphatase in vertebrates. *Biochem. Biophys. Res. Commun.* **306,** 110–115.

Klumpp, S., Faber, D., Fischer, D., Litterscheid, S., and Krieglstein, J. (2009). Role of protein histidine phosphatase for viability of neuronal cells. *Brain Res.* **1264,** 7–12.

Krieglstein, J., Lehmann, M., Mäurer, A., Gudermann, T., Pinkenburg, O., Wieland, T., Litterscheid, S., and Klumpp, S. (2008). Reduced viability of neuronal cells after overexpression of protein histidine phosphatase. *Neurochem. Int.* **53,** 132–136.

Laemmli, U. K. (1970). Cleavage of structural proteins during the assembly of the head of bacteriophage T4. *Nature* **227,** 680–685.

Linn, T. C., and Srere, P. A. (1979). Identification of ATP citrate lyase as a phosphoprotein. *J. Biol. Chem.* **254,** 1691–1698.

Luo, J., Deng, Z. L., Luo, X., Tang, N., Song, W. X., Chen, J., Sharff, K. A., Luu, H. H., Haydon, R. C., Kinzler, K. W., Vogelstein, B., and He, T. C. (2007). A protocol for rapid generation of recombinant adenoviruses using the AdEasy system. *Nat. Protoc.* **2,** 1236–1247.

Lutz, S., Hippe, H. J., Niroomand, F., and Wieland, T. (2004). Nucleoside diphosphate kinase-mediated activation of heterotrimeric G proteins. *Methods Enzymol.* **390,** 403–418.

Mäurer, A., Wieland, T., Meissl, F., Niroomand, F., Mehringer, R., Krieglstein, J., and Klumpp, S. (2005). The β-subunit of G proteins is a substrate of protein histidine phosphatase. *Biochem. Biophys. Res. Commun.* **334,** 1115–1120.

Mermi, J., Yajima, M., and Ebner, F. (1991). The control of the contraction of myocytes from guinea-pig heart by the resting membrane potential. *Br. J. Pharmacol.* **104,** 705–713.

Noiman, S., and Shaul, Y. (1995). Detection of histidine-phospho-proteins in animal tissues. *FEBS Lett.* **364,** 63–66.

Pellegrino, M., and Pellegrini, M. (1998). Modulation of Ca^{2+}-activated K^+ channels of human erythrocytes by endogenous cAMP-dependent protein kinase. *Pflugers Arch.* **436,** 749–756.

Srivastava, S., Li, Z., Ko, K., Choudhury, P., Albaqumi, M., Johnson, A. K., Yan, Y., Backer, J. M., Unutmaz, D., Coetzee, W. A., and Skolnik, E. Y. (2006). Histidine phosphorylation of the potassium channel $K_{Ca}3.1$ by nucleoside diphosphate kinase B is required for activation of $K_{Ca}3.1$ and CD4 T cells. *Mol. Cell* **24,** 665–675.

Srivastava, S., Zhdanova, O., Di, L., Li, Z., Albaqumi, M., Wulff, H., and Skolnik, E. Y. (2008). Protein histidine phosphatase 1 negatively regulates CD4 T cells by inhibiting the K^+ channel $K_{Ca}3.1$. *Proc. Natl. Acad. Sci. USA* **105,** 14442–14446.

Steeg, P. S., Horak, C. E., and Miller, K. D. (2008). Clinical-translational approaches to the Nm23-H1 metastasis suppressor. *Clin. Cancer Res.* **14,** 5006–5012.

Swanson, R. V., Schuster, S. C., and Simon, M. I. (1993). Expression of CheA fragments which define domains encoding kinase, phosphotransfer, and CheY binding activities. *Biochemistry* **32,** 7623–7629.

Swanson, R. V., Alex, L. A., and Simon, M. I. (1994). Histidine and aspartate phosphorylation: Two-component systems and the limits of homology. *Trends Biochem. Sci.* **19,** 485–490.

Wagner, P. D., and Vu, N. D. (1995). Phosphorylation of ATPcitrate lyase by nucleoside diphosphate kinase. *J. Biol. Chem.* **270,** 21758–21764.

Wagner, P. D., Steeg, P. S., and Vu, N. D. (1997). Two-component kinase-like activity of nm23 correlates with its motility-suppressing activity. *Proc. Natl. Acad. Sci. USA* **94,** 9000–9005.

Wulf, A., and Schwab, A. (2002). Regulation of a calcium-sensitive K^+ channel (cIK1) by protein kinase C. *J. Membr. Biol.* **187,** 71–79.

Wulff, H., Kolski-Andreaco, A., Sankaranarayanan, A., Sabatier, J. M., and Shakkottai, V. (2007). Modulators of small- and intermediate-conductance calcium-activated potassium channels and their therapeutic indications. *Curr. Med. Chem.* **14,** 1437–1457.

CHAPTER TWENTY-ONE

Histidine Phosphorylation in Histones and in Other Mammalian Proteins

Paul G. Besant *and* Paul V. Attwood

Contents

Abstract

The investigation of protein histidine phosphorylation has required the development of a number of methods that differ from traditional methods of phosphoprotein analysis that were developed to study phosphorylation of serine, threonine, and tyrosine, which are, unlike phosphohistidine, acid-stable. The investigation of histidine phosphorylation is further complicated by the fact that in mammalian proteins, phosphorylation appears to occur at either 1-N or 3-N positions of the imidazole ring, depending on the source of the kinase. In this review, we describe methods developed for phosphoamino acid analysis to detect phosphohistidine, including the determination of the isoform present, using chromatographic and mass spectrometric analysis of phosphoprotein hydrolysates and ^{1}H- and ^{31}P NMR analysis of intact phosphoproteins and phosphopeptides. We also describe methods for the assay of protein histidine kinase activity, including a quantitative assay of alkali-stable, acid-labile protein phosphorylation, and an in-gel kinase assay applied to histidine kinases.

School of Biomedical, Biomolecular and Chemical Sciences (M310), The University of Western Australia, Crawley, Western Australia, Australia

Methods in Enzymology, Volume 471
ISSN 0076-6879, DOI: 10.1016/S0076-6879(10)71021-1

Most of the detailed descriptions of methods are as they are applied in our laboratory to the investigation of histone H4 phosphorylation and histone H4 histidine kinases, but which can be applied to the phosphorylation of any proteins and to any such histidine kinases.

1. Introduction

While protein phosphorylation is recognized as key process in the regulation of mammalian cellular function, for the most part, attention has been focused on serine, threonine, and tyrosine phosphorylation. However, histidine phosphorylation is a well-known cellular control phenomenon in prokaryotes and in some eukaryotes (plants and fungi), being part of the "two component" histidine kinase systems (for a review see Stock *et al.*, 2000). There is now increasing recognition of the roles played by histidine phosphorylation in mammalian cells (for reviews see Besant and Attwood, 2005; Besant *et al.*, 2003).

Histidine phosphorylation in mammalian cells has been known to occur since the 1960s (Deluca *et al.*, 1963; Zetterqvist and Engström, 1967). In the 1970s there were a number of reports of histidine kinases in nuclei from regenerating liver and Walker-256 carcinosarcoma cells that phosphorylate histones, especially histone H4, where phosphorylation was reported to occur on both histidine residues (H18 and H75) (Chen *et al.*, 1974, 1977; Smith *et al.*, 1973, 1974). Later, Besant and Attwood (2000) reported the presence of histone H4 histidine kinase activity in porcine thymus nuclei, a relationship between enhanced activity of this kinase and regeneration following chemical and physical damage to liver and also enhanced activity in human hepatocellular carcinoma (Tan *et al.*, 2004). Thus, histone H4 phosphorylation is associated with enhanced cellular proliferation in liver cells and is present in proliferating thymus cells. Huang *et al.* (1991) purified a histone H4 histidine kinase from yeast, but it was not identified and fully characterized. No mammalian histone H4 histidine kinase has been characterized.

The other main focus of attention in the investigation of histidine kinases in mammals has been nucleoside diphosphate kinase (NDPK). This enzyme, whose primary cellular role is to maintain cellular nucleoside triphosphate (NTP) balance, catalyses phosphoryl transfer from NTP to NDP via the formation of an active site phosphohistidine. However, there is evidence that the enzyme can transfer the phosphoryl group from this phosphohistidine to histidines on other proteins, including G_β, leading to activation of G_S via transfer of the phosphoryl group from the phosphorylated H266 of G_β to GDP bound on G_α (Cuello *et al.*, 2003; Hippe *et al.*, 2003). NDPK has also been implicated in the histidine phosphorylation of annexin 1 in a [Cl^-]-regulated signaling system in airway epithelial cells (Muimo *et al.*, 1998, 2000; Treharne *et al.*, 2001).

In addition to the reports of histidine kinases in mammalian cells described above, histidine-phosphatases have also been found in rat liver and other tissues (Matthews and Mackintosh, 1995; Ohmori *et al.*, 1993; Wong *et al.*, 1993). Some of these phosphatases have been shown to be serine/threonine phosphatases or similar to this type of enzyme (Kim *et al.*, 1993; Matthews and Mackintosh, 1995). However, more recently a specific mammalian histidine-phosphatase has been discovered and characterized (Dombradi *et al.*, 2002; Ek *et al.*, 2002; Klumpp and Krieglstein, 2002; Klumpp *et al.*, 2002).

With the accumulating evidence of the existence and roles of histidine kinases and phosphatases, it is likely that protein histidine phosphorylation plays important roles in mammalian cell regulation. Owing to the labile nature of the P–N bond in phosphohistidine and the occurrence of both 1- and 3-isoforms of phosphohistidine, methods that have been developed to study protein phosphorylation involving the formation of phosphoserine, phosphothreonine, and phosphotyrosine have had to be modified or new methods developed. In this review, we shall describe and discuss the methods that have been developed or modified in the investigation of mammalian protein histidine phosphorylation.

2. Chemical Phosphorylation of Histone H4 Proteins and Peptides

To develop methods for the analysis of phosphorylated histone H4 or provide substrates for histidine-phosphatase studies, it has been convenient to chemically phosphorylate the protein or peptides using phosphoramidate. Potassium phosphoramidate is synthesized from phosphoryl chloride and ammonia as described by Wei and Matthews (1991). We have phosphorylated histone H4 by incubating 5 mg of the calf thymus histone with 100 mg potassium phosphoramidate in 1 ml H_2O, with the pH adjusted to 8.0 with KOH, for 3 h at room temperature (Zu *et al.*, 2007). Subsequent cleavage of the phosphorylated histone H4 with trypsin and analysis of the digest by electrospray ionization mass spectrometry showed both H18 and H75 to be phosphorylated (Zu *et al.*, 2007) (see below). Fujitaki and coworkers incubated histone H4 with phosphoramidate in a 1:1 (w/w) ratio overnight at room temperature for analysis by ^{31}P NMR (Fujitaki *et al.*, 1981). Recently, Attwood *et al.* (2010) attempted to phosphorylate the histidines in histone H4 and histidine-containing peptides based on its sequence, to study the kinetics of their phosphorylation and dephosphorylation by ^{1}H NMR. It was found that the histone H4 was very sensitive to salt concentration and precipitated quite rapidly, even with a 1:1 (w/w) ratio of histone to phosphoramidate, making the kinetic studies with histone H4 impossible. It was possible, however, to phosphorylate the histidines in the peptides and follow the kinetics of the phosphorylation reactions by ^{1}H NMR using 15:1 *M* ratios of phosphoramidate to peptide at 30 °C, pH 7.2, over a period 6–8 h (Attwood *et al.*, 2010). Formation of 1- and 3-phosphohistidine and 1,3-diphosphohistidine was observed to occur and the kinetics of their formation and hydrolysis were analyzed (Attwood *et al.*, 2010).

Phosphorylation of other histidine-containing peptides and proteins has also been performed using phosphoramidate, for example, the bacterial phosphocarrier protein, HPr (phosphoramidate:protein $>$ 10:1, w/w; incubation at 37 °C for about 5 h) (Gassner *et al.*, 1977); a peptide based on protein tyrosine phosphatase (phosphoramidate:protein $>$ 20:1, w/w; incubation at room temperature, overnight) (Medzihradszky *et al.*, 1997). In these examples, phosphorylation primarily resulted in the formation of 3-phosphohistidine. Using phosphoryl chloride, the histidine residue of a peptide based on the phosphorylation site of the bacterial histidine kinase, EnvZ, has been phosphorylated (Kleinnijenhuis *et al.*, 2007), as described by Wei and Matthews (1991).

3. Detection of Phosphohistidine—Phosphoamino Acid Analysis

Owing to the acid-labile nature of the P–N bond in phosphohistidine, the partial acid-hydrolysis of phosphoproteins that is used for the analysis of proteins phosphorylated on serine, threonine, or tyrosine is not suitable for phosphohistidine-containing proteins. There are thus two major methods of partial hydrolysis used for phosphohistidine detection, the original method involves partial hydrolysis of the phosphoprotein with strong alkali, for example, 3 *M* KOH at 105 °C for 3 h (Besant and Attwood, 2000). An advantage of this method is that any phosphoserine and phosphothreonine content of the digest is greatly depleted, as these phosphoamino acids are alkali-labile (Wei and Matthews, 1990). However, care must be taken to minimize evaporation, as this results in concentration of the KOH and subsequent hydrolysis of phosphohistidine, this can be prevented by performing the reaction under a layer of mineral oil (see Fig. 21.1) (Besant and Attwood, 1998).

We have used this method in the detection of phosphohistidine in histone H4 phosphorylated, using [γ-^{32}P]ATP, by histidine kinase from bovine thymus nuclei and yeast cells (Besant and Attwood, 2000). In these experiments, we separated the histone H4 from the other reaction

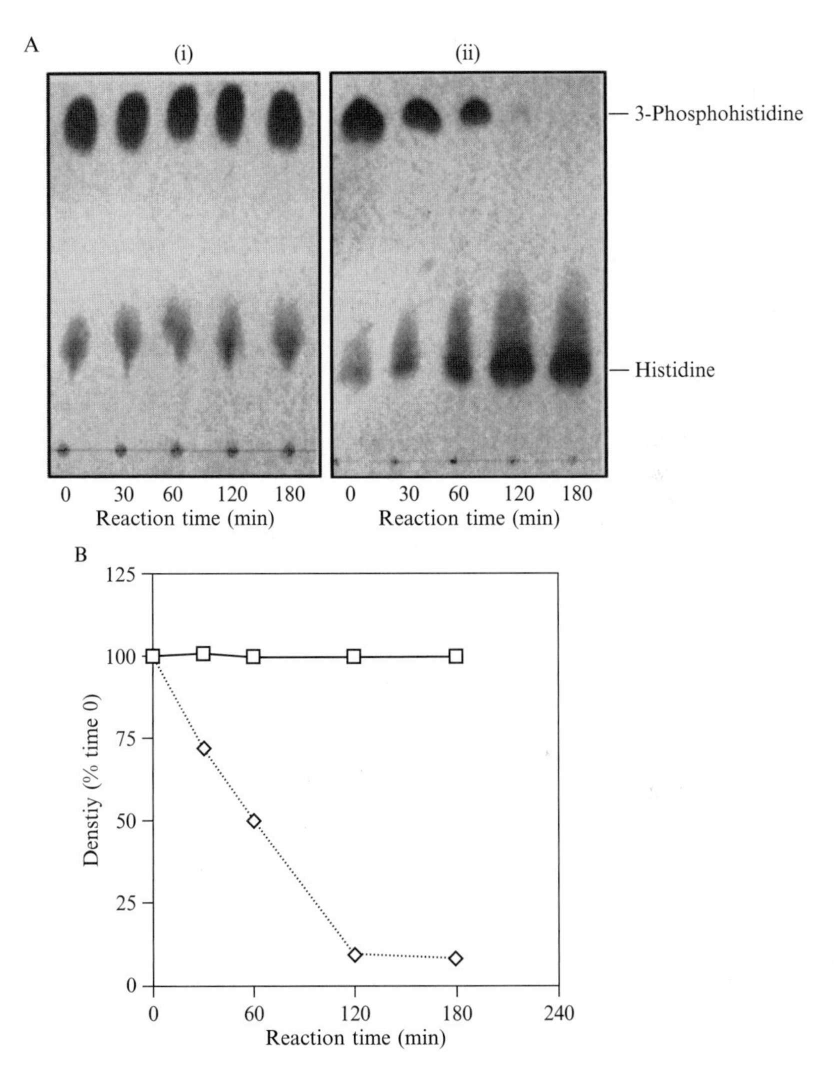

Figure 21.1 (A) TLE of 3-phosphohistidine, heated with 3 *M* KOH (with (i) and without (ii) an oil overlay) at 105 °C over a 3-h time period. (B) Density of the 3-phosphohistidine spots from (A) (i) and (ii) as a percentage of the density of the spots at time zero. □, With oil overlay; ◊, without oil overlay. Reproduced with permission from Elsevier (Besant and Attwood, 1998).

components by immunoprecipitation with a pan-histone monoclonal antibody and protein A Sepharose (Besant and Attwood, 2000). Following alkali-digestion of the immunoprecipitate with KOH, the potassium ions were precipitated from the digest by the addition of 5 μl of 100% (v/v) perchloric acid to 10 μl of the digest, followed by brief centrifugation. 1 μl of the supernatant was spotted onto a cellulose thin-layer electrophoresis (TLE) plate as quickly as possible to minimize acid-hydrolysis of the phosphohistidine. Negative control digests (minus histidine kinase), [γ-^{32}P]ATP and a phosphohistidine standard were also applied to the plate prior to electrophoresis as described by Besant and Attwood (1998) (1500 V, 150 mA, 100 W for 25 min) using 2% ammonium bicarbonate as electrophoresis buffer. Figure 21.2 shows both the phosphorimage (A) and the ninhydrin-stained image (B) of this plate (Besant and Attwood, 2000). As can be seen, there are a number of radioactive spots and streaks, but the most prominent radioactive spots that are not present in the [γ-^{32}P]ATP control correspond to the position of the 3-phosphohistidine standard.

The cellulose containing the radioactivity in these spots was scraped from the plate and placed in 200 μl microcentrifuge tubes into which had been packed a layer of glass wool. A hole was made in the bottom of the tubes with a 27-gauge needle and the tubes placed in larger 1.5 ml microcentrifuge tubes. The radiolabeled material was eluted from the cellulose by

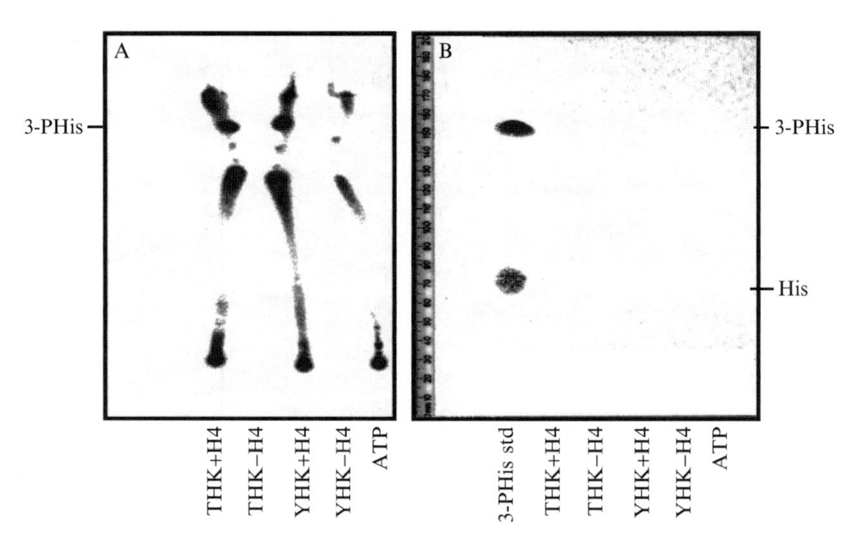

Figure 21.2 (A) Phosphorimage of thin-layer electrophoresis plate run with alkali-treated [γ-^{32}P]ATP (ATP) or alkali-hydrolyzed immunoprecipitates of histone H4 phosphorylated by either a yeast histidine kinase (YHK + H4) or putative porcine thymus histidine kinases (THK + H4). Controls (YHK − H4 and THK − H4) were performed in which histone H4 was omitted from the reaction mixtures. The spots identified as phosphohistidine have relative mobilities of 124 mm (origin to center of spot). (B) Ninhydrin-stained image of the thin-layer electrophoresis plate shown in (A), illustrating the relative mobility of a 3-phosphohistidine standard (3-PHis Std) to be 124 mm, a much lower mobility spot with a relative mobility of 45 mm corresponds to histidine (Perego *et al.*, 1994) impurity in the 3-phosphohistidine standard. Reproduced with permission from Elsevier (Besant and Attwood, 2000).

the addition of 4 × 100 μl of ethanol:ammonia:water (60:4:26), followed by brief centrifugation (Besant and Attwood, 2000). The eluates were rapidly transferred to glass tubes to minimize contact with plastic, from which contaminants leach. The 400 μl of eluates were then dried by vacuum centrifugation before being resuspended in 5 μl of 0.1% (v/v) ammonia solution. The solutions were then applied to another TLE plate, along with phosphoamino acid standards and electrophoresis performed as before (Besant and Attwood, 2000). Figure 21.3 shows the phosphorimage (A) and ninhydrin-stained image (B) of this plate (Besant and Attwood, 2000). The migration of the radiolabeled spots is clearly compatible with that of the phosphohistidine standard.

The other method of partial hydrolysis of phosphohistidine-containing proteins is by cleavage with the nonspecific protease, pronase E (*Streptomyces griseus* protease) (Tan *et al.*, 2003). In this example, histone H4 was phosphorylated by yeast histidine kinase with [γ-^{32}P]ATP and the radiolabeled phosphoprotein was then incubated overnight with 10 μg pronase E in 40 μl 20 m*M* ammonium bicarbonate at room temperature (Tan *et al.*, 2003). The digests were then dried by vacuum centrifugation and resuspended in 10 μl H_2O. 1 μl 3.6 *M* HCl was added to 5 μl of the digest and

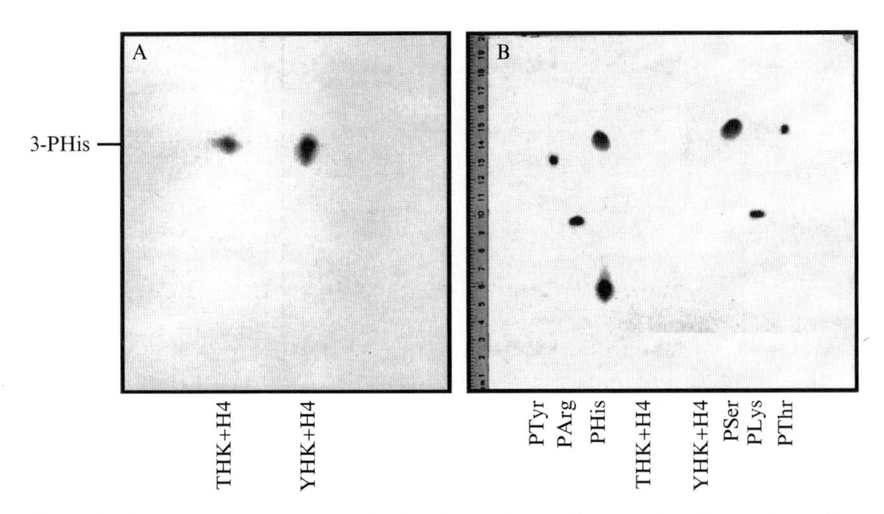

Figure 21.3 (A) Phosphorimage of a thin-layer electrophoresis plate illustrating radioactive phosphohistidine after elution from original thin-layer electrophoresis plate (Fig. 21.2B). Phosphohistidine runs with a relative mobility of 111 mm (origin to center of spot). (B) Ninhydrin stained thin-layer electrophoresis plate illustrating the relative mobility of a 3-phosphohistidine standard (PHis) to be 111 mm compared to the relative mobilities of phosphoserine (PSer); 117.5 mm, phosphothreonine (PThr); 117.5 mm, phosphotyrosine (PTyr); 101 mm, phospholysine (PLys); 69.5 mm; and phosphoarginine (PArg), 66 mm. Reproduced with permission from Elsevier (Besant and Attwood, 2000).

1 μl H_2O was added to the remaining 5 μl, both samples were then heated at 60 °C for 30 min. One microliter of each sample was then spotted on to an RP-18 F_{254S} precoated silica reverse-phase thin-layer chromatography (TLC) plate, along with phosphoamino acid standards (Tan *et al.*, 2003). The plate was developed in 72:10:9 (v/v/v) ethanol:28% ammonia solution:H_2O. Figure 21.4 shows the phosphorimage (A) and ninhydrin-stained image of the plate (B) (Tan *et al.*, 2003). In the phosphorimage there is a radioactive spot in the sample lane (lane 3) that is missing in the acid-treated sample lane (lane 4). This spot migrates to the same point as the corresponding phosphohistidine standards seen on the ninhydrin-stained image (lanes 1 and 5). Note that when the mixed phosphoamino acid standards were acid-treated as above, only the phosphohistidine spot disappeared (lane 2).

Apart from the TLE and reverse-phase TLC separation methods described above, phosphoamino acid analysis of phosphorylated histones has been performed using ascending paper chromatography (on Whatman

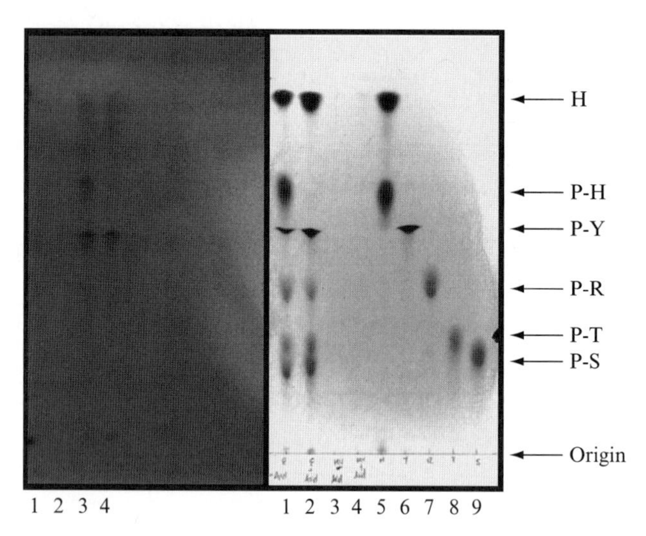

Figure 21.4 Phosphoamino acid analysis of histone H4 phosphorylated by the yeast histidine kinase preparation. A digest of the ^{32}P-labeled phosphorylated histone H4 was split in two and treated with either acid or water. The water-treated digest was run in lane 3 and the acid-treated digest was run in lane 4 of the reverse-phase TLC plate. A mixture of water-treated phosphoamino acid standards was run in lane 1 and a similar acid-treated mixture was run in lane 2. Individual phosphoamino acid standards were run in lanes 5–9: lane 5—phosphohistidine (P-H); lane 6—phosphotyrosine (P-Y); lane 7—phosphoarginine (PR); lane 8—phosphothreonine (P-T); lane 9—phosphoserine (P-S). The left hand panel shows the phosphorimage of the TLC plate while the right hand panel shows image of ninhydrin-stained plate. Positions of the phosphoamino acid standards and histidine are indicated on the right of the figure. Reproduced with permission from Elsevier (Tan *et al.*, 2003).

3MM developed with isopropanol:ethanol:water:triethylamine—30:30:39:1, v/v/v/v), where the digestion was also performed using pronase E (Chen *et al.*, 1974). Phosphoamino acid analysis of phosphorylated histone H4 was also performed using anion-exchange chromatography of the pronase digest on a Mono Q HR5/5 column (Huang *et al.*, 1991). The phosphohistidine was eluted with a linear gradient of 0.1–1.4 *M* methyldiethanolamine, pH 9.0 and radiochemical detection of [^{32}P]phosphohistidine (Huang *et al.*, 1991). Phosphoamino acid standards were used to calibrate the column and they were analyzed by postcolumn derivatization with *o*-phthalaldehyde and fluorescence detection (Wei and Matthews, 1991).

Analysis of a pronase E digest of phosphorylated histone H4 in 4.1% solution in 64% methanol (0.36 μg μl^{-1}) was performed by negative ion electrospray ionization mass spectrometry (MS) by direct infusion at 5 μl min^{-1} into an Applied Biosystems Q-STAR pulsar *i* time-of-flight (TOF) mass spectrometer (Zu *et al.*, 2007). A signal corresponding to a precursor of PO_3^- (-79 amu) was detected with an *m/z* of 234 amu that matched that of the monoisotopic singly charged negative ion of phosphohistidine (234.0285 amu). In addition, in the normal negative ion mode (MS-TOF), a signal at 234.0359 amu was observed, as well as signals corresponding to histidine and other amino acids and some dipeptides (Fig. 21.5). Positive ion MS-TOF was also performed on a digest in 10 m*M* ammonium bicarbonate and 50% methanol in which a signal corresponding to monoisotopic singly charged positive ion of phosphohistidine (236.0903 amu) was detected at 236.0832 amu, as well as signals corresponding to histidine and other amino acids and some dipeptides (Fig. 21.6A). Collision-induced dissociation (MS/MS-TOF) was performed on the putative phosphohistidine ion with a collision energy of 25 V and the product ions analyzed (Fig. 21.6B). Signals corresponding to the immonium ion of phosphohistidine (190.0376 amu), histidine (156.0690), and the immonium ion of histidine (110.0696 amu) were detected (Fig 21.6B). Although the work described above was performed on chemically phosphorylated histone H4, the method has since been successfully applied to the analysis of histone H4 phosphorylated by a thymic histidine kinase (Abzalov, 2007).

As mentioned in the introduction, both 1- and 3-isoforms of phosphohistidine occur but often phosphoamino acid analysis reported in the literature does not discriminate between these isoforms. They have been separated in alkali-digests of a bovine liver [^{32}P]phosphoprotein, later shown to be NDP kinase, using Dowex-1 × 8 anion exchange chromatography with a linear gradient of 0.2–0.8 *M* $KHCO_3$ (Walinder, 1968; Walinder *et al.*, 1968). Both 1- and 3-phosphohistidine were shown to be present in the digest. (Huebner and Matthews, 1985) used the same methodology to identify 1-phosphohistidine in histone H4 phosphorylated by a nuclear extract from *Physarum polycephalum*. The chromatography systems

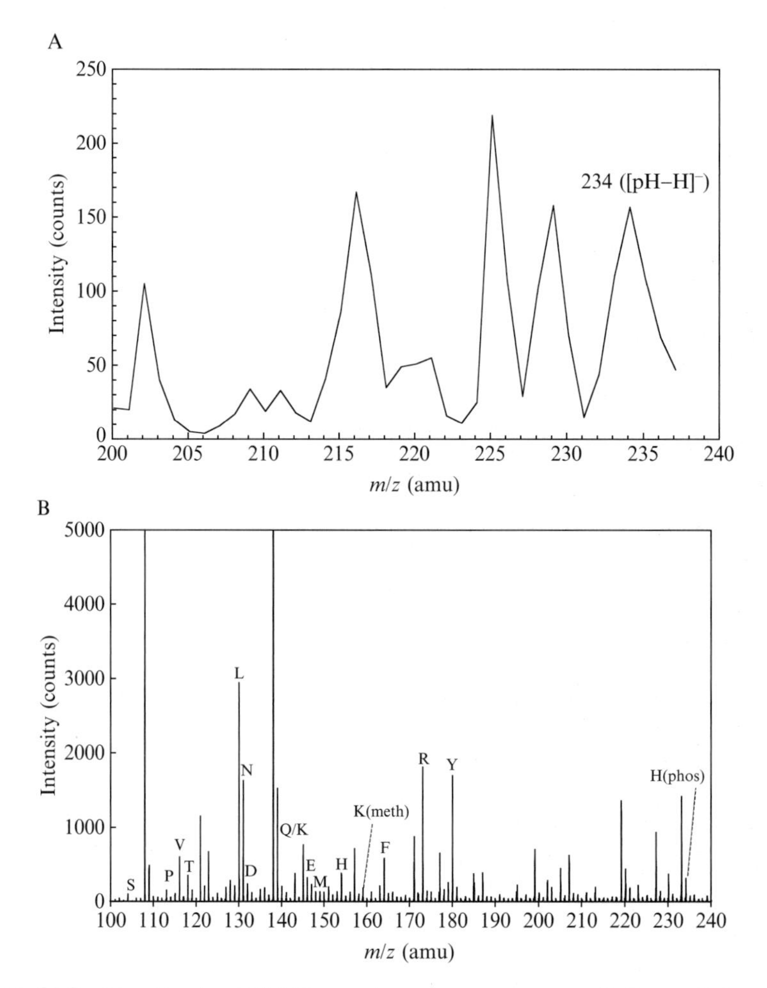

Figure 21.5 Negative ion ESI-MS experiments on the pronase E digest of phosphohistone H4. The sample was infused at 5 μl min^{-1} using a declustering potential of -50 V. (A) Precursor ion spectrum detecting the precursors of the -79 amu ion (PO_3^-). Three hundred scans were accumulated with a collision energy of -50 V. (B) MS scan of the digest where 300 scans were accumulated. Reproduced with permission from Springer (Zu *et al.*, 2007).

described by Wei and Matthews (1991), Mono Q HR5/5 (see above) and a C_{18} reverse-phase system using 0.03–0.02 *M* NaH_2PO_4 buffer, 1.25–0.25% tetrahydrofuran, and a 17–44% acetonitrile gradients can resolve the isoforms of phosphohistidine or their *o*-phthalaldehyde derivatives, respectively. In the detection of the isoforms of phosphohistidine it is necessary to have the 1- and 3-phosphohistidine standards. Often the easiest way to obtain these is to use a reaction mixture containing phosphoramidate and

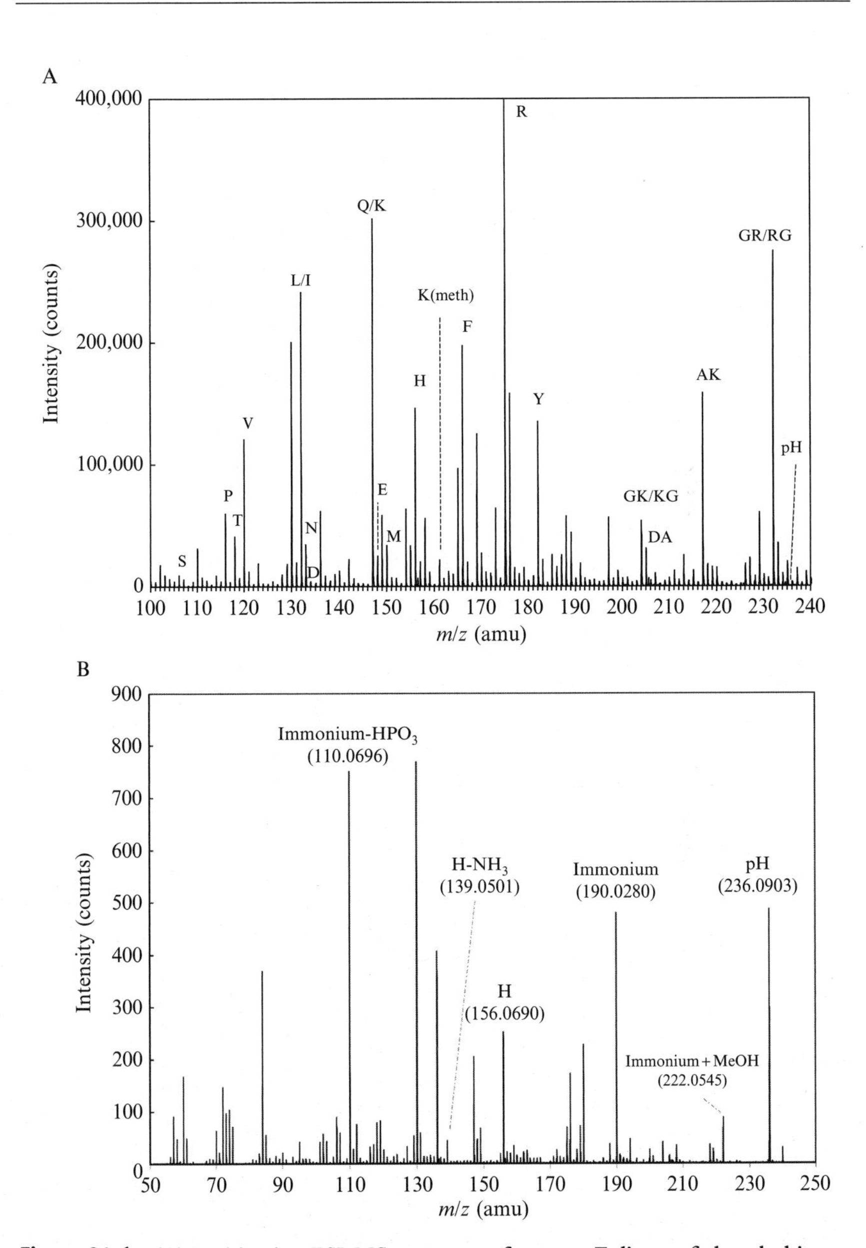

Figure 21.6 (A) Positive ion ESI-MS spectrum of pronase E digest of phosphohistone H4. The sample was infused at 1 μl min^{-1} and 150 scans were accumulated. The declustering potential was +30 V. The singly charged ion of phosphohistidine (pH) $[M + H]^+$ at 236.0903 amu is indicated along with signals corresponding to ions of amino acids and dipeptides. (B) Positive ion ESI-MS/MS spectrum of the product ions of phosphohistidine $[M + H]^+$ (236.0903 in (A)). Reproduced with permission from Springer (Zu *et al.*, 2007).

histidine where the reaction has been stopped by dilution and/or freezing after about 30–60 min (Wei and Matthews, 1991). This will contain a mixture of histidine, 1,3-diphosphohistidine and 1- and 3-phosphohistidine. Whichever separation method is employed, histidine and 1,3-diphosphohistidine should be easily identified owing to the large difference in their polarities and the use of a histidine standard. 1- and 3-phosphohistidine should migrate/elute somewhere between histidine and 1,3-diphosphohistidine. Owing to the greater instability of 1-phospho derivative, if a reaction mixture is left overnight, the predominant species of phosphohistidine is 3-phosphohistidine, thus this reaction mixture can be used as a standard to identify 3-phosphohistidine (Wei and Matthews, 1991). The different forms of phosphohistidine can be separated preparatively using silica gel chromatography, but owing to the instability 1-phospho derivative, purified 1-phosphohistidine always contains histidine (Besant *et al.*, 1998).

Another approach to detecting phosphohistidine in proteins and peptides and determining which isoform(s) is present is to study the NMR spectra of the phosphoprotein or phosphopeptide. There is a difference in chemical shift relative to phosphoric acid between ^{31}P signal of 1-phosphohistidine (about −5.1 ppm) and 3-phosphohistidine (about −4.0 ppm) (Besant *et al.*, 1998). In addition, the chemical shifts of the C2 and C4 protons of 1- and 3-phosphohistidine are also different (Besant *et al.*, 1998) (see Fig. 21.7). It should be noted that the chemical shifts of both the ^{31}P and ^{1}H signals are sensitive to pH owing to the protonation/deprotonation of the imidazole ring (Gassner *et al.*, 1977). These differences in chemical shifts and sensitivity to pH have allowed the identification of the isoforms of phosphohistidine in phosphoproteins by ^{31}P- and ^{1}H NMR analysis.

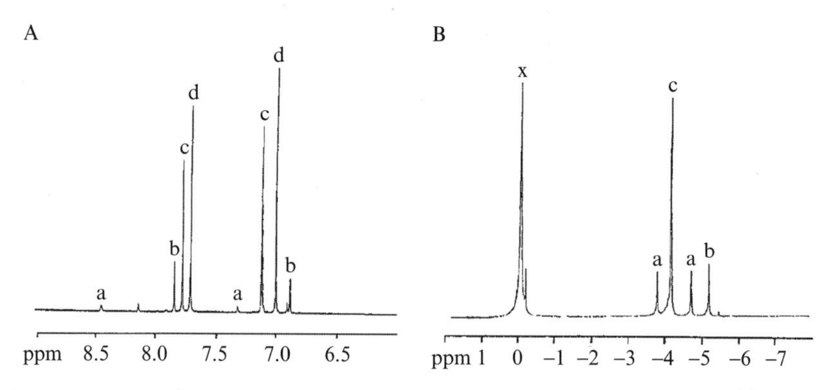

Figure 21.7 (A) ^{1}H NMR spectrum showing C2 and C4 signals and (B) ^{31}P NMR spectrum for a reaction between phosphoramidate (50 mg ml^{-1}) and histidine–HCl (21 mg ml^{-1}), pH 7.2, stopped after 15 min by adjustment of the pH 9.8. Signals observed are: 1,3-diphosphohistidine (a); 1-phosphohistidine (b); 3-phosphohistidine (c); histidine (d); 85% phosphoric acid standard (x). Reproduced with permission from Elsevier (Besant *et al.*, 1998).

^{31}P NMR was used to study histone H4 that had been either chemically phosphorylated, or phosphorylated by histidine kinases from nuclei of either regenerating rat liver or Walker-256 carcinosarcoma cells (Fujitaki *et al.*, 1981). The phosphorylated histone H4 (40–80 mg) was denatured by addition of 1% SDS and the spectra run in the presence of 0.1% SDS at pH 10. Fujitaki and coworkers were able to detect 3-phosphohistidine in the histone H4 phosphorylated by the Walker-256 carcinosarcoma kinase and 1-phosphohistidine when the kinase from regenerating rat liver was used (Fujitaki *et al.*, 1981). Both chemically and enzymically phosphorylated *Staphylococcus aureus* HPr (about 40 mg/ml at pH 9.3) was analyzed by both ^{1}H- and ^{31}P NMR (Gassner *et al.*, 1977). In the enzymically phosphorylated HPr the only histidine in the protein was phosphorylated on the N-1 position of the imidazole ring. As mentioned previously, we have recently used ^{1}H NMR to measure the kinetics of histidine phosphorylation and dephosphorylation of peptides based on histidine-containing histone H4 sequences (Attwood *et al.*, 2010).

4. Filter-Based Assay of Alkali-Stable, Acid-Labile Protein Phosphorylation (Nytran Assay)

The filter assay was originally developed by Wei and Matthews (1990) to detect alkali-stable, acid-labile phosphorylation of histone H4 but has since been modified by us to discriminate between histidine phosphorylation and possible tyrosine phosphorylation. Although this assay has been optimized for histidine phosphorylation of histone H4, it could be easily modified to test for histidine phosphorylation of other protein substrates. This assay is usually performed in conjunction with phosphoamino acid analysis (see above) to determine the nature of the alkali-stable, acid-labile residue that is phosphorylated.

The assay begins by adding 5 μl of histone H4 (5 mg/ml) to 5 μl of phosphorylation reaction mix (500 m*M* Tris–HCl, pH 7.5, 150 m*M* $MgCl_2$, 50 m*M* NaF). To this assay mix 5 μl of 2 m*M* ATP and 1 μCi [γ-^{32}P]ATP (3000 Ci/mmol) is added. To start the kinase reaction, 35 μl of a solution containing histidine kinase is added. The histidine kinase may originate from a variety of different sources depending on what tissue or partially purified extract you are testing. In the past, we have used crude nuclear extracts from regenerating liver (Tan *et al.*, 2004), porcine thymus nuclear extracts (Besant and Attwood, 2000), or partially purified yeast extracts (Tan *et al.*, 2003). The reaction is incubated at 30 °C and the length of time required for the incubation can be empirically determined by running various time points of a reaction on a 15% SDS–PAGE gel and determining the stoichiometry of phosphorylation. This assay also allows for reaction kinetics to be determined by performing this assay at various time points (Tan *et al.*, 2003). The phosphorylation reaction is terminated by adding 6 μl of stop solution (3 *M* NaOH, 2 m*M* ATP) at the desired time points. All samples are incubated at 60 °C for 30 min to deplete any phosphoserine and phosphothreonine.

To quantify the level of acid-labile phosphorylation of histone H4, the phosphorylation reactions are then divided into two. One half of each reaction is treated with acid by addition of 10 μl HCl (3.6 *M*), and 10 μl of water is added to the remaining half. These samples are incubated at 60 °C for 30 min and then returned to their original pH by the addition of 10 μl of NaOH (3.6 *M*) to the acid-treated samples and 10 μl of NaCl (3.6 *M*) to the non-acid-treated samples.

If conducting a time course of phosphorylation then the 0 min time point is established by heating the standard reaction mixture containing the stop solution together with the histidine kinase preparation to 60 °C. After this, the rest of the assay can be performed as described above.

Finally, the reaction mixtures are adsorbed onto 3 × 3 cm squares of Nytran membrane (Schleicher and Schuell) (prewashed with 10 m*M* tetrasodium pyrophosphate and dried). Once the samples have air-dried on the pretreated Nytran membrane the membranes are washed 4 × 15 min in 10 m*M* tetrasodium pyrophosphate to remove any unincorporated radiolabeled phosphate. Once dried, each square of Nytran membrane is subjected to Cherenkov counting. Activity is expressed as pmol Pi incorporated/min/mg total protein of enzyme preparation. All assays were performed in triplicate with and without histone H4 as the substrate.

The triplicate reaction in the absence of histone H4 serves as a control for any phosphorylation of endogenous protein in the sample and the acid/base treatment of each sample allows for discrimination of only the acid-labile, alkali-stable phosphorylation. To determine the nature of the acid-labile, alkali-stable phosphorylation of histone H4, the Nytran assay should also be followed up with phosphoamino acid analysis (see methods above/below). Examples of the Nytran assay being used to determine histone H4 histidine kinase activity in yeast and liver can be seen in Fig. 21.8.

5. In-Gel Kinase Assay

The use of in-gel kinase assays in identifying protein kinase activity in a complex mixture of proteins is by no means novel. However, for the purposes of identifying histidine kinases this assay has proven invaluable. The kinase assay works on the premise that the kinase substrate once polymerized into a gel will act as a substrate for protein kinases that have

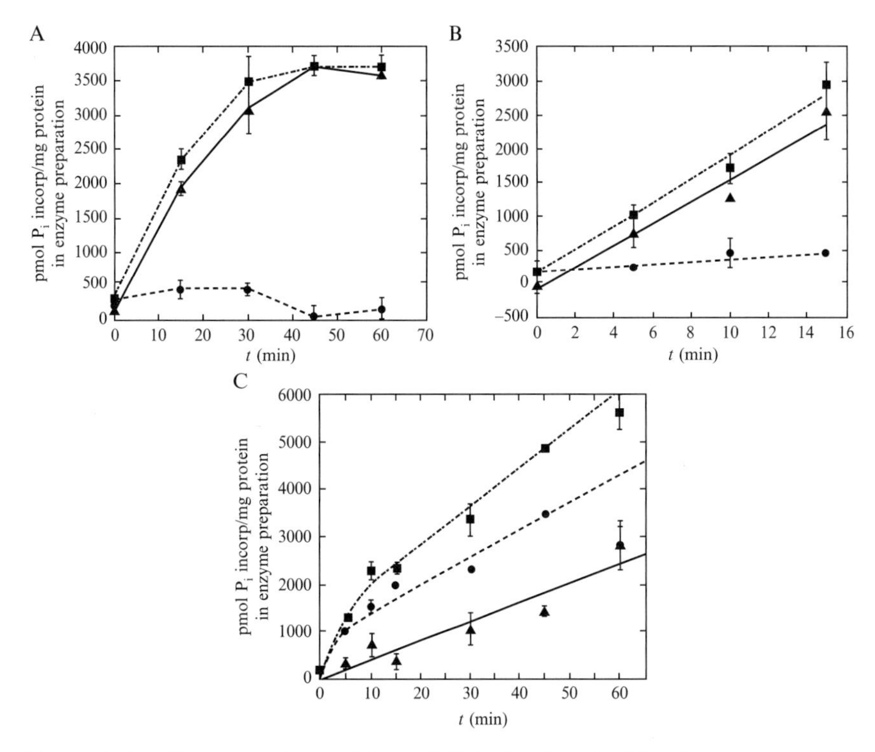

Figure 21.8 Phosphorylation of histone H4 *in vitro* by a histidine kinase preparation from regenerating rat liver. The assays were performed in triplicate and results are presented as pmol of P_i incorporated into histone H4 per mg total protein in enzyme preparation ±S.E. Time courses of alkali-stable phosphorylation (■), alkali-stable/acid-stable phosphorylation (●), and acid-labile phosphorylation (▲) are shown. Time courses are (A) 0–60 min, (B) 0–15 min, and (C) Phosphorylation of histone H4 *in vitro* by a histidine kinase preparation from yeast. The assays were performed in triplicate and results are presented as pmol of P_i incorporated into histone H4 per mg total protein in enzyme preparation ±S.E. Time courses of alkali-stable phosphorylation (■), alkali-stable/acid-stable phosphorylation (●), and acid-labile phosphorylation (▲) are shown. Lines were fitted to the alkali-stable and alkali-stable/acid-stable data by nonlinear least-squares regression analysis, fitting an equation describing a first-order exponential approach to a steady state rate. The line was drawn through acid-labile data using linear least-squares regression analysis. In the curve fits, each datum point was weighted according to the standard error of the mean associated with that datum point [weighting $> 1/(\text{S.E.M.})^2$]. Reproduced with permission from Elsevier (Tan *et al.*, 2003).

been denatured and refolded in-gel. While this method does not distinguish histidine kinases from other kinases it does serve to identify protein kinases that phosphorylate histone H4. Once separated, the radiolabeled bands on the gel can excised and identity of the phosphoamino acid determined by phosphoamino acid analysis. Also the excised bands can be rerun on a standard SDS–PAGE to separate the substrate from the kinase and the kinase potentially can be identified via proteomics techniques.

The example shown in Fig. 21.9 illustrates the detection of a yeast and porcine thymus histone H4 kinase activity that was subsequently determined to be histidine kinase activity via phosphoamino acid analysis (Besant and Attwood, 2000). The yeast histidine kinase (3 ng) and the pool containing putative porcine thymus histidine kinase (8 mg) were denatured in an equal volume of SDS sample buffer at 95 °C for 5 min. Identical samples, along with broad range molecular weight standards (Biorad) were loaded on to both halves of a 10% composite SDS–PAGE with a 3% stacking gel and then electrophoresed at 25 mA for 1 h (Biorad, Mini-Protean III).

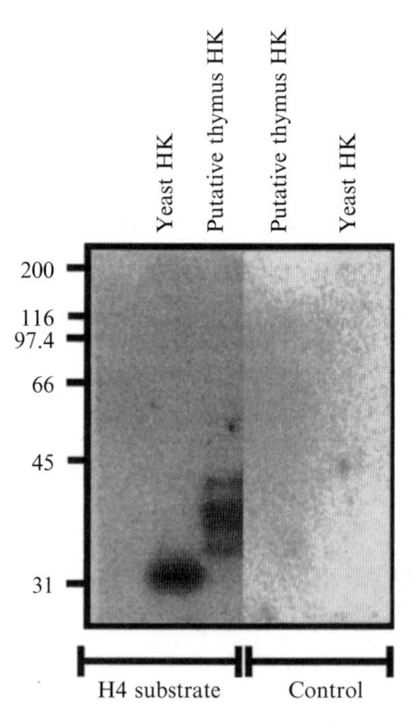

Figure 21.9 Phosphorimage of in-gel renaturation and phosphorylation using Histone H4 as a substrate of putative thymus and yeast histidine kinases as described in Materials and Methods. Control portion of the gel is identical to the histone H4 substrate portion except that no histone H4 was incorporated into the polyacrylamide matrix. Molecular weights based on standards from the coomassie-stained gel indicated on the ordinate show the yeast histidine kinase to be approximately 32 kDa and four putative porcine thymus histidine kinases to be approximately 34–41 kDa. Reproduced with permission from Elsevier (Besant and Attwood, 2000).

The method for pouring composite gels containing different substrates is described by Kameshita and Fujisawa (1996), however, if the amount of available substrate is abundant then separate gels with and without substrate can be poured. The substrate half of the gel in this experiment contained 1 mg/ml histone H4.

After electrophoresis, renaturation was essentially carried out as described by Kameshita and Fujisawa (1989) with minor modifications to the Tris buffer, which was maintained at pH 7.5 in accordance with the pH optimum for the yeast histidine kinase. After renaturation, the gel was re-equilibrated at room temperature by washing 2 × 30 min with 50 ml of 40 m*M* HEPES (pH 7.5), 2 m*M* DTT, 10 m*M* $MgCl_2$. The phosphorylation reaction was carried out by incubating the gel at room temperature for 2 h in 10 ml of 40 m*M* HEPES (pH 7.5), 0.5 m*M* EGTA, 10 m*M* $MgCl_2$, 50 m*M* ATP, and 25 mCi [γ-^{32}P]ATP (3000 Ci/mmol). To remove [γ-^{32}P]ATP and $^{32}P_i$ from the gel it was washed with 100 ml of 10 m*M* sodium pyrophosphate (pH 10), 1 m*M* ATP, 10% methanol every 30 min for 3 h, or until the radioactivity of the wash became negligible. The wet gel was then wrapped in plastic film and exposed to a phosphorimager plate overnight. The band on the histone H4 substrate gel represents renatured kinase activity. The separation of the proteins provides clues to the mass of the protein and the number of kinases in the sample. The in-gel kinase assay may not be wholly representative of every kinase as not all kinases may refold under these conditions. However, when combined with the Nytran assay and phosphoamino acid analysis, the in-gel kinase assay provides valuable information on the nature of the kinase being examined especially when assaying fractions from protein purifications columns. Owing to the high concentration of histone H4 in the gel, direct identification of the kinase in the in-gel kinase band via tryptic digestion, and mass spectrometric analysis is problematic. The excised in-gel band can be re-electrophoresed to separate and histone H4 from the kinase and if sufficient material is available, the isolated kinase can then be identified using proteomics techniques. It is important to run the kinase samples on a gel without the histone H4 substrate as this acts as a control for those proteins kinases that autophosphorylate.

6. Phosphorylation and Thiophosphorylation Site Analysis by Edman Sequencing

Edman sequencing using ^{32}P radiolabeled peptides for the identification of the site of phosphorylation has been used for serine and threonine phosphorylation but is also possible for the detection of phosphohistidine.

The only caveat here is the limitations presented by low stoichiometry of phosphorylation of the protein and the lability of phosphoramidate bond. Thus, this method is useful if the peptide being sequenced is highly phosphorylated on histidine so its identity can be clearly distinguished from background levels of $^{32}P_i$. To make this technique better suited to the study of the more labile P-N bond of phosphohistidine, a more stable ^{35}S-thiophosphorylation can be used. By replacing the double-bonded oxygen of the phosphate with a sulfur leads to a more stable N–P bond (Pirrung *et al.*, 2000).

The labeling of proteins with ^{35}S-thiophosphate and subsequent Edman sequencing was performed on G_β proteins phosphorylated by bovine NDPK (Cuello *et al.*, 2003). In this case, the proteolytic peptides were separated by HPLC and the radiolabeled peptides identified by Edman sequencing. In this study, the site histidine of phosphorylation was inferred by the presence of a histidine residue in the isolated radiolabeled peptide. In our previous work, we have taken this technique one step further by counting the fractions after each cycle of the Edman sequencing to definitively identify the site of histidine autophosphorylation of bovine NDPK (Lasker *et al.*, 1999). The NDPK protein was radiolabeled in an autophosphorylation reaction using either γ-^{32}P-ATP or ^{35}S-thio-ATP. The radiolabeled protein was digested with trypsin and the peptides separated by reverse-phase HPLC. The HPLC fractions were measured using a scintillation counter and the peak fraction was collected and immediately subjected to Edman sequencing.

The ^{32}P-phosphohistidine-containing peptide resulted in spurious sequence data with variable amounts of radioactivity released in the Edman degradation cycles, most likely due to the lability of the phosphoramidate linkage of phosphohistidine. However, the more stable thiophosphohistidine-containing peptide showed a clear signal above background for a released radioactive anilinothiazolinone (ATZ)-amino acid in position 4 (see Fig. 21.10). The only tryptic peptide derived from bovine liver NDPK with a histidine in position 4 has the sequence 115NII**H**GSDSVESAEK128. This peptide contains the predicted phosphorylated histidine residue in position 118 of NDPK from bovine liver.

For the Edman sequencing, radiolabeled phosphopeptides identified from the HPLC separation were added to 5 μl of 0.1% TFA and spotted onto a Sequelon AA membrane. The peptides coupled to the membrane according to the manufacturer's instructions PE-Biosystems, Foster City, California. The membranes were then placed into the cartridge of a Procise Protein Sequencer, Model 492 PE-Biosystems and subjected to automated Edman degradation. The released ATZ-amino acids were collected after each cycle and scintillation counted. The histogram shown in Fig. 21.10 illustrates the counts for both ^{32}P- and ^{35}S-thiophosphorylation of the peptide.

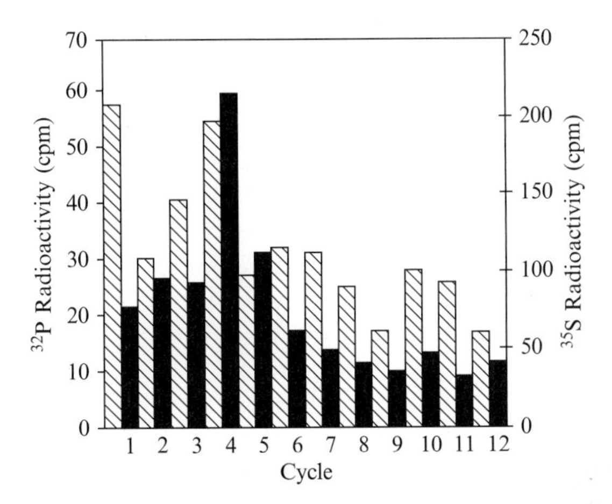

Figure 21.10 Peptide mapping analyses by RP-HPLC of autophosphorylated (▧) and autothiophosphorylated (■) NDPK tryptic digests. In-gel tryptic digests from equal amounts of NDPK were fractionated on a RP-column and 10% aliquots of each fraction were scintillation counted (Lasker *et al.*, 1999).

7. Mass Spectrometric Phosphopeptide Analysis

Mass spectrometry has also been used to identify histidine-phosphopeptides of trypsin-digested phosphohistone H4. The histidine kinase used for our analysis can be found in a commercial preparation of guinea pig liver transglutaminase 2 (TG2) but the same experiment could be done with any histone histidine kinase. The definitive identification of histidine phosphorylation of histone H4 by the TG2-kinase sample using mass spectrometry is also supported by other experimental data from Nytran assays, in-gel kinase assays, and phosphoamino acid analysis. Mass spectrometry also established the site of histidine phosphorylation of histone H4 as being H18. There was no evidence that the other histidine containing peptide (H75) was phosphorylated under the reaction conditions used. It is interesting to note that phosphorylation of H18 of histone H4 supports a study of histone H4 phosphorylation by TG2, in which the majority of phosphate incorporation occurred in the H4(1–21) peptide (Mishra *et al.*, 2006).

The kinase reaction used is similar to that used in the Nytran assay but more histone is used to improve detection of the phosphopeptides during the mass analysis. Histone H4 [50 μg] was phosphorylated in 100 μl of a reaction mixture (50 m*M* Tris–HCl (pH 7.5), 15 m*M* $MgCl_2$, 200 μM ATP) with 20 μl of guinea pig TG2 (6.6 ng) [Sigma]. The reaction was allowed to proceed for 1 h at 25 °C and then diluted to a final volume of

500 μl with 20 m*M* ammonium bicarbonate. Trypsin (5 μg) was added to the 500 μl reaction and allowed to incubate overnight at 25 °C.

The overnight digest was diluted 1:2 with 10 m*M* ammonium bicarbonate in a 75% solution of methanol and sprayed directly into the Applied Biosystems Q-STAR pulsar *i* electrospray ionization time-of-flight mass spectrometer at 2 μl min^{-1}. Data from at least 150 scan cycles were accumulated to give each spectrum, a declustering potential of 30 V was used to reduce in-source dephosphorylation and the collision energy employed in the product ion scans was 25 V or 30 V. Where *m/z* values are given, these refer to monoisotopic masses.

In the TOF-MS spectrum of the tryptic digest (Fig. 21.11) a peak corresponding to the expected mass of the singly charged ion ($[M + H]^+$, *m/z* > 392.1442) phosphodipeptide, pHR, containing pH18 was observed, however, the signal intensity was low and another peak was in close proximity, making interpretation of product ion spectra difficult. A peak with much larger signal intensity was observed, corresponding to a singly charged ion with an *m/z* of 424.1608 amu that is very close to the predicted *m/z* of 424.1704 amu for the methanol adduct of pHR ($[M + CH_3OH + H]^+$). A product ion spectrum of this species produced many ions that correspond to signature product ions expected from the pHR ion (see Table 21.1). These include methanol adducts, the c ion and the ions resulting from losses of H_2O, HPO_3, and H_3PO_4 from pHR. In addition,

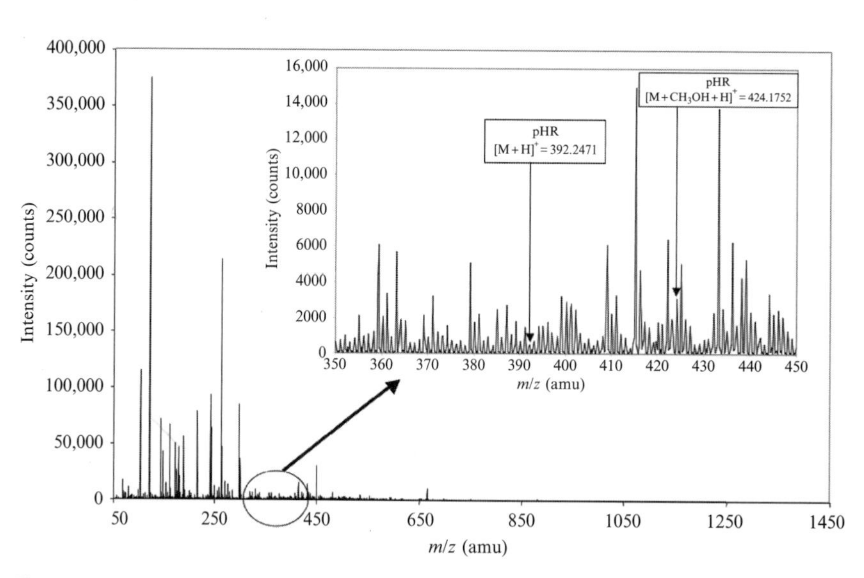

Figure 21.11 TOF–MS spectra of the tryptic digest of TG2-phosphorylated histone H4 showing the peaks for the phosphotryptic peptide pHR *m/z* > 392.2471 and the methanol adduct of pHR *m/z* > 424.1752.

Table 21.1 Product ions of the methanol adduct of the phosphorylated peptide ion (pHR, $[M + CH_3OH + H]^+$, $m/z > 424.1704$ amu) derived from the tryptic digest of TG2-phosphorylated histone H4 that correspond to predicted product ions of pHR ion ($[M + H]^+$) or their methanol adducts

Product ion	Mass (amu)
$[M + CH_3OH + H]^+ - H_2O$	406.1599
$[M + CH_3OH + H]^+ - HPO_3$	344.2140
$[M + CH_3OH + H]^+ - H_3PO_4$	326.1935
Immonium (pHis)	190.0376
Immonium (pHis) − NH_3	173.0111
Immonium (pHis) − (NH_3 + HPO_3)	93.0447
Immonium (Arg)	129.1135
b	218.0325
b − H_2O	200.0220
b − NH_3	201.0060
c	235.0591
c + CH_3CHOH	267.0853
x	201.0892
y	175.1190

Matches were positive if the observed signal was within 0.06 kDa of the theoretical mass and had minimum peak height equal to 0.125% of the maximum peak height in the spectrum (20 counts).

there are immonium ions of phosphohistidine, histidine (together with the corresponding ions showing loss of NH_3), and arginine. Also present were b, c, x, and y ions and the b ion with loss of H_2O or NH_3. There were no peaks in the TOF-MS spectrum at the expected *m/z* values of the other possible tryptic phosphatidyl phosphopeptide ion (DAVTYTEpHAK; $[M + H]^+ > 1214.5095$ amu; $[M + 2H]^{2+} > 607.7586$ amu) or at the *m/z* values of their methanol adducts. These data coupled with data from many of the other assays (Besant and Attwood, unpublished data) described in this chapter, confirmation of the presence of phosphohistidine in the phosphorylated histone H4 and suggest that a kinase in the TG2 protein preparation specifically phosphorylates H18 of the histone. It should be noted that in addition to pHR, the other phosphohistidine-containing phosphopeptide (DAVTYTEpHAK) has been detected and analyzed in chemically phosphorylated histone H4 (Zu *et al.*, 2007). In this case, the doubly charged ion was subjected to collision-induced dissociation at 20 V and six out of nine y ions of the phosphopeptide were detected, as well as a number of a, b, c, and x ions (Zu *et al.*, 2007).

In this chapter, we have presented a selection of methods that have been used to detect and analyze histidine phosphorylation in mammalian proteins and to investigate and measure the activity of mammalian histidine kinases.

Most of the methods described have been employed in our laboratory in the study of histone H4 histidine phosphorylation and mammalian histidine kinases.

ACKNOWLEDGMENTS

We gratefully acknowledge financial support from UWA Small Grant to PVA.

REFERENCES

Abzalov, A. (2007). The purification and characterisation of histone H4 histidine kinase from bovine thymus. *In* "Honours Thesis, School of Biomedical, Biomolecular and Chemical Sciences," pp. 121–134. The University of Western Australia, Perth.

Attwood, P. V., Ludwig, K., Bergander, K., Besant, P. G., Adina-Zada, A., Krieglstein, J., and Klumpp, S. (2010). Chemical phosphorylation of histidine-containing peptides based on histone H4 and their dephosphorylation by protein histidine phosphatase. *Biochim. Biophys. Acta* **1804,** 199–205.

Besant, P. G., and Attwood, P. V. (1998). Problems with phosphoamino acid analysis using alkaline hydrolysis. *Anal. Biochem.* **265,** 187–190.

Besant, P. G., and Attwood, P. V. (2000). Detection of a mammalian histone H4 kinase that has yeast histidine kinase-like enzymic activity. *Int. J. Biochem. Cell Biol.* **32,** 243–253.

Besant, P. G., and Attwood, P. V. (2005). Mammalian histidine kinases. *Biochim. Biophys. Acta* **1754,** 281–290.

Besant, P. G., Byrne, L., Thomas, G., and Attwood, P. V. (1998). A chromatographic method for the preparative separation of phosphohistidines. *Anal. Biochem.* **258,** 372–375.

Besant, P. G., Tan, E., and Attwood, P. V. (2003). Mammalian protein histidine kinases. *Int. J. Biochem. Cell Biol.* **35,** 297–309.

Chen, C. C., Smith, D. L., Bruegger, B. B., Halpern, R. M., and Smith, R. A. (1974). Occurrence and distribution of acid-labile histone phosphates in regenerating rat liver. *Biochemistry* **13,** 3785–3789.

Chen, C. C., Bruegger, B. B., Kern, C. W., Lin, Y. C., Halpern, R. M., and Smith, R. A. (1977). Phosphorylation of nuclear proteins in rat regenerating liver. *Biochemistry* **16,** 4852–4855.

Cuello, F., Schulze, R. A., Heemeyer, F., Meyer, H. E., Lutz, S., Jakobs, K. H., Niroomand, F., and Wieland, T. (2003). Activation of heterotrimeric G proteins by a high energy phosphate transfer via nucleoside diphosphate kinase (NDPK) B and Gbeta subunits. Complex formation of NDPK B with Gbeta gamma dimers and phosphorylation of His-266 in Gbeta. *J. Biol. Chem.* **278,** 7220–7226.

Deluca, M., Ebner, K. E., Hultquist, D. E., Kreil, G., Peter, J. B., Moyer, R. W., and Boyer, P. D. (1963). The isolation and identification of phosphohistidine from mitochondrial protein. *Biochem. Z* **338,** 512–525.

Dombradi, V., Krieglstein, J., and Klumpp, S. (2002). Regulating the regulators. Conference on protein phosphorylation and protein phosphatases. *EMBO Rep.* **3,** 120–124.

Ek, P., Pettersson, G., Ek, B., Gong, F., Li, J. P., and Zetterqvist, O. (2002). Identification and characterization of a mammalian 14-kDa phosphohistidine phosphatase. *Eur. J. Biochem.* **269,** 5016–5023.

Fujitaki, J. M., Fung, G., Oh, E. Y., and Smith, R. A. (1981). Characterization of chemical and enzymatic acid-labile phosphorylation of histone H4 using phosphorus-31 nuclear magnetic resonance. *Biochemistry* **20,** 3658–3664.

Gassner, M., Stehlik, D., Schrecker, O., Hengstenberg, W., Maurer, W., and Ruterjans, H. (1977). The phosphoenolpyruvate-dependent phosphotransferase system of Staphylococcus aureus. 2. 1H and 31P-nuclear-magnetic-resonance studies on the phosphocarrier protein HPr, phosphohistidines and phosphorylated HPr. *Eur. J. Biochem.* **75,** 287–296.

Hippe, H. J., Lutz, S., Cuello, F., Knorr, K., Vogt, A., Jakobs, K. H., Wieland, T., and Niroomand, F. (2003). Activation of heterotrimeric G proteins by a high energy phosphate transfer via nucleoside diphosphate kinase (NDPK) B and Gbeta subunits. Specific activation of Gsalpha by an NDPK B.Gbetagamma complex in H10 cells. *J. Biol. Chem.* **278,** 7227–7233.

Huang, J. M., Wei, Y. F., Kim, Y. H., Osterberg, L., and Matthews, H. R. (1991). Purification of a protein histidine kinase from the yeast *Saccharomyces cerevisiae.* The first member of this class of protein kinases. *J. Biol. Chem.* **266,** 9023–9031.

Huebner, V. D., and Matthews, H. R. (1985). Phosphorylation of histidine in proteins by a nuclear extract of *Physarum polycephalum* plasmodia. *J. Biol. Chem.* **260,** 16106–16113.

Kameshita, I., and Fujisawa, H. (1989). A sensitive method for detection of calmodulin-dependent protein kinase II activity in sodium dodecyl sulfate-polyacrylamide gel. *Anal. Biochem.* **183,** 139–143.

Kameshita, I., and Fujisawa, H. (1996). Detection of protein kinase activities toward oligopeptides in sodium dodecyl sulfate-polyacrylamide gel. *Anal. Biochem.* **237,** 198–203.

Kim, Y., Huang, J., Cohen, P., and Matthews, H. R. (1993). Protein phosphatases 1, 2A, and 2C are protein histidine phosphatases. *J. Biol. Chem.* **268,** 18513–18518.

Kleinnijenhuis, A. J., Kjeldsen, F., Kallipolitis, B., Haselmann, K. F., and Jensen, O. N. (2007). Analysis of histidine phosphorylation using tandem MS and ion-electron reactions. *Anal. Chem.* **79,** 7450–7456.

Klumpp, S., and Krieglstein, J. (2002). Phosphorylation and dephosphorylation of histidine residues in proteins. *Eur. J. Biochem.* **269,** 1067–1071.

Klumpp, S., Hermesmeier, J., Selke, D., Baumeister, R., Kellner, R., and Krieglstein, J. (2002). Protein histidine phosphatase: A novel enzyme with potency for neuronal signaling. *J. Cereb. Blood Flow Metab.* **22,** 1420–1424.

Lasker, M., Bui, C. D., Besant, P. G., Sugawara, K., Thai, P., Medzihradszky, G., and Turck, C. W. (1999). Protein histidine phosphorylation: Increased stability of thiophosphohistidine. *Protein Sci.* **8,** 2177–2185.

Matthews, H. R., and Mackintosh, C. (1995). Protein histidine phosphatase activity in rat liver and spinach leaves. *FEBS Lett.* **364,** 51–54.

Medzihradszky, K. F., Phillipps, N. J., Senderowicz, L., Wang, P., and Turck, C. W. (1997). Synthesis and characterization of histidine-phosphorylated peptides. *Protein Sci.* **6,** 1405–1411.

Mishra, S., Saleh, A., Espino, P. S., Davie, J. R., and Murphy, L. J. (2006). Phosphorylation of histones by tissue transglutaminase. *J. Biol. Chem.* **281,** 5532–5538.

Muimo, R., Banner, S. J., Marshall, L. J., and Mehta, A. (1998). Nucleoside diphosphate kinase and Cl(-)-sensitive protein phosphorylation in apical membranes from ovine airway epithelium. *Am. J. Respir. Cell Mol. Biol.* **18,** 270–278.

Muimo, R., Hornickova, Z., Riemen, C. E., Gerke, V., Matthews, H. R., and Mehta, A. (2000). Histidine phosphorylation of annexin I in airway epithelia. *J. Biol. Chem.* **275,** 36632–36636.

Ohmori, H., Kuba, M., and Kumon, A. (1993). Two phosphatases for 6-phospholysine and 3-phosphohistidine from rat brain. *J. Biol. Chem.* **268,** 7625–7627.

Perego, M., Hanstein, C., Welsh, K. M., Djavakhishvili, T., Glaser, P., and Hoch, J. A. (1994). Multiple protein aspartate phosphatases provide a mechanism for the integration of diverse signals in the control of development in *B. subtilis*. *Cell* **79,** 1047–1055.

Pirrung, M. C., James, K. D., and Rana, V. S. (2000). Thiophosphorylation of histidine. *J. Org. Chem.* **65,** 8448–8453.

Smith, D. L., Bruegger, B. B., Halpern, R. M., and Smith, R. A. (1973). New histone kinases in nuclei of rat tissues. *Nature* **246,** 103–104.

Smith, D. L., Chen, C. C., Bruegger, B. B., Holtz, R. M., Halpern, R. M., and Smith, R. A. (1974). Characterization of protein kinases forming acid-labile histone phosphates in Walker-256 carcinosarcoma cell nuclei. *Biochemistry* **13,** 3780–3785.

Stock, A. M., Robinson, V. L., and Goudreau, P. N. (2000). Two-component signal transduction. *Annu. Rev. Biochem.* **69,** 183–215.

Tan, E., Lin Zu, X., Yeoh, G. C., Besant, P. G., and Attwood, P. V. (2003). Detection of histidine kinases via a filter-based assay and reverse-phase thin-layer chromatographic phosphoamino acid analysis. *Anal. Biochem.* **323,** 122–126.

Tan, E., Besant, P. G., Zu, X. L., Turck, C. W., Bogoyevitch, M. A., Lim, S. G., Attwood, P. V., and Yeoh, G. C. (2004). Histone H4 histidine kinase displays the expression pattern of a liver oncodevelopmental marker. *Carcinogenesis* **25,** 2083–2088.

Treharne, K. J., Riemen, C. E., Marshall, L. J., Muimo, R., and Mehta, A. (2001). Nucleoside diphosphate kinase—A component of the [Na(+)]- and [Cl(−)]-sensitive phosphorylation cascade in human and murine airway epithelium. *Pflugers Arch.* **443** (Suppl. 1), S97–S102.

Walinder, O. (1968). Identification of a phosphate-incorporating protein from bovine liver as nucleoside diphosphate kinase and isolation of 1-^{32}P-phosphohistidine, 3-^{32}P-phosphohistidine, and N-epsilon-^{32}P-phospholysine from erythrocytic nucleoside diphosphate kinase, incubated with adenosine triphosphate-^{32}P. *J. Biol. Chem.* **243,** 3947–3952.

Walinder, O., Zetterqvist, O., and Engstrom, L. (1968). Purification of a bovine liver protein rapidly phosphorylated by adenosine triphosphate. Isolation of 1-^{32}P-phosphohistidine, 3-^{32}P-phosphohistidine, and N-epsilon-^{32}P-phospholysine from ^{32}P-labeled protein. *J. Biol. Chem.* **243,** 2793–2798.

Wei, Y. F., and Matthews, H. R. (1990). A filter-based protein kinase assay selective for alkali-stable protein phosphorylation and suitable for acid-labile protein phosphorylation. *Anal. Biochem.* **190,** 188–192.

Wei, Y. F., and Matthews, H. R. (1991). Identification of phosphohistidine in proteins and purification of protein-histidine kinases. *Methods Enzymol.* **200,** 388–414.

Wong, C., Faiola, B., Wu, W., and Kennelly, P. J. (1993). Phosphohistidine and phospholysine phosphatase activities in the rat:potential protein-lysine and protein-histidine phosphatases? *Biochem. J.* **296,** 293–296.

Zetterqvist, O., and Engström, L. (1967). Isolation of [32P]phosphohistidine from different rat-liver cell fractions after incubation with [^{32}P]adenosine triphosphate. *Biochim. Biophys. Acta* **113,** 520–530.

Zu, X. L., Besant, P. G., Imhof, A., and Attwood, P. V. (2007). Mass spectrometric analysis of protein histidine phosphorylation. *Amino Acids* **32,** 347–357.

Author Index

K

L

M

X

Y

Z

Subject Index

I

J

K

L

M

N

O

P

Q

T

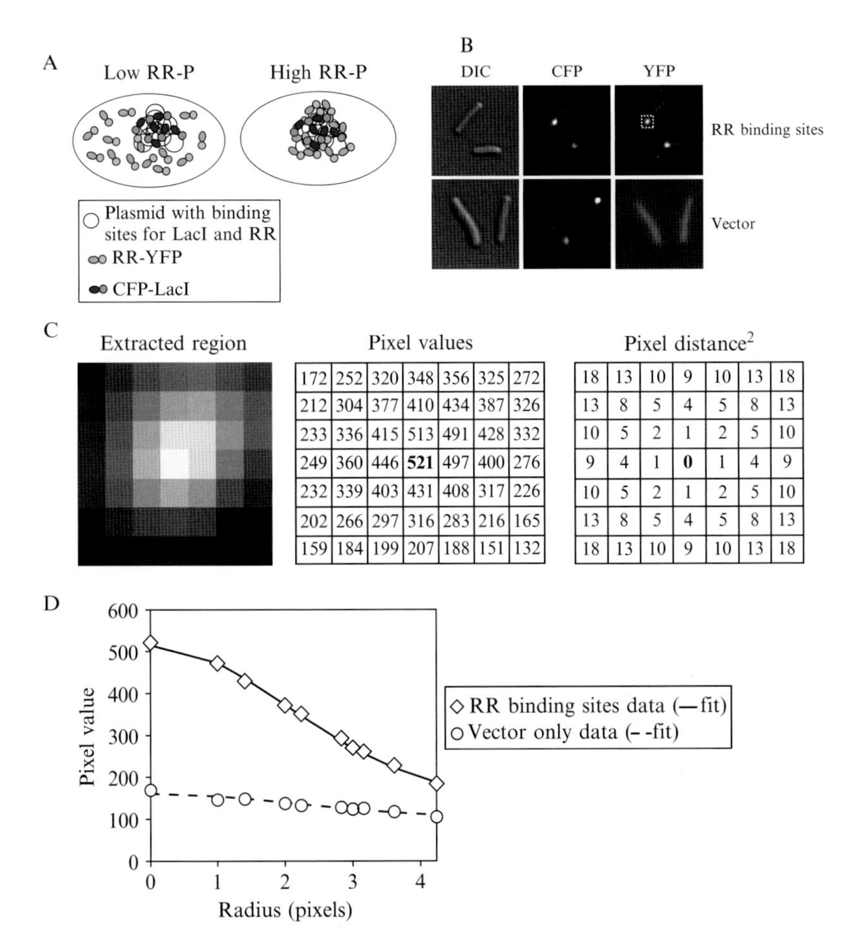

Pixel values

172	252	320	348	356	325	272
212	304	377	410	434	387	326
233	336	415	513	491	428	332
249	360	446	**521**	497	400	276
232	339	403	431	408	317	226
202	266	297	316	283	216	165
159	184	199	207	188	151	132

Pixel distance^2

18	13	10	9	10	13	18
13	8	5	4	5	8	13
10	5	2	1	2	5	10
9	4	1	**0**	1	4	9
10	5	2	1	2	5	10
13	8	5	4	5	8	13
18	13	10	9	10	13	18

Albert Siryaporn and Mark Goulian, Figure 1.3 Measurement of response regulator binding to DNA *in vivo*. (A) Schematic of the method, which uses a functional translational fusion of a fluorescent protein (e.g., YFP) to the response regulator and a plasmid containing response regulator binding sites. The plasmid also contains *lac* operators, which are bound by CFP-LacI. Plasmid clustering in the cell results in a high local concentration of binding sites, which is easily visualized as a bright fluorescent spot in the CFP channel. Response regulator binding to plasmid also results in a bright fluorescent spot in the YFP channel. (B) Images of cells expressing OmpR-YFP and containing a plasmid with OmpR binding sites (top row) or an empty control vector (bottom row). DIC—differential interference contrast image. (C) Example of a YFP spot, corresponding to the dashed square in the upper right image in (B), and the associated pixel values and distances from the center. (D) Gaussian fit to the profile of pixel values from (C). A corresponding fit to the neighborhood of a maximal YFP pixel in a cell containing the vector control is also shown.

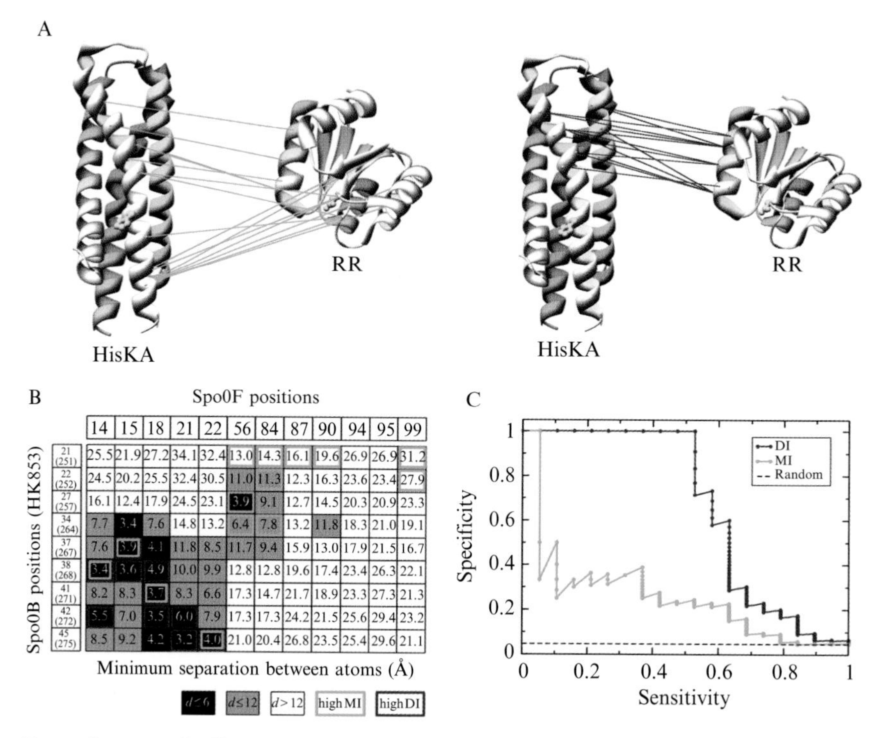

Spo0B (HK853)	14	15	18	21	22	56	84	87	90	94	95	99
21 (251)	25.5	21.9	27.2	34.1	32.4	13.0	14.3	16.1	19.6	26.9	26.9	31.2
22 (252)	24.5	20.2	25.5	32.4	30.5	11.0	11.3	12.3	16.3	23.6	23.4	27.9
27 (257)	16.1	12.4	17.9	24.5	23.1	3.9	9.1	12.7	14.5	20.3	20.9	23.3
34 (264)	7.7	3.4	7.6	14.8	13.2	6.4	7.8	13.2	11.8	18.3	21.0	19.1
37 (267)	7.6	3.9	4.1	11.8	8.5	11.7	9.4	15.9	13.0	17.9	21.5	16.7
38 (268)	3.4	3.6	4.9	10.0	9.9	12.8	12.8	19.6	17.4	23.4	26.3	22.1
41 (271)	8.2	8.3	3.7	8.3	6.6	17.3	14.7	21.7	18.9	23.3	27.3	21.3
42 (272)	5.5	7.0	3.5	6.0	7.9	17.3	17.3	24.2	21.5	25.6	29.4	23.2
45 (275)	8.5	9.2	4.2	3.2	4.0	21.0	20.4	26.8	23.5	25.4	29.6	21.1

Bryan Lunt *et al.*, Figure 2.1 Comparison of results derived by covariance analysis with direct coupling analysis. (A) The top 15 residue pairings identified by covariance analysis (left in green) and the top 15 residue pairings identified by direct coupling analysis (DCA), which includes an additional statistical inference step are mapped on exemplary structures for the HisKA domain (HK853 from *Thermotoga maritima*: PDBID 2C2A) and RR domain (Spo0F from *B. subtilis*: PDBID 1PEY). It becomes apparent that the additional inference step increases strongly the specificity of the contact residue prediction. (B) The table shows minimal atom distances in the Spo0B/Spo0F cocrystal structure between all residues that are identified by covariance analysis or DCA. The top pairings identified by covariance analysis are framed in green and identified by DCA are framed in red. Since the Spo0B helix $\alpha2$ is oriented different and cannot be aligned with regular HisKA helix $\alpha2$, residue positions 291, 294, and 298 (HK853 numbering) are ignored for this analysis. Out of the 15 pairings displayed in each of the figures in Panel (A), 14 high-MI and 10 high-DI pairings do not involve these residues, and are included into the figure. (C) Comparison of specificity/sensitivity curves of covariance analysis versus DCA, where distances below 6 Å in the Spo0B/Spo0F cocrystal structure are considered as real contacts.

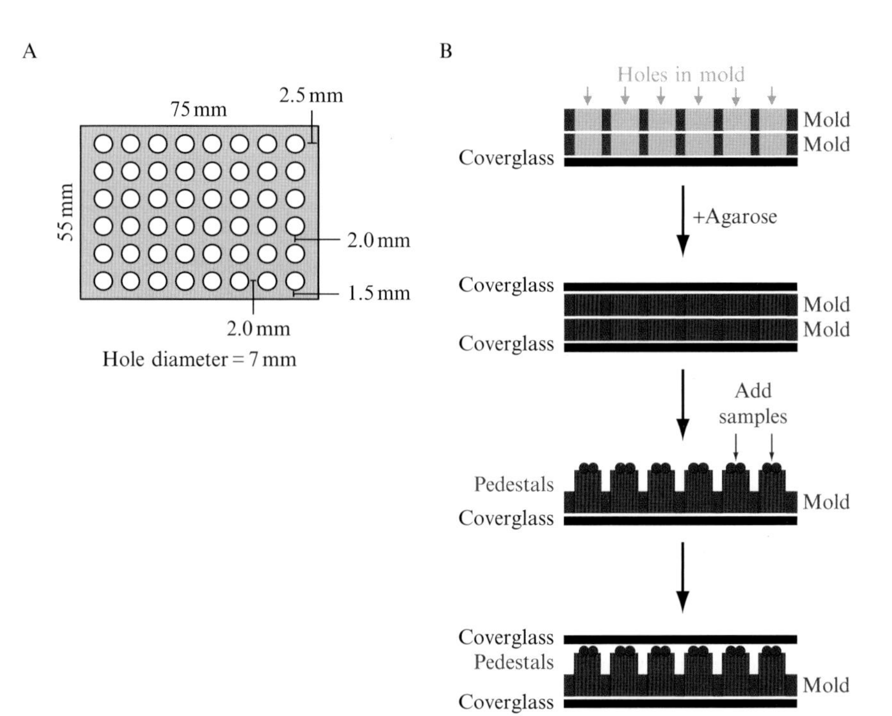

John N. Werner and Zemer Gitai, Figure 11.1 Schematic of the 48-well pedestal slide mold and pedestal slide assembly. (A) Measurements of mold are 75 mm long by 55 mm wide by 1 mm thick. The diameter of each hole is 7 mm. The width of the long border, short border, and width between holes is 1.5, 2.5, and 2.0 mm, respectively. (B) Flow chart of pedestal slide assembly as described in the text. A side view of a slice through the assembly sandwich is shown with the stainless steel molds in brown, empty holes in the mold in light gray, coverglass in black, agarose pedestals in dark gray, and culture samples in red.